Developments in Geotectonics 21

THE ORIGIN OF ARCS

FURTHER TITLES IN THIS SERIES

1. *J. AUBOUIN*
GEOSYNCLINES

2. *R.W. VAN BEMMELEN*
GEODYNAMIC MODELS

3. *A. SUGIMURA and S. UYEDA*
ISLAND ARCS

4. *A.R. RITSEMA* (Editor)
THE UPPER MANTLE

5. *C. LOMNITZ*
GLOBAL TECTONICS AND EARTHQUAKE RISK

6. *X. LE PICHON, J. FRANCHETEAU and J. BONNIN*
PLATE TECTONICS

7. *R.W. GIRDLER* (Editor)
EAST AFRICAN RIFTS

8. *S. MUELLER* (Editor)
THE STRUCTURE OF THE EARTH'S CRUST

9. *N. PAVONI and R. GREEN* (Editors)
RECENT CRUSTAL MOVEMENTS

10. *S.W. CAREY*
THE EXPANDING EARTH

11. *A.M. JOHNSON*
STYLES OF FOLDING

12. *M.H.P. BOTT* (Editor)
SEDIMENTARY BASINS OF CONTINENTAL MARGINS AND CRATONS

13. *C.A. WHITTEN, R. GREEN and B.K. MEADE* (Editors)
RECENT CRUSTAL MOVEMENTS, 1977

14. *M.N. TOKSÖZ, S. UYEDA and J. FRANCHETEAU* (Editors)
OCEANIC RIDGES AND ARCS

15. *C.E. KEEN* (Editor)
CRUSTAL PROPERTIES ACROSS PASSIVE MARGINS

16. *P. VYSKOCIL, R. GREEN and H. MÄLZER* (Editors)
RECENT CRUSTAL MOVEMENTS, 1979

17. *J.H. ILLIES* (Editor)
MECHANISM OF GRABEN FORMATION

18. *E.V. ARTYUSHKOV*
GEODYNAMICS

19. *P. MORGAN and B.H. BAKER* (Editors)
PROCESSES OF CONTINENTAL RIFTING

20. *P. VYSKOCIL, A.M. WASSEF and R. GREEN* (Editors)
RECENT CRUSTAL MOVEMENTS, 1982

Developments in Geotectonics 21

THE ORIGIN OF ARCS

Invited papers presented at the International Conference "The Origin of Arcs", held at the University of Urbino, Urbino, Italy, September 22nd–25th, 1986

Edited by

F.-C. WEZEL

Istituto di Geologia dell'Università, Via S. Chiara 27, Urbino I-61029, Italy

ELSEVIER
Amsterdam — Oxford — New York — Tokyo, 1986

ELSEVIER SCIENCE PUBLISHERS B.V.
Sara Burgerhartstraat 25
P.O. Box 211, 1000 AE Amsterdam, The Netherlands

Distributors for the United States and Canada:

ELSEVIER SCIENCE PUBLISHING COMPANY INC.
52, Vanderbilt Avenue
New York, NY 10017, U.S.A.

Library of Congress Cataloging-in-Publication Data

International Conference "The Origin of Arcs" (1986 :
Università di Urbino)
The origin of arcs.

(Developments in geotectonics ; 21)
1. Island arcs--Congresses. I. Wezel, Forese-Carlo.
II. Title. III. Series.
QE511.2.I57 1986 551.4'2 86-16747
ISBN 0-444-42688-4

ISBN 0-444-42688-4 (Vol. 21)
ISBN 0-444-41714-1 (Series)

Printed in The Netherlands

LIST OF CONTRIBUTORS

E.V. ARTYUSHKOV
Institute of Physics of the Earth
U.S.S.R. Academy of Sciences
Moscow, U.S.S.R.

M.A. BAER
Ministry of Geology of the U.S.S.R.
Moscow, U.S.S.R.

L. BECCALUVA
Dipartimento di Scienza della Terra
Via Mezzocannone 8
Napoli, Italy

V.V. BELOUSSOV
Institute of Physics of the Earth
U.S.S.R. Academy of Sciences
Moscow, U.S.S.R.

E. BIJU-DUVAL
IFREMER
66 Avenue d'Iéna
F-75763 Paris, France

R. BLANCHET
GIS Océanologie et Géodynamique
Université de Bretagne Occidentale
F-29283 Brest, France

J.-P. BOUILLIN
Laboratoire de Géologie
Méditerranéenne
Université de Toulouse
F-31400 Toulouse, France

J.-C. BOUSQUET
Laboratorie de Géologie Structurale
Université des Sciences et
Techniques du Languedoc
F-34060 Montpellier, France

J.P. CADET
Laboratoire de Géologie Dynamique
Université de Orléans
F-46067 Orléans, France

S.W. CAREY
24 Richardson Avenue
Dynnyrne, Tasmania 7005
Australia

A. CASTELLARIN
Istituto di Geologia
Università degli Studie
I-40127 Bologna, Italy

P. de CLARENS
SNEA (P)
Tour Elf
F-92078 Paris la Défense, France

S.A.P.L. CLOETINGH
Institute of Earth Sciences
University of Utrecht
3508 TA Utrecht, The Netherlands

A.R. CRAWFORD
Department of Geology
University of Canterbury
Christchurch, New Zealand

J. DEBELMAS
Geology Department
University of Grenoble
F-38031 Grenoble, France

J. DOSTAL
Department of Geology
St. Mary University
Halifax, N.S., Canada

M. DURAND-DELGA
Laboratoire de Géologie
Méditerranéenne
Université de Toulouse
F-31400 Toulouse, France

J.D. FODEN
Geology Department
University of Adelaide
Adelaide, S.A. 5001
Australia

F. FORCELLA
Dipartimento di Scienza della Terra
Università degli Studie di Milano
I-20133 Milano, Italy

F.J. HILGEN
Institute of Earth Sciences
University of Utrecht
3508 TA Utrecht, The Netherlands

A.M. HIRT
Institut für Geophysik
ETH-Hönggerberg
CH-8093 Zürich, Switzerland

L. JOLIVET
Laboratoire de Géologie
Ecole Normale Supérieure
F-75231 Paris, France

K. KOBAYASHI
Ocean Research Institute
University of Tokyo
Tokyo 164, Japan

S. LALLEMAND
Laboratoire de Géologie Dynamique
Université de Orléans
F-46067 Orléans, France

E. LOCARDI
ENEA
C.R.E. Casaccia
I-00060 Roma, Italy

W. LOWRIE
Institut für Geophysik
ETH-Hönggerberg
CH-8093 Zürich, Switzerland

A. MASCLE
IFP
F-92506 Rueil Malmaison, France

B. MERCIER DE LEPINAY
UA 215
Université de Paris-Marie Curie
Paris, France

J.E. MEULENKAMP
Institute of Earth Sciences
University of Utrecht
3508 TA Utrecht, The Netherlands

N.-A. MÖRNER
Geological Institute
University of Stockholm
S-10691 Stockholm, Sweden

S. MUELLER
Institut für Geophysik
ETH-Hönggerberg
CH-8093 Zürich, Switzerland

H. MUNSCH
Total CFP
F-92069 Paris la Défense, France

Ph. OLIVIER
Laboratoire de Géologie
Méditerranéenne
Université de Toulouse
F-31400 Toulouse, France

G.F. PANZA
Istituto di Geodesia e Geofisica
Università degli Studie
I-34100 Trieste, Italy

G. PASQUARE
Dipartimento di Scienza della Terra
Università degli Studie di Milano
I-20133 Milano, Italy

N. PAVONI
Institut für Geophysik
ETH-Hönggerberg
CH-8093 Zürich, Switzerland

H. PHILIP
Laboratoire de Géologie Structurale
Université des Sciences et
Techniques du Languedoc
F-34060 Montpellier, France

G. SERRI
Dipartimento di Scienze della Terra
Via Santa Maria 53
Pisa, Italy

J.F. STEPHAN
GIS Océanologie et Géodynamique
Université de Bretagne Occidentale
F-29283 Brest, France

A. TIBALDI
Dipartimento di Scienza della Terra
Università degli Studie di Milano
I-20133 Milano, Italy

S. UYEDA
Earthquake Research Institute
University of Tokyo
Tokyo 113, Japan

G.B. VAI
Istituto di Geologia
Università degli Studie
I-40127 Bologna, Italy

R. VARNE
Geology Department
University of Tasmania
Hobart, Tasmania 7001
Australia

L. VEZZOLI
Dipartimento di Scienzia della Terra
Università degli Studie di Milano
I-20133 Milano, Italy

F.-C. WEZEL
Instituto di Geologia
Università di Urbino
I-61029 Urbino, Italy

M.J.R. WORTEL
Institute of Earth Sciences
University of Utrecht,
3508 TA Utrecht, The Netherlands

A. ZANCHI
Dipartimento di Scienza della Terra
Università degli Studie di Milano
I-20133 Milano, Italy

AUTHOR INDEX

PREFACE

In this volume there is a collection of twenty-three papers which are to be presented as distinguished guest lectures in the International Conference on "The Origin of Arcs" which is to be held at the University of Urbino from the 22nd to the 25th of September 1986, under the joint sponsorship of the European Union of Geosciences and the Italian Geological Society.

The workshop on island and mountain arcs has been organized with the aim of increasing our understanding of the intrinsic nature of orogenic and post-orogenic processes, on the basis of empiric factual data, rather than particular theoretic models. Reality, in fact, almost always rejects affirmations and speculations built up around a table. Quite often a trivial piece of field data appears to have much more weight than many fascinating hypotheses put forward by the human mind. This seems to be much more valid in geology, where a special method is necessitated by the special nature of the geological phenomena and the time concept. Every general law deduced should be rooted in the study of the earth's development in geological time. The geological-historical method leads to the recognition of a repetition in time and space of a sequence of events. The deeper causes of these events can only be discussed in the light of a space-time system which remains quite empiric in that it must derive from geological facts and not from preconceived theory. It is the convener's opinion that there must first be an inductive picture by means of geological methods and then it must be interpreted by geophysics in the light of physical laws. The geological method must serve, besides, to test the historical credibility of geophysical theories. It is clear that these two methods, the geological-historical one and the geophysical one, must be complementary and the one must not substitute the other.

Since the problem of the structure and origin of arcs admits of several solutions, different factors being still unexplained, all opinions that have been correctly deduced have been considered by the present editor. Well aware of the many facets presented by complex geological reality, he does not consider himself entitled to reject as naive arguments and interpretations which appear apparently contrary to our present beliefs. The development of geosciences itself provides numerous examples of statements and concepts which were first said to be impossible but later proved perfectly valid. This tolerant attitude towards unorthodox ideas is not merely that of enlightenment and impartiality but reflects also a cautious wisdom in that the heresies of today may perhaps become the truths of tomorrow and the truths of today are not eternally valid dogmas. It is the conviction of the present editor that no-one knows the whole geological reality, but that each one of us, at most, presents some fragments and splinters which appear to us as tiny illuminated spaces fleetingly glimpsed through the great fog of our ignorance. Thus, it is necessary to stimulate the creativity, originality, intuition and eclecticism of young researchers, rather than block them and orientate their minds towards fixed, pre-established doctrines. As Robert M. Pirsig has well said, "We take a handful of sand from the endless landscape of awareness around us and call that handful of sand the world". The interaction amongst the different concepts could possibly, dialectically, spark off some aspects of geological reality.

The contributers to this pre-conference volume have been asked to present essential geological results, as concrete as possible, on some basic problems, such as:

- Are the island and mountain arcs primary or induced features?
- How have these orogenic festoons developed into their similar regular shapes?
- What are the relationships between "primary" active arcs and

"secondary" mountain arcs?

- What is the dominant deformational factor in the bulging of the arc?
- What is the real nature and tectonic significance of the oblique Benioff zone?

The diverse opinions expressed in their papers have been accepted by the editor without screening. Basically the aim of the Arc Conference is to hear and discuss the points of view of these guest speakers, some of whom are recognized masters of geology, selected and invited by the convener because of their preparation and experience in the topics covered by the conference.

The papers have been grouped into five more or less natural sections, of which three are defined on the basis of geography. But of course several range broadly and could have been differently grouped. The subdivision, like any classification, is quite subjective and serves only to channel the discussion in a practical way.

The generalized geotectonic concepts represent an effort to find answers to questions considered the most fundamental in a particular historical moment. It is wrong and out of place to consider them as the ultimate truths. Their usefulness can be maintained only on condition that they do not degenerate into rigid, universal schemes. The earth's arcuate belts are interpreted by S.W. Carey in the context of his expansionist hypothesis, according to which orogeny is a diapiric process analogous to that of ocean-spreading. He underlines, besides, the importance of two global torsions, viz. the Tethyan sinistral torsion and the conjugate Pacific dextral counter torsion. The similarity in the geological development between the continental orogenic areas and the island arcs is pointed out by V.V. Beloussov. Their arcuate shape is thought by him to be controlled by the contours of the stable median zones and of the subsidences bounding them. According to N. Pavoni mountain and island arcs represent the zone of convergence and shearing of the lithosphere due to large-scale movements away from the central Pacific and African antipodal plates, triggered by a bicellular convection pattern in the mantle. N.-A. Mörner pointed out that the earth's geodynamics (i.e., volcanism, seismicity and neotectonics) react very sensitively to changes in the rate of rotation, as clearly manifested by the records of the last centuries and of Holocene.

In the next section some geophysical and geochemical results considered to be of regional importance are set out. The presence of a nearly vertical massive "lithospheric root" which penetrates deeply into the asthenosphere beneath the Alpine and the Northern Apennine hinterlands, and also under the Betic Cordillera, is pointed out by St. Mueller and G.F. Panza. Are we dealing with Otto Ampferer's concept of "Verschluckung" ("down-sucking")? Some aspects of the dynamics of the subduction are analysed by M.J.R. Wortel and S.A.P.L. Cloetingh which throw light on the relation between upper mantle and tectonic processes (accretion and subduction of sediments, vertical motions, plate fragmentation). Diverse papers, in other sections too, stress the importance of rotational tectonics in shaping the active margin and island arc systems. Thus, paleomagnetic studies are of fundamental interest in verifying some geodynamic hypotheses. A method, tested by W. Lowrie and A.M. Hirt, allows us to judge if the observed curved shape of a mountain arc is an original feature or the result of oroclinal bending. Their data do not support oroclinal bending for the curvature of the Jura mountains and Umbrian-Marchean Apennines. The systematic geochemical and isotopic variations in the active volcanic arcs can provide important information on the involvement of suboceanic and subcontinental mantles in the arc magmatism. Such variation has permitted R. Varne and J.D. Foden to postulate the existence beneath the Sunda Arc of large-scale tectonic processes that are mixing the Indian Ocean and Australian mantles.

The Alpino-Mediterranean is a key region for the study of arcs, both of the mountain type and the island arc type. The present arcuate shape of the Alpidic orogenic belt is thought by E.V. Artyushkov and M.A. Baer to represent the boundary of the past deep basins with stable cratons. The rapid subsidence of continental crust, due to the destruction of lower crust and sea-floor spreading, gave rise to the deep and weak basins which were successively compressed. The subsequent formation of the topographic mountains was caused by the upwelling of a hot mantle. The data furnished by J. Debelmas indicate that the classic Western Alpine arc is in part inherited from Hercynian and Mesozoic patterns and in part exaggerated by the Insubric impingement and tha later Ligurian Apennine folding. The Southalpine and the opposite-vergent Neogene Po Plain Apennine arc systems are analysed by A. Castellarin and G.B. Vai, who explain, respectively, how the first are due to underthrusting (active A subduction) of the Padania-Adriatic crust and the second, instead, to overthrusting (passive A subduction) upon the Padania-Adriatic foredeep. The curvature of the arcs takes place around more rigid blocks, Alpine and Padanian respectively. The Mediterranean arcs form the topic of three contributions. The Gibraltar and the Calabrian tectonic arcs, at the two ends of the Maghrebide chain, are considered by J.-P. Bouillin, M. Durand-Delga and Ph. Olivier to derive their curvature both from a combination of Mesozoic paleogeographic inheritance and Tertiary motion of the internal (Alboran and Aeolian) blocks along transverse strike-slip faults. Neotectonic and microtectonic observations of J.C. Bousquet and H. Philip have pointed out the Plio-Quaternary evolution and migration of the deformation in the Calabrian Arc and the Apennines. The various recognized phases are related to the passage from an oceanic subduction to a "continental subduction" and collision. The paper of J.E. Meulenkamp and F.J. Hilgen is a good example of the importance of a detailed stratigraphic analysis in geotectonic reconstruction. It shows the existence of NW-SE parallel, prevalently sinistral, shear-zones in Calabria which are held to account for the bending of the arc. Event stratigraphy indicates that the Hellenic Arc got its final shape through the combined effects both of the southward migration of the Cretan block and Plio-Quaternary vertical movements. Finally, the paper of E. Locardi suggests the existence during the Pleistocene of two different volcanic arcs in the Italian-Tyrrhenian region. One, developed on the coastal Apennines delivering potassic magmatism, and the other, centered on the present Tyrrhenian slope, delivering calc-alkaline volcanism followed by more potassic "Mediterranean" lavas.

The Lesser Antilles Arc System is of particular importance because it is considered the prototype of a young compressional active margin where the accretion of oceanic rocks and sediments takes place. In fact, only on the Barbados Ridge margin did DSDP drilling (Leg 78A) give direct evidence of accretion. Two contributions have been invited. The paper of A. Mascle, B. Biju-Duval, P. de Clarens and H. Munsch describes the genesis of the Barbados accretionary prism and the various forms and intensities of tectonic deformation, amongst which gravitational gliding is common. An important parameter in all the tectonic processes is the occurrence of fluids and overpressured sediments. In fact, the pore fluid pressure determines the location of the main décollements, facilitates the gravitational tectonics and is a leading factor in mud diapirism. J.F. Stephan, R. Blanchet and B. Mercier de Lepinay describe the arc structures with Caribbean convexity located at the periphery of the Caribbean plate and delimited by long transcurrent fault systems. Such circum-Caribbean festoons are the result of the recent bending of the continental frame related to a relatively modest E-W shortening of the whole system ("pseudo-subductions").

The last section is devoted to the Pacific Ocean basin, the most extensive individual unit of terrestrial structure, antipodal to the Eurafrican mass. On the hypothesis of an expanding earth, A.R. Crawford postulates that the Pacific probably originated catastrophically in the Late Precambrian or Early Paleozoic, contemporaneously to the critical stage, represented by the "Pan-African Event" (750-450 Ma), when there was the first development of future continental outlines. The Pacific is conspicuously asymmetrical in having a well-defined eastern boundary and a broad western one, characterized by a complex mosaic of continental and ocean crust, with island arcs and several marginal basins. As stated by S. Uyeda in his paper, although much progress has been made in understanding the origin of active arc systems, we are still far from the solution of a number of very basic problems. The subduction models are not able to explain (1) the origin of arc volcanism and high heat flow and (2) the extensional opening of back-arc basins. Besides there is still no definitive answer to the question: why are the arcs arcuate in shape? Uyeda believes that the sphericity of the earth, as well as the collision of buoyant features, may be responsible for the curvature of the island arc. The results in the Japan Trench of the French-Japanese "Kaiko Project", as well as those of the deep dives of submersible "Nautile" are discussed by J.-P. Cadet, K. Kobayashi, S. Lallemand and L. Jolivet. They show, instead of active accretion, normal faulting and large scale sediment mass sliding, whose frontal boundary constitutes the topographic Japan Trench. Some long and steep slump scars subparallel to the trench axis were detected on the inner slope. Also, Lower Cretaceous crushed limestones were observed near the trench. The tectono--magmatic develoment of the intra-oceanic Mariana arc system constitutes the topic of L. Beccaluva, G. Serri and J. Dostal's paper. Their new chronological data seem to indicate that back-arc spreading starts during a marked decrease or even absence of arc volcanism. Also, they suggest that boninitic magmatism may be generated during different stages of the geotectonic evolution of the Mariana system, namely in Middle and Upper Eocene, Lower Oligocene and Upper Miocene. All the volcanic rocks dredged on the inner slopes of Mariana, Palau and Yap trenches are considered to be derived from island arc-interarc mantle sources; this implies a tectonic erosion of the margin since the Eocene. The complexity of the structural picture is well pointed out by G. Pasquarè, F. Forcella, A. Tibaldi, L. Vezzoli and A. Zanchi who describe a continental volcanic arc, the Trans-Mexican Volcanic Belt of Plio-Quaternary. Their accurate analysis shows the structural control of the volcanic arc, through strong and repeated reactivations of different trends, belonging respectively to the Gulf of California and Basin and Range structures. In spite of its volcanological and geochemical uniformity, this volcanic arc resulted from the control operated by different and partly diachronous geodynamic mechanisms. Finally, in a closing paper, the convener draws attention to the importance around the Pacific of the post-orogenic up-arching and collapse tectonics in the creation of the back-arc basins.

We may close this brief review noting, together with S. Uyeda, that the more our knowledge advances the more we seem to be farther away from definitive solutions. Must we then perhaps share the pessimistic philosophic view of Robert M. Pirsig who says: "Through multiplication upon multiplication of facts, information, theories and hypotheses, it is science itself that is leading mankind from single absolute truths to multiple, indeterminate, relative ones."? May we perhaps consider "the whole history of science a clear story of continuously new and changing explanations of old facts"?

I am indebted to many people. To Senator Prof. Carlo Bo, Rector of the University of Urbino, for having agreed to host the Conference. To Prof. E.R. Oxburgh, President of the EUG, and to Prof. A. Praturlon, President of the Italian Geological Society and their Councils, for having granted the sponsorship of their organizations. To Drs. David Murray, Mario Tramontana, Ennio Polidori, Pierangela Donnanno for having helped in the editing and preparation of the manuscripts. To referees and advisors too numerous to mention. To the contributors who, with few exceptions, were able to produce their manuscripts on time. Finally, to my wife and children, for having had to put up with me during the editing of the volume.

The Urbino Arc Conference, and therefore the drawing up of this volume, could not have taken place without the help and the financial support of the following institutions and organizations whom we heartily thank:

- Agip S.p.A.
- Canada Northwest (CNW) Italiana S.p.A.
- Cassa di Risparmio di Pesaro
- Consiglio Nazionale delle Ricerche
- Elf Italiana S.p.A.
- Enea
- Ministero Pubblica Istruzione
- Snia BPD S.p.A.

Urbino, May 1986　　　　　　Forese Wezel

CONTENTS

I. GEOTECTONIC CONCEPTS

DIAPIRIC KRIKOGENESIS

S. WARREN CAREY

University of Tasmania

ABSTRACT

For nearly two centuries many European and Russian geologists have regarded orogenesis as a gravity-driven diapiric process with lateral spreading dominating the upper ten km, while British and American schools attributed orogenesis to lithospheric convergence. Confirmation of ocean-floor spreading, combined with the assumption of virtually constant Earth radius, implied that ocean growth had to be balanced by lithosphere excision, which was assumed to occur at the trenches, associated with convergent orogenesis there. The diapiric alternative, with Earth expansion balancing ocean-floor growth, has been argued in German and Russian, but, until recently, has had little consideration in the English literature.

Diapiric orogenesis implies a tensional environment, and is consistent with all known constraints - seismicity, heat flux, vulcanism, telescopic stacking of nappes by hundreds of km, the essential characteristics of eugeosynclines and miogeosynclines, the circular arc pattern (krikogenesis) and common asymmetry. Benioff zones are the outer boundaries of orogenic diapirs, and bound active diapirs against passive lithosphere. Globally, orogenic zones and ocean-spreading ridges divide the lithosphere into polygonal prisms, whose size reflects the thickness of the mantle. A sinistral torsion follows the Cretaceous equatorial Tethys, and normal to it a conjugate dextral torsion separates the "land" and "water" hemispheres and passes through the Cretaceous poles, which were the centres of maximum and minimum continental dispersion.

OROGENESIS, VERTICAL OR HORIZONTAL?

From the beginning of thought about rock deformation, two opposed schools have competed: (1) orogenesis is a diapiric process with secondary lateral spreading, both gravity-driven; (2) orogenesis results from horizontal compression, caused by shrinking of a cooling Earth, or latterly, by the convergence of lithosphere plates.

Pioneers and advocates of diapiric orogenesis were Gillet-Laumont in 1799, Scrope in 1825, Schardt in 1823, Kuhn in 1836, Naumann in 1849, Herschell in 1856, Bombicci in 1882, Reyer in 1888, 1892, and 1894, Schardt in 1898, Bonarelli in 1901, Steinmann in 1905 and 1907, Anelli in 1923 and 1935, Haarmann and also Wegmann in 1930, Van Bemmelen in 1931 and several later works, Signorini in 1936, Dal Piaz in 1943 (who reviewed the literature), Ramberg in 1963 and several later works up to 1983, and an important school in Russia led through the fifties and sixties by Beloussov. Extensive model experiments, with careful attention to dimensional validity, were pioneered by Reyer, and developed by Beloussov in Russia and by Ramberg in Sweden. Every signficant school in continental Europe and Russia has contributed to diapiric orogenesis, but Britain and America are conspicuous by their absence.

The horizontal compression concept, pioneered by Sir Hames Hall in 1815, and developed by Élie de Beaumont in 1852, has been adopted almost universally in France (e.g. Goguel), Britain (e.g. Lyell, Geikie, Peach & Horne), and America (particularly by Dana, Bailey Willis, and Bucher). Hall made models, and Bailey Willis made extensive pressure-box experiments to model Appalachian foldings, but in contrast with Reyer, Belousov, and Ramberg, all of these involved shortening of the sedimentary pile but not of the supporting basement. When compelling evidnce of gross post-Jurassic spreading of the ocean floors triggered a Kuhnian revolution to plate tectonics in the nineteen-sixties, general adoption of subduction with horizontal compressional orogenesis became axiomatic faith.

This plate tectonic theory involved the marriage of three independent concepts: (1) the documented spreading of the ocean floors, (2) the assumption that the diameter of the Earth has changed litle since the beginning, and (3) that orogenesis is caused by horizontal compression. In America, (2) and (3) were axioms, hence with the advent of (1) acceptance of plate tectonics was automatic and rapid. Had ocean-floor spreading been discovered, not by Americans, but by Russians, who have long favoured diapiric orogenesis and among whom there is an important group who support gross expansion of the Earth, a very different philosophy might have emerged.

Earth expansion has had a long and diverse gestation:

(1) The concept was pioneered in Russia by Yarkovski in 1899 and 1912, Bogolepow in 1922, 1925, and 1928, and more recently by Kirillov, Neimann, and Letavin in Moscow, Blinov in Kiev, and Kuznetsov in Novosibirsk, leading up to the 1981 Moscow conference on Earth expansion and pulsation organized by Professor Milanovsky of Moscow State University.

(2) In the German literature, the Earth expansion alternative to explain the Wegener continental dispersion was developed by Lindemann in 1928, Hilgenberg in 1933, Joksch in 1955, and Jordan in 1964, and in Vienna by Keindl in 1940, and in Budapest by Egyed in 1956.

(3) Independent "loners" who argued for Earth expansion were William Lowthian Green, a Scotsman in Hawaii who preceeded even the Russians, Mantovani in France in 1909, Hixon in Boston in 1920, and Halm in South Africa in 1935, two ore geologists, the Walker brothers of Colorado, in 1954, and the Brazilian petroleum geologist Groeber in 1959.

(4) Then there was the group of globe-makers, who extended the Wegnerian Africa-to-South America closure, excised all oceans, and fitted all the continents together on a globe about half the diameter of the present Earth: Hilgenberg (mentioned above) in 1933, Brösske of Düsseldorf and Barnett of London in 1962, Creer in Newcastle in 1965, Groves of Memphis Tennessee in 1976, and Vogel of Werdau, East Germany, in 1977.

(5) Meanwhile cosmologists contemplating decline in the gravitational "constant" recognized that this implied Earth expansion: Dirac in 1937, Dicke and Brans in Princeton in 1957 and 1966, Jordon in Hamburg in 1964, Hoyle and Narlikar in 1971, and van Flandern in Washington in 1974 and 1976. Tuzo Wilson was sympathetic to this view in 1960, but switched to the subduction model later.

(6) Important reviews, supporting Earth expansion from a tectonic viewpoint, were published by Fairbridge of Columbia in 1964, and Holmes in 1965. Heezen suggested Earth expansion in 1957, but subsequently shifted to the plate-tectonics model.

(7) Dearnley, of the British Institute of Geological Sciences, reconstructed Precambrian orogenic belts on an Earth 69% of the present radius in the Proterozoic, which had expanded to 94% by the Cambrian; by contrast, Waterhouse in 1967, then of Toronto, argued for slow expansion until the

Mesozoic, but very rapid expansion since the Mid-Cretaceous.

(8) Meservery, of the Massachussets Institute of Technology, concluded in 1969 that the generation of the present configuration of the continents from Pangaea was topologically impossible without Earth expansion, a conclusion confirmed by Owen of the British Museum by precise cartography, and recently by Perry of Wyoming by systematic computer modeling.

Space does not allow discussion here of this literature or the many paradoxes and dilemmas inherent in subduction and in the assumption of constant Earth radius. These have already been fully discussed by me (Carey, 1958, 1973, 1976, 1983, 1986).

The question of Earth expansion should shortly be resolved finally by the current NASA intercontinental chord measurements. The rate of change of a single inter-continental chord can be interpreted either as "plate" movements or Earth-radius change. But Dr. W.D.Parkinson has shown that when three stations lie on a single great circle, three *simultaneous* equations can be written of the form $dC/_{dt} = 2\ dR/_{dt} \sin {}^{\varnothing}/2 + dD/_{dt} \cos {}^{\varnothing}/_{2}$ in which the derivatives of C, R, D, and ø are the rates of change of chord, Earth radius, surface arc, and angle subtended at the Earth's centre. Given chord changes, these simultaneous equations can be solved uniquely for the rate of change of Earth radius. This is pure geometry, irrespective of any theory of subduction, expansion, or whatever. On the data so far published by NASA the rate of increase of Earth radius comes out at 2.4 ± 0.8 cm/year. This rate of radius increase accounts for *all* ocean floor growth since the Early Cretaceous, without *any* subduction at all.

The avalanche of the plate-tectonics revolution has established compressional orogenesis in the minds of most. However my task herein is to examine the alternate combination of ocean-spreading, Earth-expansion, and diapiric orogenesis, and to show, not only that it remains consistent with all the new data, but is in fact the only consistent solution.

PARADIGM FOR A SIMPLE DIAPIRIC OROGEN

The stages in the development of an orogen are shown in Figure 1. The initiation of primary stretching in continental crust leads first to "necking" or thinning, so that the top and bottom surfaces of the continental crust converge towards zero at some five km below sea-level, by which time the mantle below it has already risen 30 km. Thus, although the surface of the thinning continental crust subsides steadily, the bottom and the mantle diapir below it rises, and continues to do so throughout orogenesis. Crustal thinning caused by the expanding interior, resulting in gravity drive towards isostatic equilibrium, causes all the motions.

Orthodox compressional theory agrees that during the geosynclinal stage of orogenesis the continental crust must stretch and thin, because otherwise there is no possibility of maintaining even approximate isostatic balance through the millions of years involved in this stage. The fact that, after the melting of the ice sheets of the last ice age, isostatic balance has been nearly completely restored in only 10,000 years in cold inactive crust of Canada and Scandinavia, implies that, in the much hotter profile of an orogenic zone, isostatic balance would be closely maintained during this extensional stage. But orthodox tectonics then reverses from crustal extension to gross shortening. By contrast, in the expansion model, extension persists through all stages, and the gravity-driven subcrustal diapiric motion is *upward at all times.*

At the stage when the continental crust has thinned to zero (middle diagram of Figure 1) there are two contrasting sites for deposition of sediment. In the axial zone is a eugeosyncline, which is deep, with an active and unstable floor, with many faults and rifted troughs; volcanic lavas come up from the hot rising diapir below; sediment accumulates rapidly without much reworking by wave action; earthquakes trigger turbidity flows; whole slabs of sediments, several hundred metres thick and several square kilometres in area, creep slowly down-slope, gliding on deeper sand layers (which have become quicksand because of excess water pressure), and may become contorted in the process.

These eugeosynclinal sediments contrast with sediments in the miogeosyncline, which subsides because the continental crust below it is thinning, and isostatic balance therefore requires that the surface falls below sea-level, deepening in the direction of thinning, that is, towards the main trough (Figure 1, middle diagram). But the continental floor is thicker and much more stable than that below the eugeosyncline; heat flux and temperatures are lower, with no vulcanism; seas are shallower so wave action winnows out rock fragments and weak mineral grains; marine life is abundant, so limestones are common.

A ridge of basement commonly rises as a rim separating the miogeosyncline from the eugeosyncline, because the density profile below the eugeosyncline, from the surface right on down, is less than normal (because of steeper temperature gradient), and the consequent upward drive is resisted somewhat laterally by the strength of the crust. As regional isostatic equilibrium is approached, a positive gravity anomaly over the ridge balances a negative anomaly over the axial zone. This basement ridge separates two distinct

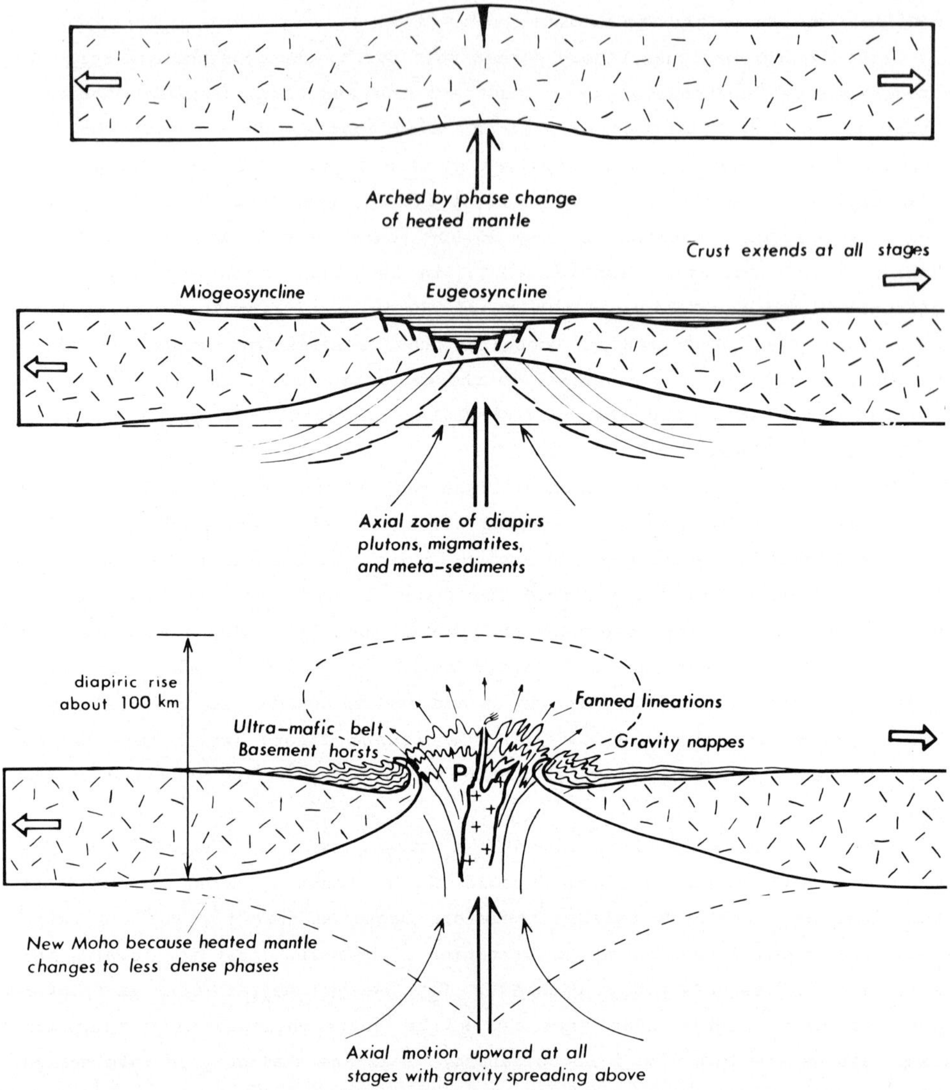

FIG. 1. Model of the development of a geosycline and orogen during continuous crustal extension and continuous diapiric rise of about 100km.

facies. On the eugeosynclinal side, not only are there contemporary volcanics, and injections of granites during the subsequent folding, but also serpentine belts and ophiolites derived directly from the mantle, interlarded with the eugeosynclinal sediments.

After the pre-orogenic continental crust has thinned to zero, the continuing and accelerating ascent of the deep diapir begins to drive out the new sediments, regurgitating the eugeosynclinal gut, which then spreads laterally at the surface. Consider the point marked **P** in the middle of the rising orogen at the general level of the surface outside the orogen (Figure 1). It is being driven upward by the rising diapir below, but pressing down on it is the weight of the pile above. So it is compressed between the two vertical forces, which squeeze it to spread sideways. The rate at which it spreads depends on how heavy is the overburden load, and the "viscosity" of the materials. If this rate is less than the rate of the rising diapir, the orogen rises higher (which increases the overburden load at the level of **P**) and continues to rise in this way until the overburden load at **P** is sufficient to drive the sideways flow at the rate that the diapir is pushing up from below. The *surface* of the orogenic zone then ceases to rise, but the orogen continues to spread laterally and will continue to spread as long as the diapir rises, piling nappe on nappe.

The root zone of an orogenic belt (Figure 1) would be something like 100km wide. To be conservative, let us assume only 50. The vertical rise of the driving mantle column would also be something like 100km, but again to be conservative let us assume only 50. So the rising ram of the orogenic axial zone would be 2500 sq km in cross-section. If this extruded at the surface as a sheet 1000 metres thick (an average thickness for nappes), the sheet would flow out for 2500km. But such a sheet would be halted by increasing bottom friction before traveling far, so the rest of the sheet overides the first part, so that nappe stacks on nappe, or thrusts into or below earlier nappes. But, because of the continuing rise of the diapir root, the cumulative sum of overthrusting would still add up to the 2500km. So the total nappe overthrusting observed in the Alps, Himalayas, or Appalachians, or amy other orogen, which according to orthodox dogma implies crustal compression by several hundred km, does not imply *any crustal shortening whatever.* How wrong can they be? Far from crustal shortening of several hundred kilometres, the Alps probably represent crustal *widening* of a few tens of kilmetres!

Within the orogenic zone, the lineations and thrust surfaces are all in the direction of the flow, the inner zones always over-riding their flanking neighbours, very steeply near the centre, but becoming flatter and flatter outwards, as nappes over-ride the miogeosyncline. The soles of these nappes may even come to slope downward when the nappe pile extrudes over the miogeosyncline beyond the positive anomaly zone separating the eugeosyncline from the miogeosyncline, but if the sole is traced back to its source it originates from the steep upward drive from the diapir. Compare this with the outflow of the crystaline salt nappe from an Iranian salt diapir (Figure 2).

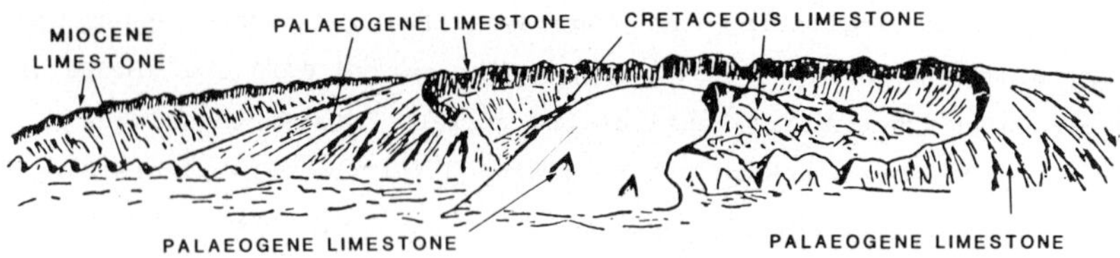

FIG. 2. Salt "glacier" nappe extruding from Kuh-i-Anguru salt diapir, Iran. (From a field sketch by Dr.G.M.Lees)

Observe, in the bottom diagram of Figure 1, that the thinned-out edge of the original continental crust is turned up and overturned to overthrust the miogeosyncline, where its driving pressure, and the pressure from higher nappes over-riding it, push the miogeosynclinal sediments forward, producing more thrust sheets and trains of concentric folds.

In the folded frontal belt, all the manifestations of horizontal compressional tectonics develop. The miogeosynclinal zone *is* foreshortened. The microscopic fabric of minerals *do* conform to compressional patterns. Folds *are* "flattened" transverse to the axial surfaces of the folds. But none of these phenomena involve shortening in the basement, nor narrowing of the orogenic belt as a whole, which actually widens during orogenesis.

The Moho is defined by a sudden jump in the velocity of seismic compressional waves from 6.7-7.2 km/s in the lower crust to 7.6-8.6 km/s at the top of the mantle. In the top diagram of Figure 1 the Moho is the base of the continental crust. In the bottom diagram, it is shown as a broken line plunging down under the orogen to a depth of several hundred kilometres. The material in the rhomboid region above this new Moho is former mantle material originally far below the Moho, but it has risen a hundred kilometres or so with the diapir. As a guide to the amount of vertical rise, a thin broken line completes a broad arch to connect the back-turned edges of the original base of the continental crust.

This material is still at a relatively high temperature, but under very greatly reduced pressure, so it undergoes a phase change from dense eclogite (or equivalent material) with a seismic velocity above 8.0 to less dense gabbro with a seismic velocity well below 8.0 (or other similar phase change). A considerable volume of water in the gas phase (i.e. above its critical temperature) is also introduced, which results in serpentinization, with reduction in density and in seismic velocity. So a seismologist finds a deep root under the orogen. Likewise a geophysicist measuring gravity finds densities of about 3.0 extending down deep under the orogen instead of the

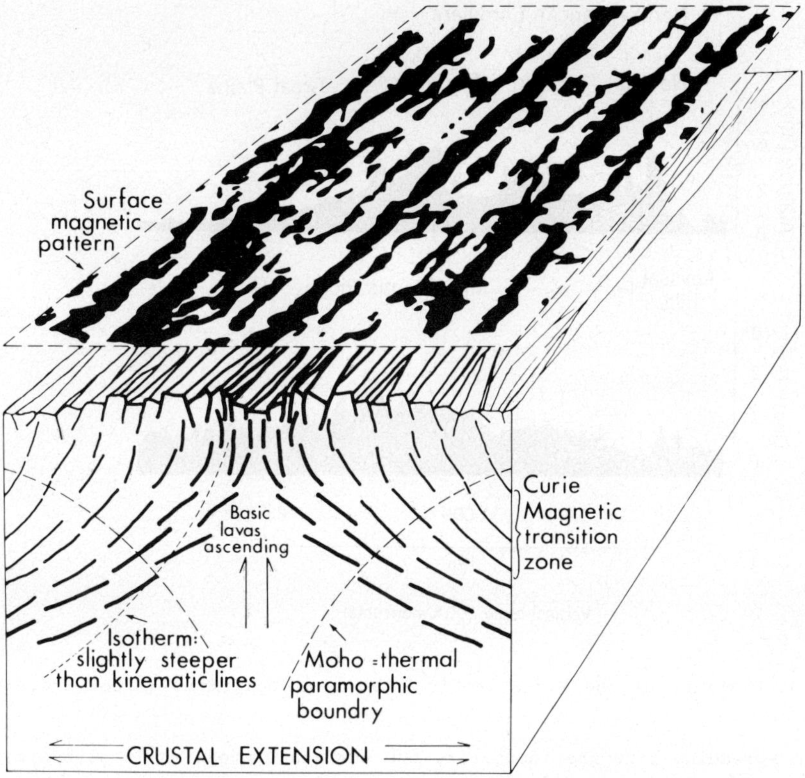

FIG. 3. The diapiric root of an oceanic spreading ridge is genetically identical with an orogenic root (Fig. 1).

higher densities of the mantle, so he confirms that the orogen has a deep root.

Those who believe the orogen is the locus of intense crustal compression interpret the depression of the Moho as evidence that the crust has been squeezed together and doubled in thickness to form the root. In this regard, comparison should be made with the mid-ocean spreading ridges (Figure 3). Under them the Moho, as determined by the compressional seismic velocity, plunges down exactly as in Figure 1, and so does the density distribution as determined by the gravity profile. No continental crust exists there to squeeze down as a root, and in any case everybody agrees that these are zones of extension. In the subsurface, orogens and spreading ridges are identical seismicly and gravity-wise. It will be shown later that both link up into the extensional mosaic of the earth's expanding surface. They differ because orogens incorporate continental crust and a large volume of sedimentary material. Because the underlying distribution of mass below an orogen and an

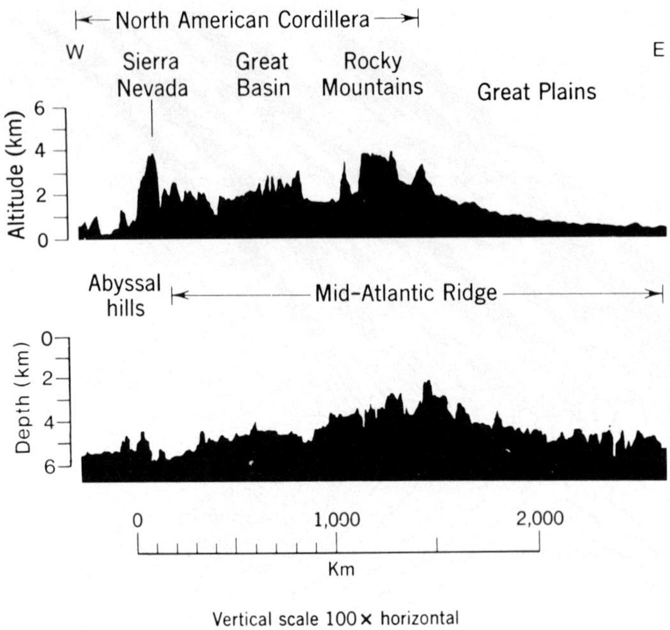

FIG. 4. Profiles of the Mid-Atlantic Ridge and Rocky Mtns.(After Sheperd).

oceanic spreading rift are identical, the surface expression is also similar (Figure 4).

The genetic similarity of spreading ridges and orogens has been recognized by geologists not blinded by compressional creed. Thus Zonenshayn at the 1972 International Geological Congress crystallized the wisdom of many Russian geologists that, fundamentally, geosynclines are the sites where initially new oceanic crust is created: *"The origin of a geosyncline is accompanied by tension, by moving apart of lithosphere plates and by creation of a new oceanic floor....Eugeosynclines appear to be analogous in their initial stage to ocean ridges, and in the mature stage to island arcs. On the whole, the evolution proceeds, possibly, from conditions resembling mid-ocean ridges to an island arc and further to an orogenic zone....Geosynclines (and eugeosynclines especially) are deep-rooted structures in which the energy and substance of the upper mantle rise up toward the earth's surface".*

The granitic cores of orogens have commonly been assumed to extend down to great depths, indeed, the name bathylith (Greek βαθψζ deep, λιθοζ rock) implies this. However, evidence has been increasing that true granite forms only the hood of such plutons. Ross (1985) has inferred, from a study of the most deeply eroded parts of the Sierra Nevada bathylith and of the fragments brought up in the rising magma, a downward sequence from granite to 10km, through tonalite, then sillimanite replaces andalusite, then migmatite, then hornblende granulite to 25 km, and two-pyroxene granulite at 30km.

APPLICATION TO REAL OROGENS

This model of an orogen is over simplified: (1) It is bilaterally symmetrical, whereas orogens are commonl asymmetric (like the salt nappe of Figure 2). (2) The model considers only one orogenic cycle of sedimentation and folding. (3) The rate of orogenesis increases through geological time, so that ancient orogens may differ from later ones. Although each orogenic zone has its own individuality, the basic principles outlined above apply to all. Bearing this in mind, let us now compare with the basic model three orogens, Appalachians, Alps, and Himalaya.

Appalachians

In the mid-west of the United States the pre-Appalachian basement crops out as in the bottom diagram of Figure 1, thence eastward to flat-lying strata of the miogeosyncline in the Allegheny synclinorium, which thicken progressively toward the concentric folds and nappes of the Valley-and-Ridge zone, which are thrust forward on flat or shallowly-dipping surfaces. Next basement rocks of the Blue Ridge are thrust over the miogeosynclinal strata. Beyond are the complexly folded eugeosynclinal facies, with granites and ophiolitic rocks, then the Piedmont, where the Appalachian rocks disappear under the younger cover of the Mesozoic.

The front against the miogeosyncline is bowed into a series of arcs. The overthrusting is most intense where the arcs bow outward, but much less intense where the arcs bow inward towards the crystalline core. This is what should be expected from a chain of diapirs, but in the compression model, the overthrusting should be most intense where the craton advanced most.

King (1959) has depicted four independent sediment cones spreading northwest from the Appalachian axial zone (Figure 5). In the Mid Ordovician an orogenic tumour rose actively in North Carolina and its rapid erosion shed a fan of marine clastic sediments 2400m thick over Kentucky, while less than 1000m accumulated elsewhere along the Appalachian front. In the Late Ordovician, an active tumour regurgitated in Maine and spread a thick fan of sediments across Vermont and far into Quebec. In the Late Devonian it was New York's turn, and a clastic fan up to 2700m thick spread across West Virginia, Pennsylvania and upper New York State while only 600m were deposited in Tennessee and about 30m in Alabama. But Alabama had its turn in the

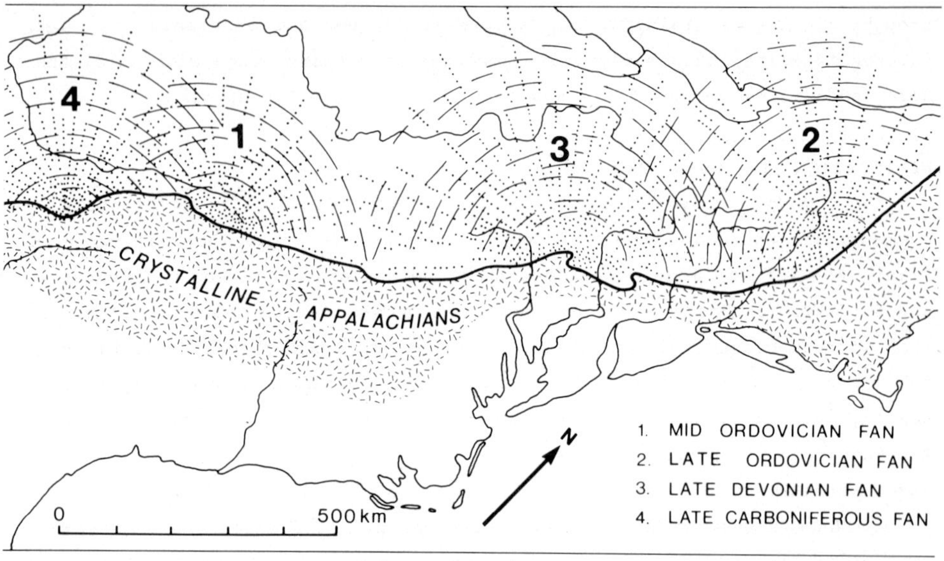

FIG. 5. Clastic fans from Appalachian diapirs. (After P.B. King).

Late Carboniferous when a regurgitating tumour rose in Gerogia to spread a fan of sediments up to 3000m thick.

This succession of events is logical in the diapiric model, but in the convergent-continents model a block of the foreland a couple of hundred km wide would have to close on the Appalachians in the Mid Ordovician towards Kentucky, a separate block would have to converge on Maine in the Late Ordovician, still another against New York in the Late Devonian, and another in the far south in Late Carboniferous times.

Even though the diapiric process may have been initiated more or less equally all along the trough, equality would be transient, because any slight inequality which raises the isotherms is favoured by feedback, because the rate of yield under a given load increases exponentially with absolute temperature. Hence diapirism always tends toward rounded tumours (krikogens).

Alps

The Vosges, Schwarzwald, and Bohemian massifs represent pre-Alpine basement. Lapping onto them are flat-lying Permian and Mesozoic in miogeosynclinal facies, which, farther to the southeast, start to buckle in the concentric folds of the Jura, underlain by flat-lying thrust surfaces as in the Valley-and-Ridge zone in front of the Appalachians. The Swiss plain (a sagging zone caused by continuation of the regional extension represented by the Alps themselves), intervenes between the Jura and the Alps. The van of the Alps are the Helvetic nappes, a stack of flat overthrusts of limestone and shale of the miogeosyncline facies, then upturned pre-Alpine basement, represented by Mt Blanc, Aiguilles Rouge, and Aar massif, which signals change of facies to the schistes lustrés of the Pennine nappes, then the zone of roots, and the ophiolites of the Ivrea zone. All this is consistent with the diapiric model.

It is interesting to compare the movement patterns of the Alps with the Malaspina Glacier where it flows on to the Alaskan coastal plain (Figure 6). The scale of the Alps is five times larger, and as gravity severely flattens upward movement, Figure 7 shows the Malaspina lobe compressed three-to-one for comparison with the Alps. Imagine now a narrow axial zone of "roots", whence "nappes" rise steeply out of the "pre-orogenic trough". One by one they turn over to the NW as great "nappes", over-riding above, or driving into the backs of others. The outermost "Pre-Alps" are the farthest travelled. A few lobes turn over toward the SE. Although not true in detail, the Malaspina pattern certainly mimes the Alpine structures. Yet orthodox geologists tell us that the Alps represent great shortening, and that the overthrust belt has been compressed to one eighth. But we know that the Malaspina lobe produced similar geometrical and structural patterns while dilating to 30 times its width! Can there be any question that the structures of the Alps could be produced by the upward extrusion of the contents of the geosyncline, spreading where the pile lacked lateral support? The jaws of the extruding zone may have remained fixed while the material squeezed out, or they could themselves have dilated.

Himalaya

Figure 8 shows two profiles across the Himalaya, the first by Gansser, and the second in accordance with diapiric orogenesis. As in the model, Mesozoic and Tertiary strata of the miogeosyncline thin out onto the pre-Himalaya basement SSW of the Indo-Gangetic lowland and thicken toward the synclinorium of the miogeosyncline. The first folds are the Siwaliks, (**si**), concentric folds in the youngest strata of the miogeosyncline. Thence the miogeosynclinal strata of the Lesser Himalaya, (**ls**), continue to the "Great Boundary Thrust" (**B**) where the Lesser Himalaya are thrust forward, then a succession of thrusts until the Great Central Thrust (**C**), the beginning of the High Himalaya, including Mt. Everest (**E**). The Main Central Thrust is where the pre-Himalayan basement first turns up to the surface, and is carried forward as a great nappe, most of which has already been eroded away leaving isolated klippen on the miogeosynclinal sediments. These klippen are upside down - they are less strongly metamorphosed at the base than higher up. This is as should be expected, because following back on the broken line of the thrust

FIG. 6. Flow pattern of the Malaspina Glacier, Alaska.

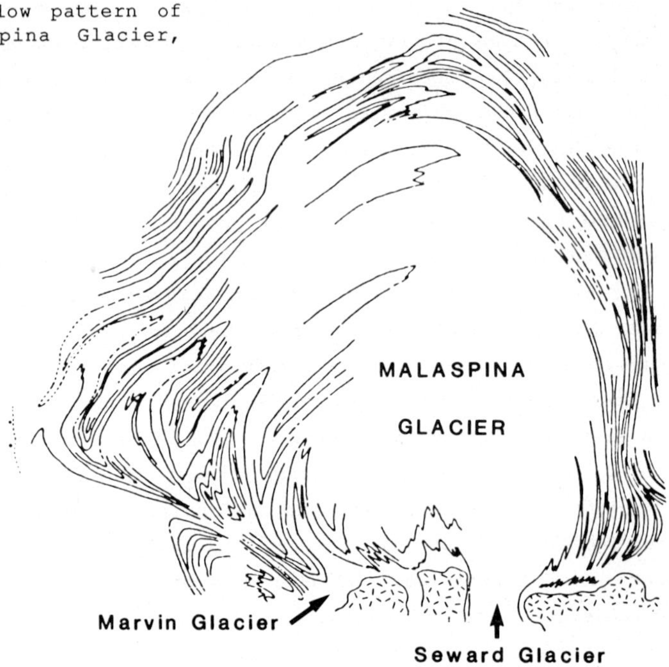

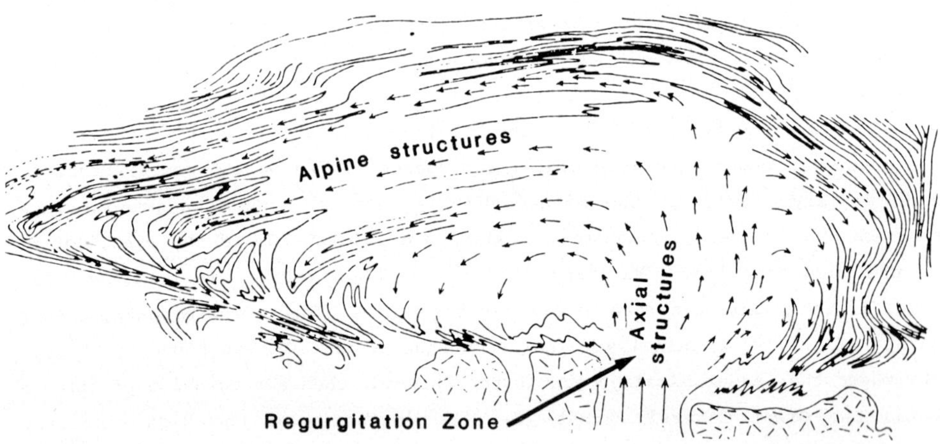

FIG. 7. Flow of Seward Glacier flattened 3:1 simulating extruding orogenic diapir flattened by its own weight. Compare the actual pattern of the Alps.

surface to its root leads back to the thinned and overturned basement. The miogeosyncline continues until the next major thrust at the Indus suture (S), and changes there to eugeosynclinal sediments. Mt Everest consists of Mesozoic

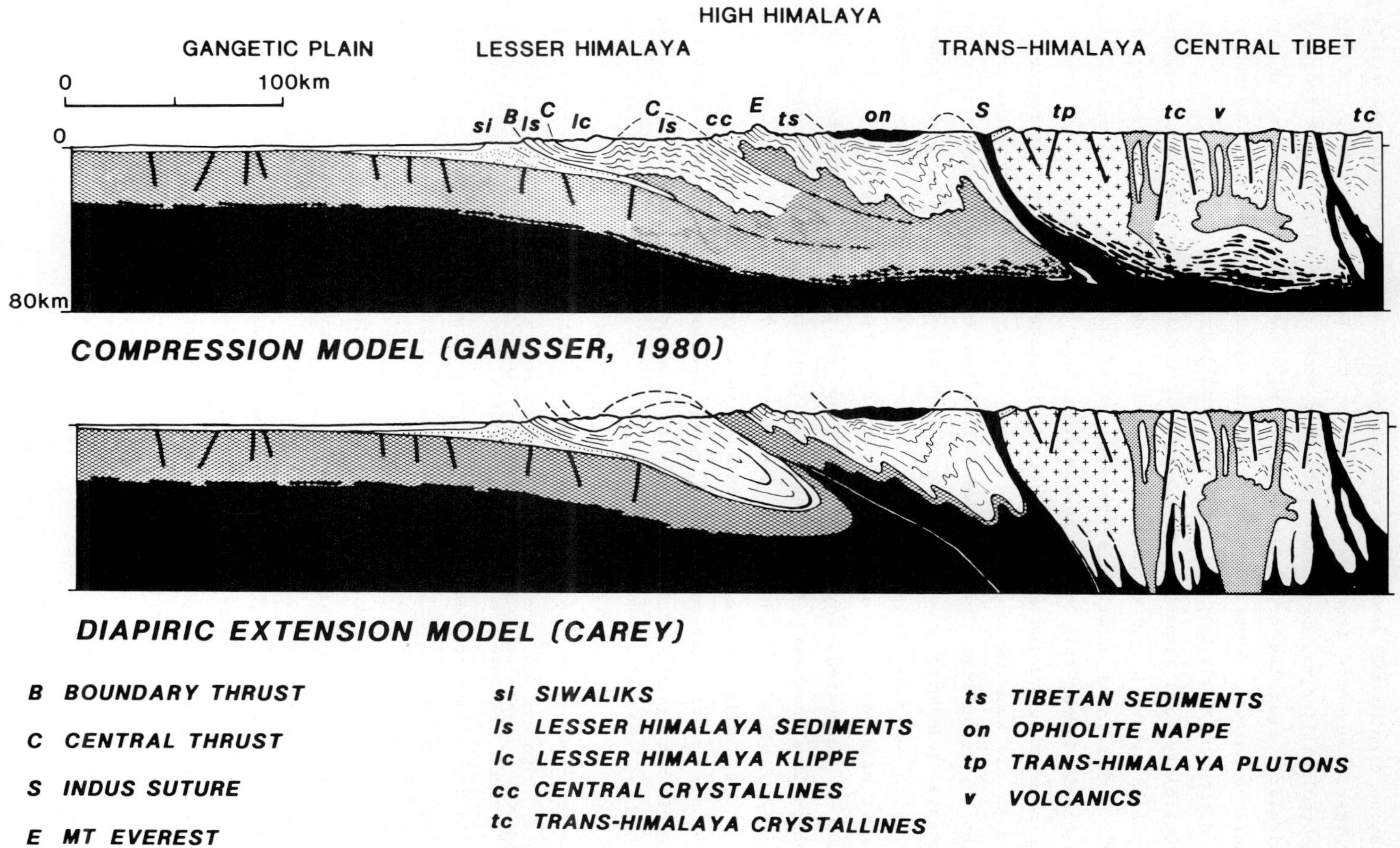

FIG. 8 Compressional and diapiric models of the Himalaya.

limestone dipping NNE toward the deepest part of the miogeosyncline, with a conformable sequence from the Cambrian to the Eocene. Ophiolites, of the eugeosynclinal facies, do not occur in this zone. Décollements are frequent particularly near the crystalline floor, with much back-folding as in the calcareous Helvetic Alps. Deep rift troughs developed near the Indus suture zone during the Early Tertiary, and rapidly filled.

Ophiolites and serpentinites rising diapirically from the Indus suture are pushed laterally 100 km SW as a great nappe (**on**) over the miogeosynclinal strata. The Indus suture has the long straight outcrop characteristic of megashears. Indeed this suture is the main shear of the Tethyan torsion (of which more later). Large exotic blocks of limestone foreign to that region are found in the zone SW of the Indus suture; their source is therefore to be sought in Afghanistan. The megashear along the Indus suture complicates the structure, producing steeply-dipping slices of various rocks, and frequent ophiolitic mélanges. Mélanges are a common feature of megashears, indeed, whenever mélanges occur along a long straight line of valleys, a megashear should be suspected. Beyond the Indus suture the facies changes to the eugeosynclinal sediment (**tc**) of central Tibet, intruded by large bodies of granite, tonalite, and gabbro (**tp**), and extensive Tertiary volcanics (**v**).

The Himalaya is nearly 400 million years younger than the Appalachian orogen and is still rising diapirically and spewing out its nappes. After the lapse of that amount of time, erosion will have cut ten kilometres or more off the highest parts of the Himalaya, reducing it to a peneplain, which will have arched up again to maintain isostatic balance because of its remaining roots, and this second-generation epeirogenic Himalaya will have been dissected by new rivers. Himalaya will then resemble more closely the present Appalachians. I say "resemble" because the tempo of orogenesis has accelerated significantly, and Late Tertiary orogenesis differs in degree from Late Paleozoic orogenesis, although conceived and delivered from similar orogenetic wombs.

GENETIC KINSHIP OF OROGENIC AND OCEAN-SPREADING DIAPIRS

As Zonenshayn (1973) has emphasized, orogens and ocean-spreading zones are genetically the same process. The Earth's surface is made up of eight primary polygons, each a few thousand km across, which meet each other along tectonically active zones (Figure 9). Plate theory calls them plates, some 100km thick. I call them polygonal prisms, extending down the full depth of the mantle, 3000km. Plate theory calls the spreading ridges mid-oceanic ridges,

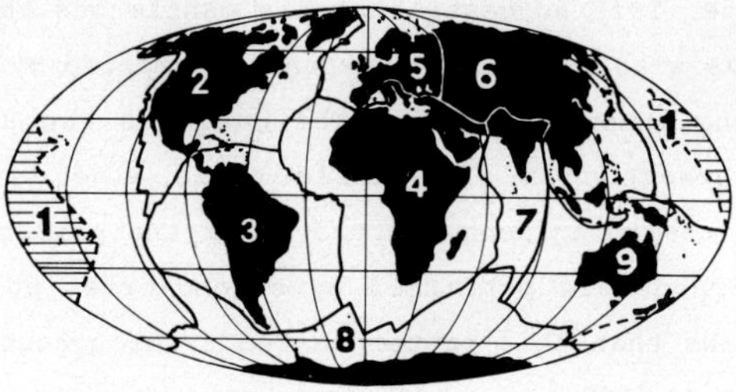

FIG. 9. Primary polygons of the Earth's surface.

although they are really peri-continental and only mid-oceanic for continent to continent separations. I combine them with the orogenic zones as peri-continental spreading zones.

The primary inhomogeneity of the Earth is a fluid core of nearly half its radius, overlain by solid mantle. Surely we should expect that this primary inhomogeneity would find most obvious expression at the surface, as indeed it does. The primary polygons, 3000km thick are a few thousand km across. If the mantle were only 1000 km thick the surface would express this with twice as many primary polygonal prisms. Observe on Figure 9 that each primary polygon consists of a continent surrounded by its accreted oceanic crust added during the last 100 million years. If that growth is removed, the polygons are reduced

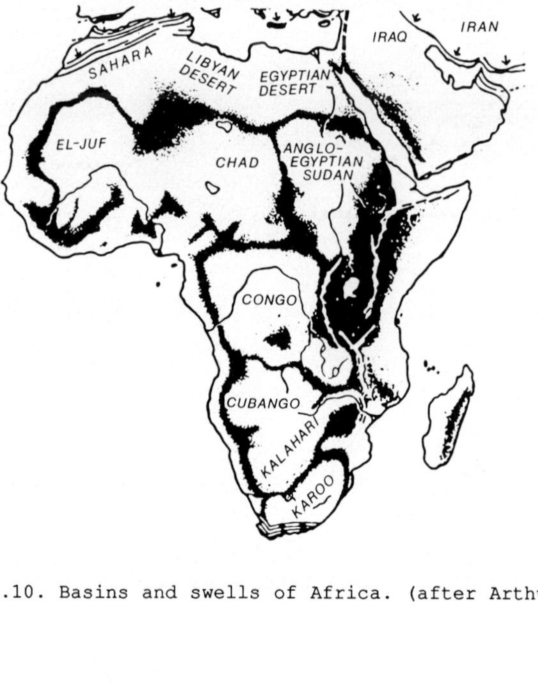

FIG.10. Basins and swells of Africa. (after Arthur Holmes).

to about half the size. This suggests that the mantle was then only 1500 km or so thick, which agrees with the rate of expansion implied by other data.

The second-ranking inhomogeneity of the Earth is the asthenosphere which separates the lithosphere from the mantle. This also is expressed at the surface, as the next most prominent feature after the primary polygons (Figure 10). The primary polygons are patterned by second-order polygonal basins and swells. Figure 11 shows that this pattern extends throughout the Earth surface without any significant difference between continent and oceanic lithosphere.

Looked at globally, by far the greatest share of the heat flux emerges at the spreading zones between the primary polygons, coming up all the way from the core. Likewise the second-order polygons, but less. The highest heat flux is along the swells. Indeed, that is precisely why they are swells. Most mineral substances exist in different forms according to temperature and pressure. Thus silica exists as quartz under shallow conditions, coesite deeper, and as much denser stishovite at still greater depths. Likewise the graphite-diamond and gabbro-eclogite transitions. Such transitions extend right through the core. Temperature has the opposite effect to pressure. So, if two places have the same kinds of rock composition below them, the depth of the many density transitions will be deeper at the hotter site, which swells upward in gravity equilibrium. Thus the swells rise higher because the

FIG.11. The second-order basin-and-swell pattern extends similarly over continental and oceanic crust.

temperature below them is higher, and the basin floors remain low because the temperature below them is lower. Primary polygon boundaries have the greatest share of earthquakes. Second order swells are also seismically active, to a second degree. Faulting associated with swells may develop into major rift valleys, like the rifts which frame the Lake Victoria basin in Africa.

As the Earth expands, the first adjustment to decreasing surface curvature occurs at the primary spreading ridges, but if this were all, the curvature of the primary polygons would remain too much. The lithosphere is not strong enough to support a major departure from isostasy, so the surface of the primary polygons adjusts to average out the new curvature by the second-rank basins and swells. If this were all, the unbalanced weight would still exceed rock strength, so adjustment continues the extension hierarchy to the third rank, the fourth, and so on. The thicker lines on Figure 12 indicate polygons tens of km across, which are the locus of minor seismicity. Within them are still lower rank polygons, five km or so across, which tilt independently with

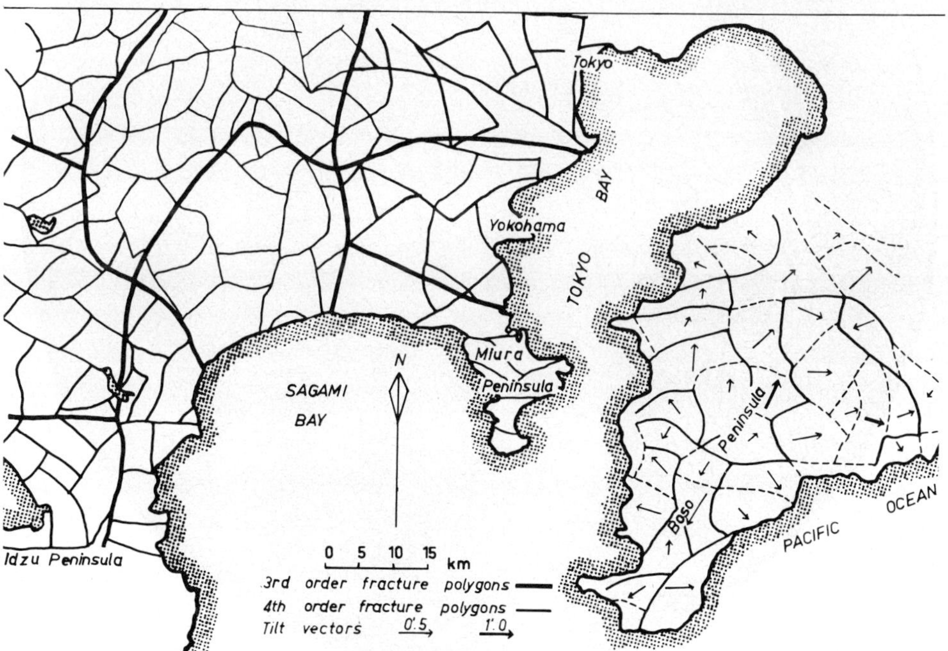

FIG. 12. Tilt blocks in Central Honshu. (After Miyabe, 1931).

a still lower level of micro-seismicity; and so on down to the master joints, a few hundred metres apart, and within them in turn systematic joints, by which all rocks are broken, and which allow the final adjustment to the changing curvature. All regions, except the youngest sediments which still yield rather than fracture, acquire two sets of joints nearly at right angles with their intersections nearly vertical - epeirogenic joints, the final adjustment to the decreasing curvature.

FIG. 13. Vogel's primitive globe within a modern outer globe shows that continents mover radially outward.

RADIAL MOVEMENT OF PRIMARY CONTINENTAL PRISMS

Klaus Vogel, of Werdau, German Democratic Republic, assembled all continents on a globe nearly half the size of the equivalent Earth, then enclosed his globe in an outer globe, to show relations to the present Earth (Figure 13). He found that the continents had moved *radially* outward (1983).

Schmidt and Embleton (1981), investigating Proterozoic polar-wander paths by moving continents until the polar curves coincided, were surprised to find that paths for North America, Greenland, Africa, and Australia coincided within error limits *with these continents in their present positions on the globe,* which suggested that their angular positions with respect to the Earth's centre now, were about the same as they were a billion years ago, their present separation being due to their movement out *radially* as the Earth expanded.

Dr.Ken Perry of Wyoming has demonstrated with geometrical precision what Vogel found with his globes. He set up a program based on matrix algebra and a hidden-line algorithm, so that continents may be moved radially from the centre of the Earth and translated using one centre of coincidence and one rotation pole (Figure 14). His program generated successive positions of spreading ridge, fracture zones, and magnetic anomaly lineations, and from these he calculated the amount of radial expansion implied by each anomaly. Dr. Perry also demonstrated by direct computation that the geometry of the Mid-Atlantic

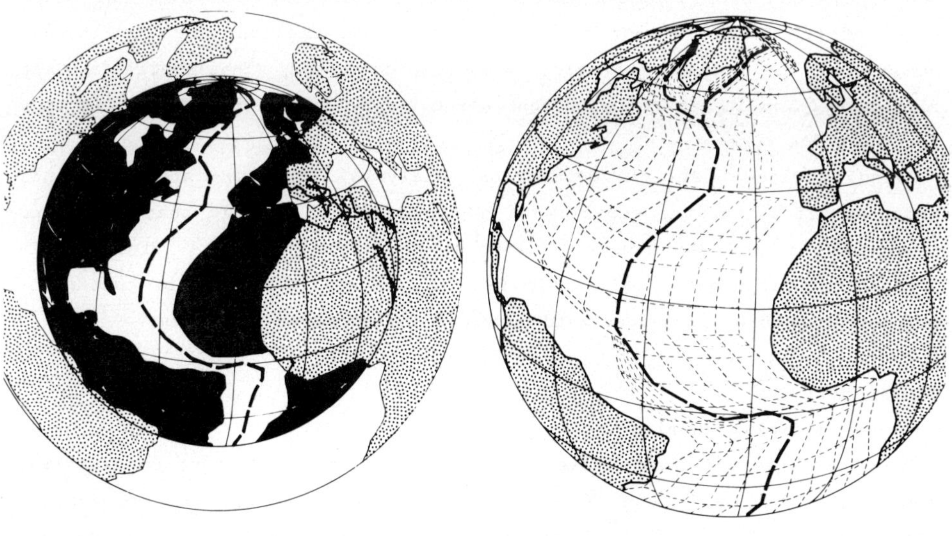

FIG. 14. Perry's radial translation of the continents, and generation of the Mid-Atlantic Ridge.

Ridge is compatible *only* with a radially expanding Earth.

Continents are welded to their own mantle

One of the surprises of the early 'sixties was that the rate of heat flow through continents is statisticly the same as through the ocean floors. Dr.V. V. Belousov, of the Russian academy, and Dr. Uwe Walzer, of the East German Academy, have each pointed out that this general equality of heat flux between continents and oceans contradicts plate tectonics. The radiogenic heat yield of continental rocks exceeds that of oceanic rocks by an order of magnitude. If continental lithosphere moved over passive mantle on a yielding asthenosphere, the heat yield below the asthenosphere should be generally uniform, so continental heat flux should be significantly greater than oceanic, whereas if the continents derived their radiogenic elements by differentiation of the mantle *fixed* below them, equilibrium heat flux would be more or less constant everywhere irrespective of the degree of differentiation, which is found to be so.

Dziewonski and Woodhouse (1984) investigating the transmission of large earthquakes recorded at thirty or more observatories in order to measure the temperature distribution below the surface, prepared global temperature maps at depths of 100km and 340km, and found that the cold temperatures below continents persisted down through the 100 and 340km maps, and even through 500km, the greatest depths studied. This agrees with expansion theory, but is contrary to plate theory, which requires the detachment of the plates from the mantle and free relative movement at the asthenosphere. Even the older parts of the Pacific floor formed more than 100 million years ago remain attached to the underlying mantle. Woodhouse commented, "That's still a great difficulty. Now it has to be thought about." The "great difficulty" is only for the plate concept.

Vogel's globes, Perry's geometrical analysis, Schmidt's and Embleton's polar-wander paths, the equality of continental and oceanic heat flux, and Dziewonski's and Woodhouse's temperature distribution - five wholly independent techniques - all indicate that the separation of the continents has been caused by radial outward movement during Earth expansion.

GLOBAL TORSIONS

Tethyan Sinistral Torsion

All Laurasian continents have moved westward in relation to Gondwanaland opposites (Figures 15 and 16). When the Americas are fitted back against Africa in the "Bullard fit", North America has to be moved 1800km further than South America. This means that during the separation of the Americas from

Africa, sinistral shearing of this amount occurred through Central America. This sinistral offset shows up in the magnetic spreading stripes in the Atlantic and Perry's computer regression of them (Figure 14). North of the Caribbean the spreading ridge is "mid-Atlantic", but south of the Caribbean spreading on the west side of the ridge is less than on the eastern. Likewise, on the western side of the Americas, spreading south of the Tethyan torsion is symmetrical east and west, but north of it significant spreading only occurs on the west side. In the Atlantic, cental eastern Pacific, and in the western Pacific, the computer plot (Figure 21) indicates the torsion zones by plotting the growth-lines very close together.

Figure 15 (upper) shows that the Andesite Line, the nominal boundary of the western Pacific, is offset 5,500 km from Samoa to New Guinea. Right around the Tethyan girdle, blocks broken from the continents and local orogenic belts are rotated sinistrally: Spain 35°, Corsica-Sardinia 90°, Italy 110°, India 70°, Seram 100°, New Guinea 30°, Mesozoic Mexico 130°, Colombian orogens 40°-60°, Greater Antilles 35°-45°, Newfoundland 25° (Figure 15, upper). All of these were published by me on tectonic grounds long before they were confirmed palaeomagnetically. Indeed, the first recognition of the Tethyan torsion was when I wrote in my 1938 doctoral thesis: "New Guinea has been sheared westward under a colossal shear system on a scale grander than has been demonstrated anywhere else on the globe....The stresses which are responsible for this great westerly displacement are of continental dimensions. They are probably related to the main architectural pattern of the globe." Recently, Pigram *et al.* (1985) reported that the stratigraphy of the Sula platform indicated its westward displacement 2,500km from near central New Guinea.

An orogenic belt forms an **S**-shape when dragged sinistrally, a **Z**-shape when dragged dextrally. Several **S** drags occur along the Tethyan torsion zone, e.g.: from the Atlas of North Africa, through Sicily, Italy, and the Alps; from the Zagros of Iran, through Baluchistan, Kashmir, and the Himalaya; from Sumatra, through Java, the Banda loop, SE and NE Sulawesi, the Sula spur, and New Guinea; the loop from northern Venezuela, Trinidad, the Lesser Antilles, and the Greater Antilles is the southern half of such an **S**, the upper half being represented by a sinistral megashear through the Cayman Trench and Motagua.

When North America is rejoined to the bulge of Africa according to the Bullard fit, Europe has to be pushed westward to allow Newfoundland to wrap around Africa as far as Oran. This is indicated independently by the structure of Europe, where the existing complex of oroclinal bends and paleomagnetic rotations are automatically resolved by reversing the offset of the Tethyan torsion and the opening of the Mediterranean (Figure 17).

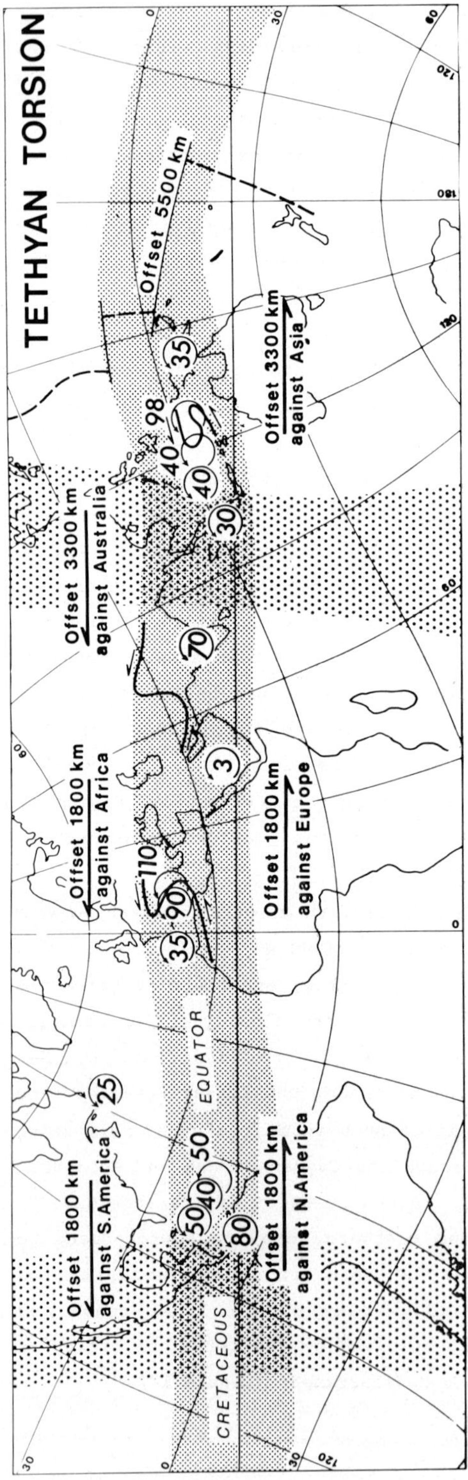
TETHYAN TORSION
Offset 5500 km
Offset 3300 km against Asia
Offset 3300 km against Australia
Offset 1800 km against Africa
Offset 1800 km against Europe
Offset 1800 km against S.America
Offset 1800 km against N.America
EQUATOR
CRETACEOUS
35
98
40
40
30
70
3
110
90
35
25
50
50
40
80

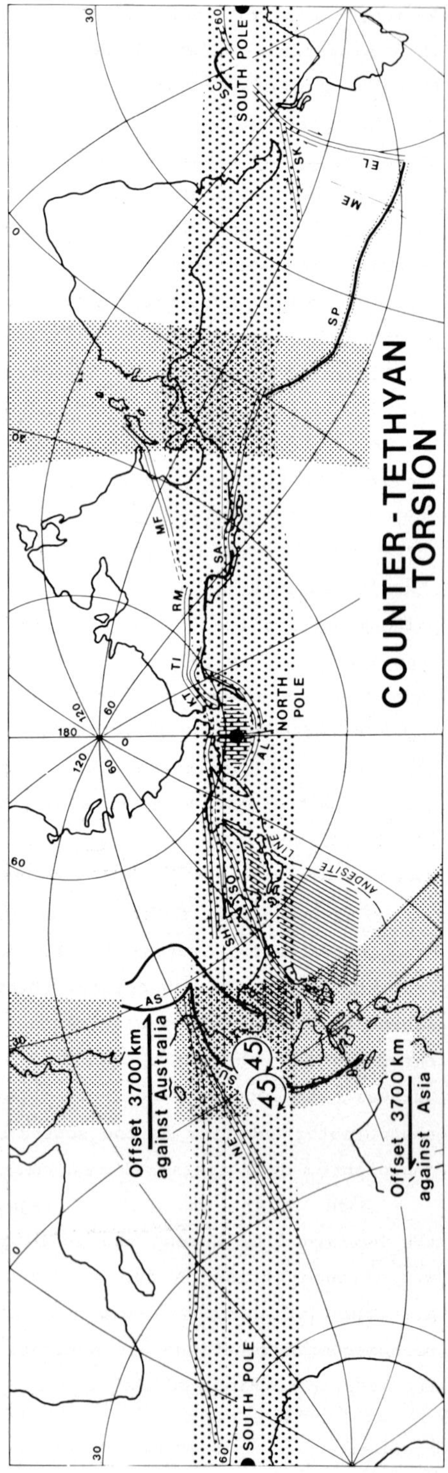
COUNTER-TETHYAN TORSION
SOUTH POLE
NORTH POLE
SOUTH POLE
ANDESITE LINE
Offset 3700 km against Australia
Offset 3700 km against Asia
SC
SK
EL
ME
SP
MF
SA
RM
TI
KT
AL
SO
SH
AS
NE
SU
45
45

Fig. 15 (p. 24): Tethyan sinistral torsion (fine stipple) and dextral counter torsion (coarser stipple) on oblique Mercator projections normal to each other.

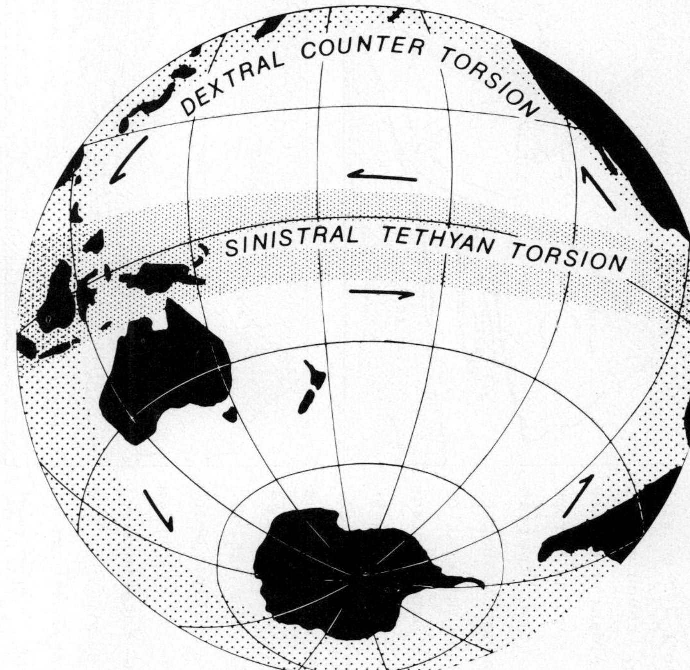

Fig. 16 (above): Oblique equal-area projections of the "land hemisphere" and "oceanic hemisphere", showing the Tethyan sinistral torsion (fine stipple) and the dextral counter torsion (coarser stipple).

Western Europe has been dragged westward about 700 km with respect to Africa, concurrently with about 700 km widening of the Mediterranean zone. Pushing Europe westward rotates the Iberian peninsula 35°, opening the Bay of Biscay. The westward movement of Spain with the Betic orogen drags the Riff against the Moroccan megashear, forming the Gibralar, Tangiers, Riff orocline. The development of the **S**-coupled oroclines through the Atlas, Sicily, Appenine, and Alps opens the Ligurian and Tyrhenian sphenochasms. The rotations of Spain, Corsica, Sardinia, and Italy, and the non-rotation of Sicily have been confirmed palaeomagnetically. The Rhône and Rhine grabens are in tensional directions of the sinistral shear

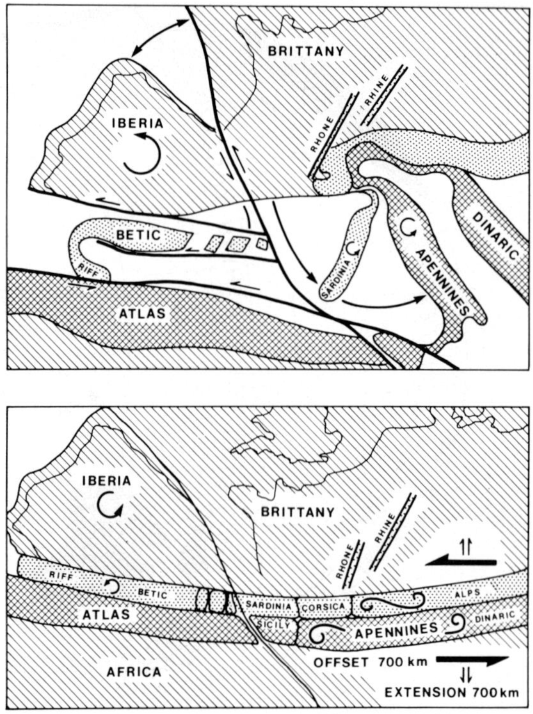

FIG. 17. Tectonic evolution of the Mediterranean by 700km sinistral shear plus 700km of transverse extension.

In a single stroke, the Tethyan torsion with transverse extension reduces the complex tectonic confugurations of western Europe to simplicity.

Dextral Counter Torsion.

A conjugate dextral counter torsion crosses the Tethyan sinistral torsion at right angles at the East Indies and the West Indies (Figures 15 and 16). The Tethyan torsion follows the Cretaceous equator, and the counter torsion passes through the Cretaceous poles (Figure 15). Large scale dextral shearing has long been well known along the west coast of North America from Mexico to Alaska, and includes the San Andreas system, the Rocky Mtn. Trench, and the numerous terranes, which have been shown by facies, fauna and palaeomagnetic latitudes to have been displaced northward long distances. Irving (1983) showed that the palaeomagnetic latitudes of much of British Colombia indicated northward displacement relative to stable North America of 1000km or more. Further north Gabrielse (1985) estimated dextral shift of

900km with a further 300km west of it. This dextral offset continues along the Aleutian trench, as shown by the 700km dextral offset of the Zodiac fan (Carey 1976).

The coupled Mendocino and Idaho oroclines (Figure 18) have the typical **Z** form, indicating dextral drag with a shift of about 800 km. Just as the Tethyan torsion integrates the diverse structure of Western Europe, so the dextral torsion integrates in one motion the Basin-and Range province, the Cascadan province, the Columbia Plateau basalts, the Rocky Mtn. Trench, the San Andreas system, the coupled Mendocino and Idaho oroclines, and the many palaeomagnetic rotations in Washington and Oregon.

Matching the Mendocino Idaho oroclinal couple are the Assam and Sunda coupled oroclines which offset the Tethyan front 34° southward from Assam to the Java Trench, again with the typical **Z** dextral drag pattern. Figure 19 shows the offset of the dextral torsion through Australasia. 66° of new crust has been inserted between Antarctica and India, but only 32° between Antarctica and Australia. The 34° difference is the same as the offset of the Assam-Sunda oroclinal couple, and hence is the southward displacement of Australia with respect to Asia. Figure 19 also shows that between northwest Australia and China there are a series of extensional basins, some still empty and some (like the Ordos Basin) already filled with seven km or more of sediment. These basins also add up to 34° and thus show the trail of Australia's southward drag by the dextral torsion. The series of small seas of East Asia, all the way up to the Aleutians, also show this dextral drag (Carey 1958). This 34° offset also appears in the evolution of the Pacific Ocean (Figure 20). The stippled area is

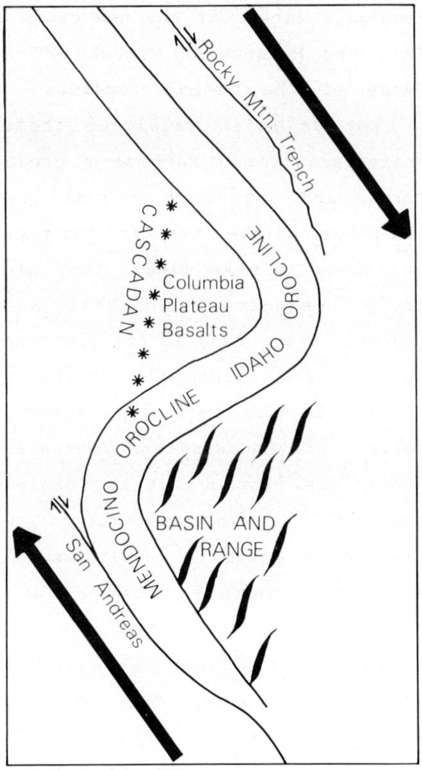

FIG. 18.

the Pacific crust inserted since the Mid Cretaceous from the present spreading ridge indicated by the double line. When Australia is moved north 34° as indicated above, the Melanesian plateau from New Guinea to Samoa fits against the Hawaian Ridge. The combination of the two torsions closes the Shatsky Sphenochasm, bringing together the Emperor, Mariana, and Palau Ridges against the Philippines.

Dr Perry has extended to the Pacific his computer program by radial regression of the continents inwards, which throws up both the Tethyan and counter torsions. (Compare Figures 20 and 21). Reversal of the dextral torsion, which moves Australasia northward, closes **X** to **X'**, and reversal of the Tethyan torsion, which moves Australasia westward to bring **Y** to **Y'**, then neatly closes the Early Pacific. Note that the computer plot generates no ocean growth east of the spreading ridge north of the Tethyan torsion, and generates little if any new crust along the transcurrent zones of east Asia **XX'**, the Melanesian Plateau **YY'**, and southwest of Panama.

Cause of the Global Torsions

Let us first catalogue their characteristics. Each is a great circle and strictly normal to the other, as the two Mercator maps of Figure 15 make clear. They are truly conjugate, sinistral and dextral. They cross in the East and West Indies, which are remarkably similar Tethyan achipelago systems between the Gondwana and Laurasian continents. The sinistral torsion follows the Tethys along the Cretaceous equator. The dextral torsion separates the "land hemisphere" from the "water hemisphere" (Figure 16). The dextral torsion passes, not only through the East and West Indies, but through the the Cretaceous poles, one in the Bering Sea in longitude 180° and about 30° from the present north pole, and the other in longitude 0° and about 30° the present south pole. I have previously

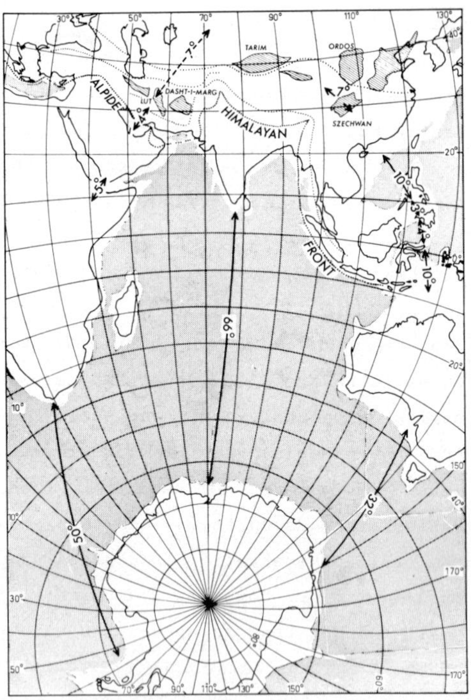

FIG. 19.

shown that the latter point is the centre of maximum dispersion of the continents and the former is the centre of minimum dispersion (Carey 1970).

The centre of gravity of a continent in isostatic equilibrium is more than two kilometres higher than the centre of gravity of oceanic crust down to the

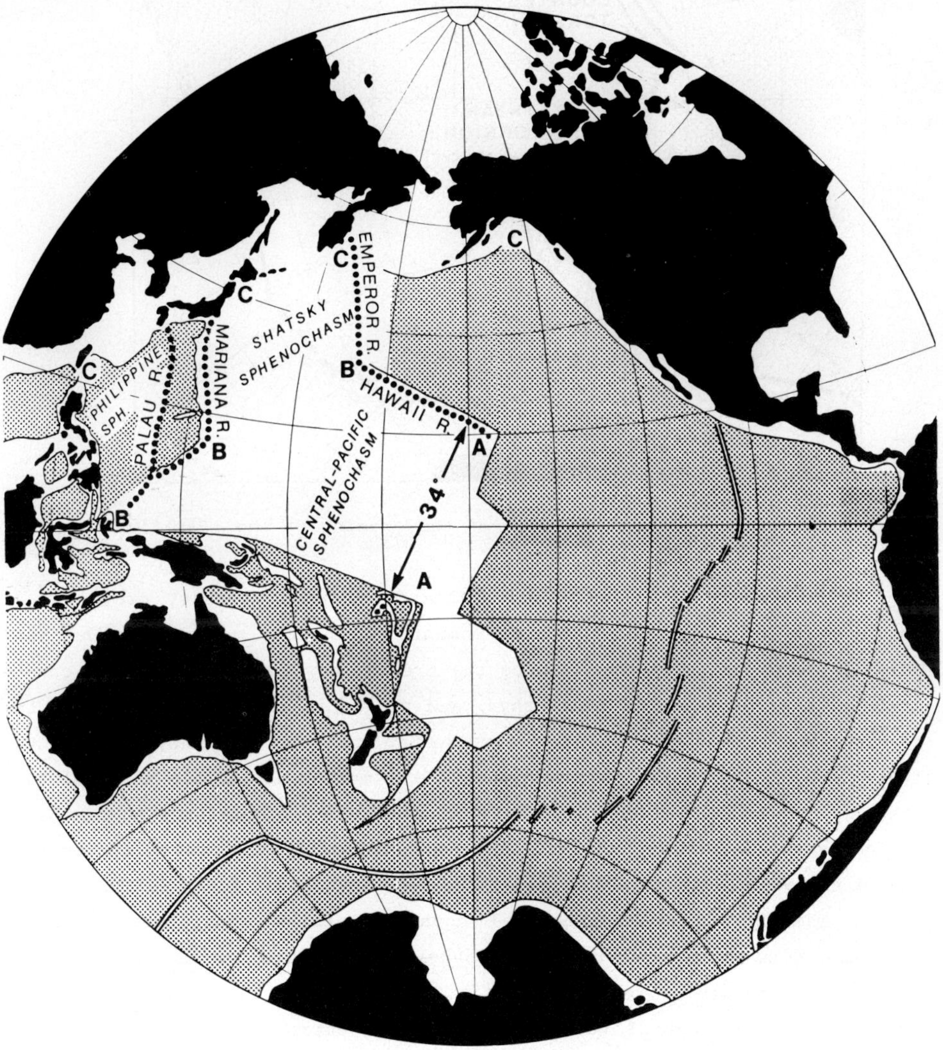

FIG. 20. Opening of the Pacific Ocean.

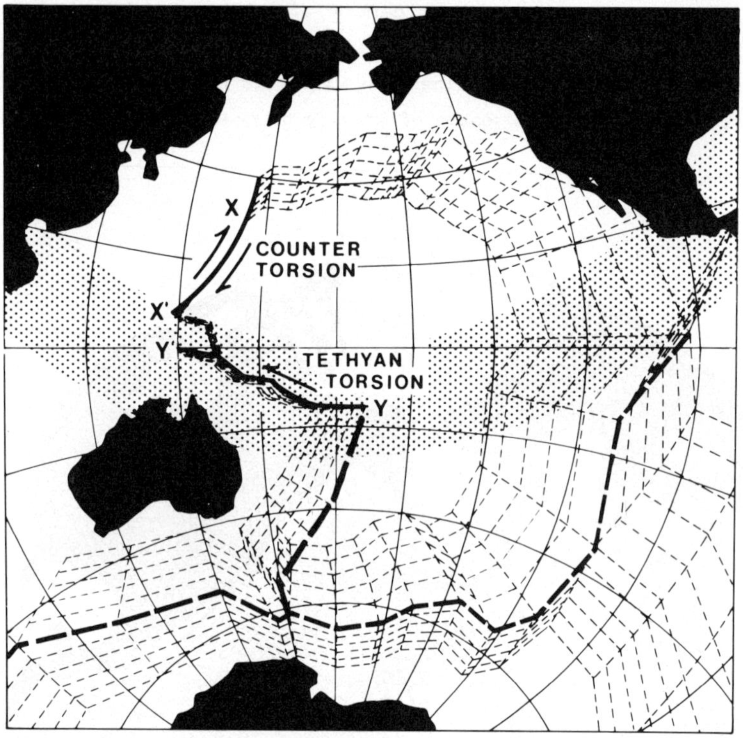

FIG. 21. Perry's computer plot of the opening of the Pacific.

level of the base of the continent. Therefore, when isostasy prevails, a continent contributes more to the moment of inertia of the Earth than equivalent oceanic crust does. Since the Cretaceous, much more oceanic crust has been inserted south of the Tethys, and the ratio of continental crust to oceanic crust has become progressively greater north of the Tethys than south of the Tethys. This means that increase of the moment of inertia north of the Tethys has exeeded the increase south of the Tethys. Therefore the northern continents would tend lag behind their southern neighbours in their rotation towards the east. This is exactly what the Tethyan torsion does.

The fact that the dextral counter torsion separates the land hemisphere from the oceanic hemisphere suggests that this torsion also results from the interaction of rotation inertia with differential moment of inertia. The poles of the dextral torsion are in the centre of maximum moment of inertia (in the centre of the land hemisphere) and the centre of the minimum moment of inertia (in the centre of the water hemisphere). However no-one has yet mastered the dynamics of zonal motions on a rotating gravitating expanding spheroid. Also, the Cretaceous EoPacific was quite small, although the poles of maximum and minimum moments of inertia would still have passed though the centre of Pangaea and the centre of the EoPacific.

This tendency of continental crust to lag westward with respect to oceanic crust shows up on various scales. The Tethyan torsion between hemispheres expresses it on the global scale. Next it shows up in the the asymmetry of the Pacific. From the Aleutians to Anarctica the western Pacific margin shows a chain of small seas, whereas the east Pacific coast, from the Aleutians to Antarctica, has none. On a smaller scale, where oceanic crust separates continents north and south of it, the oceanic crust tends to lie east relative to the continents lagging westward. Hence between the Americas the Caribbean bows eastward into the Lesser Antilles arc, between Asia and Australia the Banda Sea bows east into the Banda arc, and between South America and Antarctica the Scotia Sea bows eastward into the Scotia arc. Equivalent westward bowing does not occur. The east-west asymmetry of the Pacific results in successive orogenic belts from Late Proterozoic to Quaternary on the western side of the Pacific to be spaced, but in the Cordillera on the eastern Pacific margin they are superimposed (Figure 22).

The computer plot of Figure 21 shows the Tasman Sea growing eastward to the Kermadec Trench, but no growth east of this trench. This agrees with the situation found there (Figure 23). Superficially, the Lord Howe, Norfolk, Three Kings, Lau, and Tonga submarine ridges resemble successive moraines left behind by a retreating glacier, except that the moraine nearest the glacier is the last left behind, whereas in the Tasman Sea the Lord Howe ridge through into the South Island of New Zealand was the first left behind by Austalia in the Cretaceous. Vulcanism and seismicity have long since left the Norfolk ridge, but are currently very active along the outer ridge.

The spreading of east Asia is similar, but an orogen does not grow uniformly along its length, but tends to form diapiric foci some six or seven hundred km apart (determined by the thickness of the lithosphere). If conditions were symmetrical, funnel-shaped diapirs would develop, appearing as ring-shaped orogens at the surface (Wezel's krikogens). Because of the prevailing east-west asymmetry along the east Asian coast, oceanic crust grew eastward forming a row of asymmetric krikogen basins with rifted continental crust on the western side, a newly-risen floor of oceanic crust which has grown from west to east, and an orogenic arc on the eastern side. The spreading orogenic nappes over-ride toward the east. Similarly, the Cordilleran nappes, all the way from Alaska to Chile, over-ride toward the

FIG. 22. Cordilleran orogenies migrated little since the Proterozoic, but along west Pacific the orogens migrated westward.

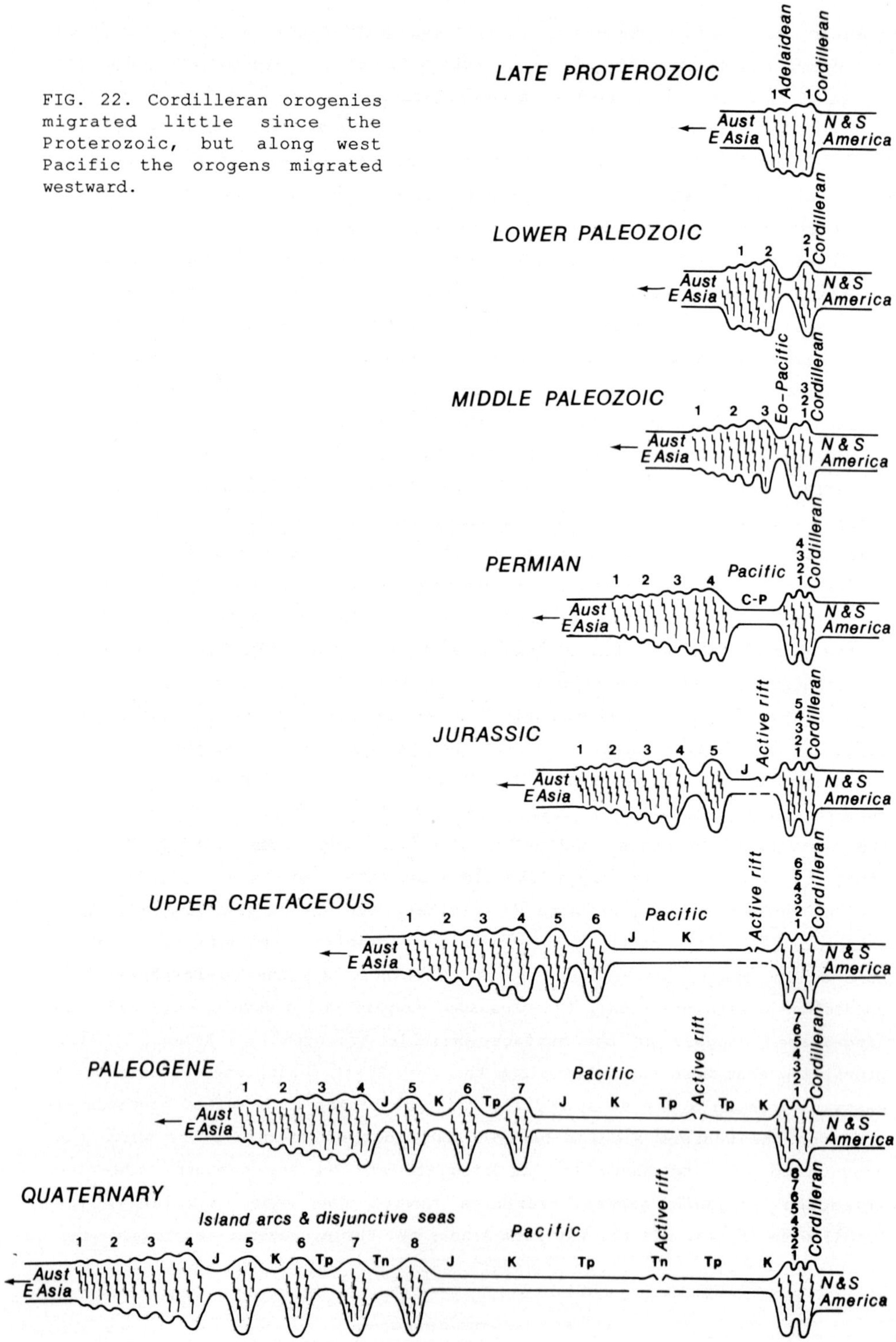

east, even though this is a west-facing coast. The east Asian krikogens are further modified by overprinting by the dextral torsion.

Southeast Asia between the Australian and Asian continents has suffered the interaction of a series of diapiric krikogens, 34° of dextral torsion, a similar amount of Tethyan sinistral torsion, and the pervasive east-west asymmetry. This complex my be unraveled as follows (Figure 24):

1) Excise the new crust which had formed the many small basins in the region (such as the South China, Sula, Celebes, Flores, and Banda Seas), which for example, closes Borneo and Palawan against South China.

(2) Shorten appropriately the orotaths (stretched orogens): For example, the Sunda-Banda orogen becomes progressively more stretched along the arc from Sumatra, Java, Bali, Lomboc, Sumbawa, Flores, Lamblen, Pantar, Alor, Wetar, Romang, Damar, Banda, Turtle, Lucipara, thence via a continuous submarine ridge to Salajar, and southwest Sulawesi. Likewise the continuous ridge through Sumba, Savu, Roti, Timor, Moa, Barbar, Tanimbar, Kai, Gorang, Seram, Buru, Tukangbesi, Butung, to southeast Sulawesi. Such orogens are elongated partly by fault rift gaps between them and partly by flow attenuation. If an orogen standing 1,000m above sealevel were streched to twice its length, isostatic equilibrium would leave it as a submarine ridge rising to 1500m below sealevel. Such submarine ridges and island chains indicate what points on major islands have to be brought closer, and the degree of necessary shortening.

(3) Straighten the oroclines, such as the Assam-Sumatra couple, and the loops around the Banda Sea.

(4) Check palaeomagnetic orientations, rotations, and palaeolatitudes. It was found that the preceding steps had automatically corrected these. For example, McElhinny *et al.* (1974) reported that in the Early Cretaceous, the Malay Peninsula lay a few degrees north of the equator elongated east-west, which is precisely where the above procedures had left it (Figure 24). Likewise the 180° rotation of Seram. However in a region as contused as the East Indies with the crossing of the dextral and sinistral torsions, further data may require adjustments to this 1976 reconstruction, but these will be minor, because (1) (2) and (3) above determine the gross pattern.

(5) Check the large transcurrent faults associated with the Tethyan and counter torsions because quite large translations may have occurred along these. Again it was found that the foregoing operations had largely corrected for them.

(6) Check what had happened to the many faunal and palaeogeographic links, such as Wallace's Line, the many disrupted faunal distributions, and the tight

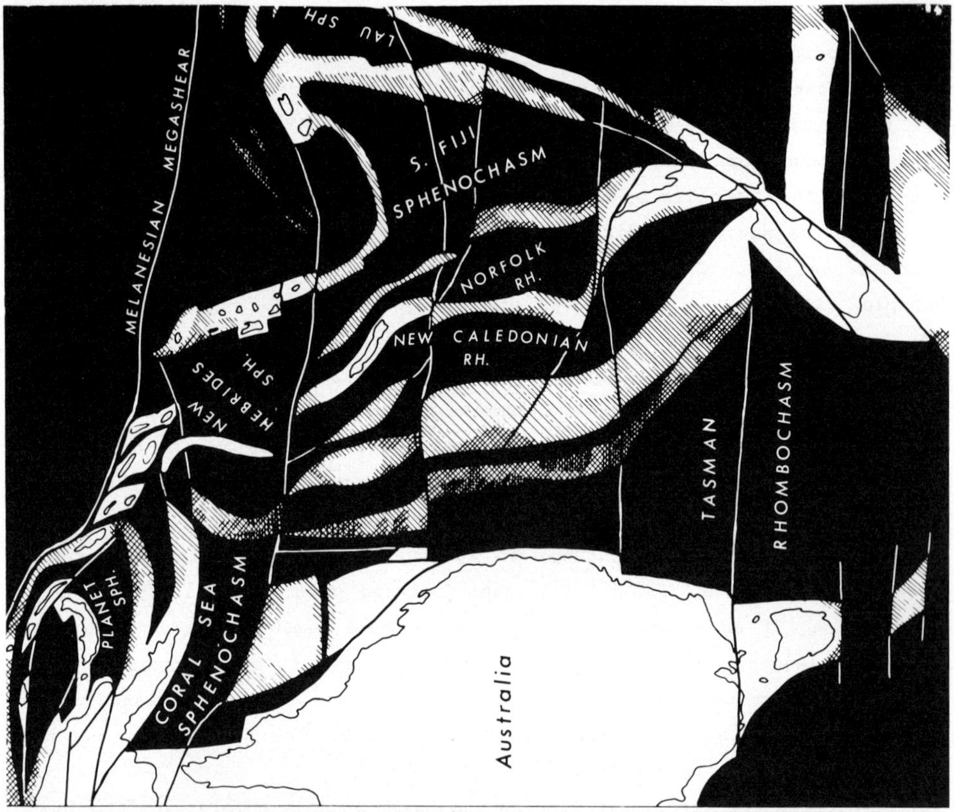
MELANESIAN MEGASHEAR
LAU SPH.
S. FIJI
SPHENOCHASM
NORFOLK
RH.
NEW CALEDONIAN
RH.
NEW
HEBRIDES
SPH.
PLANET
SPH.
CORAL SEA
SPHENOCHASM
TASMAN
RHOMBOCHASM
Australia

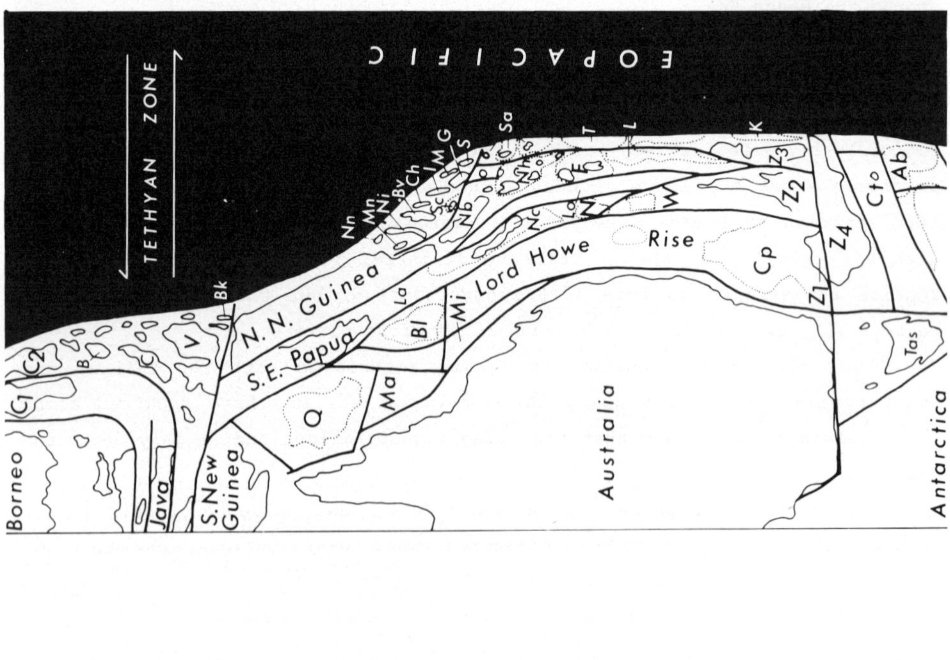
TETHYAN ZONE
EOPACIFIC
Borneo
Java
S.New
Guinea
N. N. Guinea
S.E. Papua
Lord Howe Rise
Australia
Antarctica
Q
Ma
Bl
Mi
Cp
Z1
Z2
Z3
Z4
Cto
Ab
Tas
C1
C2
V
Bk

FIG. 23 (opposite page):
Tertiary drift of Australia away from the Early Cretaceous margin of the EoPacific, leaving successive ridges like retreat moraines separated by younger rhombochasms and sphenochasms.
C1, SW Sulawesi; C2, SE Sulawesi; B, Buru: C, Ceram; V, Vogelkop; Bk, Biak; Nn, Ninigo Group; Mn, Manus; Ni, New Ireland; Bv, Bougainville; Ch, Choiseul; I, Santa Isobel; Sc, New Georgia; M, Malaita; G, Guadalcanal; S, San Christobal; Sa, Samoa; T, Tonga; L, Lau; K, Kermadek; Nb, New Britain; Nh, New Hebrides; F; Fiji: La, Louisiade Archipelago; Nc, New Caledonia; Lo, Loyalty Group; N, Norfolk Ridge; Z1, Z2, Z3, Z4, New Zealand; Q, Queensland Plateau; Bl, Bellona Plateau; Cp, Campbell Plateau; Ct, Chatham Rise; Ab, Antipodes, Bounty: Nz, New Zealand Plateau; Aukland Island.

FIG. 24. Jurassic reconstruction of Southeast Asia. Numbered asterisks indicate shortened orotaths. Modern coastlines are shown for identification.
1, Andaman-Nicobar; 2, Nicobar-Sunda; 3, Timor-Ceram; 4, Banda Arc; 5, Sula Spur; 6, Sangi Kawaio; 7, Talair; A, Aru; B, Buru; C, Ceram; J, Japen; L, Luzon; M, Misool; Mi, Mindoro; Mn, Mindanao; N, Negros; Ni, Nias; O, Obi; P, Peleng; Pa, Palawan; Pn, Panay; R, Roti; S, Samar; Su, Sumba; Sl, Sula; Si, Siberut; T, Timor; Ta, Tanimbar; V, Vogelkop; W, Waigeo.

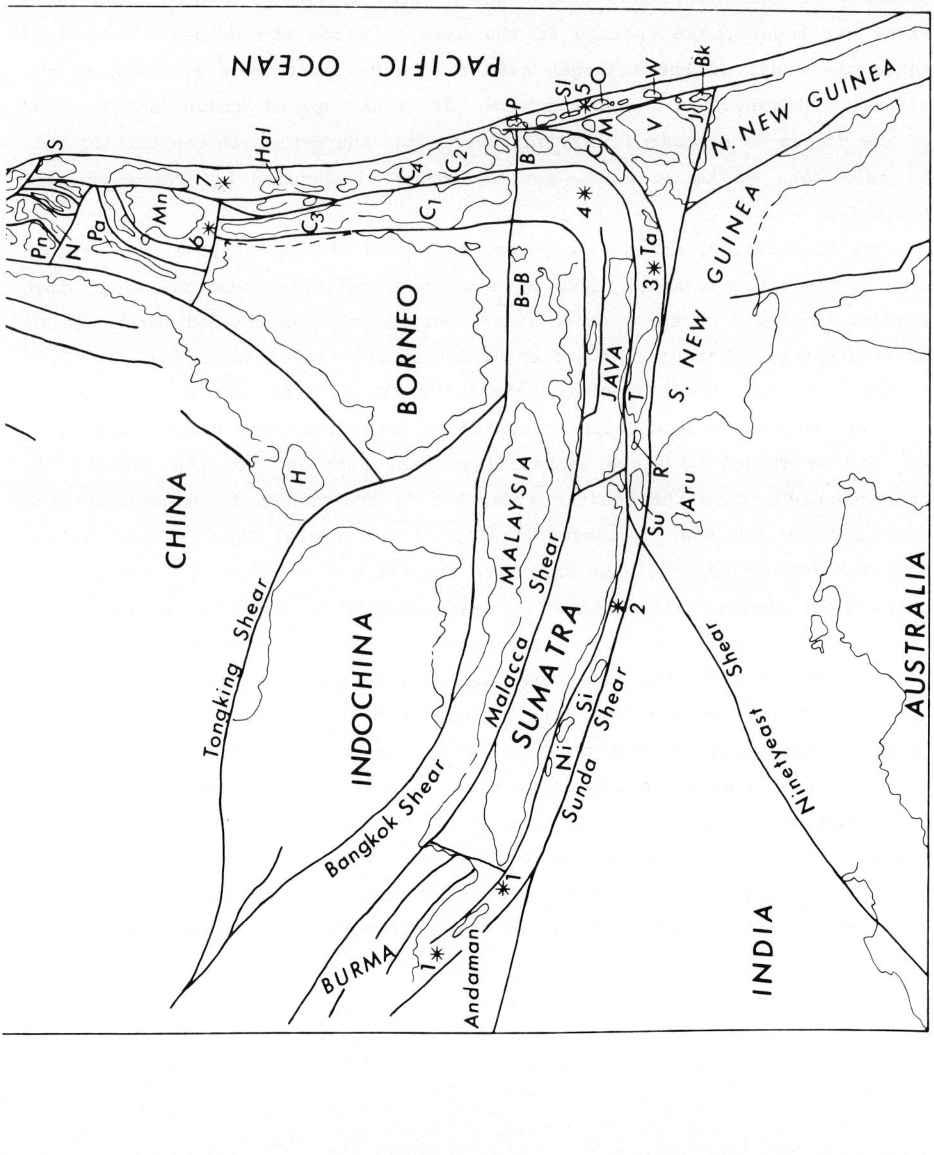

facies and provenance links reported forcefully by Umbgrove (1938) and more recently by Pigram *et al.* (1984), Stait & Burrett (1985), and several others. Here again the basic operations had already met the requirements.

BENIOFF ZONE

The expansion and subduction models agree that the Benioff Zone is a zone of shearing fracture, with the orogen side moving *relatively* upward and over the oceanic side, and this is confirmed by the sense of first motion deduced from the earthquakes which propagate from the fractures. But the subduction model claims that the oceanic lithoshpere is thrusting downward beneath the orogen, and has done so for thousands of kilometres. In contrast, in the expansion model the oceanic crust is stationary, and the Benioff zone is the boundary of the upward-thrusting diapir, which flares in a bell-shape as it rises (krikogen). The spacing of the diapirs along the orogenic belt (and hence the radii of the krikogen arcs at the surface) is a function of the effective thickness of the lithosphere. The total upward motion in the center of the orogen is a hundred or so kilometres and the motion at the Benioff zone is only tens of km at most, not thousands as assumed by the subduction hypothesis.

The question is, which side moves, and which side is passive, the oceanic side or the orogenic side? Continuous seismic profiling shows relatively thin regular sediments on the oceanic side, monotonously undisturbed for thousands of kilometres, right up to the trench. But as the trench is crossed, tectonic violence of all kinds erupts: seismicity, thrusting, and gross slumping towards the trench repeatedly triggered by the continuous oversteepening by the rising orogen to slope instability. Heat outflow from the interior is consistently low on the oceanic side, but as the trench is crossed it more than doubles, and even increases locally up to tenfold. This is incompatible with the subduction model, in which the overall trench-orogenic arc system is where cold oceanic lithosphere is descending deep into the mantle. This requires that the total heat flux in the region should be much *less* than average. By contrast, the expansion model asserts that the whole orogenic zone is where hotter mantle material is forced upward and outward by the expanding interior; this requires that the total heat flux in the orogenic region should be significantly *more* than average, which is precisely what is found.

Everything in the orogenic zone moves *up*. The isotherms are tens of km higher than average. In the axial zone, volcanoes bring up lavas from the partially melted mantle; magmas rise to crystallize in plutons; metamorphic rocks from deep below the floor of the geosyncline are forced up a dozen kilometres or more, to crop out high in the mountains; serpentinites and peridotites, originally even deeper below these crystallines, do likewise; gneiss domes intrude diapirically upward into the geosynclinal strata; diapirs pierce upward in a viscosity peck-order - gneiss domes, migmatite diapirs, magmatic plutons, volcanic lavas, and nuées ardentes. Overthrusts are numerous, always with the axial side riding up over the trench side, on very steep thrust surfaces near the axial zone, but on progressively flatter slopes farther from the axial zone.

What forces all these bodies up? Not buoyancy, because except for the magmas they are denser than the rocks they penetrate, and they are driven to higher altitudes than their density would warrant. They are forced up by the ascending mantle below them, which changes phase to less dense paramorphs as higher temperatures are brought to levels of lower confining pressure.

From the axial zone right out to the Benioff zone is a single complex diapiric unit. The Benioff zone is the boundary of the ascending diapir against the stationary oceanic lithosphere. The axial zone, where temperatures are highest and "viscosity" least, ascends most rapidly under a given driving force. Outward from the axial zone, the temperature gets progressively lower, the "viscosity" progressively higher, and the rate of flow under the same stress-difference gets slower. Farther out, with still lower temperature and still higher "viscosity", the rate of flow of the rocks becomes too slow to accommodate the rising diapir, so, as the stress-difference is still above the shear strength, fractures occur--hence the Benioff zone, which is analogous to the boundary of a salt diapir against the passive sediments (Figure 25). Where the rate of ascent of the diapir is relatively fast, the Benioff zone of fractures goes deeper to higher temperatures. Slower upward movement can be accommodated by flow in relatively cooler rocks, so the Benioff fractures only go down a couple of hundred kilometres, even though the ascending diapir originates much more deeply than this.

The rocks just outside the Benioff zone are held in a state of elastic strain, at a stress level just below what would break them. They are the constraining wall of the diapir. Shear fracture could not occur without such a constraint, because an unconstrained shear stress produces rotation, not shear fracture as in the Benioff zone. Experiments on rocks held under shear stress below fracture level show that they transmit sound and seismic waves five per cent faster than when not stressed. The seismic velocity in the zone underneath the Benioff zone has been found to be higher than normal. This has been interpreted as evidence for a cold descending lithosphere slab, whereas such anomalous velocities are inevitable in the constraining rocks of the ascending diapir, just below the threshold of fracture.

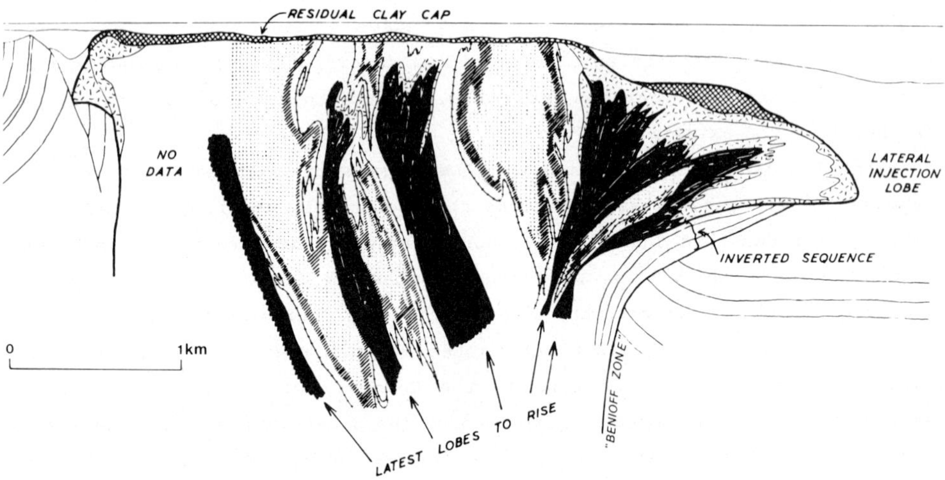

Fig 25. Successive salt diapirs within the Heide salt dome. (After Bentz). 1, diluvium; 2, Tertiary; 3, Late Cretaceous; 4, Senonian; 5, Turonian; 6, Cenomanian; 7 Albian; 8, Aptian; 9, Hauterivian; 10, Lias; 11, Rhaetic; 12, Middle Keuper; 14, Upper Muschelkalk; 15; Middle Muschelkalk; 16, Lower Muschelkalk; 17, Red Beds; 18, Bunter; 19, Residual clay cap; 20, Anhydrite cap; 21, oldest salt diapir; 22, Red salt clay; 23 potash bed; 24, younger salt diapir; 25, youngest salt diapir.

THE BLINKERS OF FAITH

Since the Kuhnian revolution to plate tectonics in the 'sixties, the validity of this dogma is taken for granted by most "respectable" scientists. When a new fact appears, it is automatically interpreted in terms of this ruling faith, even though it may be better explained otherwise. Indeed, if it were explained in terms of Earth expansion, it would be sent back by the referees for rewriting, if not rejected outright as naïve. American journals were the worst offenders during the 'thirties,'forties, and 'fifties, in rejecting outright anything smelling of what is now called plate tectonics, and since the revolution, have been the worst offenders in rejecting as naïve anything based on any alternative model.

The magmatic-differentiation sequence from the trench to the volcanoes of the orogenic arc, the so-called "paired metamorphic belts", the ophiolite zones, the beryllium-ten anomaly, and the petrological succession of elemental abundances, pillars of the subduction theory, are more simply explained under Earth expansion.

An extreme example is the interpretation of the so-called "blue schists" (metamorphic rocks containing the blue amphibole glaucophane, with lawsonite, aragonite, and ubiquitous quartz) which occur near the outer edge of the orogen, nearest to the trench. Laboratory experiments have produced this mineral assembly under relatively low temperature and very high confining pressure. To achieve these conditions, the subduction theory drags trench sediments down to great depths, then brings them back for tens of kilometres *against the direction of flow* of the coldest part of their adjacent down-going slab! If I find an erratic boulder in a glacier, I know for certain that it has come from further up the glacier. There is no possibility whatever of it having been brought there from *down* the glacier. Nor is there any possibility of the blue schists ascending several tens of kilometres up against the flow of the down-going slab. To increase the absurdity, this relatively low-temperature part of the lithosphere is certainly a million times as viscous as glacier ice.

Plate tectonicists have reached this absurd position because of their faith, and through misinterpretation of the environment for crystalization of blue schists. The laboratory experiments involved *static* confining pressure. But the blue schists are *schists.* The name itself implies, not static confining pressure where the stress-difference is zero, but very large stress-difference. High confining pressure with no stress-difference produces granulite, tectonically isotropic marbles, and quartzites, with minerals typical of that environment, such as garnets, pyroxenes, calcite, and quartz. The same rock in an environment of large stress-difference recrystalizes as gneiss or schist, made up of *schistophile* minerals, different from those of the granulite facies. Schistophile minerals, such as micas, amphiboles, and aragonite, commonly are denser in one crystal direction and less dense in another direcion. The laboratory experiment to produce the blue-schist mineral assembly used very high pressure with low temperature. But this is *not* their environment. They really form in conditions of relatively low temperature but large stress-difference, and not necessarily with deep burial. These are exactly the conditions which apply in the diapiric orogen inward from the Benioff zone where, according to the expansion theory, stress-difference is maximum. As the motion there is upward at all times, there is no absurdity of bringing these rocks up against the flow.

REFERENCES

Carey, S.W., 1958: The tectonic approach to continental drift. In: S.W.Carey (Editor): Continental drift - a symposium. University of Tasmania, 177-363.

Carey, S.W., 1970: Australia, New Guinea, and Melanesia in the current revolution in concepts of the evolution of the Earth. Search 1(5):178-189.

Carey, S.W., 1975: The expanding Earth - an essay review. Earth Sci.Rev., 11(2): 105-143.

Carey, S.W., 1976: The Expanding Earth. Elsevier, Amsterdam, 488 pp.

Carey, S.W., 1981: The necessity for Earth expansion. In: S.W.Carey (Editor): The expanding Earth - a symposium. University of Tasmania, 367-396.

Carey, S.W., 1986: Philosophy of the Earth and Universe. Stanford University Press, (in press).

Dziewonski, A. and Woodhouse, J., 1984: A map that gets under the Earth's skin. Science (June): 6.

Gabrielse, H., 1985: Major dextral transcurrent displacements along the Northern Rocky Mtn. Trench and related lineaments in north-central British Columbia. Geol. Soc. Am., Bull. 96: 1-14.

Irving, E., 1983: Fragmentation and assembly of the continents, Mid-Carboniferous to the Present. Dominion Observatory, Ottawa, Geophys, Surv., 5: 299-333.

King, P.B., 1959: The Evolution of North America. Princeton Univ. Press, 1-190.

McElhinny, M.W., Haile, N.S., and Crawford, A.R., 1974: Palaeomagnetic evidence shows Malay Peninsula was not part of Gondwanaland. Nature, 252: 641-645.

Miyabe, N., 1931: Notes on the block structure of the Earth's crust. Earthquake Res. Inst., Tokyo Imp. Univ., Bull. 9: 256.

Pigram, C.J., Surono, and Supandjono, J.B., 1985: Origin of the Sula Platform, Eastern Indonesia. Geology, 13: 246-248.

Ross, D.C., 1985: Mafic gneissic complex (bathylithic root?) in the southernmost Sierra Nevada, California. Geology, 13: 288-291.

Schmidt, P.W. and Embleton, B.J.J., 1981: A geotectonic paradox: has the Earth expanded? Journ. Geophys., 49: 20-25.

Stait, B.A., and Burrett, C.F., 1985: South East asia as a part of an Ordovician Gondwanaland - a palaeobiogeographical test of a tectonic hypothesis. Earth and Plan. Sci. Lett., 75: 184-190.

Umbgrove, J.H.F., 1938: Geological history of the East Indies. Am. Assoc. Petrol. Geol., Bull., 22(1): 1-70.

Vogel, K. 1933, Global models and Earth expansion. In: S.W.Carey (Editor): The expanding Earth - a symposium. University of Tasmania, 17-28.

Zonenshayn, L.P., 1973: The evolution of central Asiatic geosynclines through sea-floor spreading. Tectonophysics, 19: 213-232.

ARCUATE CRUSTAL STRUCTURES

V.V. BELOUSSOV

Institute of Physics of the Earth, USSR Academy of Sciences, Moscow

ABSTRACT

The orogenic ridges on continents and island arcs have a similar history of development and character of tectonic deformations. They also have much in common with the linear ridges. The arcuate shape of orogenic uplifts is associated with long geological history, during which the consecutively appearing structures overlaid one another or inherited from one another. The shape of island arcs is presumably the result of their former geological development.

GENERAL REMARKS

The numerous large arcuate topographical formations on the Earth's surface are seen at a glance on the geographical map of the world. On continents they feature mountain ridges, for example, Western Alps, the Carpathians, Pamir, Verkhoyansk Ridge, Himalayas, and island arcs outside the continents. The continental arcuate ridges are mostly grouped in Eurasia;the larger part of island arcs is situated in the transition zone from Asia and Australia to the Pacific Ocean. They resemble one another morphologically, because among the continental ridges and the island arcs there are slightly curved long structures and steeply curved short ones (Fig.1). Not only their shape, but to a certain extent their geological history and tectonic structure are similar.

Before analysing this analogy, however, our concept of an island arc should be defined in clearer terms. In the transition zone in the Western Pacific there are actually two types of islands grouped in arcs (Beloussov and Ruditch,1961).The first type includes groups of large islands with the typically continental crust and policyclic geosynclinal development: the Japanese Islands, New Guinea, Philippines, New Zealand. They are, as it were, small continents with all the peculiarities of continental geological history. The second type of island arcs includes the chains of volcanic islands

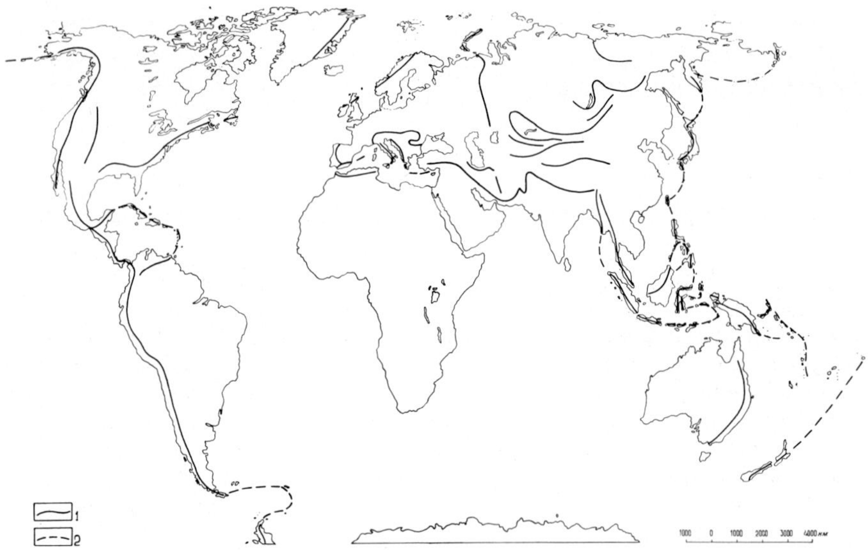

Fig. 1. Orogenic and island arcs. 1- orogenic arcs; 2- island arcs.

located either on small blocks of the reduced continental crust, or on the oceanic crust. Among them are the Aleutian, Kuril, Mariana, Tonga, Kermadec and other islands. In some places, the island arcs of the second type are as if superimposed on the islands of the arcs of the first type and continue within their area as mountain ridges crowned by volcanoes. For example, the continuation of the Kuril arc of the second type can be seen in the volcanic ridges of the Hokkaido Island, which belongs to the arc of the first type;the Marianas arc running north crosses the Honshu Island and forms there the peculiar volcanic structure Fossa Magna; the chain of volcanoes of the "Tonga and Kermadec arc" continues on the Northern Island of New Zealand; and so on. It is actually the island arcs of the second type with their continuations on the islands of the arcs of the first type that are considered as"island arcs" in contrast to the continental arcs. The islands of the arcs of the first type are just blocks of continental crust with corresponding regularities of geological evolution. Further in the text, when using the term "island arcs", we shall mean the arcs of the second type.

From the viewpoint of geological history,the continental arcuate ridges are similar to island arcs: like the latter, they are geologically young. Both these forms of topography appeared during the so-called neotectonic epoch, i.e., Neogene and Quaternary. This was the time of overall tectonic activisation on the Earth, when not only arcuate ridges and island arcs rose, but in general the entire modern high relief was raised on the Earth's surface. The linear Caucasus in its present-day form was shaped at that time, as also the Cordilleras and Andes. The ancient fold zones, such as the Appalachians, Caledonides of Scandinavia, Urals were likewise activated. Therefore, the age of arcuate continental ridges and island arcs is the same as that of all other present-day ridges.

During the up-lift the ridges on continents and the island arcs in transition zones experienced volcanic processes and eruption of lavas of predominantly andesitic and andesite-basaltic composition. The only difference was that the amount of erupted material was much larger on island arcs and in the Pacific zone in general than on continents. But neotectonic intensive andesitic volcanism was active, for example, in the Caucasus, in the Carpathians. At that time, andesitic effusions occurred in the Alps.

As regards tectonic deformations, all modern ridges on continents, including the arcuate ones, are the results of the orogenic stage of development. Many of them were formed as soon as the geosynclinal development was accomplished (for example,the Carpathians and Alps). These structures are epigeosynclinal orogens. Others have risen after a more or less long lapse of time since the termination of the geosynclinal stage (for example, the Tien Shan ridges, where the platform regime has set in since the end of the Paleozoic). These structures are the epiplatform orogens. In all cases, however, the uplift of ridges took place after the termination of folding and regional metamorphism, whereas displacements acquired block character. The ridges on continents are structures composed of blocks of various sizes lifted vertically to different heights and divided in most cases by vertical faults. The linear and arcuate ridges do not differ in this respect as well.

There are hardly any data on the pre-history of those major parts of island arcs of the second type, which lie outside the island arcs of the first type. Thick accumulations of volcanogenic--sedimentary formations on some of them, however, allow us to suppose that before the uplift of the modern arcs there were subsidences in the crust in their place where these formations accumulated. In those parts, where the island arcs of the second type are superimposed on the islands of arcs of the first type, the typically epigeosynclinal orogenic structures have appeared. These structures, as also the other parts of the island arcs of the second type and the continental ridges, are composed of blocks. They likewise have neither folded deformations, nor any evidence of regional metamorphism. The folding, observed on island arcs of the first type, was formed at an earlier period, in the pre-orogenic stage, i.e. during the stage of geosynclinal development.

Therefore, the nature of tectonic deformations of the modern continental ridges, including the arcuate ones, is quite similar to that of the island arcs. The only important difference probably is that the young ridges on continents are often associated with large intrusions of alkalic potassic granites unknown on island arcs.

Taking into account this difference and also the common points, we may use the data on continental arcuate ridges to reach a general understanding of the nature of not only these ridges, but of the island arcs as well. There are, obviously, much more geological opportunities for the study of the continental mountain ridges than for the study of island arcs. We shall, therefore, start our analysis with continental orogenic arcs.

CONTINENTAL OROGENIC ARCS

It seems natural to try to find the solution of the problems of the origin of orogenic arcs not in the so-called "actualistic" theoretical models, but in the historical approach traditional in geology. We are necessarily restricted to just a few selected examples.

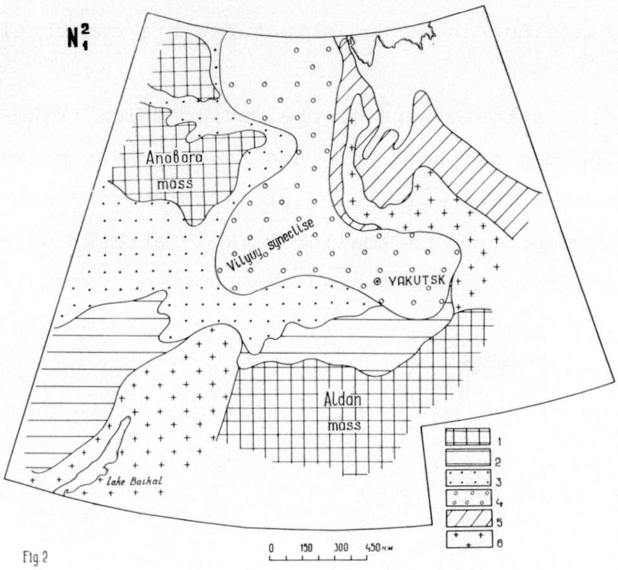

Fig. 2. North-East of the USSR. Paleotectonic scheme for the Middle Miocene (according to A.P.Vinogradov, ed., 1967). 1-shields; 2-anteclises; 3-syneclises; 4-deep Vilyuy syneclise; 5-moderate orogenic uplifts; 6-high orogenic uplifts.

The first example is related to the northern part of Eurasia, the Verkhoyansk and Baikal-Patomsk mountain arcs (Fig.2).

In their modern aspect, these arcuate ridges have constituted a part of the Siberian relief since the Miocene. The gently curving arc of the Verkhoyansk Ridge stretches along the Lena River. The steeper arc of the Baikal-Patomsk highlands is situated north of the Baikal Lake. The gently curving blocklike folds of the Verkhoyansk Ridge were formed at the end of the Mesozoic in place of the Verkhoyansk geosyncline of the Cimmerian cycle. Metamorphism and granitisation processes were weak. The Baikal-Patomsk geosyncline developed during the Baikal cycle, i.e., in the Late Proterozoic; crumpling into folds, regional metamorphism and granitisation occurred in that region in the interval between the Proterozoic and Phanerozoic. Consequently, the time period from the end of the geosynclinal development to recent orogenic uplift is quite different for these two cases: it is not more than a hundred million years in the Verkhoyansk region and about five hundred million years on the Baikal-Patomsk highlands.But these two structures are connected historically.

Let us analyse the paleotectonic maps of the Soviet Union (Vinogradov ed.,1967-1969). The map of the Middle and Late Riphean shows that the outlines of the two then existing regions of crustal subsidence, the vast Verkhoyansk and the smaller Baikal-Patomsk, were slightly rounded, their convex sides facing one another (Fig.3).

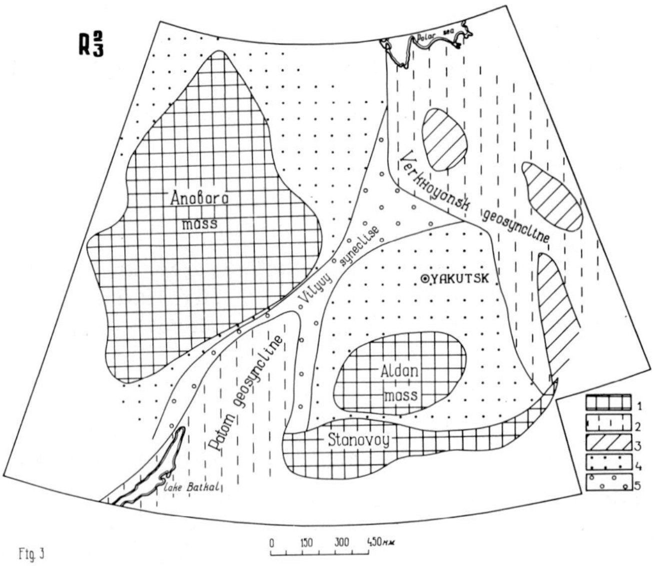

Fig. 3. North-East of the USSR. Paleotectonic scheme for the Middle and Late Riphean (according to A.P.Vinogradov, ed., 1968).
1-shields; 2-geosyncline; 3-median massifs; 4-syneclises; 5-deep Vilyuy syneclise.

The thickness of the Riphean deposits in the two subsidences amounts to many thousands of meters. The interesting point,however, is that both subsidences were linked together by a depression of the deep platform syneclise type. This is the Vilyuy syneclise and its Riphean deposits are up to 3000 m thick. The syneclise was located exactly on the continuation of the axes of both these large subsidences. It formed a kind of reduced pericline common to the two large subsidences. The whole system, therefore, can be regarded as one depression composed of two oval widenings with maximum subsidence ("geosynclinal" scale) and the less subsided contracted zone between them ("platform" scale). As shown in Fig.2, this contraction was situated between two very old uplifts (shields): the Anabar (forming the nucleus of the Siberian platform) and the Aldan. On the whole, this system can be described as the intersection of two zones with opposite tectonic tendencies: the uplift zone, stretching from south-east to north-west, and the subsidence zone with the southwest to north-east strike. The subsidence zone is superimposed on the uplift zone, which results in considerable narrowing of the depression and in the reduction of its amplitude. There are two alternative interpretations: either the causes of the uplifting and subsiding processes are located at different depths, or the transformation of a more ancient deep process, causing the uplift, and its substitution by the process, causing the subsidence,are opposed by the "conservative tendencies" of the first process. At the present state of studies, no suggestions could be offered about the physical nature of these deep processes.

As far back as in the Cambrian, the Baikal-Patomsk geosyncline having accomplished its cycle of development, changed to the orogenic regime. In the Verkhoyansk geosyncline, at the same stage of development, only the central part of the "subsidence oval" experienced tectonic inversion, deformations, metamorphism, and uplift and formed the Kolyma median massif, which since then was repeatedly slightly lifted (Fig.4). The displacements of mostly block character were also not large in the Kolyma median massif. At that time, between the Kolyma massif and the edge of the Verkhoyansk "subsidence oval", an arcuate depression appears as the residual structure.

During the subsequent periods of geological time, the Kolyma median massif expanded, contracted and disintegrated into several small massifs. The Patomsk orogen remained lifted, whereas the Vilyuy syneclise still subsided and changed its contours and the rate of subsidence from time to time.

In the beginning of the Cretaceous, as the result of the inversion of the tectonic regime, the Verkhoyansk Ridge rose the

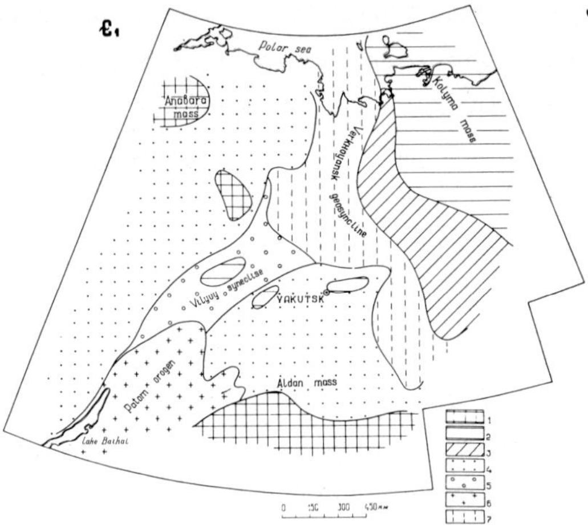

Fig. 4. North-East of the USSR. Paleotectonic scheme for the Early Cambrian (according to A.P. Vinogradov ed., 1968).
1-shields; 2-median massif; 3-slope of median massif; 4-syneclises; 5-deep Vilyuy syneclise; 6-orogenic uplift; 7-geosyncline.

first time from the arcuate depression bounding the Kolyma median massif. In the Neogene, after the Late Cretaceous and Paleogene erosional levelling, it again became active and acquired its present-day aspect. In the process, it inherited the arcuate form of the residual depression. The Baikal-Patomsk orogen became active as well and its shape was also inherited from the old geosynclinal subsidence oval. The Vilyuy syneclise, however, did not experience an inversion and is still the zone of slight crustal subsidence (Fig.2).

Another example is the large and complicated Urals-Mongolian-Okhotsk geosynclinal belt, which stretches in a wide arc convex to the south across the larger part of Asia. The inner part of the belt covers the basement of the modern West Siberian lowland, Central Kazakhstan, Altai, and Sayan. This territory had manifestations of diastrophism during the Baikalian, Caledonian and Hercynian cycles and, except West Siberian lowland, experienced recent activisation (Fig.5).

In that belt, peculiar arcuate structure was formed in Central Kazakhstan. The limbs of the Caledonian and Hercynian geosynclines,

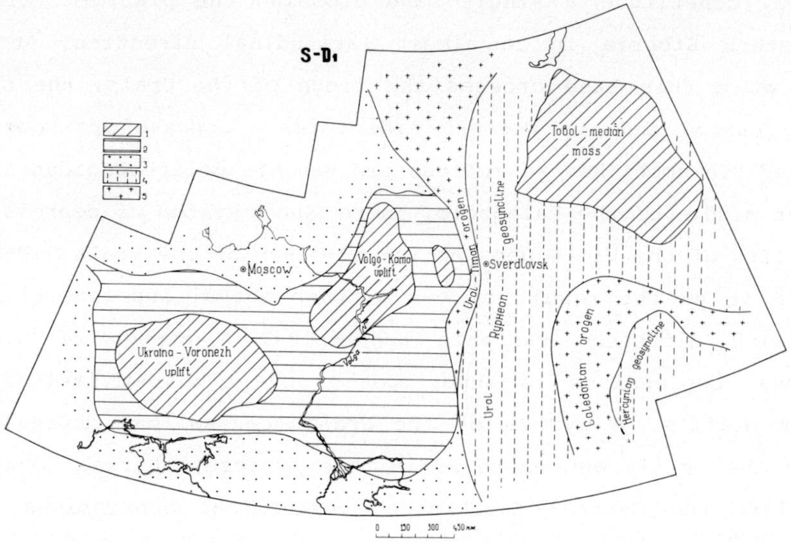

Fig. 5. Paleotectonic scheme of the eastern part of the East European platform, the Urals and Central Kazakhstan (according to A.P. Vinogradov, ed.,1969). 1-high uplifted anteclises; 2-moderately uplifted anteclises; 3-syneclises; 4-geosynclines; 5-orogenic uplifts.

reaching from the east and generally curving in the Balkhash Lake region from the north-western to the meridional strike, at latitude 53-55°N close up and do not continue further north. Moreover, the Caledonian structures form an arclike boundary to the younger Hercynian structures. As shown in Fig.5, the orogenic uplifts, having terminated the Caledonian cycle, surrounded in a semicircle the Hercynian geosyncline, which is only just appearing (Zaitsev, 1984).

The history of this region shows that in the Proterozoic the Urals with Timan and the whole West Siberian lowland belonged to the large geosynclinal area of the Baikalian cycle where the Tobolsk folded uplift was formed by the beginning of the Cambrian and became the median massif. This uplift was probably the decisive factor, which hindered the Paleozoic geosynclines from expanding further north from Central Kazakhstan.

The Tobolsk median massif in the Trans-Urals and the Ukrainian-Voronezh and Volga-Kama, old uplifts on the East European platform, obviously constitute a single band crossing the platform, Urals, and Western Siberia in an almost latitudinal direction. At the point, where this band crosses the ridge of the Urals, the ridge is very narrow and its western side forms a concave arc bounding so-called Ufa amphitheatre and the old Bashkir uplift.Incidentally, till the middle of the Paleozoic, there was a system of depressions and uplifts of the Urals geosyncline in place of the Urals Ridge.

Could it be the result of yet another mutual superposition of zones with different tectonic tendencies? A number of almost meridional depressions is superimposed on the almost latitudinal zone of uplifts. A system of the Urals geosynclinal depressions crossed the uplift and narrowed in this process, whereas south of the uplift the Central Kazakhstan geosynclinal depressions have completely closed up and at their closure formed peculiar arcuate structures still discernible in the morphology of the area.

Sharply outlined arcuate curves form the Alpine folded chains in Baluchistan and Makran in the south-east of Iran and in the south of Pakistan. In these regions, unusual meridional strikes appear against the background of the general latitudinal strike of the orogenic structures of the Tethys (Fig.6). This area lies in the zone of the so-called Urals-Oman lineament, which stretches along the Urals, and the Turgay "strait", Turanian young platform, western offspurs of Tien Shan, and the ridges of Eastern Iran (Goriachev, 1980). The lineament is a wide zone of deep faults of the meridional or almost meridional strike. The faults were formed long ago. They were already in existence in the Middle and Late Paleozoic, when the Urals geosyncline was developing. Anyway, they appeared long before the formation of the latitudinal geosyncline of the Alpine Tethys. Consequently, we may suggest, that at the place where this young Alpine geosyncline was superimposed on the meridional lineament, the crust of which was greatly fragmented and had high penetrability, the shape of the geosyncline was distorted: the vertical crustal movements in this area adjusted to the meridional strike of old faults. This distortion resulted in the formation of the arcs of Baluchistan and Makran.

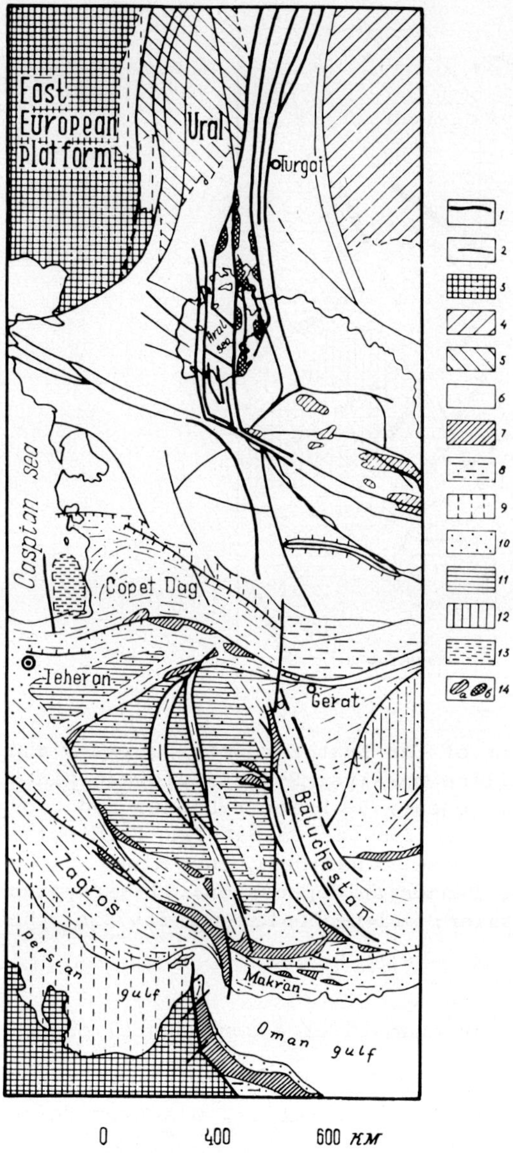

Fig. 6. The Urals-Oman deep fault zone (according to A.V.Goriachev, 1980). 1-main deep faults; 2-secondary deep faults; 3-East-European platform; 4-Caledonian Kazakh foldbelt; 5-Hercynian Urals fold-belt; 6-Turan young platform; 7-outcrops of the Turan platform basement; 8-foldbelts of the Alpine age; 9-fore-deeps; 10-Alpine belt; 11-Central-Iranian median massif; 12-Central-Afghan median massif; 13-South-Caspian median massif; 14-ophiolites (a-cretaceous; b-pre-Mesozoic).

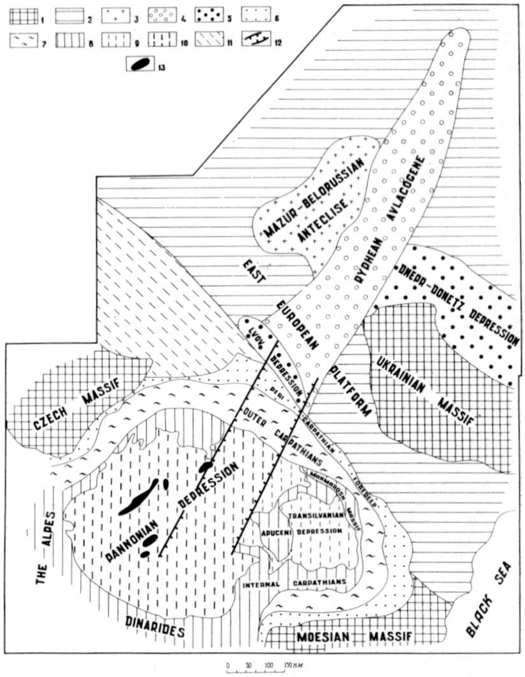

Fig. 7. Tectonic scheme of a part of the East-European platform and the Carpathian region. 1-crystalline massifs; 2-East-European platform; 3-anteclise; 4-Riphean aulacogene; 5- Dnieper-Donetz and Lvov depressions; 6-Carpathian fore-deep; 7-outer Carpathian fold-belt; 8-internal Carpathian fold-belt, the Dinarides, the Alps; 9-the Transylvanian depression; 10-the Pannonian depression; 11-Hercynian foldbelt; 12-transverse depression; 13-tectonic uplifts in the Pannonian depression.

The Carpathians are yet another example of a continental orogenic arc. Their arc bounds the Pannonian depression (Fig.7). The latter, according to the classical interpretation, was considered as a typical median massif inside the Alpine geosyncline. Now we know that the structure of this depression is more complicated than was supposed earlier (Körössy, 1964; Wein, 1969; Balogh and Körössy, 1974, etc.). The eastern part of the depression (the Apuseni Mts. and the Transylvanian subsidence) was formed in the end of the Miocene and was superimposed over the inner zone of the folded belt of the Carpathians. It is the result of the recent expansion of the Pannonian depression. To the west of this area, there is a structure which has all the features of the median massif. The basement of the depression has evidence of diastrophism of Baikalian time. The last, general distrophism, however, was Hercynian. The Sudeten phase of folding between the Early and Middle Carboniferous was the major and final one in the history of the basement. The metamorphism and the granitic intrusions also occurred at that time. It is universally known that the end of the early and the beginning of the Middle Carboniferous is the usual time for inversion in the Hercynian geosynclines. The Hercynian structures had north-eastern and north-western strikes, the former dominating. In the end of the Paleozoic, the Pannonian massif rose up and was denuded.

Not the entire massif was lifted, however. Initially, in its inner part the Igal-Bukk graben of the north-eastern strike remained as the subsidence zone, where sediments accumulated till the end of the Triassic. Later, during the Mesozoic, new grabens appeared on the massif and collected the Jurassic and Cretaceous sediments. The graben of the Middle Hungarian mountains was rather stable and subsided till the end of the Cretaceous.

Since in the Western Carpathians the continental sediments are found to contain the flora of the Late Carboniferous, Permian and Early Triassic, we may deduce that the subsidence continued also in this zone, which divides the inner part of the Pannonian massif and the Czech ancient massif (Planderova and Vozarova, 1982). This zone had the north-eastern strike as the mentioned grabens of the inner part of the Pannonian massif. According to S.G. Rudakov's data (Rudakov,1980), similar continental deposits of the Late Carboniferous and Permian (coal-bearing and redstones) are also observed in the Eastern Carpathians on the Marmarosh massif. In this place, the strike of all tectonic zones is north-western and parallel to the edge of the ancient East European platform.

In the Southern Carpathians, the continental deposits are also dated from the Upper Carboniferous and Permian; they overlie the crystalline Baikalian basement (Oncescu, 1959). The crystalline rocks of Baikalian age compose the basement of the Moesian platform

(Bonchev, 1971).

These data provide grounds to suggest that at the end of the Paleozoic the vast uplift, locally complicated by grabens, was formed in the inner part of the modern Pannonian depression. This uplift, in place of the modern Western, Eastern and Southern Carpathians, till the end of the Paleozoic was bounded by a depression, which can be classified as the fore-deep by the sedimentary formations of that depression. The fore-deep bounds the inner Pannonian uplift by an arc, which on the outer side was confined by the ancient massifs: the Czech massif in the north, the East European platform in the west, the Moesian massif in the south. Starting from the Triassic, the Carpathian geosyncline actually developed on the basis of this fore-deep and inherited its shape; the termina tion of its development was manifested by the rise of the mountain ridge of the Carpathians in the end of the Cretaceous and, particularly, in the Miocene and later the ridge also, inherited the arcuate shape of the preceding structures.

In other words, as in the Verkhoyansk region, in this area in the Paleozoic there was the "subsidence oval", where by the end of the Paleozoic, in the process of Geosynclinal development during tectonic inversion, a large and complicated uplift was formed; in the periphery, the residual depression still remained in its historical association with the Carpathian mountain arc. The modern Pannonian depression is the manifestation of the recent process superimposed on all the previous processes and evidently unconnected directly with them. This process is a part of a different endogenous regime not analysed here.

The position and shape of the Carpathian arc is apparently associated with the position of the deep Riphean avlacogene exactly opposite the northernmost part of the arc stretching into the ancient platform from Lvov through Belorussia almost to Novgorod. At a higher stratigraphical level, the through-going Pannonian-Volyn depression is identified, the interesting feature of which is that it crosses the Carpathians. Along its entire length all the folds of the Carpathians are depressed, whereas the Paleozoic Lvov and the Jurassic Stryj depressions are located on the platform; further on there is the deep commissure between the uplifts of the Ukrainian shield in the south-east and the Mazury-Belorussian anteclise in the north-west. In the south-east the Marmarosh massif bounds the depression in the Carpathians and the Apuseni Mts. bound the depression in the Pannonian region (Glushko, 1968). Perhaps, this band of platform subsidences is the attenuating pericline of the Hercynian (and more ancient) Pannonian "subsidence oval", which repeats the interrelations observed in the Verkhoyansk oval and the Vilyuy syneclise. Probably, they form the geosynclinal oval, which passed through the stages of tectonic inversion,folding and uplift, and its periclinal slightly depressed part, which survived as a group of platform syneclises.

To conclude, a better insight into the nature of continental orogenic arcs should be achieved not by constructing theoretical models, but by reconstruction of their geological history. Here yet again we observe that the modern structural mosaic of the Earth's crust is the result of a complicated history, whose progress was stipulated by previous events and during which the different structures were consecutively superimposed and affected one another.

The analysed examples featured the "subsidence ovals".Many other examples can be also found. In actual fact, the subsidence ovals can be identified as the initial structures, which later experienced distortion by subsequent processes. The existence of the structures of that shape, predominantly in the ancient Precambrian, is evidence that the deep processes were primarily manifested in the rounded structures. During the Precambrian the rounded granite and granite-gneiss domes were dominant; they were bounded by arcs and rings of metamorphic rocks crumpled into complicated folds (Fig. 8). At later stages, the linear structures appear with the formation of deep faults in the crust. The tendency to rounded forms, however, has more or less remained till recent geological time. Probably, this shape is more easily acquired by the vertical flows of matter in the Earth's mantle causing subsidences and uplifts on the surface.

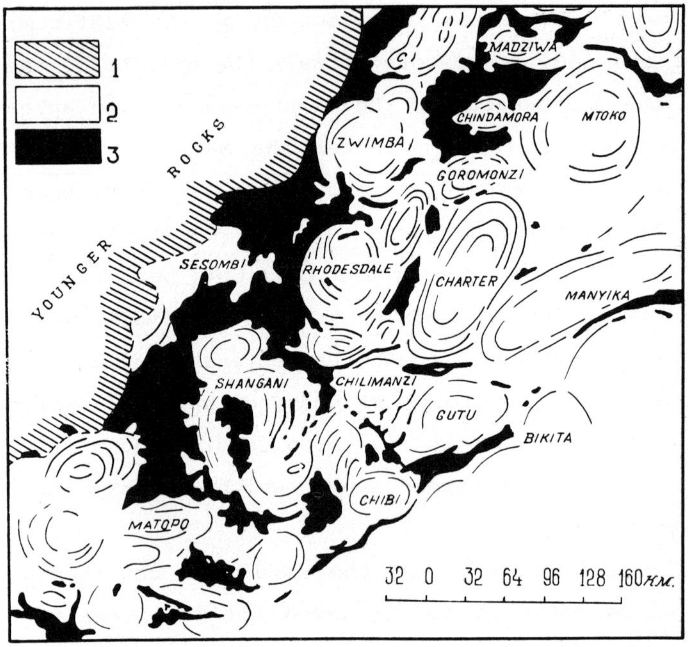

Fig. 8. Ancient dome-like structures in East Africa (Rhodesia). 1-younger rocks; 2-granites; 3-schists.

ISLAND ARCS

The common features in the history of development and in the character of displacements of continental and island arcs described in the beginning of the paper imply that the shape of island arcs is also controlled by the preceding geological history. It is impossible now to prove this since the relevant historical-geological data is unavailable.

There are still certain doubts, however, whether the popular concepts about the origin of island arcs are plausible. These concepts in fact solve the problem of arcs as the problem of the origin of marginal ("back-arc") seas.

The marginal seas are situated partly on the continental, but mostly on the oceanic crust. From drilling data we know that their age is different (Beloussov, 1984). The Tasman Sea appeared in the interval between the Cretaceous and Paleocene. The crust of other seas is of Eocene to Pliocene age (Fig.9).

Therefore,the time of formation of marginal seas is not directly

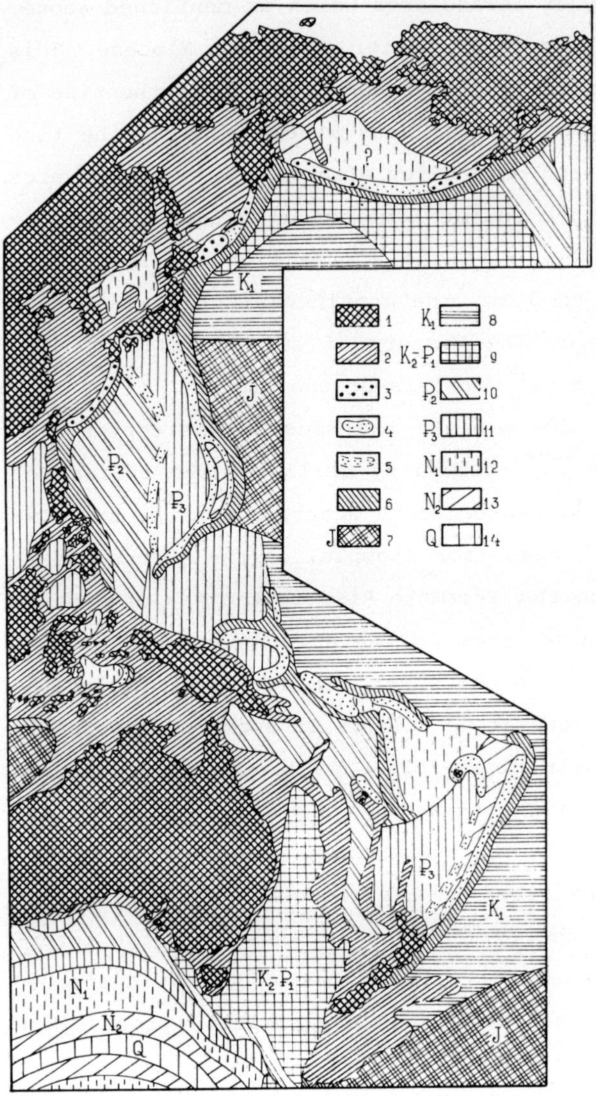

Fig. 9. Transition zone between the Pacific Ocean and the Asian and Australian continents. 1-continents and island arcs of the first (continental) type; 2-marginal seas continental crust; 3-island arcs of the second type on reduced continental crust; 4-island arcs on oceanic crust; 5-submarine ridges in marginal seas; 6-trenches; 7-14 - ages of the most ancient sediments in marginal seas: 7-Jurassic; 8-Early Cretaceous; 9-Late Cretaceous and Paleocene; 10-Eocene; 11-Oligocene; 12-Miocene; 13-Pliocene; 14-Quaternary.

connected with the history of island arcs which, as mentioned above, acquired their modern aspect not earlier than the Miocene. This means that the larger part of the seas appeared before the rise of the arcs of the second type. Several seas were formed at the time of the uplift of the arcs and only a few after the rise of the arcs.

In marginal seas, located on the continental crust, the thickness of the crust diminishes with the increase of the sea depth. At the sea depth from 2 to 3 km, the continental crust is wedged out and changes to oceanic. The history of all marginal seas on continental or on oceanic crust consists of the process of the subsidence and deepening of the sea. In the seas on the continental crust, where the basement was reached by drilling, the surface of the basement was eroded and the sediments started from the shallow--water or even continental ones (for example, on the Queensland or New Zealand plateau, or on the Yamamoto Rise in the Sea of Japan). In the seas on the oceanic crust, however, the change of the character of sediments from bottom upwards in the section implies consistent deepening. Many submarine uplifts in such seas initially rose above sea level. We can, therefore, assume that the thinning of the continental crust and its transition into the oceanic is historically associated with its subsidence, i.e., according to geophysical parameters, the crust in the process of "oceanisation" gradually transforms from the typically continental and the intermediate to the typically oceanic crust. I have repeatedly described my understanding of the mechanism of the oceanisation process and will, therefore, omit it now (Beloussov, loc. cit.).

The "plate tectonics" also ascribes the secondary origin to marginal seas, but associates it with "scattered spreading" (Karig, 1971).

It is natural to wish to verify the probability of this origin of marginal seas by reconstructing the situation before spreading. If the marginal seas did not exist previously and appeared as the result of spreading, then the material of all island arcs of the Pacific Ocean should have been concentrated initially near the Asian and Australian continents. The band of this material, conse-

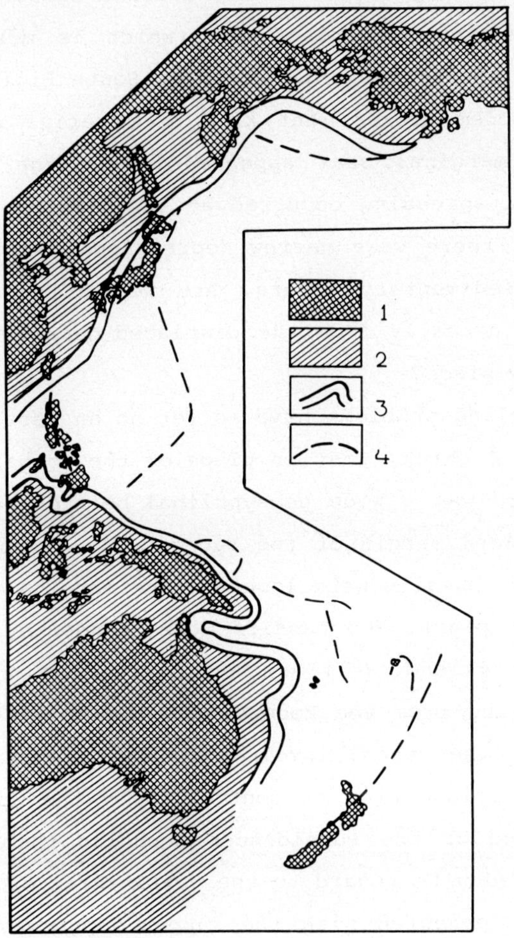

Fig. 10. Transition zone between the Pacific ocean and the Asian and Australian continents in supposed state before spreading in marginal seas. 1-continents; 2-marginal seas on continental crust; 3-supposed initial location of island arcs of the second type; 4-present-day location of the same island arcs.

quently, should invariably be extremely meandering in accordance with the windings of the edge of the continental crust (Fig.10). How could such regular arcs with similar morphology be produced from this winding band by scattered spreading? Furthermore, how in certain places could such straight lines of islands be produced from this winding band? For example, the straight line of the archipelago of the Tonga and Kermadec islands should have appeared as the result of the summarised and (which is most important!) differently aged spreading in the Tasman, South Fiji and Lau seas. Besides, the problem arises what kind of material was displaced. If most of the marginal seas appeared before the uplift of the island arcs, then spreading occurred at the time when, in place of the future arcs, there were narrow depressions, which were filled by volcanogenic-sedimentary layers. Are we to believe, that the island arcs rose actually from the displaced depressions when they reached their new place?

All these puzzling problems have as yet no answers. It is reasonable to suggest, I think, that in place of the transition zone, in the Mesozoic there was a wide geosynclinal belt on the continental crust in the western margin of the Pacific. In the middle of this belt, large median massifs were located. The marginal seas have now appeared in their place. The massifs were bounded by narrow mobile zones, in which island arcs were formed after the subsidence terminated. Starting from the Eocene, the median massifs were gradually, one after the other, involved in the oceanisation process. The shape of the island arcs is controlled by the contours of the median massifs and of the subsidences bounding them, in the same way as was observed with regard to the continental orogenic arcs.

The problems, connected with the supposed process of oceanisation, for example, with the effect of the Benioff zones or the nature of deep-sea troughs, are beyond the frame of this Symposium. To some extent,they were discussed in my earlier papers (Beloussov, loc. cit.).

With time,the progress of science and technology will reach such a stage, when the geological history of marine areas can be treated not from the position of "models", but on the basis of factual data compatible in reability with the data of continental geology. No doubt, continental geology has many mysterious aspects as well, but in that case we at least can use the historical method, always keeping in mind that geology is the natural - **historical** discipline.

REFERENCES

Balogh, K. and Korössy, L., 1974. Hungarian Mid-Mountains and adjacent areas. In: M. Mahel (Editor), Tectonics of the Carpathian-Balkan regions. GUDS Bratislava: 391-403.

Beloussov, V.V. and Ruditch, YE.M.,1961. Island arcs in the development of the Earth's structure (especially in the region of Japan and the Sea of Okhotsk). J. Geol., 69 (6): 3-23.

Beloussov, V.V., 1984. Certain problems of the structure and evolution of transition zones between continents and oceans. Tectonophysics, 105: 79-102.

Boncev, Ek., 1971. Problems of geotectonics of Bulgaria. Sofia: 204 pp. (in Bulgarian).

Glushko, V.Ya., 1968. Tectonics, and oil and gas distribution in the Carpathians and in adjoining regions. Nedra, Moscow: 264 pp. (in Russian).

Goriatchev, A.V., 1980. About structure and structural position of the Ural-Oman zone of deep faults. Geotektonika, 1: 56-71 (in Russian).

Karig, D.E., 1971. Origin and development of marginal basins in the Western Pacific. J. Geophys. Res., 76: 2542-2561.

Korössy, L., 1964. Tectonics of the basin areas of Hungary. Acta Geol. Acad. Sci. Hung., 8: 377-394.

Oncescu, N., 1959. Geologia Republicii Populare Romine, Editia a II-a, Bucuresti: 520 pp.

Planderova, E. and Vozarova, A., 1982. Biostratigraphic correlation of Late Paleozoic Formations in the West Carpathians, JGCP, Project No 5, Newsletter, No 4: 67-71.

Rudakov, S.G., 1980. Pre-Mesozoic history of the Marmarosh massif, East Carpathians (Soviet Part). J.G.C.P., Project No 5, Newsletter, No 2: 49-52.

Sollogub, V.B. and Chekunov, A.V., 1980. Structure and age of the basement of the crust in Central and South-East Europe. Geotektonika, 1: 17-29 (in Russian).

Vinogradov, A.P. (Editor-in-Chief), 1967-1969. Atlas: lithologo-paleotectonic maps of the USSR, I-IV.

Wein, Gy., 1969. Tectonic review of the Neogene-covered areas of Hungary. Acta Geol. Acad. Sci. Hung., 13: 399-436.

Zaitsev, Yu.A., 1984. Evolution of geosynclines. Nedra, Moscow: 208 pp. (in Russian).

REGULARITIES IN THE PATTERN OF MAJOR FAULT ZONES OF THE EARTH AND THE ORIGIN OF ARCS

N. PAVONI

Institut für Geophysik, ETH-Hönggerberg, CH-8093 Zürich, Switzerland

ABSTRACT

The considerable tectonic mobility of the earth as demonstrated by the Mesozoic-Cenozoic evolution of the oceanic lithosphere argues against the existence of regular, rhegmatic shear systems of global extent. Arc structures are characteristic features of the tectonically mobilized belts of convergence of the earth's crust and lithosphere. Emphasis is given to horizontal crustal deformation within the Eurasian and circum-Pacific orogenic belts as revealed by the pattern of Cenozoic strike-slip faults. Distinct regularities in the orientation of maximum horizontal crustal shortening and compression are observed. The following geotectonic model is proposed: The orogenic belts represent the zone of convergence and shearing of the lithosphere situated between two major geotectonic units, the Pacific unit and the Gondwana unit. In the Pacific hemisphere, large-scale diverging movements of plates away from the central Pacific region toward the margins of the Pacific are documented by the breakup of the paleo-Pacific lithosphere and the growth of the Pacific plate. In the anti-Pacific hemisphere, large-scale diverging movements away from Africa are demonstrated by the breakup of Gondwana. During collision twisting movements will be generated within the orogenic belt. Fan-shaped fault patterns, comparable to the pattern of slip-lines formed in deformation of rigid-plastic solids, are associated with the formation of mountain arcs.

1. INTRODUCTION

Mountain arcs and island arcs are characteristic features of the tectonically mobilized belts of convergence of the earth's crust and lithosphere. The origin of arcs is interrelated with the processes of collision and subduction of lithosphere. The recognition of distinct fault patterns associated with arcs might contribute to the knowledge of the geotectonic frame within which arc structures evolve. At the invitation of the convenor of this symposium, I was asked to review and discuss evidence and arguments concerning global fracture and shear patterns, and to report about possible regularities in the arrangement of major fault systems.

In the following, I will (1) summarize some major results of geotectonic research of the last three decades, (2) shortly present some differing hypotheses concerning the nature of the lithosphere and correlated mountain building processes, (3) try to give an evaluation of these hypotheses in the light of the new findings of geotectonic research and discuss the consequences thereof with respect to the nature of crustal fracture patterns, (4) have a look at the patterns of strike-slip faulting as related to Cenozoic orogenic belts and (5) discuss the evolution of mountain arcs and associated faults in the case of continent-continent collision.

2. SOME MAJOR RESULTS OF RECENT RESEARCH IN GEOTECTONICS

The impressive results of world-wide geophysical, geological and geodetic research of the last three decades have convincingly demonstrated the great tectonic mobility of the earth's lithosphere. They have provided distinct answers to many longstanding, fundamental questions in geotectonics:

- The system of active oceanic ridges and the system of young fold mountains, both of world-wide extension, represent the two major belts of tectonic activity and transformation.
- New oceanic crust and lithosphere are being generated at the oceanic ridges by sea-floor spreading. Old lithosphere is being "destroyed" within the belts of converging lithosphere by subduction. Both spreading and subduction rates are similar in magnitude, varying between 1 and 18 cm per year. It appears reasonable to assume that an approximately constant surface area of the earth is maintained.
- The construction of new lithosphere, the large-scale horizontal tectonic transport of lithospheric plates, and the final destruction of lithosphere are but the external manifestation of a continuous endogenic process: the circulation of mantle material.
- During the Mesozoic-Cenozoic period, i.e., during the Alpine orogenic cycle, plate movements and deformations have been taking place at varying rates but, in principle, in a continuous way.
- During the last 200 million years, i.e., geologically speaking, in a relatively short time span, new oceans, such as the Atlantic Ocean and the Indian Ocean, as well as the whole floor of the Pacific Ocean have been newly formed by sea-floor spreading. Thus, more than half of the earth's

lithosphere is of young age.

- Large-scale horizontal rotations of lithospheric plates are demonstrated by the results of paleomagnetic and paleogeographic research.
- There is a subdivision of the lithosphere into oceanic and continental lithosphere. This fundamental heterogeneity of the lithosphere is of upmost geotectonic importance. The continental lithosphere is less dense than the oceanic lithosphere. Due to this density difference the continental lithosphere cannot be subducted deep into the mantle. It remains at the earth's surface and so does not take part in the mantle recycling process. By accretion of crustal material the continental lithosphere is slowly increasing in volume.
- Most of the continental lithosphere is much older than the oceanic lithosphere. For this reason, it preserves much older structures, fault patterns, etc., than the oceanic lithosphere.
- The whole lithosphere is in nearly perfect isostatic equilibrium.

3. GEOTECTONIC HYPOTHESES AND POSSIBLE EVOLUTION OF CRUSTAL SHEAR PATTERNS

For a long time, geologists have been convinced that the lithosphere was a solid shell, and that the single parts, such as the continents, were constrained from lateral movement by the finite strength of the material. In this fixist picture of the lithosphere, orogeny and crustal shortening are explained by shrinking of the earth through cooling. The most rigid portions of the lithosphere glide over the substratum along a surface in the asthenosphere and the shortening is concentrated in the zones of weakness, the orogenic belts. The geologists of the 19th century, such as Dana (1847), Heim (1878), and Suess (1885), were proponents of the contraction hypothesis. In our century and up to the present day the hypothesis finds well-known earth scientists, such as Bucher (1933, 1957), Cloos (1936), Jeffreys (1959), Kober (1942), Stille (1924), among its adherents. With respect to the evolution of global rhegmatic fracture systems the assumptions of the contraction hypothesis are favorable. However, serious objections have been raised against the contraction hypothesis. Since the detection of radioactivity and the respectable amount of radioactive heat produced in the crust, cooling and contraction have become questionable. In contrast to the contraction hypothesis an expanding earth has been postulated (Hilgenberg, 1933; Jordan, 1966; Egyed, 1957; Carey, 1958; Owen, 1981). As compared with the contraction hypothesis an expanding earth allows a higher degree of relative movements and displacements of lithospheric plates. A similar statement about relative plate movement holds true for the undation and oscillation hypotheses (Haarmann, 1930; van Bemmelen, 1932, 1935; Belousov, 1980).

The highest degree of tectonic mobility of lithospheric plates is granted by the convection hypotheses, in which deep-seated convection currents act as motor of lithospheric movements (Ampferer, 1906; Schwinner, 1919; Holmes, 1929; Pekeris, 1935; Hales, 1936; Griggs, 1939; Vening Meinesz, 1952).

Large-scale horizontal displacements are a part of the hypotheses of continental drift (Wegener, 1912; Taylor, 1910; du Toit, 1937; Argand, 1922; Staub, 1928). The explanations of the mechanism and the motor of continental movements remain inadequate. The processes by which large-scale horizontal lithospheric movements are realized have been described by the hypotheses of sea-floor spreading (Hess, 1962; Vine and Matthews, 1963) and plate tectonics (see e.g. Graham, 1981). The motor of plate movements is still a matter of debate.

Joints and small faults are to be found in practically all competent rocks exposed at the surface. It is a remarkable fact that the rocks of the continental crust, and especially the crystalline rocks of the basement, are intensively cut by several sets of joints and faults into an assemblage of polyhedral blocks. In old rocks, the origin and age relationship of joint sets are extremely difficult to establish or are not known. There is abundant field evidence that joints have developed at practically all ages in the history of rock. It is extremely unlikely that all joints observed in an exposure of old rock are the result of a single mechanism. Incorrect assumptions regarding the ages of joints may easily be made and invalidate the conclusions of the analysis (Price, 1966).

In sufficiently large exposures of competent rocks subvertical and inclined joints and faults will be found. Locally and with respect to the subvertical joints, four to six main orientations, or sets of joints and faults are generally observed. In many cases, two sets are orthogonally oriented to each other, forming a joint "system". The orientations of major joint sets remain remarkably constant and parallel over large areas. Since the earth's crust can to a certain degree transfer stresses and stress changes (variations) without failure, the number of different orientations of joint sets and fault sets at one locality is limited (angles of 20 - 30 degrees between two joint sets).

The constraints about the number of possible orientations of joint and fault sets at a single site, as well as the observed regional constancy of joint and fault patterns, led to the concept, held by many earth scientists, that regular, continent-wide stationary fracture systems exist. For the process causing the regular joint and fault patterns Sonder (1947) has introduced the term rhegmagenesis (often written regmagenesis).

The evolution of stationary fault systems of continents has been described by Burtman (1979). According to this author stationary fault systems, which are of old origin, have propagated from older consolidated areas of the continent to the adjacent younger tectogenes. Very little is known about the mechanisms and forces by which the regmatic grid pattern is produced or about the age sequences of the sets and the mechanisms of expansion from one area to the neighboring one. In this regard the deep-seated faults or deep faults (Belousov, 1980) must be mentioned. These faults represent dislocations of old origin and of great vertical and horizontal extent which have undergone a long, polyphase development. Also, the zones of folding and orogeny are interpreted to be intimately related to the extent and pattern of deep faults (Ashgirei, 1963). Based upon the results of continent-wide studies, Sonder (1947), Moody (1966) and others have postulated the existence of regular, rhegmatic fracture systems of global extent.

The formation of regular rhegmatic shear fault systems of global extent presupposes the presence of a solid lithospheric shell as in the contraction hypothesis. In a recent paper, Belousov and Dmitriyeva (1984) have reported about a comprehensive study of orientations of folded structures in continental areas. They found that the same preferential trends, a north-south one, an east-west one, and two diagonal ones, are present in all continents. According to these authors, these preferential trends have been maintained with reference to the present geographic grid from Archean to Cenozoic times, and thus, argue against any mobilistic concept in geotectonics. It should be mentioned here that planet-wide fault and lineament systems have been described for Moon, Mars and Mercury (Fielder et al., 1976; Harp, 1976; Katterfeld, 1976).

With respect to the origin of arcs, it is of interest to note that both de Sitter (1956) and Jacobs et al. (1959) have described the Eurasian and circum-Pacific fold belts as representing the two fundamental world-embracing active fracture belts caused by contraction of the earth. The island arcs and mountain arcs are interpreted as surface expressions of conical thrust zones reaching several hundred kilometers into the contracting sphere (see also Benioff, 1954).

Finally, it should be mentioned that satellite imagery has become a powerful aid in recognizing and mapping linear features over vast areas. The interpretation of satellite pictures has given a new impetus to the study of lineament and fault patterns of continents. Interesting results are reported in the proceedings of the four Conferences on the New Basement Tectonics (Hodgson, 1976; Podwysocki and Earle, 1979; O'Leary and Earle, 1981; Gabrielsen et al., 1984). However, in many cases, the interpretation is controversial and the true nature of lineaments observed on satellite pictures remains enigmatic.

That the continental crust is intensively cut by joint and fault systems is of upmost importance with respect to its deformation under changing stress conditions. A gradual, relative change of the orientation of the stress field will not create new fault sets, but will instead use the existing ones and preferentially reactivate those which are optimally oriented with respect to the new applied stresses.

Considering rotations of crustal blocks we are faced with a fundamental difficulty. In the case of a fully developed joint and fault pattern, a horizontal rotation of 20 - 30 degrees will represent a pattern which, relative to the geographic grid, is very similar to the one before the rotation. Since the origin and relative age sequence of the fault sets usually are not known, the sets are not identifiable with certainty after the rotation. The fundamental difficulty of identification of single joint and fault sets renders it impossible to really prove the existence of a global rhegmatic fracture pattern.

In addition to the commonly observed subvertical joint and fault systems, geological and geophysical studies reveal the fundamental importance of repeated, large-scale thrust movements and detachments at midcrustal levels (Burchfiel et al., 1980). These processes tend to destroy older fault patterns. Thus, one should avoid speaking of primordial fault patterns.

The considerable tectonic mobility of the earth as demonstrated by the Mesozoic-Cenozoic evolution of the oceanic lithosphere argues against the existence of a global rhegmatic fault pattern.

4. REGULARITIES IN THE PATTERN OF MAJOR CENOZOIC FAULT ZONES

A different approach to the understanding of large-scale regularities in the arrangement of faults and fault systems, also with respect to the origin of arcs, is based on the study of present tectonic processes and on the reconstruction of the Mesozoic-Cenozoic evolution of the lithosphere. The study of active faults and their relationship to neotectonic deformation is promising, as are seismotectonic studies, i.e., the investigation of seismicity, focal mechanisms and centroid-moment tensor solutions of earthquakes and their relationship to local tectonics (Fig. 1). The advantage of such an approach is that the observation and description of modern tectonic processes and deformations are far more comprehensive and certain than the statements based on kinematic analyses of old structures which have undergone repeated cycles of deformation.

Fig. 1. The arc of the Western Alps. Characteristic, fan-shaped pattern of maximum horizontal crustal shortening (MHS, dashed lines), as derived from a kinematic analysis of late-Cenozoic folds (dotted) and faults of the Jura Mountains and the Chaînes Subalpines (after Carte tectonique de la France 1:1 000 000, 1980), and orientations of P-axes (thick bars) from focal mechanisms of earthquakes (Bossolasco et al., 1974; Frechet, 1978; Pavoni, 1980). B: Bern, L: Lyon, M: Marseille, T: Torino.

The following discussion is mainly based on Cenozoic faulting and fault patterns observed within the continental crust. Special attention is given to faulting within the fold belts of Alpine orogeny since arc structures typically originate within the zones of convergence and collision of the lithosphere. Emphasis is given to strike-slip faults and their patterns. There are several reasons to look particularly at strike-slip faults:

- Major strike-slip fault zones are of considerable extent (up to 1700 km) and are relatively accessible over long distances.
- As a rule, the sense of displacement, dextral or sinistral, remains the same along the whole length of the fault zone.
- The sense of displacement remains the same over long periods of time, i.e., over several millions of years.
- The amount of accumulative lateral displacement may be considerable, reaching tens or hundreds of kilometers, and is greater by an order of magnitude than the vertical displacement.

These particular qualities of strike-slip faulting are due to the fact that the earth's lithosphere is in isostatic equilibrium. Horizontal displacements of rock masses do not involve a change of potential energy. Strike-slip faulting is in a definite and well-known way related with normal faulting, reverse faulting, thrust faulting, and folding. Structural analyses of regional patterns of strike-slip faults and associated faults provide valuable information about large-scale deformation and geodynamic processes.

For a long time, the importance of strike-slip faults in geotectonics was underestimated by the geologists. One reason is that in the framework of the contraction, oscillation and undation hypotheses, horizontal displacements are of secondary importance. Another reason is that in outcrops and vertical profiles, lateral components of displacement are difficult to observe.

The increasing use of air photographs and satellite imagery in geological prospecting has revealed the importance of horizontal movements. Last, it was nature itself that demonstrated the significance of horizontal displacements in the numerous surface breaks observed along active faults during major earthquakes. Hill and Dibblee (1953), Wellman (1955), Moody and Hill (1956), Rod (1956), St. Amand (1957), Vroman (1957), Lensen (1960), Pavoni (1961), Allen (1962), Crowell (1962), and Burtman et al. (1963) have emphazised the importance of strike-slip faulting in regional tectonics. In the concept of plate tectonics, strike-slip faults play an important role. They have usually been interpreted as transform faults (Wilson, 1965).

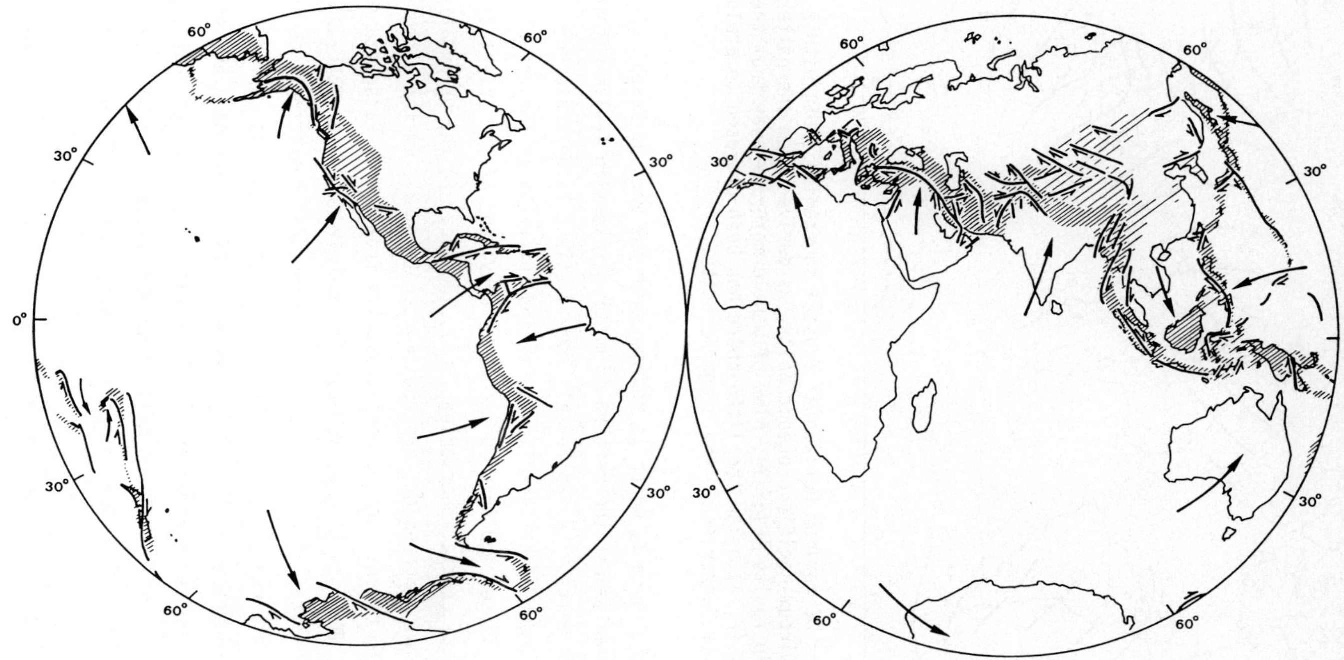

Fig. 2. Pattern of major strike-slip fault zones within the Alpine orogenic belt (hatched area) and related horizontal movements of the lithosphere (arrows).

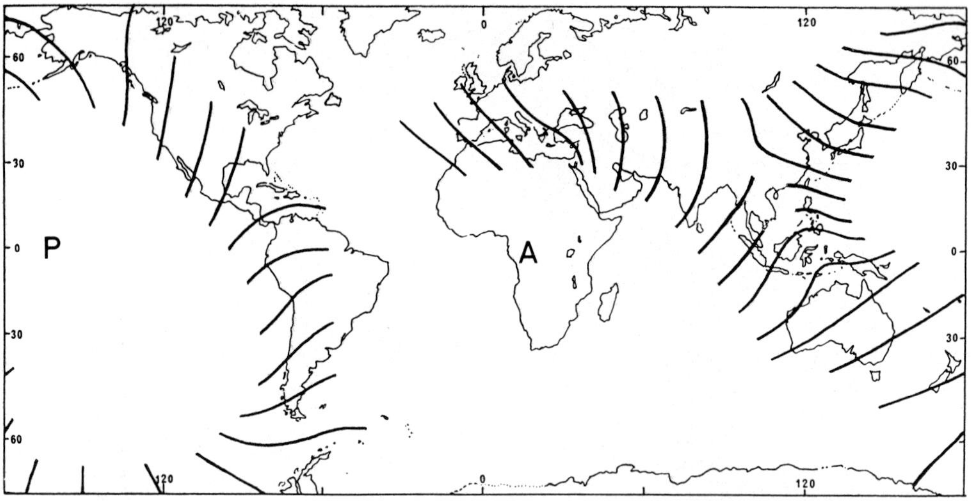

Fig. 3. Orientations of maximum horizontal crustal shortening (MHS) within the Mediterranean and Circum-Pacific orogenic belts, as derived from the pattern of Cenozoic strike-slip faults (Fig. 2). Flow of mantle material away from the two geotectonic centers (P, A) below the lithosphere has been proposed to explain the pattern of MHS trajectories in the orogenic belts.

The world map of Figure 2, which is redrawn from the work of Pavoni (1962), shows in a simplified manner the pattern of major strike-slip faults within the young orogenic fold belts. The map reveals some distinct large-scale regularities in the arrangement of strike-slip faults and interrelated horizontal crustal movements. These regularities become more evident if the orientations of maximum horizontal crustal shortening (MHS), identical to the orientations of maximum horizontal compression, are constructed (Fig. 3). In the circum-Pacific belt the MHS orientation rotates gradually from E-W in Patagonia to N-S in California, NW-SE in Alaska, WNW-ESE in southern Japan, and ENE-WSW in New Zealand. The MHS trajectories encircle a hypothetical geotectonic center (P) in the central Pacific region. Similarly, a center (A) in Africa, in antipodal position to P, is encircled by the MHS trajectories of the Eurasian and southern, circum-Pacific fold belts. Further studies have led to the concept that the Eurasian and circum-Pacific orogenic belts represent zones of crustal shearing and convergence between two major geotectonic units, the Pacific unit and the Gondwana unit.

It is of interest to note, that the orientations of P-axes of fault-plane solutions (Ritsema, 1960; Balakina et al., 1967) and mean orientations of P-axes of centroid-moment tensor solutions (Dziewonski et al., 1981) of earthquakes located in the fold belt correspond well with the MHS orientations.

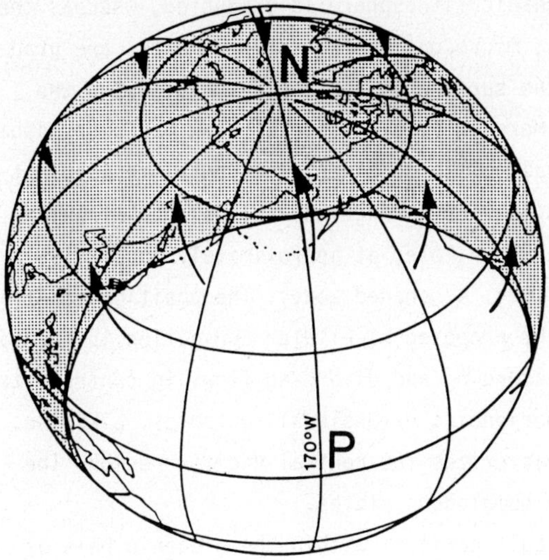

Fig. 4. Idealized representation of the bicellular flow pattern in orthographic horizontal projection. P: Pacific pole at 170°W/0°N. Light areas: Diverging lithosphere. Shaded areas: Zone of converging lithosphere.

Convergence and subduction of lithosphere in the fold mountain belts and island arcs require tectonic transport of lithospheric plates from the neighboring regions toward the belts, and consequently, lithospheric spreading and diverging movements within the Pacific unit and/or within the Gondwana unit. In the Pacific hemisphere, diverging movements of plates away from the central Pacific are documented by the breakup of the paleo-Pacific lithosphere. With respect to the Pacific plate, which is centered at 170°W/0°N (P), the Kula plate moved toward the northwest and was subducted under Asia and Alaska, the Farallon plate moved toward the northeast and was subducted under North America and South America, and the Phoenix plate moved toward the southeast and was subducted under South America and Antarctica. At the same time the Pacific plate was growing to its present size (Larson and Pitman, 1972; Hilde et al., 1976). In the anti-Pacific hemisphere, which is centered in Africa at 10°E/0°N (A), diverging movements away from Africa are documented by the breakup of Gondwana.

In order to explain these large-scale diverging movements, a bicellular convection system in the mantle has been proposed with ascending flow under the central Pacific and African plates. Underneath the lithosphere the ascending flow diverges horizontally away from the centers. At 70 - 90 degrees distance from the centers the flow converges and descends into the mantle (Fig. 4). In the zone of converging flow, the oceanic lithosphere is subducted, whereas the continental lithosphere, continents, microcontinents, and sediments, are gradually accumulated and accreted at the surface in a circular belt. Below the oceanic ridges, unidirectional and more or less horizontal flow has been assumed to exist (Pavoni, 1969, 1981). Actually, the continents of South America, Antarctica, Australia, Asia, and North America arrange themselves in a circular zone, extending over North Pole and South Pole, at approximately equal distances from both pole P and pole A (Fig. 4, shaded zone). The position of Africa is different. In the framework of the proposed bicellular convection in the mantle, Africa is situated above the ascending and diverging flow. In contrast to the other continents, Africa is experiencing extensional tectonics. Likewise, an extensional tectonic regime characterizes the central Pacific region. The African plate and Pacific plate are homologous plates.

The interpretation of Cenozoic fault patterns within the orogenic belt as well as the Mesozoic-Cenozoic evolution of the system of oceanic ridges reveal a fundamental Pacific/anti-Pacific bipolarity in the structure of the earth's mantle, defined by pole P and pole A. It is of significance to note that the same Pacific/anti-Pacific bipolarity is evident in the global distribution of residual geoid heights and hotspots (Chase, 1979; Crough and Jurdy, 1980), and in the distribution of lateral heterogeneities of P-wave velocities in the lower mantle (Dziewonski, 1984). That the same bipolarity is evident in the global data sets of three independent fields of earth science research points to a common origin of the anomalies and the bipolarity (Pavoni, 1985a,b). The geophysical and geodetic data strengthen the arguments in favour of mantle-wide or thermally coupled convection in the mantle with ascending flow beneath both the central Pacific region and the African plate, and descending flow at 70 - 90 degrees distance from the spreading centers P and A.

The proposed model of bicellular convection in the mantle constitutes, in my opinion, the geotectonic frame within which arc structures evolve. The young fold mountain belts and island arcs are situated in the zone of convergence of lithosphere between the Pacific unit and the Gondwana unit. Their location, within the convergence zone, is strongly influenced by the distribution of large, relatively rigid, and irregularly accreted blocks. Therefore, their actual position may deviate considerably from their "regular" position with respect to the geotectonic model. Preferentially, they extend along the margins of continental blocks.

Accumulation and collision of continental lithosphere, as well as the converging flow in the mantle beneath the lithosphere, produce complicated pat-

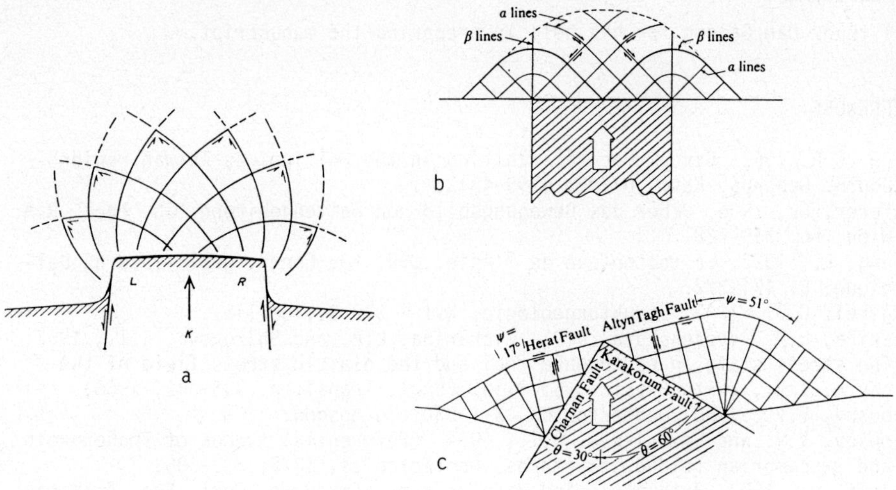

Fig. 5. Slip-line fields in rigidly indented rigid-plastic solids.
(a) After Pavoni, 1961; (b) and (c) after Tapponnier and Molnar, 1976.

terns of indented crustal blocks and wedges, mutually protruding into the neighboring crust by thrust faulting and lateral faulting. At the same time, horizontal twisting movements and vortex-like patterns may evolve. Also, vortex-like patterns are observed with subduction zones (e.g. Fig. 2, New Hebrides/Tonga arcs).

In this regard the fan-shaped fault patterns produced in the process of continent-continent collision must be mentioned. These fault patterns are comparable to the slip-line fields in rigidly indented rigid-plastic solids (Fig. 5). They are typically observed in the external parts of mountain arcs and in the adjacent continental foreland. Their origin is related with active protrusion of the arc, relative to the foreland. They are an indication of intensified deformation and tectonic mobilization of the crust. The comparison of fan-shaped strike-slip fault patterns with plane slip-line fields was first used in the description and interpretation of fault and fold patterns of the Jura Mountains and the Chaînes Subalpines of the Western Alps (Pavoni, 1961), and has since been successfully applied to the interpretation of continental tectonics in Asia (Pavoni, 1964; Tapponnier and Molnar, 1976), and other areas, such as the Slave-Churchill boundary zone in Canada (Gibb, 1978).

ACKNOWLEDGMENT

I thank Dan Galson for his help in preparing the manuscript.

REFERENCES

Allen, C.R., 1962. Circum-Pacific faulting in the Philippines-Taiwan region. Journ. Geophys. Res., 67/12: 4795-4812.
Ampferer, O., 1906. Ueber das Bewegungsbild von Faltengebirgen. Jb. Geol. R.A. Wien, 56: 539-622.
Argand, E., 1922. La tectonique de l'Asie. C.R. 13e Congr. géol. intern. Belgique, 5: 171-372.
Ashgirei, G.D., 1963. Strukturgeologie, XVI + 572 S., Berlin.
Balakina, L.M., Vvedenskaya, A.V., Misharina, L.A. and Shirokova, E.I., 1967. The stress state in earthquake foci and the elastic stress field of the earth. Izv., Earth Physics, 6: 3-15. (Engl. Transl. p. 333-342, 1968).
Belousov, V.V., 1980. Geotectonics, X + 330 p., Moscow.
Belousov, V.V. and Dmitriyeva, B.I., 1984. Preferential trends of Phanerozoic and Precambrian folded structures. Geotectonics, 18/5: 383-386.
Benioff, H., 1954. Orogenesis and deep crustal structure. Geol. Soc. America Bull., 65: 385-400.
Bucher, W.H., 1933/1957. The deformation of the earth's crust. Reprint of the original edition of 1933, XIV + 518 p., New York 1957.
Burchfiel, B.C., Oliver, J.E. and Silver, L.T., eds., 1980. Continental tectonics. Studies in Geophysics. National Academy of Sciences, XII + 197 p., Washington.
Burtman, V.S., 1979. Das stationäre Bruchnetz der Paläozoiden und der Mobilismus. Z. geol. Wiss., 7/10: 1257-1267, Berlin.
Burtman, V.S., Peive, A.V. and Ruzhentzev, S.V., 1963. Main wrench faults in Tien-Shan and Pamir. Akad. Nauk USSR, Geol. Inst., 80:152-172.
Caire, A., 1974. Tectonique spirale en Méditerranée centrale. C.R. Acad. Sc. Paris, t. 278, Série D, 3165-3167.
Carey, S.W., 1958. The Tectonic Approach to Continental Drift. Symposium, Univ. Tasmania, p. 177-209. Hobart (Tasmania) 1958.
Chase, C.G., 1979. Subduction, the geoid, and lower mantle convection. Nature, 282: 464-468.
Cloos, H., 1936. Einführung in die Geologie. XII + 503 S., Berlin.
Crough, S.T., Jurdy, D.M., 1980. Subducted lithosphere, hotspots, and the geoid. Earth and planet. Sci. Lett., 48: 15-22.
Crowell, J.C., 1962. Displacement along the San Andreas Fault, California. Geol. Soc. Am., Spec. Paper 71, 63 p.
Dana, J.D., 1847. On some results of the earth's contraction on cooling. Am. Journ. Sci., 2/3: 176-188.
De Sitter, L.U., 1956. Structural geology. VIII + 552 p., New York.
Du Toit, A.L., 1937. Our Wandering Continents. An Hypothesis of Continental Drifting. 366 p., London.
Dziewonski, A.M., 1984. Mapping the lower mantle. Determination of lateral heterogeneity in P velocity up to degree and order 6. J. Geophys. Res., 89: 5929-5952.
Dziewonski, A.M., Chou, T.-A. and Woodhouse, J.H., 1981. Determination of earthquake source parameters from wave-form data for studies of global and regional seismicity. J. Geophys. Res., 86: 2825-2852.
Egyed, L., 1957. A new Dynamic Conception of the Internal Constitution of the Earth. Geol. Rdsch., 46:101-121., Stuttgart.
Fielder, G., Fryer, R.J., Gash, P.J.S., Whitford-Stark, J.L. and Wilson, L., 1976. Lineament patterns on the Moon, Mars and Mercury. In: R.A. Hodgson et al. (eds.), Proc. First Intern. Conf. on The New Basement Tectonics,

379-388, Salt Lake City.
Gabrielsen, R.H., Ramberg, I.B., Roberts, D. and Steinlein, O.A. (Editors), 1984. Proceedings of the Fourth International Conference on Basement Tectonics.
Gibb, R.A., 1978. Slave-Churchill collision tectonics. Nature, 271: 50-52.
Graham, A.L., 1981. Plate tectonics. In: Cocks, L.R.M. (Editor): The evolving earth, 165-177, Cambridge.
Griggs, D.T., 1939. A Theory of Mountain-Building. Amer. Journ. of Science, 237: 611-650.
Haarmann, E., 1930. Die Oscillationstheorie. Ferdinand Enke Verlag, Stuttgart.
Hales, A.L., 1936. Convection currents in the Earth. Geoph. suppl. Monthly Not. Astr. Soc., 3: 372-379.
Harp, E.L., 1976. Fracture Systems of Mars. In: R.A. Hodgson et al. (Editors), Proc. First Intern. Conf. on The New Basement Tectonics, 389-408, Salt Lake City.
Heim, A., 1878. Untersuchungen über den Mechanismus der Gebirgsbildung. 2 Bände, 346 S. und 246 S., Basel.
Hess, H.H., 1962. History of ocean basins. In: A.E.J. Engel, H.L. James and B.F. Leonard (Editors), Petrologic studies: a volume in honor of A.F. Buddington: Geological Society of America, 599-620.
Hilde, T.W.C., Uyeda, S. and Kroenke, L., 1976. Tectonic history of the western Pacific. In: C.L. Drake (Editor), Geodynamics: Progress and Prospects, AGU, Washington, D.C., 1- 15.
Hilgenberg, O.C., 1933. Vom wachsenden Erdball, 56 S., Berlin.
Hill, M.L. and Dibblee, T.W. Jr., 1953. San Andreas, Garlock and Big Pine Faults, California. Bull. Geol. Soc. Am., 64: 443-458.
Hodgson, R.A. (Editor), 1976. Proceedings of the First International Conference on the New Basement Tectonics. 636 p.
Holmes, A., 1929. Radioactivity and earth movements. Transactions of the Geological Society, 18: 559-606.
Jacobs, J.A., Russell, R.D. and Wilson, J.T., 1959. Physics and Geology. XII + 424 p., New York.
Jeffreys, H., 1959. The Earth. Its Origin, History and Physical Constitution. 420 p., Cambridge.
Jordan, P., 1966. Die Expansion der Erde. Folgerungen aus der Diracschen Gravitationshypothese. Die Wissenschaft, 14, 180 S., Braunschweig.
Katterfeld, G.N., 1976. Global and regional systems of lineaments on the Earth, Mars and the Moon. In: R.A. Hodgson et al. (Editors), Proc. First Intern. Conf. on The New Basement Tectonics, 369-378, Salt Lake City.
Kober, L., 1942. Tektonische Geologie. 492 S., Berlin.
Larson, R.L. and Pitman, W.C., 1972. World-wide correlation of Mesozoic magnetic anomalies, and its implications. Bull. geol. Soc. Amer., 83: 3645-3662.
Lensen, G.J., 1960. Principal horizontal stress directions as an aid to the study of crustal deformation. Publ. Dominion Obs. Ottawa, XXIV/10: 389-397.
Moody, J.D., 1966. Crustal shear patterns and orogenesis. Tectonophysics, 3, 6: 479-522, Amsterdam.
Moody, J.D. and Hill, M.J., 1956. Wrench-fault tectonics. Bull. Geol. Soc. Am., 67: 1207-1246.
O'Leary, D.W. and Earle, J.L. (Editors), 1981. Proceedings of the Third International Conference on Basement Tectonics. 404 p.
Owen, H.G., 1981. Constant dimensions or an expanding earth? In: L.R.M. Cocks (Editor). The evolving earth, 179-192, Cambridge.
Pavoni, N., 1961. Faltung durch Horizontalverschiebung. Eclogae geol. Helv., 54: 515-534.
Pavoni, N., 1962. Rotierende Felder in der Erdkruste. Abh. deutsch. Akad. Wiss. Berlin, Kl. Bergbau, Hüttenwesen, Montangeologie, Jg. 1962, Nr. 2, 1. Internat. Symposium über rezente Erdkrustenbewegungen, Leipzig, 257-270.

Pavoni, N., 1964. Aktive Horizontalverschiebungszonen der Erdkruste. Bull. Ver. Schweiz. Petrol.-Geol. u. Ing., 31: 54-78, Basel.

Pavoni, N., 1969. Zonen lateraler horizontaler Verschiebung in der Erdkruste und daraus ableitbare Aussagen zur globalen Tektonik. Geol. Rdsch., 59: 56-77.

Pavoni, N., 1985a. Pacific/anti-Pacific bipolarity in the structure of the earth's mantle. Eos, 66/25: cover, June 18, 1985.

Pavoni, N., 1985b. Die pazifisch-antipazifische Bipolarität im Strukturbild der Erde und ihre geodynamische Deutung. Geol. Rdsch., 74/2: 251-266.

Pekeris, C.L., 1935. Thermal convection in the interior of the earth. Monthly Geoph. suppl. Not. R. Astr. Soc., 3: 343-367.

Podwysocki, M.H. and Earle, J.L. (Editors), 1979. Proceedings of the Second International Conference on Basement Tectonics. 595 p.

Price, N.J., 1966. Fault and joint development in brittle and semi-brittle rock. X + 176 p., Oxford.

Ritsema, A.R., 1960. Further focal mechanism studies at de Bilt. Publ. Dominion Obs., Ottawa, XXIV/10: 355-358.

Rod, E., 1956. Strike-slip faults of northern Venezuela. Bull. Am. Ass. Petr. Geol., 40/3: 457-476.

Schwinner, R., 1919. Vulkanismus und Gebirgsbildung. Ein Versuch. Z. f. Vulkanologie, 5: 175-230, Berlin.

Sonder, R.A., 1947. Discussion of shear patterns of the earth's crust by F.A. Vening Meinesz. Trans. Amer. Geophys. Union, 28: 939-946.

Sonder, R.A., 1956. Mechanik der Erde. 291 S., Stuttgart.

St. Amand, P., 1957. Circum-Pacific orogeny. Publ. Dominion Obs., Ottawa, XX/2: 403-411.

Staub, R., 1928. Der Bewegungsmechanismus der Erde. Borntraeger, Berlin.

Stille, H., 1924. Grundfragen der vergleichenden Tektonik. 443 S., Berlin.

Suess, E., 1909. Das Antlitz der Erde. 3 Bde., 2778 S., Prag und Leipzig 1885.

Tapponnier, P. and Molnar, P., 1976. Slip-line field theory and large-scale continental tectonics. Nature, 264: 319-324.

Taylor, F.B., 1910. Bearing of the Tertiary mountain belt on the origin of the Earth's plan. Bulletin of the Geological Society of America, 21: 179-226.

Van Bemmelen, R.W., 1932. De Undatie theorie, hare afleiding en toepassing op het westelijk deel van de Soenda boog. Nat. Tijdschr. v. Ned. Indië, 92/I: 85-242.

Van Bemmelen, R.W., 1935. The Undation theory on the development of the earth's crust. Proc. 16th Int. Geol. Congr. Washington, 2, 965-982.

Vening Meinesz, F.A., 1952. Convection-currents in the earth and the origin of the continents. 1. Proc. Kon. Ak. v. Wetensch., Amsterdam (B), 55/5: 527-554.

Vine, F.J. and Matthews, D.H., 1963. Magnetic anomalies over oceanic ridges. Nature, 199: 947-949, London.

Vroman, A.J., 1957. Strike-slip movements, their associated features and their occurence in Israel. XX. Congr. Geol. Intern., Mexico, Sec. V/2: 399-408.

Wegener, A., 1912. Die Entstehung der Kontinente. Geol. Rdsch., 3: 276-292.

Wellman, H.W., 1955. New Zealand Quaternary tectonics. Geol. Rdsch., 43/I: 248-257.

Wilson, J.T., 1965. A new class of faults and their bearing on continental drift. Nature, 207: 343-347, London.

Addenda:

Bossolasco, M., Eva, C. and Pasquale, V., 1974. On seismotectonics of the Alps and Northern Apennines. Riv. It. Geofis., 23: 57-63.

Frechet, J., 1978. Sismicité du sud-est de la France et une nouvelle méthode du zonage sismique. Thèse de 3ème cycle, Univ. Grenoble.

Pavoni, N., 1980. Crustal stresses inferred from fault-plane solutions of earthquakes and neotectonic deformation in Switzerland. Rock Mechanics, Suppl. 9, 63-68.

GLOBAL NEOTECTONICS, ARCS AND GEOID CONFIGURATION

NILS-AXEL MÖRNER
Paleogeophysics & Geodynamics, Geological Institute, University of Stockholm, S-10691 Stockholm, Sweden

ABSTRACT

The direction and motion of plates have sometimes undergone rapid shifts. This happened, for example, at around 38 Ma. The mid-oceanic ridges have not experienced rapid volumetric changes to account for the rapid sea level falls in the Exxon "Eustatic" Curve. These rapid regressions are dismissed. Arc volcanism varies with time and is closely related to the slip rates. The plate motions and geodynamics of subduction zones (i.e. volcanism, seismicity and neotectonics) are sensitively reacting to variations in the Earth's rate of rotation. This is manifested in records of the last centuries. The 1900 AD change in rotation is, for example, well expressed in the seismicity of Venezuela and Fennoscandia. The Mediterranean seems to have experienced a change in dominant tectonic regiment at about 2.5 Ma; the underthrusting by the African plate then became dominant and large areas of Europe were suddenly uplifted. In South America, intense uplift of the active western margin strongly contrasts with the passive eastern margin. The understanding of converging plate margins needs multiple analyses of different processes and, maybe especially, of different time units (both short and long periods).

INTRODUCTION

Converging plate margins in the form of island arcs, collision zones and continental arcs constitute the well-known Circum-Pacific Belt and Alpine-Himalayan Zone. Along these zones, sediments, sedimentary rocks and oceanic lithospheric rocks of one plate are forced down (subducted) beneath the other plate giving rise to intensive metamorphic and magmatic processes, large contrasts in mass and density, disturbed thermal and pressure gradients, and high stress and strain values. It is therefore quite natural that we find, concentrated in these zones, intensive seismic activity, large neotectonic movements, intensive volcanism and large variations in the geoid height.

The rate of plate motion varies considerably along the converging plate margins. The common rates are in the order of a few to several cm/year; the plate motion off the Sunda Trench is, for example, 7 cm/year (Beaudry and Moore, 1985). The present rates of plate motion have, of course, varied with time; both in the long-term range (millions of years) and in the short-term range (thousands of years and shorter time periods as discussed below).

Yonekura (1983) investigated the general vertical uplift along the Circum-Pacific Belt. He found that the uplift is larger (1-10 mm/year) in continental arcs and collision zones than in island arcs (below 0.5 mm/year).

The subduction zones may also change position and characteristics with time. In Japan, the underthrusting at the Boso Peninsula (eastern Japan) may be on its way to shift to the western side of Japan (Nakamura, 1983). In Ecuador, there may be a shift from an original underthrusting from the east to the west in late Cretaceous and early Cenozoic time to an underthrusting from the west to the east (Garner, 1983).

Heavy geoid variations are generally recorded over the trench and arc zones (e.g. Fig. 1 of Cheney et al., 1984). In the northeastern Mediterranean, there is an exceptionally large and steep geoid anomaly (Brennecke et al., 1985) due to heavily irregular distribution of mass and/or density in association with the underthrusting of the African plate under the Eurasian plate.

PLATE MOTIONS AND GEODYNAMICS ON THE LONG-TERM RANGE

The morphology and geodynamics of the converging plate margins are primarily controlled by the rate and direction of plate motions, which we try to determine via the rate of sea floor spreading according to the magnetic anomaly pattern of the oceanic crust, the displacement of volcanism over hot spots, the subsidence of the ocean floor in response to the increasing age and distance from the spreading center, and the changes in geographic relation to the mutual VGP migration curve. The volumetric changes of the mid-oceanic ridges have also been used to establish the general activity of the sea floor spreading mechanism.

Rates and direction of plate motions

The magnetic anomaly patterns in the ocean floors have been used to build up Magnetic Polarity Time Scales (Heirtzler et al., 1968; Ness et al., 1980; Berggren et al., 1984). With a fixed time scale, the actual oceanic spreading rates can be calculated.

The age/distance relationship - i.e. the rate of plate motions - away from spreading centers, of course, differ over the globe. With a constant spreading rate, the age/distance relationship should form a straight line. This is not the case, either in local and regional details, or between different oceans. A major discrepancy between the Atlantic, Pacific and Indian Oceans exists at about 38 Ma; at this point there is a significant break in the age/distance relation in one ocean if it is kept linear in the other, and vice versa. Many other changes at about the same time speak for a major change in the Earth's plate tectonic regiment at about 38 Ma.

The Emperor-Hawaiian chain makes a 55°-bend at about 40.5 Ma (e.g. Bargar and Jackson, 1974; Shaw et al., 1980; Moberly and Cambell, 1984) indicating a rapid shift in the direction of plate motions. The Greenland-Svalbard region opened (Eldholm and Talwani, 1977) and/or was flooded, allowing free exchange of water between the Atlantic and Arctic Basin expressed as a drastic signal in the oxygen isotope records (Buchardt, 1978) and other oceanographic effects (cf. Mörner, 1980). The shift from an easterly to a westerly subduction in Ecuador (Garner, 1983) occurred at about the same time, too.

All this seems to indicate that the Earth experienced a general shift or change in plate tectonics at about 38 Ma. Probably the axis of rotation and twisting of the plates (e.g. Heirtzler, 1968; Dewey, 1972) was abruptly displaced.

Variations in the mid-oceanic ridge growth

In 1977, the Exxon group (Vail et al., 1977) presented a supposed sea level curve for the last 200 Ma, known as the Exxon "Eustatic" Curve. It is characterized by numerous very abrupt (not to say instantaneous) regressions of high amplitudes and rates. Many authors have tried to explain these rapid regressions and continuous transgressive rises in terms of variations in the mid-oceanic ridge growth (which controls the sea floor spreading and hence the plate motions). Mörner (1982, 1986a) objected to this and claimed that the regressions - if real - could not represent globally synchronous events (as claimed by the Exxon group) but must primarily represent deformations of the geoid configuration.

In 1985, the Exxon group (Vail, 1985; Posamentier, 1985) revealed that they intended to revise the original curve themselves as to shape, amplitudes, timing and number of events. This means that we can happily forget about the previous "sudden regressions of global validity". This is good. They were physically impossible (Mörner, 1982, 1986a). The fact that, in the meantime, so many authors have claimed that they had identified (and verified) the sudden regressions of Vail et al. (1977) remains a sad example of strong phycological influence on the interpretation of geological observation.

For the plate motions and arc dynamics, this means that there are no reasons left for sudden, synchronous, changes in the ridge growth. Instead, we have reasons to expect that the general pattern is smooth alternations between accelerations and decelerations in the rate of plate motions and subduction, and that this varies over the globe. (This does not preclude a few more abrupt events; like the one suggested at about 38 Ma.)

Variations in arc volcanism

The volcanic activity in relation to subduction zones is likely to vary with the rate of plate convergence. According to Kay (1980), there is no simple and strait forward relationship, however. Wadge (1984) investigated the volcanic production in the 300, 10,000 and 100,000 period time ranges, respectively, for the subduction zones along the eastern coast of Central America and along the Lesser Antilles. Along the Central American zone, the rate of volcanic production per km along the zone ($km^3.Ma^{-1}.km^{-1}$) increased from 31 to 62 from the 100,000 year period to the 300 year period (i.e. from 50% to 100% of the production of the last centuries). Along the Lesser Antilles zone, it increased from 3 to 4 to 5 for the 100,000, 10,000 and 300 years period, respectively (i.e. from 60% to 80% to 100% of the production of the last centuries). If not a function of lost resolution with increasing time in the past, this may record a general increase in volcanic production with time along these two subduction zones.

The relationship (here given in percent) in volcanic production between the Central American and Lesser Antilles zones remained fairly constant at about 80-90% to 20-10% from the 100,000 period (92 to 8%) to the period of the last 300 years (88 to 12%). The tectonic slip rates for the last 3 Ma were about 7-9 cm/year and 2 cm/year (i.e. about 75% to 25%), respectively. The corresponding seismic slip rates for the last 80 years were 2.2 cm/year and 0.2 cm/year (i.e. about 92% to 8%), respectively. This seems to suggest that the relationship in approximate geodynamic activity between the two zones was in the order of about 80-90% to 20-10%, respectively.

This is taken up as an example of semi-quantitative estimates of the geodynamic activity along subduction zones and the approximate relationship between slip rates and volcanic activity.

SHORT-TERM CHANGES IN SEISMICITY AND PLATE MOTIONS

Anderson (1974) proposed that there was a causal correlation between the change in the Earth's rate of rotation at around 1900 AD and a general increase in volcanism and seismicity at about the same time. Courtillot et al., 1978) showed that rotational changes at 1900 and 1967 (and 1840, too) were linked to significant geomagnetic field changes at about the same time. Because Båth (1978) had shown that the seismic energy release of earthquakes in Sweden and Fennoscandia as a whole had changed drastically at about 1900-1910 AD, and there seemed to be a significant change in the rate of uplift at about the same time (Mörner, 1979), Mörner (1981) concluded: "Consequently, at about 1900-1910, there was a shift in the rate of rotation, a geomagnetic shift, a change in the seismic energy release and a change in the rate of uplift in Norrland. This may indicate a causal connection among all four variables. If this is correct, the Earth's geodynamics are very sensitively reacting to minor changes in the rate of rotation."

Aggarwal (personal communications, 1983) has provided a diagram of cumulative moment release of earthquakes in Venezuela from 1585 to 1967 (Fig. 1). At 1765 and 1900 there are major breaks (minor breaks may occur at 1810, 1837 and 1888). The break at 1900 is important as it corresponds to the break in seismic energy release in Fennoscandia (Båth,

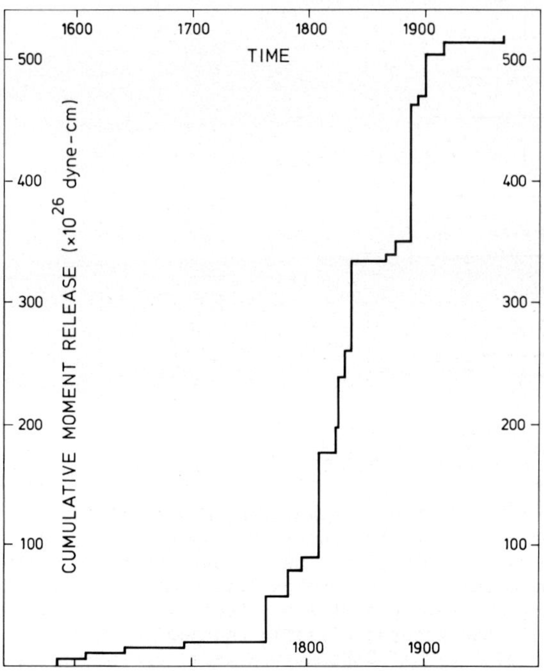

Fig. 1. Seismic activity (cumulative magnitudes) in Venezuela from 1585 to 1967 (Aggarwal, personal communications, 1983). At about 1765 and 1900 AD, there are major changes in the activity. These breaks are also well established in diagrams of cumulative numbers of earthquakes.

1978). If the cratonal Fennoscandian shield and the active plate boundary in Venezuela simultaneously experience a significant change at the same time as the Earth's rate of rotation undergoes a major shift; "this indicates that the Earth's entire plate system was affected by this rotational change" (Mörner, 1984, p. 497).

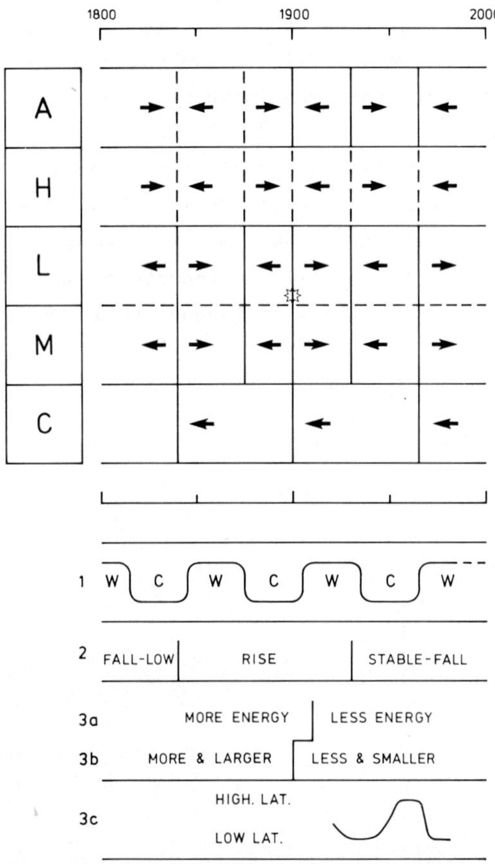

Fig. 2. Differential rotation during the last centuries (from Mörner, 1986b). A = atmosphere, H = hydrosphere, L = lithosphere, M = mantle, C = core. Arrows pointing right denote deceleration and arrows pointing left acceleration of angular momentum. Star denotes time when the viscous lithosphere/mantle coupling has set up a stress affecting the seismicity. Below are given: (1) the general climatic changes in northwest Europe due to the pulsation of the Gulf Stream (w = warm, c = cold), (2) the main eustatic sea level changes in northwestern Europe, (3a) the seismic energy release in Fennoscandia, (3b) the seismic activity in Venezuela (Fig. 1), and (3c) the latitudinal time distribution of earthquakes over the globe from 1930 to 1965 according to Mogi (1979).

The Earth is not a simple solid body but consists of a gaseous atmosphere, a liquide hydrosphere, a solid lithosphere and mantle coupled via a viscous astenosphere, a liquid outer core and a solid inne core with an electromagnetic coupling between the core and the mantle. Because of this, it experiences a differential rotation where the rate of rotation of the different layers and sublayers increases and decreases with respect to each other in order to keep the momentum constant (Mörner, 1984, 1986b, 1986c).

Because of the large differences in mass between the mantle and the lithosphere, they should have experienced differential rotation with respect to each other, if they had not been coupled together via the asthenosphere. The asthenosphere is a partially melted, low-viscosity (10^{18}-10^{21} Poise) layer. Therefore, large stress variations are likely o be created in the asthenosphere (and transferred into the lihosphere) in connection with accelerations and decelerations of the mantle-lithosphere system. These stress variations "are likely to affect earthquakes, volcanism, plate tectonics, phase transitional boundaries, etc." (Mörner, 1986b; 1984, Fig. 10).

It seems significant in this respect that the rates of plate motions (Solomom et al., 1975) and structural patterns of plate tectonics (Weijermars, 1984) exhibit an axial symmetry about the Earth's axis of rotation.

Mogi (1979) analyzed the geographical distribution of earthquakes for the period 1930 to 1978. From 1930 to 1960 the major earthquake occurrence was displaced from a more global and low-latitude spread to a more restricted high-latitude spread. After 1969 this tendency reversed and the spread of earthquakes was displaced towards the equatorial region again. This correlates well with the Earth's deceleration from about 1930 to 1965, and acceleration after 1965. During a deceleration period the mass would tend to move polewards and so would the main earthquake activity, following the "expanding" areas. During an acceleration period the situation would be reversed, and the main earthquake activity would move towards low latitudes. This is in perfect agreement with the finding of Mogi (1979; cf. Taira, 1981) and was therefore used by Mörner (1986c) as a supporting evidence in his theory of differential rotation and its multiple geophysical effects.

In order to explain the correlations found between variations in the Gulf Stream activity, short-term changes in climate in Scandinavia and eustatic sea level variations in northwestern Europe, Mörner (1984; cf. 1986b, 1986c) proposed a gravitational-rotational-oceanographic model based on differential rotation and redistribution of mass and energy over the globe. He noted that the Kuro Siwo Drift seemed to have undergone parallel changes judging from the correlations found by Taira (1981) between the climatic and eustatic records in Japan and in northwestern Europe.

According to Taira (1981, 1982, 1983, 1984) periods of geoid deformations (as proposed by Mörner, 1976) in Japan correlated with cooling events due to changes in the Kuro Siwo Drift, and increased plate tectonic motions and volcanism in the subduction zones.

All this speaks for a very high (much higher than generally assumed) sensivity of plate motions and subduction processes - i.e. volcanism, seismicity and neotectonics - to variations in the Earth's rate of rotation. This is of utmost importance for the understanding (and prediction) of the seismic and volcanic activity associated with subduction zones and arc belts.

Fig. 2 summarizes some of the data discussed above (from Mörner, 1986c).

THE EAST/WEST VERTICAL DISPLACEMENT IN SOUTH AMERICA

South America has a passive continental margin in the east and an active plate margin in the west where the Nazca plate is subducted under the continent with the formation of high continental arcs with intense volcanic and seismic activity.

Along the eastern coasts, there is a high marine Miocene level of about +100-200 m, a long period of continental deposition and lack of high marine terraces, a last interglacial level at about +5-7 m, and a Holocene maximum at about +2-5 m.

Along the western coast, the marine Miocene level is lifted up to about 700-1000 m, and is followed by a sequence of elevated beaches all from the Pliocene up to the last interglacial (generally at about +100 m, occasionally at +225 m) and the Holocene maximum (generally at about +10-20 m, occasionally at +33 m).

This means that the western side is strongly uplifted whilst the eastern side has remained more stable and intermittently, maybe, even has subsided (Fig. 3). Because of the mass and density redistributions associated with the geodynamic processes, there must have been considerable changes in the geoid configuration. Real vertical uplift and/or deformations of the geoid configuration cannot be separated.

PLIOCENE CHANGES OF THE MAJOR TECTONIC REGIMENT OF THE EUROPEAN PLATE

The dominant Tethyan-Alpine tectonics of the Cenozoic of Europe seem to have changed drastically in Pliocene time at about 3.0-2.0 Ma. It is proposed that this change in predominant tectonic regiment represents the first general and strong sign of the underthrusting of the African plate.

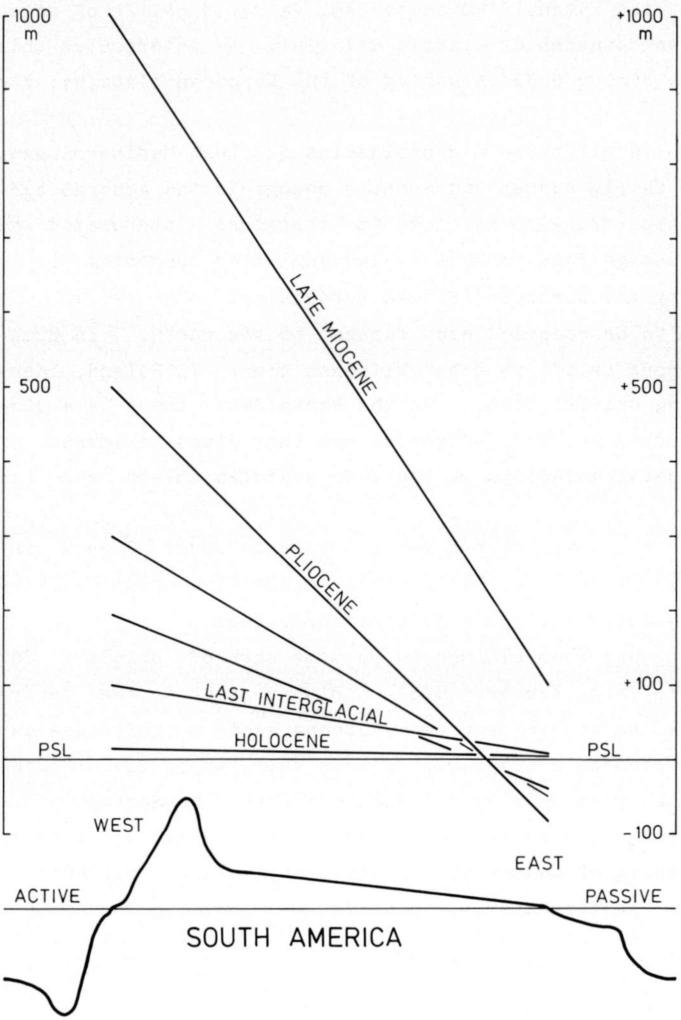

Fig. 3. Generalized profile across South America (at about Lat. 40^0S) from the active plate margin in the west to the passive continental margin in the east. Approximate shorelines from the Miocene to the Holocene illustrating the strong uplift in the west compared to the passivity in the east. PSL = Present Sea Level.

In Spain, there are structural and stratigraphic signs of a new, southernly, direction of thrusting (Goy and Zazo, 1983) occurring sometime in the upper Pliocene.

In Italy, there is an erosional phase, the "Acquatraversan", at about 2.5 Ma (Arias et al., 1979). It represents a period of major uplift with a sudden downwash of clastic material.

In Greece at about 3.0-2.0 Ma (Kirittopoulos, 1984), a new tectonic regiment was induced with intensified tectonism, a rapid uplift of several 100 m and a sudden downwash of clastic material. We interpreted this as a clear sign of a strong underthrusting of the European plate by the African plate.

This means that in all three big peninsulas in the Mediterranean, there are signs of a fairly sudden and drastic change in the general tectonic regiment at about 3.0-2.0 Ma. It is therefore interpreted as marking the final change from Tethyan-Alpine dominated tectonism to a tectonism dominated by the European/African subduction.

This change seems to be recorded even further to the north. In Rumania, there was a strong uplift in Upper Pliocene time. In Poland, there was a period of strong uplift, too. In the North Sea, there is a distinct angular unconformity of Mid-Pliocene age that gives evidence of sudden and large crustal movements of the Fennoscandian shield and the surrounding areas.

With respect to the very strong geoid gradients over Greece and southern Italy (Brennecke et al., 1984), this intensified underthrusting must have generated significant geoid deformations, too.

The Mediterranean was once claimed to be some sort of standard for Quaternary sea levels (e.g. Zeuner, 1945). With respect to the geodynamic processes operating and the extremely strong geoid relief, one can hardly find a worse area as a standard. It is therefore quite logical that Hey (1978) had to give this up and conclude that "the Mediterranean shorelines can no longer be regarded as a means of obtaining precise values for the altitudes of Quaternary eustatic stillstands" but "can be used as evidence for recent crustal movements"; and geoid deformations, one may add.

CONCLUSIONS

(1) The Earth's geodynamics are a complex interaction of different processes and forces.

(2) The entire plate system may sometimes undergo major changes and shifts; such as that at about 38 Ma.

(3) The long-term changes in plate motions and arc dynamics are characterized by variations between accelerations and decelerations. Previous hypotheses of very abrupt changes in the mid-oceanic ridge volumes (seen as rapid falls in sea level) are not tenable.

(4) The volcanic and seismic activity of the subduction zones are related to the rates of plate motions and to the actual slip rates. These change with time.

(5) Variations in the Earth's rate of rotation are, in the short-term range, found to be a very sensitive controller of plate motions and the geodynamics of the subduction zones; i.e. volcanism, seismicity and neotectonics.

(6) This high sensivity to changes in the Earth's rate of rotation seems well established for the records of the last centuries and for the Holocene period. Naturally, it applies for the longer periods, too.

(7) In the Upper Pliocene at about 3.0-2.0 Ma, there was a general shift in the Mediterranean region from a Tethyan-Alpine dominated tectonism to a tectonism dominated by the European/African subduction. In general, Europe seems to have undergone a rapid uplift at about this time. Real crustal uplift and geoid lowering cannot be distinguished, however.

(8) The active western and passive eastern coasts of South America exhibit drastically different sea level records. In the west, they are highly elevated, whilst they in the east are at much lower elevation (the marine Miocene level) or absent and below the present sea level (the Pliocene to second last interglacial levels). This implies drastically different tectonic and/or gravitational changes across the continent.

We have, with different examples, tried to show that the Earth's lithospheric plates and their converging margins are highly active zones that need to be understood by multiple analyses of different processes and, especially, different time units; i.e. both the short-term and the long-term periods.

REFERENCES

Andersson, L.D., 1974. Earthquakes and the rotation of the Earth. Science, 186: 49-50.

Arias, C., Bigazzi, G., Bonnadonna, F.P. and Urban, B., 1979. Tentative correlations among Quaternary series of central and southern Europe. Acta Geol. Acad. Sci. Hungaricae, 22: 81-88.

Bargar, K.E. and Jackson, E.D., 1974. Calculated volumes of individual shield volcanoes along the Hawaiian-Emperor chain. U.S. Geol. Survey J. Res., 2: 545-550.

Båth, M., 1978. Energy and tectonics of Fennoscandian earthquakes. Tectonophysics, 50: T9-T17.

Beaudry, D. and Moore, G.F., 1985. Seismic stratigraphy and Cenozoic evolution of West Sumatra Forearc Basin. A.A.G.P. Bull., 69: 742-759.
Berggren, W.A., Kent, D.W., Flynn, J.J. and van Couvering, J.A., 1984. Paleogene and Neogene Geochronology and Chronostratigraphy. Preprints, 1984.
Brennecke, D., Lelgeman, D., Reinhart, E., Torge, W., Weber, G, and Wenzet, H.-G., 1983. A European Astro-Gravimetric Geoid. Deitsche Geodät. Komm., Reihe B, 269: 1-115.
Buchardt, B., 1978. Oxygen isotope paleotemperatures from the Tertiary period in the North Sea Area. Nature, 275: 121-123.
Cheney, R.E., Marsh, J.G. and Martin, T.V., 1984. Applications of satellite altimetry to oceanography and geophysics. Marine Geoph. Res., 7: 17-32.
Courtillot, V., Ducruix, J. and Le Mouel, J.-L., 1978, Sur une accélération récente de la variation séculaire du champ magnétique terrestre. C.R. Acad. Sc. Paris, 287D: 1095-1098.
Dewey, J.F., 1972. Plate tectonics. In: J.T. Wilson (Editor), Continents adrift and continents aground, Scientific American 1976, pp. 34-45.
Eldholm, O. and Talwani, M., 1977. Sediment distribution and structural framework of the Barents Sea. Geol. Soc. Am. Bull., 88: 1015-1029.
Garner, H.F., 1983. Large-scale tectonic denudation and climatic morphogenesis in the Andes mountains of Ecuador. In: R. Gardner and H. Scoging (Editors), Mega-geomorphology, Oxford Univ. Press, pp. 1-17.
Goy, J.-L. and Zazo, C., 1983. Pleistocene tectonics and shorelines in Ameira (Soain). Bull. INQUA Neotectonics Comm., 6: 9-13.
Heirtzler, J.R., 1968. Sea-floor spreading. In: J.T. Wilson (Editor), Continents adrift and continents aground. Scientific American 1976, pp. 76-86.
Heirtzler, J.R., Dickson, G.O., Herron, E.M., Pitman III, W.C. and Le Pichon, X., 1968. Marine magnetic anomalies, geomagnetic field reversals, and motions of the ocean floor and continents. J. Geophys. Res., 73: 2119-2136.
Hey, R.W., 1978. Horizontal Quaternary shorelines in the Mediterranean. Quaternary Res., 10: 197-203.
Kay, R.W., 1980. Volcanic arc magmas: Implications of a melting-mixing model for element recrystaling in the crust-upper mantle system. J. Geol., 88: 497-522.
Kirittopoulos, P., 1984. Neogene lacustrine deposits in the inland basins of northern Greece: Stratigraphy, paleoenvironment, paleomagnetism and industrial usefulness. Medd. Stockholm Univ. Geol. Inst., 261: 1-96.
Moberly, R. and Campbell, J.F., 1984. Hawaiian hotspot volcanism mainly during geomagnetic normal intervals. Geology, 12: 459-463.
Mogi, K., 1979. Global variation of seismic activity. Tectonophysics, 57: T43-T50.
Mörner, N.-A., 1976. Eustasy and geoid changes. J. Geol., 84: 123-151.
Mörner, N.-A., 1979. The Fennoscandian uplift and Late Cenozoic geodynamics: geological evidence. GeoJournal, 3: 287-318.
Mörner, N.-A., 1980. Earth's movements, paleoceanography, paleoclimatology and eustasy: major events in the Cenozoic of the North Atlantic. Geol. Fören. Stockh. Förhandl., 102: 261-268.
Mörner, N.-A., 1981. Crustal movements and geodynamics in Fennoscandia. Tectonophysics, 71: 241-251.
Mörner, N.-A., 1982. Sea level changes as an illusive "geological index". Bull. INQUA Neotectonics Comm., 5: 55-64.
Mörner, N.-A., 1984. Planetary, solar, atmospheric, hydrospheric and endogene processes as origin of climatic changes on the Earth. In: N.-A. Mörner and W. Karlén (Editors), Climatic Changes on a Yearly to Millennial Basis, Reidel, Dordrecht, pp. 483-507.

Mörner, N.-A., 1986a. Eustasy, unconformities and a revision of the Exxon "Eustatic" Curves. In: C.K. Wilgus, E.C. Pitman III and C.G.S.C. Kendall (Editors), Sea level changes - An integrated approach, (Proc. Houston, 1985, conference), SEPM Spec. Publ., in press.

Mörner, N.-A., 1986b. Short-term paleoclimatic changes. Observational data and a novel causation model. In: Proc. "A climatic symp. in honour prof. R.W. Fairbridge", New york, in press.

Mörner, N.-A., 1986c. Dynamic sea surface changes in the past and redistribution of mass and energy. Submitted.

Nakamura, K., 1983. Accretion of the northeast Japan microplate and formation of a new converging boundary along the eastern Japan Sea at 1-2 Ma. Bull. Earthq. Res. Inst., Univ. Tokyo, 58: 711-722.

Ness, G., Levi, S. and Cough, R., 1980. Marine magnetic anomaly time scale for the Cenozoic and Late Cretaceous. A précis, critique, and synthesis. Rev. Geophys. Space Phys., 18: 753-770.

Posamentier, H., 1985. Eustatic control on clastic depositional stratal patterns. Lecture, SEPM Res. Symp. "Sea level changes - An integrated approach", Houston 1985.

Shaw, H.R., Jackson, E.D. and Bargar, K.E., 1980. Volcanic periodicity along the Hawaiian-Emperor chain. Am. J. Sci., 280A: 667-708.

Solomon, S.C., Sleep, N.H. and Richardson, R.M., 1975. On the forces driving plate tectonics: Inferences from absolute plate velocities and interplate stress. Geophysa. J. Roy. Astr. Soc., 42: 769-801.

Taira, K., 1981. Holocene tectonism in eastern Asia and geoidal changes. Paleogeogr. Paleoclim. Paleoecol., 36: 75-85.

Taira, K., 1982. The effect of tectonism on the climate of the last 1200 years. Paleogeogr. Paleoclim. Paleoecol., 39: 165-169.

Taira, K.,1983. Accelerated Pacific plate movements and climatic changes on time-scales of 10^2-10^3 years in the Late Quaternary: A synthesis. Paleogeogr. Paleoclim. Paleoecol., 44: 203-214.

Taira, K.,1984. Neotectonics, plate movements and variations in the Earth's orbital geometry. Bull. INQUA Neotectonics Comm., 7: 18-19.

Vail, P., 1985. Sea level from sequence stratigraphy. Lecture, SEPM res. Symp. "Sea level changes - An integrated approach", Houston 1985.

Vail, P., Mitchum, K.G., Thomson III, S., Todd, R.R., Sangree, J.B., Vidmier, J.M., Bubb, J.N. and Hatlelid, W.G., 1977. Seismic stratigraphy and global changes in sea level. A.A.O.G. Memoir, 26: 49-212.

Wadge, G., 1984. Comparison of volcanic production rates and subduction rates in the Lesser Antilles and Central America. Geology, 12: 555-558.

Weijermars, R., 1984. New structural theorems on new global tectonics. Terra Cognita, 4: 175-181.

Yonekura, N., 1983. Late Quaternary vertical crustal movements in and around the Pacific as deducted from former shoreline data. In: W.C. Hilde and S. Uyeda (Editors), Geodynamics of the Western Pacific-Indonesian Region, Am. Geophys. Union, Geodynamics Series, 11: 41-50.

Zeuner, F.E., 1945. The Pleistocene period, its climate, chronology and faunal succession. Roy. Soc. London, Monogr. 130. 322 pp.

II. GEOPHYSICS AND GEOCHEMISTRY

EVIDENCE OF A DEEP-REACHING LITHOSPHERIC ROOT UNDER THE ALPINE ARC

S. MUELLER[1] and G.F. PANZA[2]

[1]Institut für Geophysik, ETH-Hönggerberg, CH-8093 Zürich (Switzerland)

[2]Istituto di Geodesia e Geofisica, Università degli Studi, I-34100 Trieste and International School for Advanced Studies, I-34014 Trieste (Italy)

ABSTRACT

The Alps are a deep-reaching crust-mantle structure in Europe situated at the northernmost tip of the Adriatic promontory of the African plate which presently pushes against the Eurasian plate in a NW-SE direction. A detailed regional analysis of seismic surface-wave dispersion and of P-wave travel time residuals as well as long-range seismic refraction measurements and the interpretation of "stripped" residual gravity anomalies have revealed a rather anomalous structure within the uppermost mantle under the Alpine arc. As a result of the massive continent-continent collision parts of the lithosphere have been delaminated - a process which has led to "flaking" in the upper crust, combined with a thickening of the entire crust and the formation of a pronounced relatively cold, dense and slowly subsiding "lithospheric root" beneath the mountain chain. In this still ongoing plate collision the lower parts of the lithosphere apparently have penetrated into the asthenosphere to a depth of 150 to 220 km in a nearly vertical zone of "subfluence" ("Verschluckung"). On this scale the continuing uplift of the Alpine chain (by about 1 mm/year) is a secondary effect due to isostatic rebound of less dense crustal masses which previously had been forced to greater depths. A configuration of this type generates regionally a compressive dynamics of its own on which - within a wider framework - rotational processes may be superimposed.

INTRODUCTION

The ongoing geodynamic processes in the Alpine-Mediterranean region are primarily governed by the relative movement of the African lithospheric plate against the Eurasian plate resulting in a predominantly compressional regime along the zone of contact. A gradual counterclockwise rotation of the African plate caused by a higher spreading rate of the ocean floor in the South Atlantic (~40 mm/year) as compared to the Central Atlantic (~30 mm/year) causes presently a roughly northward directed push leading to a lithospheric shortening in the Mediterranean-Alpine belt of about 5 to 10 mm/year. With a northwest-southeastward oriented spreading in the North Atlantic (at a rate of 18 mm/year) the zone of plate contact is, therefore, to a large extent under compression thus providing the framework for mountain building processes accompanied by active rifting in old preconditioned zones of lithospheric weakness (Mueller, 1982).

Several sets of geophysical data indicate that the Alpine arc is a deep-reaching continent-continent collision structure situated at the northernmost tip of the Adriatic promontory, a spur-like microplate which has moved with the African plate since the Early Mesozoic thereby eventually creating this spectacular mountain range. Seismological and gravimetric data have shown that the continuing collision process must have led to a considerable thickening of the crust under the Alps accompanied by the formation of a cold, dense and slowly subsiding "lithospheric root" which substantially amplifies the compressive regime in the Alpine region (Mueller, 1984).

GEOPHYSICAL EVIDENCE

Regional surface-wave analysis

During the past three decades a sizable number of long-period seismograph stations in the European-Mediterranean area have been recording teleseismic events. Suitable station combinations have permitted the measurements of fundamental mode phase velocities of

Rayleigh waves from various azimuths in a wide period range across this seismograph network. Since the observed phase velocity dispersion is caused primarily by the geometric and physical properties of the lithosphere-asthenosphere system, this frequency-dependent dispersion can be used to determine the average thickness and shear-wave velocity of the lower lithosphere or lid, once the crustal structure is known. Most of the European-Mediterranean region is sufficiently covered by phase velocity "profiles", which provide the basis for a uniform "hedgehog" inversion of the dispersion observations. In order to apply this method it must be assumed that a representative regionalization of the phase velocity dispersion for seismic surface waves can be derived from the available observational data (Panza et al., 1980). In this procedure a systematic search in a multidimensional parameter space allows models to be determined that are compatible with a given range of observational uncertainty.

The results of this comprehensive inversion procedure are summarized in Fig. 1a, which shows a schematic map of the lithosphere-asthenosphere system for most parts of the European-Mediterranean area (Panza, 1985). When interpreting this map, it is important to realize that it represents only an approximate solution to the inverse problem and is subject to inherent uncertainties (e.g. 15 to 20 km in lithospheric thickness) due to the limited resolution of long-period surface waves. Significant deviations from the "normal" lithospheric thickness of 90 to 100 km in central and western Europe are, for instance, found in the Tyrrhenian and Balearic Basins of the Western Mediterranean Sea where the lithosphere thins to only about 30 km.

Another, striking lithospheric feature which is conspicuously delineated in Fig. 1a, is the "Central European Rift System". It reaches from the North Sea to the western Alps with an average lithospheric thickness of about 50 km and markedly lowered shear-wave velocities. Also noteworthy is the continuous structure which extends from North Africa through the Ionian Sea into the

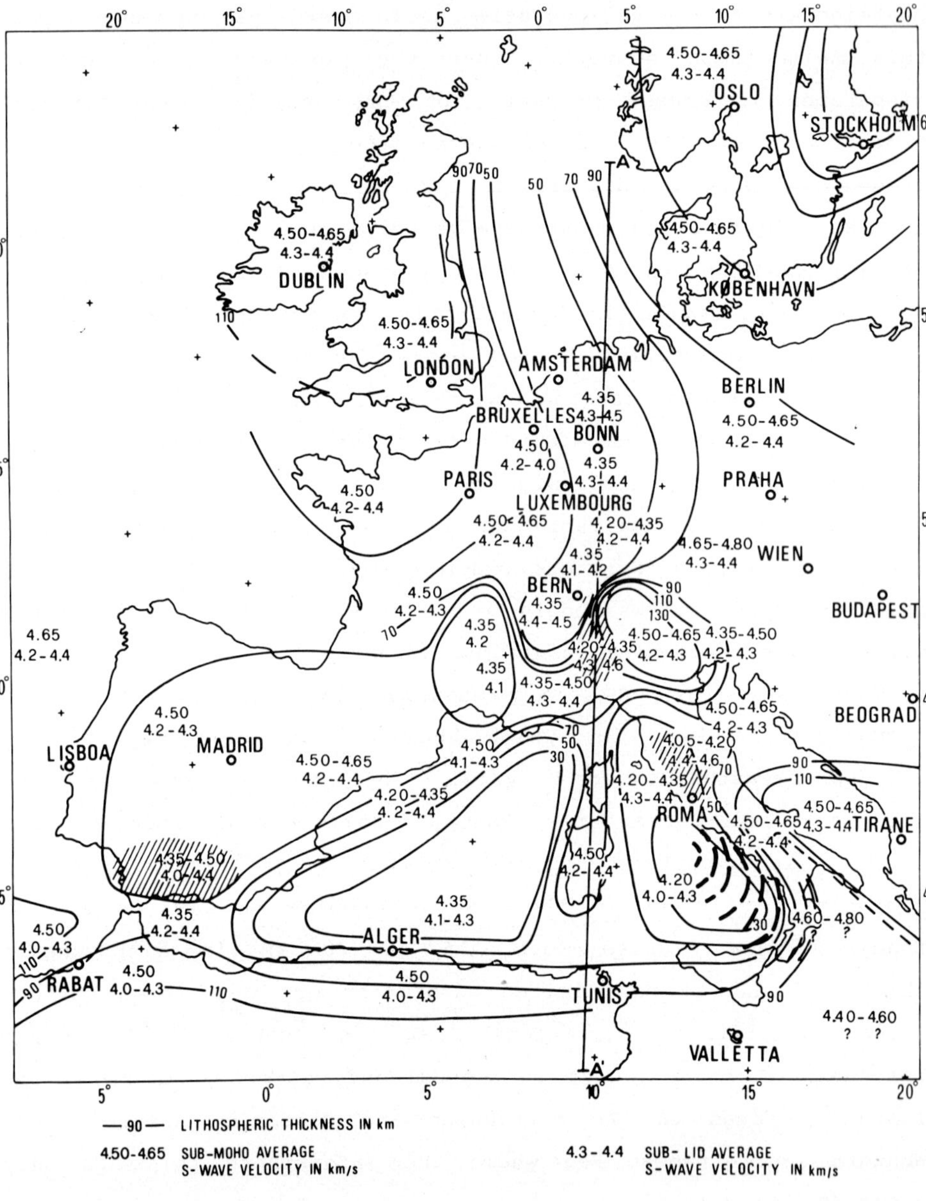

Fig. 1a. Map of lithospheric thickness (in km) in the European-Mediterranean region, deduced from the regional dispersion analysis of seismic surface waves (after Panza, 1985). Representative shear-wave velocities are given for the lower lithosphere or lid (row of upper numbers) and for the upper asthenosphere (row of lower numbers). The three shaded areas indicate the postulated presence of "lithospheric roots" to depths of about 200 km, while the thick dashed lines define the area of intermediate and deep focus earthquakes in the Tyrrhenian sea.

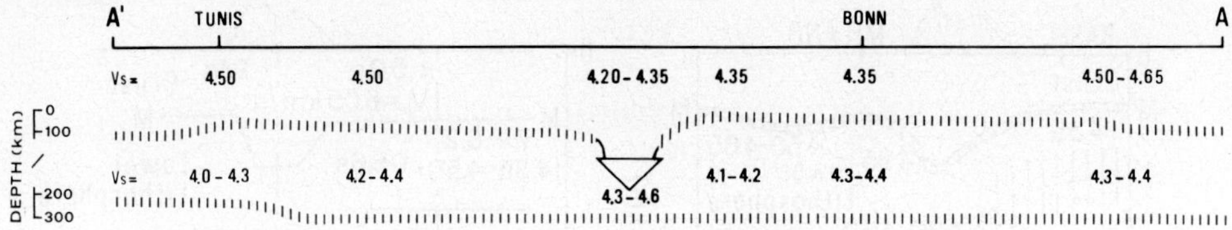

Fig. 1b. Cross section through the crust-mantle system from the North Sea (A) to Southern Tunisia (A') showing representative shear-wave velocities (V_s, in km/s). The vertical "fence" hachures indicate the uncertainty in depth of the upper and lower bounds of the upper asthenosphere.

Adriatic Sea. There the thick lithosphere seems to have a clear African affinity in accordance with the concept of an Adriatic promontory of the African plate (Channell and Horváth, 1976). Particularly interesting is the region of the Alps where the lithospheric thickness increases to 130 km and attains in places even greater values. These so-called "lithospheric roots" are found under mountain belts in zones of continent-continent collision, such as the Betic Cordillera in southern Spain, the Tuscany region in central Italy, and beneath the Alps (see the shaded areas in Fig. 1a). A north-south cross section (A-A') through the lithosphere-asthenosphere system from the eastern part of the North Sea to southern Tunisia showing the shear-wave velocity distribution in the uppermost mantle, is displayed in Fig. 1b. The change in lithospheric structure from the margin of the Baltic Shield across the "Central European Rift System" to the neighbouring Alps, the island chain of Corsica and Sardinia, and finally to the North African craton can be clearly seen.

The conspicuous thickening of the Alpine crust and the asymmetric shape of the crust-mantle boundary (Mueller et al., 1980) are an indication of a much deeper reaching structural anomaly beneath the Alps involving the entire lithosphere-asthenosphere system. The detailed regional analysis of seismic surface-wave dispersion and P-wave travel time residuals in combination with long-range seismic refraction observations in the northern Alpine foreland and along the strike of the Alps have

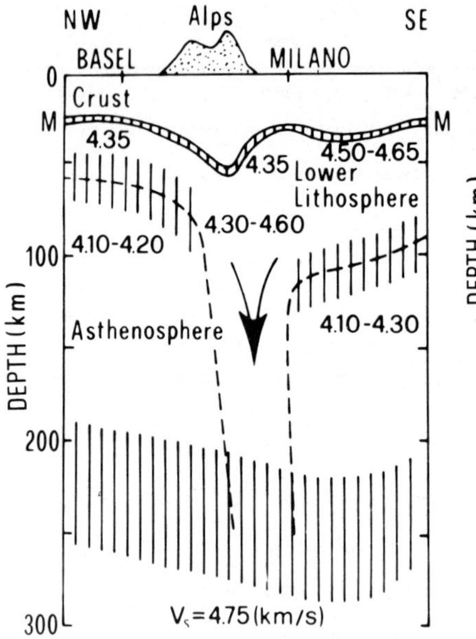

Fig. 2. Crust-mantle cross section derived from the simultaneous inversion of available surface waves dispersion data (after Panza and Mueller, 1979). Hatched areas indicate the range of uncertainty of the bottom of the crust (M), of the lithosphere and of the upper asthenosphere. In the collision process the lower lithosphere has penetrated into the asthenosphere thus forming a nearly vertical zone of "subfluence" ("Verschluckung").

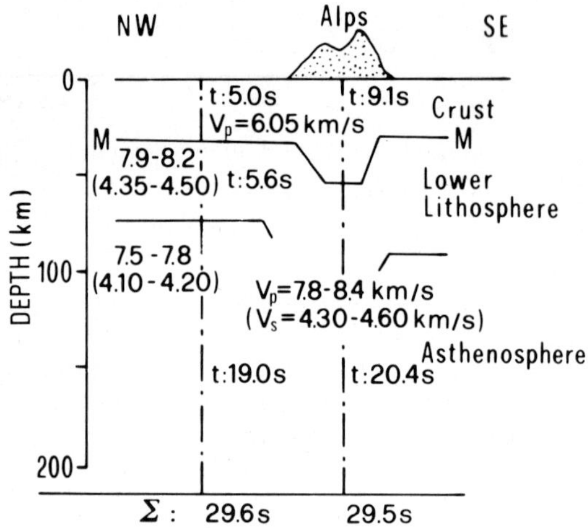

Fig. 3. Schematic crust-mantle cross section through the Central Alps along the Swiss Geotraverse (compare with Fig. 2, after Panza and Mueller, 1979). The one-way travel times of P-waves propagating vertically upwards from the bottom of the upper asthenosphere to the surface corroborate the hypothesis that the time delay caused by the low-velocity crustal root must be compensated by the underlying high-velocity "lithospheric root" which is sinking into the asthenosphere (after Baer, 1980). In the model a ratio V_p/V_s of 1.82 has been assumed.

revealed a rather anomalous distribution of P- and S-wave velocities with depth.

On a large scale the existence of a high-velocity "block" within the upper mantle under the Alps could be deduced from the dispersion of Rayleigh waves (Panza and Mueller, 1979). Fig. 2 shows a schematic cross section of the lithosphere-asthenosphere system along a profile from Basel to Milano which roughly coincides with the Swiss Geotraverse (Rybach et al., 1980). The hatched areas in this figure mark the range of uncertainty for the bottom of the crust (M), of the lithosphere and of the upper asthenosphere. The center part is characterized by a high-velocity, subducted lower lithosphere penetrating into the asthenosphere to depths of 150 to 220 km. This lithospheric "block" of higher velocity probably corresponds to the two slabs of lower lithosphere subducted to the south and to the north during the plate collision process ("subfluence" or, in German, "Verschluckung") forming the Alps (Panza and Mueller, 1979).

Teleseismic P-wave delays

A crust-mantle model of this type not only provides a reasonable solution for the deposition of the excess lithospheric material (of about 400 to 500 km in width) which must have been subducted during the shortening of the Alpine lithosphere, but it also gives an explanation for the apparently non-existing difference in the P-wave travel time residuals of teleseismic events observed at seismograph stations in the northern Alpine foreland and in the Alps (Baer, 1980). Fig. 3 demonstrates that the proposed model with a high-velocity "block" in the uppermost mantle under the Alps will provide the required compensation for the extra time delay caused by the much thicker crust beneath the central portion of the Alps. It should be noted that the zone of near-vertical subduction or "subfluence" (corresponding to the "Verschluckungszone" after Ampferer, 1906) is not symmetrical with respect to the central zone of the Alps, but appears to be somewhat displaced southeastwards (see Fig. 2), i.e. towards the inner side of the Alpine arc.

Deep seismic sounding

Deep seismic sounding experiments have been carried out in the Alps since 1956 and are still going on. A number of important results have so far been obtained which can be summarized as follows (Miller et al., 1982): The Alpine crust is characterized by an asymmetric crust-mantle boundary (M = Mohorovičić

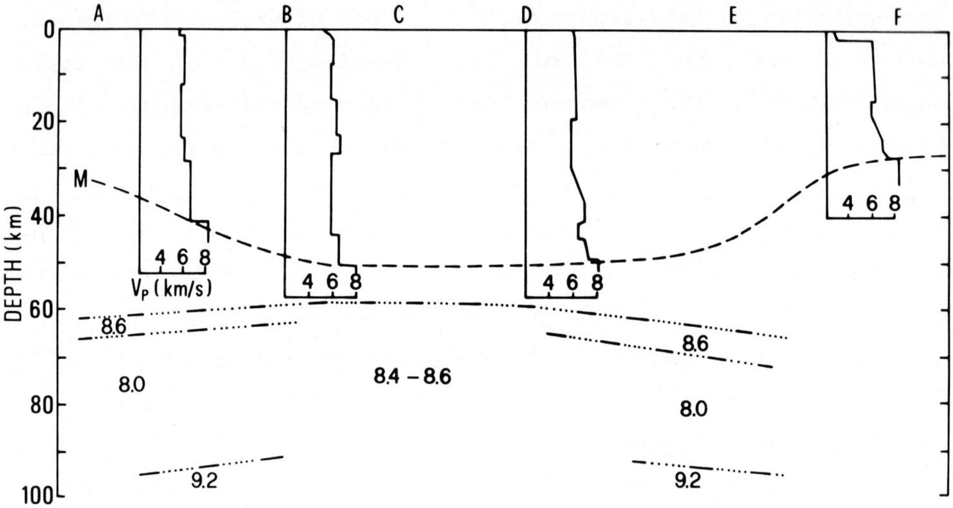

Fig. 4. Schematic structural section through the crust and uppermost mantle along the axis of the Alps from Aix-les-Bains (A) in France to the Pannonian Basin (F) in Hungary (after Miller et al., 1979 and 1982). Crustal structure is represented by selected P wave velocity-depth functions. The dashed line indicates the crust-mantle boundary (M). Note the high-velocity "block" (8.4 - 8.6 km/s) under the central portion of the Alps which extends from a depth of 60 km to at least 100 km.

discontinuity) which reaches its greatest depth of some 55 km south of the central region and rises rather steeply towards the inner arc side of the Alps. The internal structure of the crust differs markedly between the Eastern and the Western Alps with more pronounced lateral velocity variations existing in the Central and Western Alps, which might be an indication that crustal stack-interfingering was the dominant process forming the present-day crust, since crustal shortening was greater in the Central and Western Alps than in the Eastern Alps.

About 15 years ago it became apparent that specially designed experiments were necessary in order to gain knowledge on more detailed crustal and upper mantle structure sufficiently accurate for the development of geodynamic models. To this end international cooperative measurements along the "Alpine Longitudinal Profile 1975 (ALP 75)" were carried out between

Aix-les-Bains in France and the Pannonian Basin in Hungary (profile A...F in Fig. 4). This long-range experiment, besides probing detailed crustal structure in the axial region of the Alps, also yielded information on the structure of the lower lithosphere beneath the Alps.

Long-range seismic refraction measurements in the northern Alpine foreland and along the strike of the Alps revealed the existence of a layer with a markedly high P-wave velocity of up to 8.6 km/s in the uppermost mantle (Fig. 4). North of the Alps this layer has a thickness of the order of 5 to 10 km and lies between the crust-mantle boundary (M) and the relatively shallow base of the lithosphere located at depths of 50 to 55 km (Ansorge et al., 1979; Panza et al., 1980). As the Alps are approached this layer of high velocity (8.6 km/s) is found at slightly greater depths (60 to 70 km). Following the Alpine chain from west to east, refraction seismic evidence (Miller et al., 1982) shows that roughly between the Mont Blanc massif in the west and the eastern end of the Tauern window the mean P-wave velocity from a depth of 60 km down to at least 100 km amounts to about 8.5 km/s (see center of Fig. 4). In contrast to the marginal regions of the Alps in the west and east, where in the same depth range of the uppermost mantle a distinct layering has been found, no such structure could be detected beneath the central part along the axis of the Alps. The dimensions of this anomalous "block" of high-velocity material are estimated to be about 500 km in length (west-east) and 100 to 150 km in width (north-south) extending to a depth of at least 100 km.

GEOLOGIC-TECTONIC CONCEPTS

This new geophysical model of the crust-mantle system in the Alps is in full accord with geologic-tectonic schemes which have been suggested earlier. Ever since the late Eocene, i.e. for the last 40 Ma, the continuing push of the African plate against the Eurasian plate has led to a considerable shortening of the a

b

c

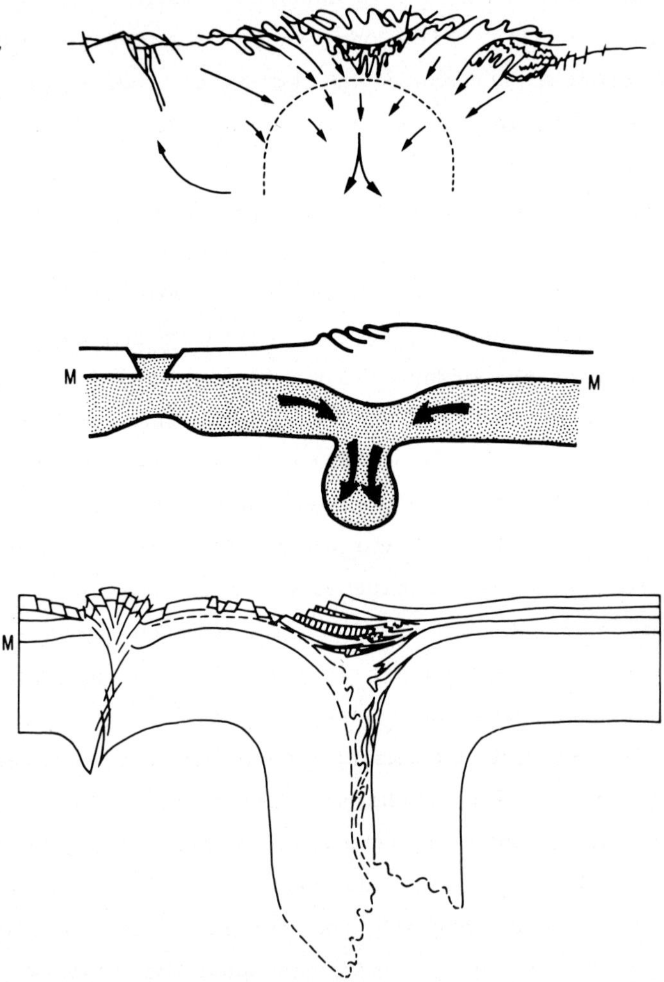

Fig. 5. Geologic-tectonic concepts of the deep structure beneath the Alps: (a) Schematic model of the Alps as symmetrical "double orogen" with lithospheric material flowing downwards into the mantle (after Kraus, 1931). This model basically corresponds to a nearly vertical zone of "subfluence". (b) Sketch illustrating the process leading to the formation of a cold, denser "lithospheric root" under a mountain range (after Fleitout and Froidevaux, 1982). The creation of a graben structure at the (northern) margin may result from lateral variations in the mechanical properties of the lithosphere. (c) Schematic cross section through the crust-mantle system of the Alps (after Laubscher, 1974). Caused by the plate collision process the subducting northern and southern lithospheres form a nearly vertical zone of "subfluence" in the Alpine region which penetrates deeply into the asthenosphere.

lithosphere in that collision zone. The geophysical evidence presented indicates that after the closure and subduction of the South Penninic Ocean relatively dense and cool lithospheric material has been pushed almost vertically into the upper mantle ("Verschluckungszone" after Ampferer, 1906) down to depths of about 200 km.

An anomalous mantle zone of this type under the Alps had already been postulated 55 years ago (Kraus, 1931). The sketch in Fig. 5a illustrates the schematic model of a "double orogen" that was envisaged with lithospheric material "flowing down into the mantle" from both sides. In the foreland extensional structures associated with rifting and voicanism may thus be created. As a consequence of the described deep-reaching lithospheric subduction slivers of upper crustal material were peeled off ("flakes") and piled up (Laubscher, 1974) thus creating the complex nappe edifice of the Alpine mountains (Fig. 5c). Deeper parts of the crust were shoved into each other (Mueller et al., 1980) and in this manner formed the less dense crustal root (see Fig. 5b) above the much wider and deeper-reaching "lithospheric root" beneath the Alps (Fig. 5c). A crust-mantle model of this type most easily explains where the excess lithospheric material must have gone during this plate collision process.

The tectonics of the Alps are usually displayed in the form of two-dimensional cross sections. As the arc of the Western Alps has a radius of only about 100 km which is of the same order as the thickness of the lithosphere, it is necessary to view the actual situation in all three dimensions (Laubscher, 1985). The pile of nappes which has been mapped on the surface consists of rocks derived from sediments or the uppermost more or less 5 km of crystalline basement (Fig. 6a). It is apparent that the surficial masses were delaminated (or "decoupled" in a low-velocity zone) from the main body of the crust and lower lithosphere which must have been subducted (Mueller et al., 1980). Since upper crust of both the African and Eurasian plates are involved in the Alpine nappe edifice, both African and Eurasian lithosphere must have been subducted. The simplest solution would be by double or "bivergent" subduction as illustrated in Fig. 6b (Laubscher, 1985).

Between the shallow and deep lithospheric masses there exists a depth range of profound disharmony which makes it virtually impossible to combine the visual surface geology (Fig. 6a) with the deep structures deduced by geophysical methods (Fig. 6b). Smoothing out the surficial nappes in the Western Alps proves shortening both in a north-south and in an east-west direction. Consequently, keeping the Eurasian plate (E) in a fixed position, the Adriatic promontory of the African plate (A) must have moved relatively northwestwards. The motion at the plate boundary must, therefore, have had a compressive and a strike-slip component which at depth looks quite different from what can be mapped at the surface.

Field evidence supports the concept of "flake tectonics" which requires zones of weakness within the lithosphere which are identical with depth ranges of lowered seismic velocities (Mueller, 1977). Three depth levels have been identified where slivers of crustal material can be sheared off more easily: (1) at the top of the crystalline basement beneath the sedimentary cover ("Jura" type), (2) near the lower boundary of the upper crustal (or "sialic") low-velocity zone ("Central Massif" type), and (3) immediately above the M discontinuity ("Ivrea" type). The Ivrea Body (Iv in Fig. 6b) is a prime example where the obducted "flake" consists of lower crustal material and even thin slices of upper mantle rocks.

MODELING OF THE "LITHOSPHERIC ROOT"

A synthesis of the available seismic and gravimetric information provides a unique data set, which permits a rough modeling of the lithospheric structure along a north-south profile across the Central Alps (Schwendener, 1984; Schwendener and Mueller, 1985). Any reasonable model will have to satisfy the

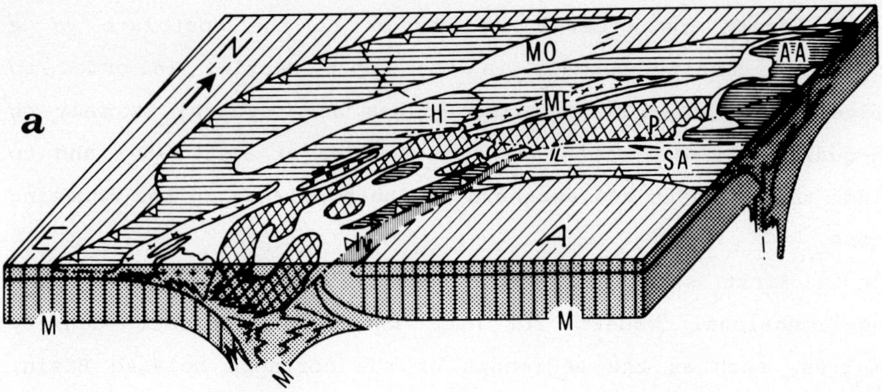

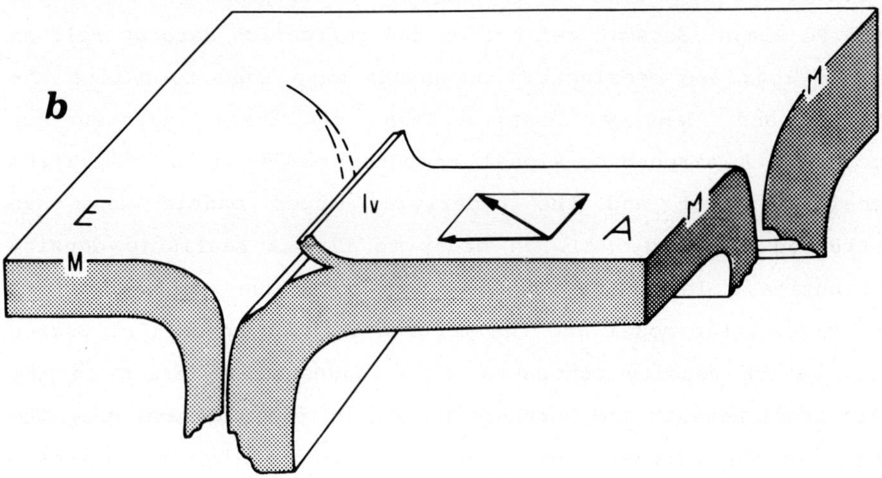

Fig. 6. Perspective view of the lithosphere system in the arc of the Western Alps (after Laubscher, 1985): (a) Surface geology and simplified crustal structure. Note that only the upper part of the crust is involved in the Alpine nappe tectonics, except for the zone of Ivrea (Iv) where a whole crustal "flake" seems to have been obducted. (b) Schematic block diagram of the lower lithosphere (uppermost mantle). Since both the African (A) and the Eurasian (E) upper crusts make up the Alpine nappe edifice, both lower lithospheres must be involved and somehow subducted at depth. A "bivergent" (double) subduction zone is the simplest and most reasonable model to explain this deep-reaching collision structure.

following three boundary conditions for the center of the Alps: (1) a maximum crustal thickness of some 50 km, (2) a minimum Bouguer gravity anomaly of -180 mgal, (3) an average topographic relief of some 2500 m. Combining these conditions leads either to an anomalously low crustal density or to the postulate of a deep-seated density anomaly in the upper mantle. In order to resolve this problem by modeling it has been crucial to rely on good-quality seismic data (see e.g. Mueller et al., 1980) and to include the northern and southern Alpine forelands in the modeling process.

In a first step the Bouguer anomalies were reduced with three-dimensional models for all known near-surface density anomalies, such as the sediments of the northern Molasse Basin, the Quaternary sediments of the Alpine valleys and the sediments of the Po Plain. Seismic reflection and refraction data as well as borehole logs and geoelectric soundings were used to define the geometry and density distributions of these near-surface structures. A three-dimensional computer model of a simplified three-layer crust and the underlying upper mantle was then constructed (see Fig. 7c). In order to find a realistic density distribution a least-squares inversion technique was applied to this crust-mantle model and the reduced Bouguer anomalies. Bodies with negative density contrasts were suggested to exist in the middle crust beneath the northern margin of the Alps and near the surface in the northern Apennines as indicated by the gravity, seismic, or geological data.

As a result of this "stripping" procedure a smooth positive residual gravity anomaly of about 500 km in width has been obtained with a maximum amplitude of close to +100 mgal which is situated in the region of the Southern Alps (Fig. 7a). The good coverage with high-quality seismic data, on which the construction of this gravimetric model was based, precludes an intracrustal source for this positive residual anomaly. It must, therefore, originate in the upper mantle in concordance with the findings from seismic surface-wave studies (Figs. 1 and 2) and travel time investigations (Fig. 3). The shape of the anomalous body in the upper mantle (not shown) was chosen according to the temperature anomalies (maximum decrease of -400° to -500°C compared to the ambient temperature) calculated with models of the Alpine orogeny

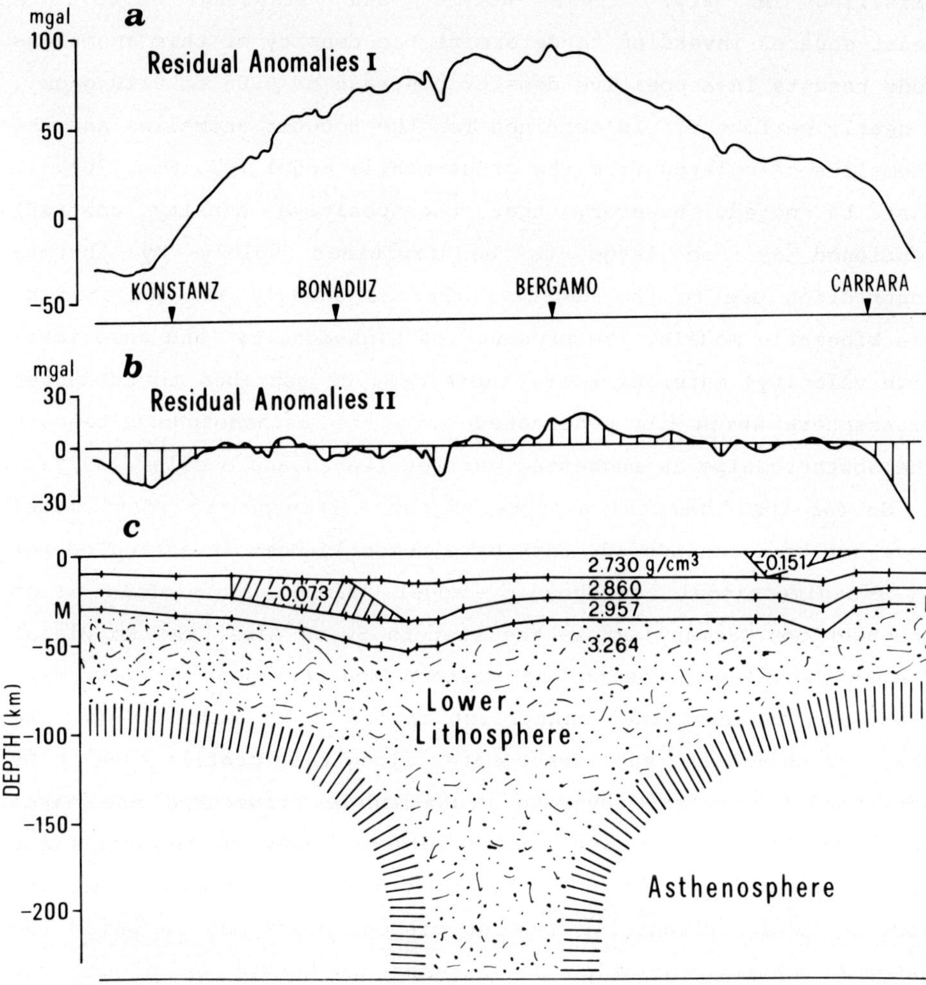

Fig. 7. Interpretation of a gravimetric crust-mantle profile from the northern foreland of the Alps to the Ligurian Sea (after Schwendener, 1984; Schwendener and Mueller, 1985): (a) Pronounced positive residual gravity anomalies obtained after the mass effect of a three-dimensional crustal model (see cross section in Fig. 7c) has been subtracted from the Bouguer anomalies. (b) Residual gravity anomalies after subtraction of the mass effect caused by the crust and the "lithospheric root" in the upper mantle. (c) Simplified model of the three-layered crust and the anomalous "block" in the upper mantle with the densities obtained by inversion. Local anomalous masses in the middle crust under the northern margin of the Alps and in the upper crust of the Northern Apennines with their respective negative density contrasts are indicated by shading.

(Kissling et al., 1983; Werner and Kissling, 1985). A least-squares inversion to determine the density of this anomalous body results in a positive density contrast of 0.05 to 0.10 g/cm^3. A nearly perfect fit is obtained for the Bouguer anomalies and the anomalies calculated from the crust-mantle model (cf. Fig. 7b). It must be noted, however, that the positive density contrast mentioned is too large to be explained solely by thermal contraction due to the negative thermal anomaly as suggested by the kinematic models. The presence of high-density (and associated high-velocity) material must, therefore, be ascribed to subducted lithosphere which has penetrated into the asthenosphere beneath the Southern Alps as suggested earlier (Panza and Mueller, 1979).

So far the dynamical aspects of the "lithospheric root" model have not been considered. They have recently been investigated for a two-dimensional geodynamic model along a profile from southwestern Germany across the eastern Swiss Alps, the Po Plain, the northern Apennines to the western Mediterranean Sea (Werner, 1985). As an additional constraint in the model calculations the observed uplift and subsidence rates along this profile have to be satisfied. Since only average crust-mantle structures are taken into account, a two-dimensional treatment seems to be justified. The method of calculation is based on the Navier-Stokes equation applied to a horizontally layered viscous halfspace in which the lateral density variation is assumed to behave as a harmonic function (Fleitout and Froidevaux, 1982; Werner, 1985).

Two sets of input parameters have to be extracted from observations or plausible assumptions have to be made: (1) the "load" represented by the density distribution, and (2) the viscosity which is dependent on depth (i.e. on pressure and temperature). Three types of "load" along the profile (Fig. 8a) have to be considered: (I) the topographic "load", including sedimentary basins and sea water, (II) the varying crustal thickness (Mueller et al., 1980; Miller et al., 1982), (III) the "lithospheric root", characterized by relatively high seismic wave velocities (Panza and Mueller, 1979; Baer, 1980), whose increased

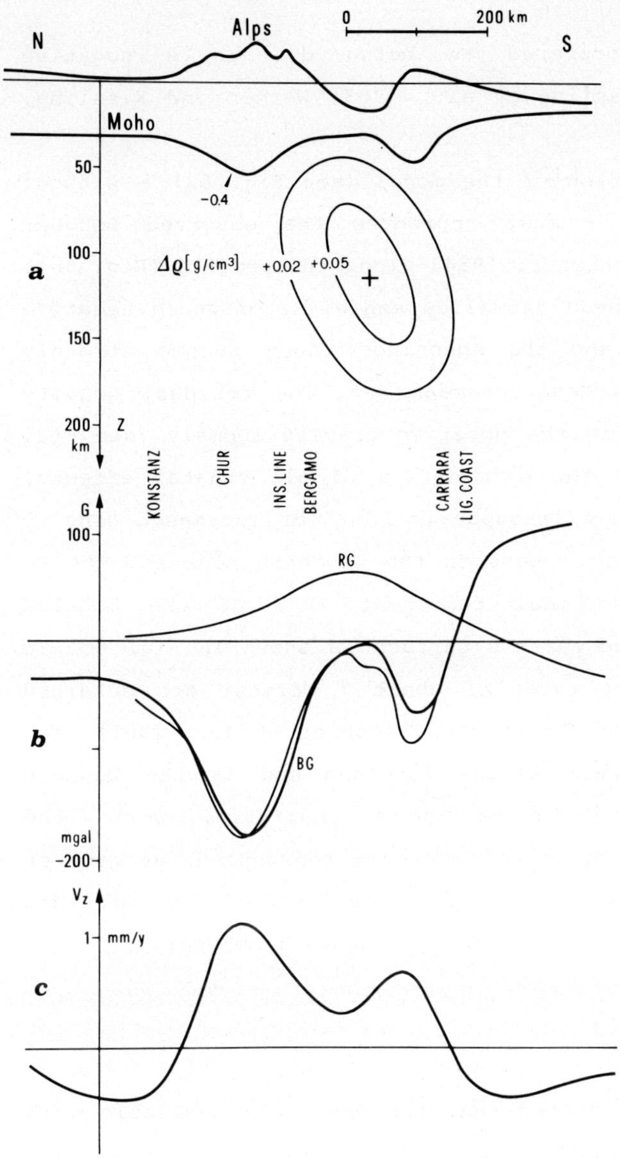

Fig. 8. Geodynamic model cross section across the Alps (after Werner, 1985): (a) Model structure consisting of the topographic "load", the varying crustal thickness and the "lithospheric root" with increased density. (b) Residual gravity effect (RG) due to the mantle anomaly sketched in Fig. 8a. The curves BG show the combined gravity effect caused by the mantle anomaly and the density jump at the crust-mantle boundary (MOHO) - thick line -, as compared to the observed gravity anomalies after near-surface density disturbances have been eliminated (Schwendener, 1984) - thin line. (c) Resulting vertical movement at the earth's surface, which reacts rather sensitively to both the "loading structure" and the viscosity-depth distribution.

density has been interpreted as being due to a negative temperature anomaly (Kissling et al., 1983; Werner and Kissling, 1985).

The density distribution of the model (see Fig. 8a) - without the topographic "load" - must reproduce the observed Bouguer gravity anomalies (Schwendener, 1984) along the profile (Fig. 8b). As is to be expected, these anomalies exhibit pronounced negative values under the Alps and the Apennines; they become strongly positive in the Western Mediterranean Sea. The residual gravity anomaly (RG in Fig. 8b) is the positive gravity anomaly (see Fig. 7a) which remains after the elimination of all crustal effects, i.e. it is caused by the "lithospheric root" of increased density residing in the upper mantle beneath the Southern Alps and the Po Plain. Finally, the vertical component (V_z) of the modeled displacement field at the earth's surface is shown in Fig. 8c. In the Alps relative uplift rates of about 1 mm/year are obtained which agree with geodetic observations (Gubler et al., 1981). The subsidence in the northern Alpine foreland and in the Western Mediterranean Sea seems to be quite realistic, while the foundering of the Po Plain is not properly reproduced. At greater depths the model postulates the "lithospheric root" to sink into the asthenosphere at a present rate of roughly 1 mm/year.

CONCLUSIONS

In summary it can be stated that the presently available data strongly support the hypothesis for the presence of a massive "lithospheric root", i.e. a deep-reaching anomalous zone residing in the mantle beneath the inner side of the Alpine arc. Based on this evidence, it must be concluded that a large-scale plate-tectonic mechanism, such as the northwest-southeast directed compression caused by the push of the African plate against the Eurasian plate through the Adriatic promontory is the fundamental driving force ultimately responsible for the formation of the Alps. The collision of the two continental plate margins will not only lead to a thickening of the crust, but will also produce a cold, dense and slowly subsiding "lithospheric root" beneath the mountain belt. This process in turn will induce strong regional compressive stresses (see Fleitout and Froidevaux, 1982) which are able to maintain the mountain building process without requiring sustaining forces transmitted from far away. A positive feedback system results that is inherently unstable; whenever it passes through a climax episodic phases of strong compression will be observed accompanied by accelerated uplift (see e.g. Kissling et al., 1983). On this scale, the still ongoing uplift of the Alpine mountain chain is a secondary effect mainly due to the isostatic adjustment of the heavily deformed upper part of the lithosphere which has been considerably thickened in this severe collision process of two major continental plates.

ACKNOWLEDGMENTS

The synopsis presented in this paper would not have been possible without the help and support of many individuals and institutions who cannot all be named here; their contribution is gratefully acknowledged. Sincere thanks for many stimulating discussions are due to the authors' colleagues and collaborators. The authors are indebted to H.P. Laubscher (Basel) and D. Werner (Zürich) for granting permission to reproduce Figs. 6 and 8 from the "Proceedings of the Second Workshop on the European Geotraverse Project". Special thanks go to Mrs. I. Galante for the patient and accurate typing for camera-ready reproduction, to Mr. G. Cavicchi and Mr. D. Crozzoli for the drawings.

This paper is Contribution No. 506 of ETH-Geophysics, Zürich (Switzerland). Financial support from Italian M.P.I. (60% and 40% funds) is gratefully acknowledged.

REFERENCES

Ampferer, O., 1906. Ueber das Bewegungsbild von Faltengebirgen. Jahrbuch k.k. Geol. Reichsanstalt, Wien, 56: 539-622.

Ansorge, J., Bonjer, K.-P. and Emter, D., 1979. Structure of the uppermost mantle from long-range seismic observations in Southern Germany and the Rhinegraben area. Tectonophysics, 56: 31-48.

Baer, M., 1980. Relative travel time residuals for teleseismic events at the new Swiss seismic station network. Ann. Géophys., 36: 119-126.

Channell, J. and Horváth, F., 1976. The African-Adriatic promontory as a paleogeographic premise for Alpine orogeny and plate movements in the Carpatho-Balkan region. Tectonophysics, 35: 71-101.

Fleitout, L. and Froidevaux, C., 1982. Tectonics and topography for a lithosphere containing density heterogeneities. Tectonics, 1: 21-56.

Gubler, E., Kahle, H.-G., Klingelé, E., Mueller, St. and Olivier, R., 1981. Recent crustal movements in Switzerland and their geophysical interpretation. Tectonophysics, 71: 125-152.

Kissling, E., Mueller, St. and Werner, D., 1983. Gravity anomalies, seismic structure and geothermal history of the Central Alps. Annales Geophysicae, 1: 37-46.

Kraus, E.C., 1931. Die Alpen als Doppelorogen. Geolog. Rundschau, 22: 65-78.

Laubscher, H.P., 1974. The tectonics of subduction in the Alpine system. Mem. Soc. Geol. Ital., 13: 275-283.

Laubscher, H.P., 1985. The tectonics of the Western and Southern Alps: Correlation between surface observations and deep structure. In: D.A. Galson and St. Mueller (Editors), Second EGT Workshop - The Southern Segment. European Science Foundation, Strasbourg, pp. 93-101.

Miller, H., Ansorge, J. and Mueller, St., 1979. Evidence of a high-velocity zone in the upper mantle under the central part of the Alps (abstract). EOS Trans. AGU, 60: 594.

Miller, H., Mueller, St. and Perrier, G., 1982. Structure and dynamics of the Alps - a geophysical inventory. In: H. Berckhemer and K. Hsü (Editors), Alpine-Mediterranean Geodynamics. AGU-GSA Geodynamics Series, Washington, 7: 175-203.

Mueller, St., 1977. A new model of the continental crust. In: J.G. Heacock (Editor), The Earth's Crust, Geophys. Monograph Series, American Geophysical Union, Washington, 20: 289-317.

Mueller, St., Ansorge, J., Egloff, R. and Kissling, E., 1980. A crustal cross section along the Swiss Geotraverse from the Rhinegraben to the Po plain. Eclogae geol. Helv., 73: 463-485.

Mueller, St., 1982. Geodynamic aspects of the Mediterranean-Alpine region. Revista de Geofisica, Madrid, 38: 103-111.

Mueller, St., 1984. Dynamic processes in the Alpine arc. Annales Geophysicae, 2: 161-164.

Panza, G.F. and Mueller, St., 1979. The plate boundary between Eurasia and Africa in the Alpine area. Mem. Sci. Geol. (Univ. Padova), 33: 43-50.

Panza, G.F., Mueller, St. and Calcagnile, G., 1980. The gross features of the lithosphere-asthenosphere system in Europe from seismic surface waves and body waves. Pure and Appl. Geophys., 118: 1209-1213.

Panza, G.F., 1985. Lateral variations in the lithosphere in correspondence of the southern segment of EGT. In: D.A. Galson and St. Mueller (Editors), Second EGT Workshop - The Southern Segment. European Science Foundation, Strasbourg, pp; 47-51.

Rybach, L., Mueller, St., Milnes, A., Ansorge, J., Bernoulli, D. and Frey, M., 1980. The Swiss Geotraverse Basel-Chiasso - a review. Eclogae geol. Helv., 73: 437-462.

Schwendener, H., 1984. Ein gravimetrisches Krusten-Mantel-Modell für ein Profil vom nördlichen Alpenvorland bis an die Ligurische Küste. Geodätischgeophysikalische Arbeiten in der Schweiz (Schweizer. Geodät. Kommission), Kümmerly und Frei, Bern, Band 36, 160 pp.

Schwendener, H. and Mueller, St., 1985. New evidence for a density anomaly in the upper mantle below the Southern Alps. In: D.A. Galson and St. Mueller (Editors), Second EGT Workshop - The Southern Segment. European Science Foundation, Strasbourg, pp. 115-120.

Werner, D., 1985. A two-dimensional geodynamic model for the southern segment of the EGT. In: D.A. Galson and St. Mueller (Editors), Second EGT Workshop - The Southern Segment. European Science Foundation, Strasbourg, pp. 65-69.

Werner, D. and Kissling, E., 1985. Gravity anomalies and dynamics of the Swiss Alps. Tectonophysics, 117: 97-108.

ON THE DYNAMICS OF CONVERGENT PLATE BOUNDARIES AND STRESS IN THE LITHOSPHERE

M.J.R. WORTEL and S.A.P.L. CLOETINGH
Vening Meinesz Laboratory
Institute of Earth Sciences, University of Utrecht
P.O. Box 80.021, 3508 TA Utrecht, The Netherlands

ABSTRACT

The evolution of active continental margins and island arcs is closely related to the process of subduction of (predominantly) oceanic lithosphere. In this paper we deal with some aspects of the dynamics of the subduction process. Understanding the dynamics is taken to be a prerequisite for improving our insight into the relation between upper mantle processes and structures or processes near the surface. A major force associated with the subduction of oceanic lithosphere is the slab pull resulting from the density contrast between the cold and dense descending slab and the surrounding warmer mantle. This force is generally assumed to be compensated by resistive forces acting on the slab. Here we emphasize that such a compensation may be valid for a plate's convergent boundary as a whole but certainly not always locally for each segment of the trench system. Transmission of stress from the subducted slab towards the attached lithospheric plate at the surface is possible, and taking this into account provides insight into the relation between upper mantle processes and tectonic processes near the surface (e.g. accretion of sediments at trenches, and uplift, subsidence and tilting of continental margins).

INTRODUCTION

In the context of the tectonic evolution of island arcs overlying subduction zones, the subduction process may be considered to be the basic process in which many geodynamic processes at or near the surface find their origin and cause. How, exactly, these processes are related to subduction is not yet (fully) understood. To gain more insight into this problem it is essential to work with a model of the subduction process as complete as possible. The fundamental dynamics of the process are expected to provide keys to the complexity of subduction zones. In this paper we address some aspects of subduction zone dynamics, with special emphasis on the role of the slab pull.

Already in early studies of the dynamics of plate motion the slab pull was recognized as a major force responsible for and maintaining the motion of lithospheric plates (Elsasser, 1967; McKenzie, 1969). This force arises from the temperature contrast - and, hence, via thermal contraction the density contrast - between the cold descending slab in a subduction zone and the warmer upper mantle surrounding the slab. Thus, the dense subducted slab may be conceived to sink under its own weight, pulling the attached plate towards the trench.

Subsequent quantitative investigations by Forsyth and Uyeda(1975), Chapple and Tullis(1977) and Backus et al.(1981) have confirmed this early assessment. Another outcome of these studies, however, was that the slab pull is balanced - or very nearly so - by the resistive shearing forces acting on the downgoing slab. Such a balance would imply that, albeit the high magnitude of the slab pull, very little of this force is transmitted to the attached horizontal part of the plate. Consequently, within this framework the slab pull would not have a significant effect on geodynamic processes near the surface.

Several authors, including Forsyth and Uyeda(1975) and (especially) Chapple and Tullis (1977), have pointed out that the compensation of the slab pull by the resistive forces should not be taken to apply for each segment of a convergent plate boundary (subduction zone), but rather for a subducting plate boundary as a whole. Nevertheless, local compensation of the slab pull is generally assumed in many studies and very little attention has been paid to the possible transmission of stress from the subducted slab to the upper horizontal part of the plate. Insight gained by studying the role of lithospheric age in the subduction process (Vlaar, 1975; Vlaar and Wortel, 1976; Wortel and Vlaar, 1978; Wortel, 1980, 1982, 1984; England and Wortel, 1980) has clearly brought out the significance of a (partially) uncompensated or over-compensated slab pull force. The extent to which the slab pull force is compensated may well vary considerably along a convergent plate boundary, thus giving rise to laterally varying boundary forces acting on the (non-subducted) lithosphere. This, in turn, strongly affects the stress field in the lithosphere.

In this paper we will show that taking into account an unbalanced slab pull contributes to a better understanding of several tectonic processes at or near the surface, in particular in the vicinity of the trenches along convergent plate boundaries.

DYNAMICS OF SUBDUCTION

To place the dynamics of the subduction process into the perspective of the dynamics of the lithosphere as a whole, and to convey the importance of the age of the descending lithosphere for the dynamics of the subduction process two short sections on plate tectonic forces and age-dependent subduction are included.

Plate tectonic forces

The main forces acting on the lithosphere are indicated in Fig. 1, which is a schematic cross section through a spreading ridge and a subduction zone.

The forces which drive the plates are the ridge push F_{rp} and the slab pull F_{sp}. The ridge push force is not a boundary force acting at the ridge axis but a pressure gradient integrated over the area of the plate, with contributions from all parts of the plate where differences in elevation exist (Lister, 1975). This pressure gradient results from the cooling and densification of the lithosphere. The other driving force, the slab pull F_{sp}, results from the density contrast between the cold descending slab and the surrounding warmer upper mantle (e.g. Elsasser, 1967; McKenzie, 1969).

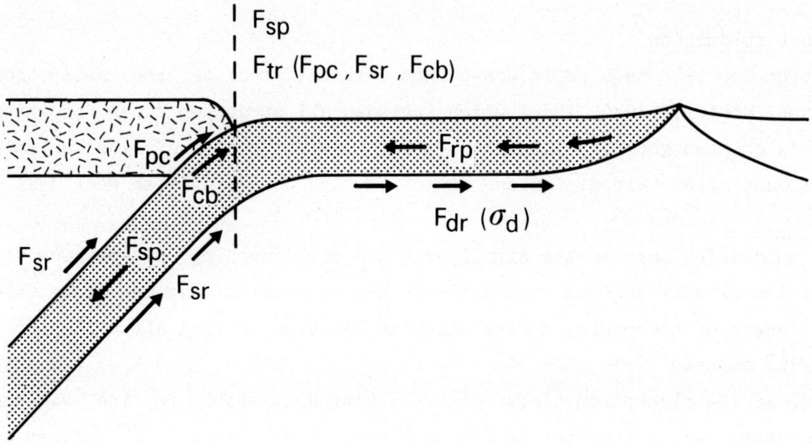

Fig. 1 - Schematic representation of the main forces acting on a subducting oceanic plate.
Driving forces: F_{rp} (ridge push) and F_{sp} (slab pull).
Resistive forces: F_{dr} (drag force at the base of the plate; σ_{dr} is shear stress), F_{tr} (resistance at the trench and in the subduction zone) . F_{tr} is composed of three parts: F_{cb} (compositional buoyancy), F_{pc} (resistance at the plate contact) and F_{sr} (slab resistance by viscous shearing forces along the slab/upper-mantle interfaces).

The resistive forces shown in Fig. 1 are the drag force acting at the base of the lithosphere (that is the shear stress σ_b integrated over the area of the base of the plate) and the resistive forces F_{tr} acting on the downgoing slab in the trench region and in the upper mantle part of the subduction zone. The latter forces comprise the compositional buoyancy force F_{cb} and resistive shearing forces (F_{pc}, the integrated value of the shear stress σ_{pc} along the plate contact with the overriding plate, and F_{sr} , the integrated value of σ_{sr}, the resistance to slab penetration in the upper mantle by viscous shearing forces). F_{cb} is associated with the petrological stratification of the oceanic crust and upper mantle created during the spreading process (Vlaar and Wortel, 1976; Oxburgh and Parmentier, 1977). The density structure of the layering contributes a gravitationally stable component to the lithosphere-asthenosphere system. Upon subduction this is felt as a force (F_{cb}) counteracting the gravitationally unstable component resulting from the cooling and densification of the lithosphere which is represented by F_{sp}.

Depending on the specific properties of a plate, some other forces (not indicated in Fig. 1) may be relevant, like drag beneath the continents and the forces associated with continental collision (see Cloetingh and Wortel, 1985, 1986).

Age-dependent subduction

Information for this section is drawn from a global study of the subduction process (see Wortel, 1980, 1982) which included all subducton zones for which adequate data on slab geometry (length, dip and maximum depth of the seismic zone), relevant plate kinematics and age of the the downgoing slab were available.

For the subduction process the significant age of downgoing lithosphere is the age at the time it arrived at the trench and started to descend, since this equals the length of the period during which it has been cooling after formation at a spreading centre.

Fig. 2 shows the resorption times of the lithospheric slabs in the various subduction zones as a function of age. The resorption time t_{res} is defined as the downdip length of the seismic zone (S_{sz}, measured from the surface to the depth of the deepest earthquake foci) divided by the convergence rate normal to the plate contact (v_c). The graph shows a strong increase in resorption time with the age of the downgoing lithosphere. This can be approximated by a linear relationship for the resorption time:

$$t_{res}=(0.12\pm0.03).age \tag{1}$$

Thus, the downdip length of the seismic zone is given by

$$S_{sz}=(0.12\pm0.03).v_c.age \tag{2}$$

Note that the maximum observed resorption time is 15 Ma.

Thermal model calculations have been carried out (Wortel, 1980, 1982) on the basis of McKenzie's (1969) approach in which a slab is assumed to sink through a mantle with an adiabatic temperature gradient. The adiabatic temperature distribution in the upper mantle below the lithosphere corresponds to a constant potential temperature T_m^p here taken as 1200°C. Potential temperature may be regarded as the real temperature minus the effect of adiabatic compression and the contribution of latent heat from phase changes (McKenzie, 1970). In Fig. 2 the values of $x_{max}^{(i)}/v_c$, where $x_{max}^{(i)}$ is the maximum downdip penetration of an isotherm of potential temperature, are plotted for the 600°C and 750°C isotherms. It follows that the downdip ends of seismic zones in subducted slabs are characterized by minimum potential temperatures between 600°C and 750°C.

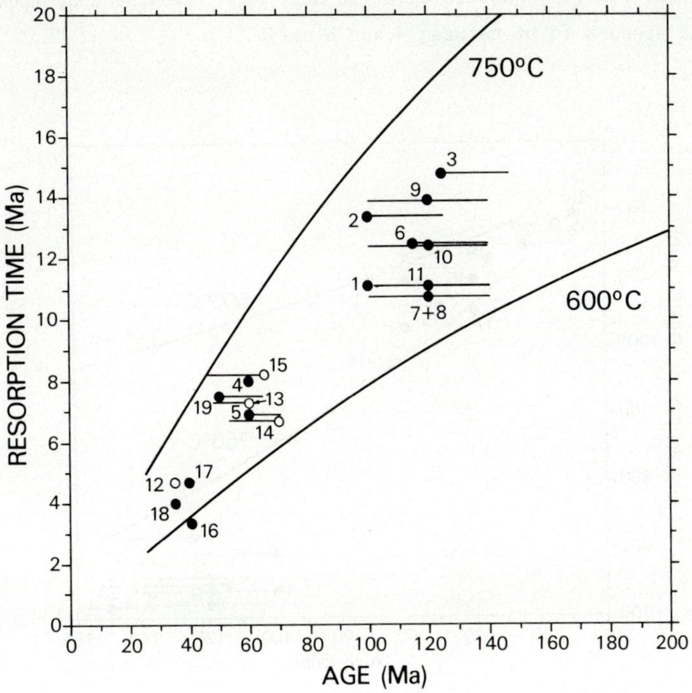

Fig. 2 - Resorption times of subducted oceanic lithosphere as a function of lithospheric age. The symbols indicate the resorption times for the following zones: 1, Kuriles; 2, N. Honshu; 3, Izu-Bonin; 4, Ryukyus; 5, Sumatra; 6, Java; 7, Tonga (20°S); 8, Tonga (24°S); 9, Kermadec (28°S); 10 and 11, New Zealand; 12, Chile (36-42°S); 13, Chile (27-33°S); 14, Chile (18-26°S); 15, North-central Peru; 16, Central America 87°W); 17, Central America (90°W); 18, Alaska (150°W); 19, Aleutians (180°W). Bars on only one side of the symbols indicate temporal variations in age: the symbols are plotted at the estimated age at which the deepest part of the subducted slab started to descend. The other end of the bar indicates the present age at the trench. Two-sided bars are used for some zones in which the age of the descending lithosphere may vary between 100 and 140 Ma. The two curves indicate the resorption time (that is, the maximum downdip penetration divided by the convergence rate) of two isotherms of potential temperature.

Fig. 3 displays the maximum focal depths of earthquakes in subducted slabs as a function of lithospheric age. This relation has been established by Vlaar and Wortel (1976). Only subduction zones in which lithosphere older than 70-100 Ma is consumed exhibit deep focus earthquakes. The depth-age relations for the 600°C and the 700°C isotherms were calculated for an average sinking rate v_z of 2.5 cm/yr. All data points in Fig. 2, with the exception of point 16, are bounded by the two isotherms. In Fig. 3 this is only the case for ages below 70 Ma. This means that the sinking rate v_z=2.5 cm/yr is appropriate for these ages but clearly too low for ages above 70-100 Ma. For lithosphere of these ages the sinking rate appears to be between 4 and 6 cm/yr.

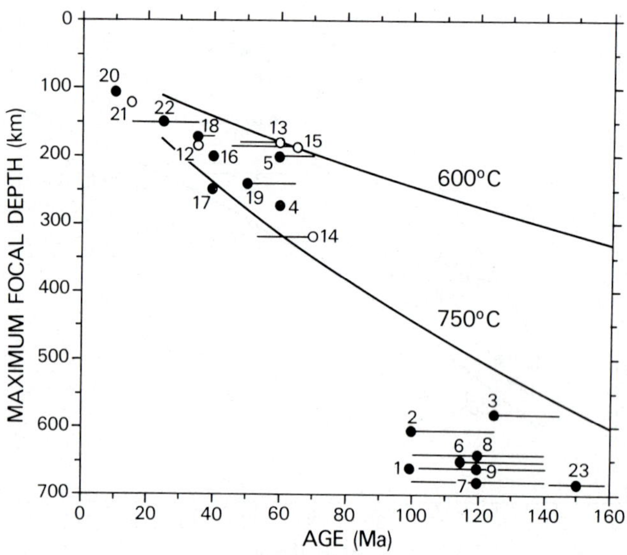

Fig. 3 - Maximum focal depth of earthquakes in subduction zones as a function of the age of the subducted lithosphere. Symbols and numbers are the same as in Fig. 2, with the addition of 20, Cascades; 21, Chile (42 - 46 ^{o}S); 22, Mexico (102 ^{o}W); 23, Marianas. The penetration depths of the 600 ^{o}C and the 750 ^{o}C isotherms of potential temperature are plotted for a vertical velocity v_z=2.5 cm/yr.

Compensation of the slab pull: partial or complete?

To answer the above question we consider the dependence of the slab pull on the various pertinent parameters.

The slab pull F_{sp} is the downdip component of the gravitational body force acting on the dense subducted slab. Per unit of length in the direction of the trench this can be expressed as:

$$F_{sp}=g \sin\phi \int_0^L \int_o^{S_{sz}} \Delta\rho \, dx dz \qquad (3)$$

Here, g is the gravitational acceleration, ϕ the dip and L the thickness of the downgoing slab, and $\Delta\rho$ the density contrast between slab and surrounding mantle.The x- and z- directions are downdip and vertical to the upper (dipping) interface between the slab and the upper mantle, respectively. The downdip integration is limited to the end of the seismic zone (S_{sz}), which is reasonable in view of the probable changes in rheology and subsequent deformation beyond this point (see Wortel, 1986).

Using McKenzie's (1969) thermal model we find

$$F_{sp} = \frac{4g\alpha\rho_m T_m L^3}{\kappa\pi^4} v_c \sin\phi \, [1-\exp(-\frac{\pi^2 \kappa S_{sz}}{v_c L^2})] \tag{4}$$

The parameters α, ρ_m, T_m and κ are the coefficient of thermal expansion, the density and the temperature of the asthenosphere, and the thermal diffusivity of the asthenosphere/lithosphere. For the present purpose, all these parameters may be taken to be constant. The thickness L of the lithosphere increases with age t according to a square-root-of-age relation, at least for t<70 Ma (see Wortel, 1980 for a review). Using this age-dependence of L, the relation (2) and $\kappa=0.9\times10^{-6}\ m^2s^{-1}$ we obtain

$$[1-\exp(-\frac{\pi^2 \kappa S_{sz}}{v_c L^2})]=0.29$$

Thus, the factor between brackets in (4) does depend neither on v_c nor on the age of the lithosphere. Putting $v_c \sin\phi = v_z$, it is found that

$$F_{sp} \propto v_z L^3 \tag{5a}$$

Taking into account the above-mentioned age-dependence of L gives

$$F_{sp} = c_{sp} v_z t^{3/2} \tag{5b}$$

where c_{sp} is a constant.

The resistive (viscous shearing) stress σ_{sr} along the slab/upper-mantle interface is taken to be proportional to the relative velocity of the slab and the upper mantle. Integrated along the downdip extent (S_{sz}) of the slab, this yields (again per unit width):

$$F_{sr} \propto v_c S_{sz} \tag{6a}$$

and, with (2):

$$F_{sr} = c_{sr} v_c^2 t \tag{6b}$$

where c_{sr} is a constant of proportionality.

Combining the forces on the slab, all per unit width in the direction of the strike of the trench, we obtain the net force in the trench region exerted by

the slab on the horizontal part of the plate. This net force, designated by F_{nt}, acts on the cross section of the plate indicated by the dashed line in Fig. 1. Positive values of F_{nt} - and of the other forces - correspond with forces directed from the ocean towards the trench. Hence, negative values correspond with resistance to continuing subduction.
Thus, the net force is

$$F_{nt} = F_{sp} - (F_{cb} + F_{pc} + F_{sr}) \tag{7a}$$

or

$$F_{nt} = c_{sp}\, v_z\, t^{3/2} - (F_{cb} + F_{pc} + c_{sr}\, v_c^2\, t) \tag{7b}$$

It should be noted that v_z can vary with the age of the lithosphere (according to the relation discussed for Fig. 3), whereas v_c can not. v_c depends on the full dynamics of the two converging plates: it cannot simply adjust to the age of the downgoing slab at a certain segment of the trench.

Sofar, the compositional buoyancy force F_{cb} and the shear resistance at the plate contact F_{pc} are taken to be independent of age. For the former force this is a sound assumption because the density variations due to composition are not likely to vary significantly with time and age (see also England and Wortel, 1980). As to the latter force, there is strong observational evidence (from seismological studies) that the coupling between the downgoing plate and the overriding plate decreases with the age of the slab (Ruff and Kanamori, 1980). This would imply a decrease in F_{pc} with age. From (7b), and in particular from the difference in the age-dependences of the driving and the resistive forces, it follows that the net force is clearly a function of the age of the downgoing lithosphere. Starting from a value smaller (in magnitude) than the resistive forces, the driving force grows faster with age than the resistive forces do. The probable decrease of F_{pc} with age even strengthens this trend. Thus, qualitatively , F_{nt} is negative for young lithosphere (over-compensation, implying resistance to continuing subduction)) and positive for old lithosphere (partial compensation). This relationship can be quantified further by using the results of England and Wortel (1980).

The account given above serves to bring out the basic acpects of the dynamics of subduction. A more detailed discussion, however, is beyond the scope of the present paper. Instead we will now present the results of some investigations which clearly and quantitatively demonstrate the significance of age-dependent forces. In conclusion, on the basis of the analysis of the slab pull force and the forces counteracting the slab pull it appears that complete compensation should be considered as an exceptional situation rather than a common one.

STRESS FIELD IN THE LITHOSPHERE

General aspects

The plate tectonic forces considered above contribute to the state of stress in the lithosphere. In fact, their contribution generally outweighs those of other origins (see Cloetingh, 1982). An exception to this are bending stresses in the lithosphere, for example near trenches along convergent plate boundaries (see below).

If F_{sp} were exactly balanced by resistive forces one could infer the stress in the oceanic lithosphere from a cross section such as shown in Fig. 1. Along these lines, and using their main result - an almost complete compensation of the slab pull F_{sp} by the viscous shearing force F_{sr} (see Fig. 1) -, Forsyth and Uyeda (1975) predicted that the horizontal part of oceanic plates would be in a state of weak compressive stress owing to the ridge push F_{rp} counteracted by the drag force F_{dr} and the resistance at the plate contact F_{pc}. It will be shown below that the stress field in oceanic lithosphere is in general more complicated than that.

As Chapple and Tullis (1977) realized, the complexity of the net subduction force (here denoted by F_{nt}) exerted on a plate from a long convergent plate boundary may give rise to intraplate stresses of considerable magnitude. On the basis of the force modelling discussed in the previous chapter we will illustrate the importance of the slab pull for the intraplate stress field by presenting results of numerical models for the Nazca (and the Farallon) plate and the Indo-Australian plate .

With a rheological model for the lithosphere the stress field in the lithosphere associated with the plate tectonic forces can be calculated numerically. This field, which is called regional stress field, represents the non-hydrostatic state of stress in the lithosphere. Because the stress values are to be compared with the strength distribution in the lithosphere - which is expressed in terms of differential stress - the hydrostatic part of the stress field (depth-dependent pressure) can be omitted.

To ensure mechanical equilibrium, the net torque exerted by the forces on a plate is required to vanish:

$$\sum_{N} \int_{P} \overline{r} \times \overline{F} \, dp = 0 \tag{8}$$

where the sum is over the total set of N forces and the integration over the boundaries or the area of the plate P, depending on the nature of the force involved.

On the basis of their established age-dependence the driving forces F_{rp} and F_{sp} can be calculated. Several unknown parameters associated with the resistive forces can be determined from eq.(8) (see Wortel and Cloetingh, 1981, 1983 and Cloetingh and Wortel, 1985, 1986).

The plates are modelled as being elastic, with a reference thickness of 100 km, Young's modulus $E = 7\times10^{10}$ Nm^{-2} and Poisson's ratio $\nu = 0.25$. From model results for this rheology the stresses in a plate with a more realistic rheology (see Goetze and Evans, 1979) can simply be derived.

For a comparison of calculated stress fields with the stress directions inferred from focal mechanism data we refer to Cloetingh and Wortel(1986).

Fragmentation of plates

Studies of marine magnetic anomalies have clearly shown that lithospheric plates can break up, thus giving rise to the formation of new spreading centres. A well-documented example of such a process is the fragmentation of the Farallon plate into two plates (the Cocos plate and the Nazca plate) which took place approximately 25-30 Ma ago (Hey, 1977).

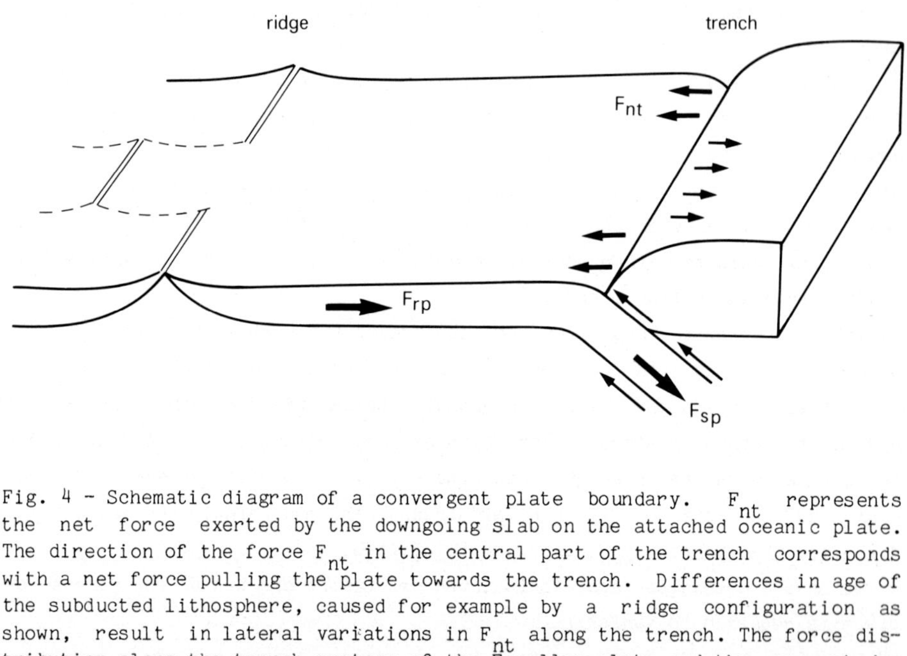

Fig. 4 - Schematic diagram of a convergent plate boundary. F_{nt} represents the net force exerted by the downgoing slab on the attached oceanic plate. The direction of the force F_{nt} in the central part of the trench corresponds with a net force pulling the plate towards the trench. Differences in age of the subducted lithosphere, caused for example by a ridge configuration as shown, result in lateral variations in F_{nt} along the trench. The force distribution along the trench systems of the Farallon plate and the present-day Nazca plate ressembles the situation shown in this figure.

Using finite element methods and a reconstruction of regional plate boundaries Wortel and Cloetingh (1981, 1983) investigated the state of stress in the Farallon plate approximately 25-30 Ma ago, to see whether the stress field provided a clue towards understanding the fragmentation. On the basis of a reconstructed age pattern in the Farallon plate and the established age-dependence of the driving forces F_{rp} and F_{sp}, and with the use of eq.(8), the forces acting on

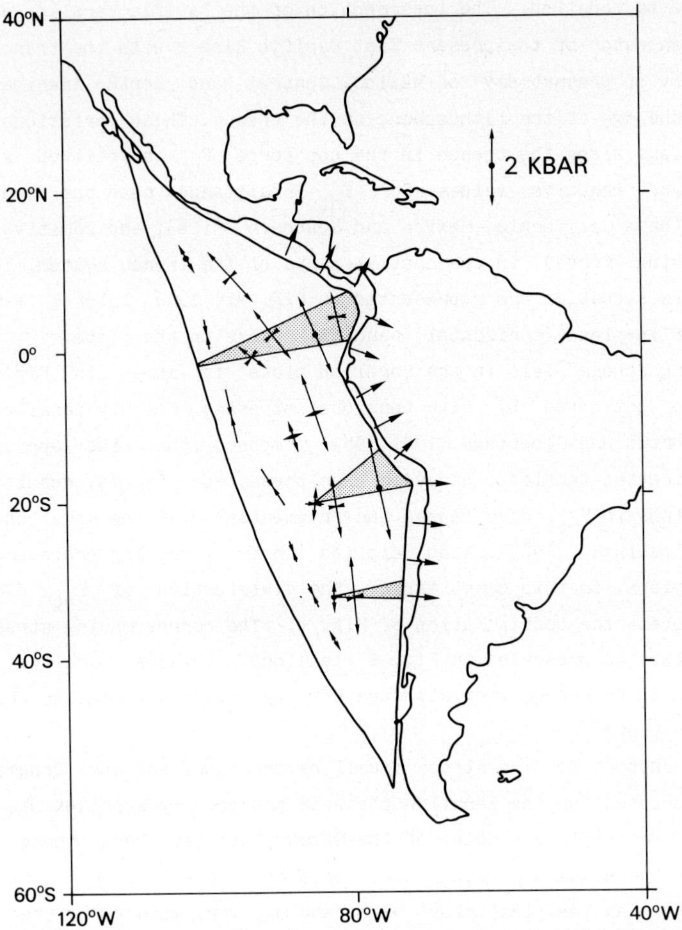

Fig. 5 - The calculated stress field in the Farallon plate under the reconstructed conditions of 25-30 Ma ago, prior to the break-up into the Cocos plate and the Nazca plate. The stress values are displayed for a uniform elastic plate with a reference thickness of 100km. Net slab forces pulling the plate towards the trench ($F_{nt} > 0$) are indicated by arrows. In the trench segments without arrows resistive forces act on the Farallon plate ($F_{nt} < 0$). Three zones exhibiting tensional phenomena are indicated. The northernmost one, near Panama, developed into the Cocos-Nazca spreading centre (Galapagos Rift).

the plate could be modelled. The intersection of the Pacific-Farallon spreading centre - the ancestor of the present East Pacific Rise - with the trench system (along the coast of present-day of Mexico, Central and South America) imply variations in the age of the lithosphere at the trench. These variations lead to lateral variations along the trench in the net force F_{nt} associated with the subducting slab: negative values for F_{nt} (resistance) near the trench-ridge intersections (Baja California, Mexico and Southern Chile), and positive values for F_{nt} (driving force) in the central parts of the trench system. The basic features of this situation are represented in Fig. 4. Such lateral variations cause a sort of in-plane (horizontal) bending process in the plate.

The resulting stress field in the Farallon plate is shown in Fig. 5. It appears to be dominated by high tensional stresses, roughly parallel to the trench axis. Wortel and Cloetingh (1981,1983) proposed that the approximately North-South directed tensional stresses near present-day Panama, resulting from lateral variations in F_{nt}, have caused the fragmentation of the Farallon plate. Wortel and Cloetingh (1983) also applied their modelling procedure to the present Nazca plate. In this case, again, the distribution of F_{nt} along the trench ressembles the configuration of Fig. 4. The corresponding stress field in the Nazca plate is presented in Fig. 6: tensional stresses parallel to the trench can again be recognized, although the magnitudes are smaller than those of the Farallon plate.

Very strong support for the stress modelling procedure and our fragmentation hypothesis formulated for the Farallon plate is provided by Warsi et al. (1983). In their marine geophysical studies of the Nazca plate off Peru these authors discovered that the Mendana Fracture zone, oriented perpendicularly to the Peruvian trench, is in an incipient stage of spreading, very much along the lines we suggested for the origin of the Cocos-Nazca spreading centre (Wortel and Cloetingh, 1981, 1983).

In conclusion, lateral variations in the age of the descending lithosphere, and hence variations in F_{nt}, may be the source of significant tensional stresses in the plate to which the slab is attached. They constitute a possible cause for fragmentation of plates, in general, and of the Farallon plate, in particular.

Accretion and subduction of sediments

The process of sediment accretion at trenches has been recognized to contribute strongly to the evolution of island arcs and continents. A puzzling feature of observational results obtained sofar is the extremely wide variety of structures and tectonic processes encountered in the trench regions. Wortel and Cloetingh (1985) addressed this problem by considering a prominent example of a non-uniform arc or trench system, the Peru-Chile Trench. Schweller et al. (1981) have established that this trench exhibits significant lateral variations,

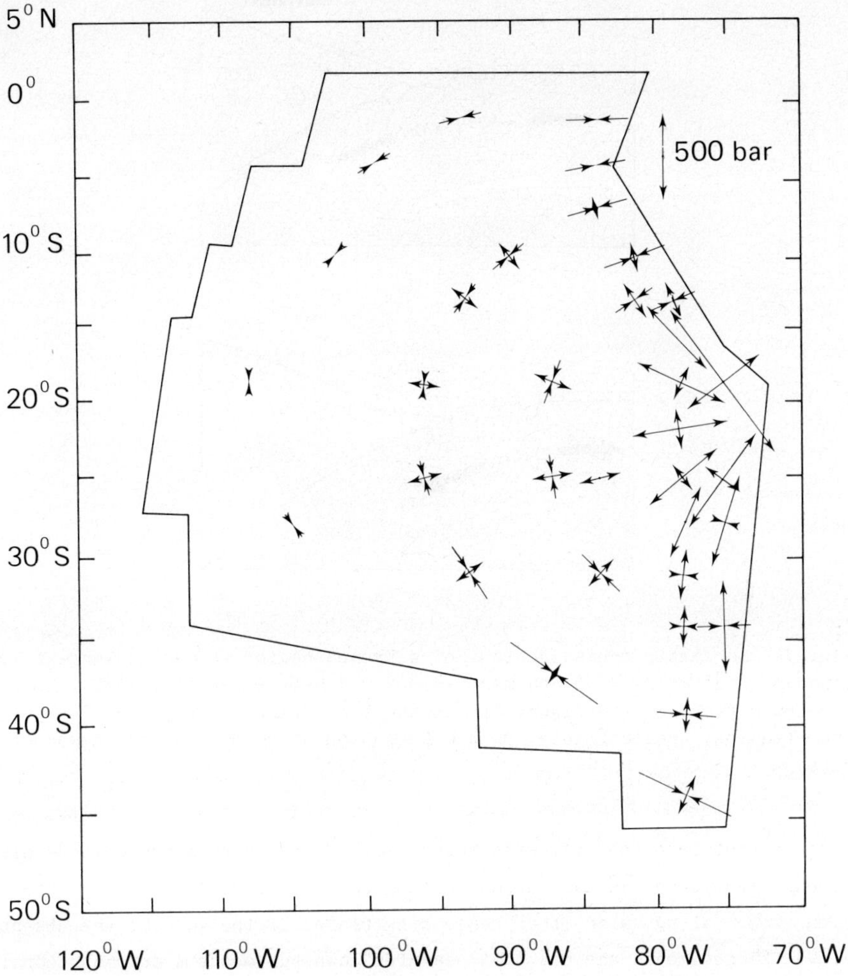

Fig. 6 - Regional stress field in the Nazca plate. The solid line indicates the boundaries of the Nazca plate. The (stress) scale is indicated in the upper right-hand corner. The stress values are displayed for a uniform elastic plate with a thickness of 100 km. After Wortel & Cloetingh (1985).

implying sediment accretion in the latitude ranges 6°-19.5°S (Peru) and 27°-45°S (Central-South Chile) and sediment subduction in the range 19.5°-27°S (North Chile).

These authors also noted a distinct correlation between graben development on the seaward trench slope and the fate of the sediments in the trench region (see Fig. 7). This led them (following Hilde and Sharman, 1978, see also Hilde,

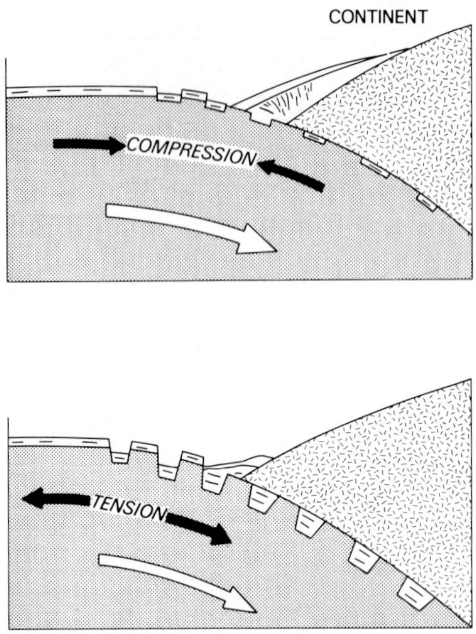

Fig. 7 - Schematic cross sections of a trench region at a convergent plate boundary (after Hilde & Sharman, 1978 and Schweller et al., 1981). The upper and lower parts of the figure illustrate the effect of regional compression and tension, respectively, on the formation of graben on the seaward trench slope.

1983), to postulate the following mechanism: If the grabens are well-developed, the trench sediments can be trapped in them and are carried down into the subduction zone along with the descending plate. If the grabens are absent (or shallow) the sediments are not (all) carried down but accrete onto the continent or island arc. Thus, the controlling factor would be the ratio between the transport volume of the grabens and the sediment supply, both per unit of time or rather averaged over a certain period of time.

Using finite element methods Wortel and Cloetingh (1985) investigated the physical background of the graben formation, that is the stress distribution in the downbending Nazca plate. This stress distribution is dominated by two contributions: stresses associated with the bending of the plate and the regional stress field (see Fig. 6).

To bring out the regional stress situation (Fig. 6) along the Peru-Chile Trench more clearly two stress components of the regional field are graphed in Fig. 8 for the near-trench region of the Nazca plate: $\sigma_{\perp}$ is the normal stress component in the direction normal to the trench (as a normal component it acts on a vertical plane parallel to the trench axis); $\sigma_{\parallel}$ is a normal stress component in the direction parallel to the trench axis, acting on a vertical plane perpendicular to the trench axis. The $\sigma_{\perp}$ and $\sigma_{\parallel}$ -curves are shown for three models A, B and C, which differ somewhat in the distribution of the resistive forces acting on the plate. All three models exhibit the same interesting feature (for $\sigma_{\perp}$: compression normal to the trench off Peru, tension normal to the trench off North Chile and again compression along the southern part of the trench system. Such lateral changes are a direct consequence of the variations in the net subduction force F_{nt} along the trench!

The right-hand side of Fig. 8 gives the division of the Peru-Chile Trench according to Schweller et al. (1981), based on the shallow structure of the trench axis and the seaward trench slope: in provinces 1,2 and 4,5 - where grabens are absent or poorly developed - accretion occurs, whereas in province 3, with deep grabens, sediments are subducted, without any significant amount of accretion. Note that the latitude range with tension normal to the trench (positive $\sigma_{\perp}$) coincides remarkably well with the extent of province 3, where the formation of grabens is most prominent. For all stress values in Fig. 8 (as well as in Figs. 5 and 6) it should be kept in mind that they are displayed for a uniform elastic plate with a reference thickness of 100 km.

Along the Peru-Chile Trench neither the curvature nor the age of the Nazca plate is constant. The age variations are adequately covered by considering three cross sections with oceanic lithosphere of 15 Ma (minimum age), 40 Ma and 70 Ma old (maximum age). In Fig. 8 (and in the inset in Fig. 9) it is indicated where these ages are encountered. In calculating the regional stress field (Figs 6 and 8) we were essentially interested in the stress integrated over the plate's thickness. For this purpose a uniform elastic plate is adequate. If one is interested in the stress variation with depth, however, as in the case of bending, a realistic depth-dependent rheology must be incorporated. Therefore, a depth- and temperature-dependent rheology (after Goetze and Evans, 1979) was implemented in the models. The variations in curvature of the plate are taken from Schweller et al. (1981). Fig. 9 shows the strength distribution and the stress distribution with depth for lithosphere of the three selected ages. Two parameters of the stress distributions are of particular interest to the formation of grabens: the depth of the neutral plane (D_{NP}) and the depth down to which brittle failure takes place (D_{BF}), that is the maximum depth at which the stress equals the brittle strength in the upper part of the plate. For comparison the stress distribution in plates with the same curvature, but without regional stress field, are also graphed in Fig. 9.

As the 15 Ma cross section represents only a small part of the entire trench, the differences between the 70 Ma and the 40 Ma cross sections are most relevant. It appears that the regional stress field strongly amplifies the rela-

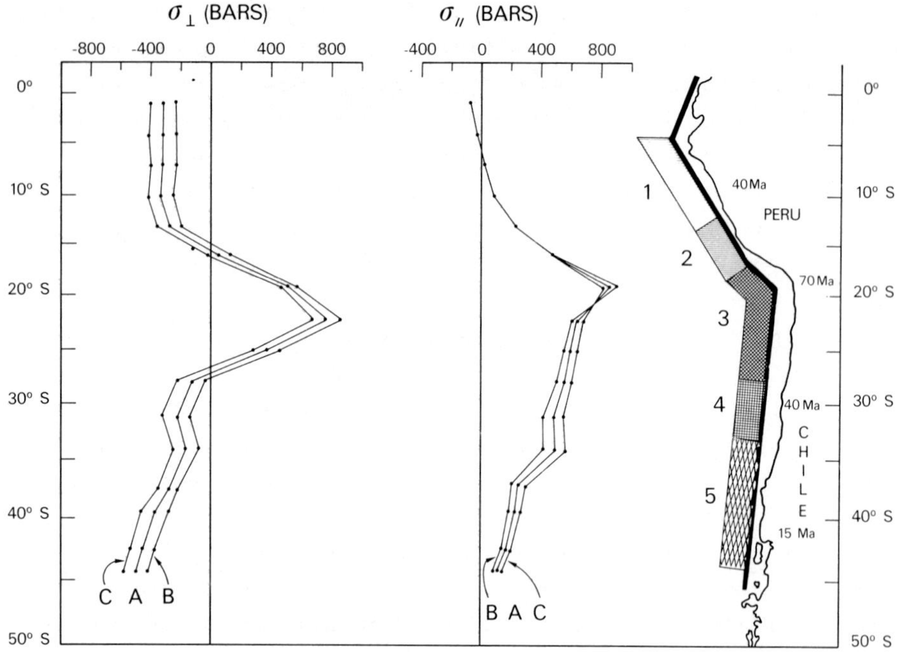

Fig. 8 - $\sigma_\perp$ - and $\sigma_{\parallel}$ - values for the region near the eastern boundary of the Nazca plate along the Peru-Chile trench. Positive values indicate tension, negative values compression. $\sigma_\perp$ is the normal stress component acting in the direction perpendicular to the trench axis on a vertical plane parallel to the trench axis. $\sigma_{\parallel}$ is the normal stress acting in the direction parallel to the trench axis on a vertical plane perpendicular to the trench axis. Values are given for a uniform elastic plate with a reference thickness of 100 km. The curves A, B and C are labelled according to the force model used in the calculation. Model A is the preferred model. $\sigma_{\parallel}$ is of interest to the possible fragmentation of the Nazca plate, along fracture zones oriented roughly perpendicularly to the trench axis (see Warsi et al., 1983). The right-hand side of the figure shows the division of the Peru-Chile trench into five provinces, as made by Schweller et al.(1981). The thick black line schematically indicates the trench axis. Extensional faulting and formation of grabens (on the seaward trench slope) are most prominent in province 3. After Wortel & Cloetingh (1985).

tively small differences in D_{BF} due to variation in curvature and (age-dependent) thermo-mechanical structure of the downbending plate: D_{BF} is 20 km and 15 km for the 70 Ma and 40 Ma sections without regional stress field, respectively, and 26 km and 12 km for the same sections with regional stress field. It is of course due to the change of sign in $\sigma_\perp$ (see Fig. 8) that the role of the regional field is so pronounced. Regional tension, superimposed on the tensional bending stresses in the upper part of the plate, promotes the for-

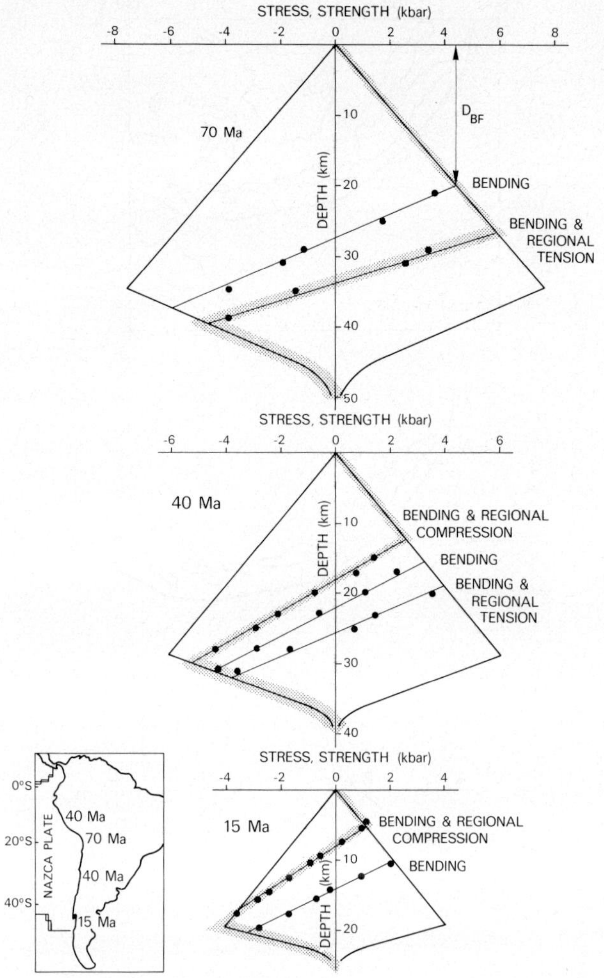

Fig. 9 - Strength distributions and stress distributions in cross sections through lithosphere of 70 Ma (top), 40 Ma (middle) and 15 Ma (bottom), respectively. Sign convention: positive values indicate tension, negative values compression. Stress and also strength refer to differential stress. Solid outer lines indicate the strength envelopes. Where the stress is less than the strength the dots represent the calculated stress values. The stress distribution with the shading are the ones appropriate for the conditions along the Peru-Chile Trench. The regional stress field is taken from Fig. 8 ($\sigma_{\perp}$, model A). In addition, the stress distributions for the same sections, but without regional stress field are graphed for comparison. The inset shows where lithosphere of the selected ages is encountered along the Peru-chile Trench. In this figure D_{BF} is indicated only for bending of 70 Ma old lithosphere without regional stress field.

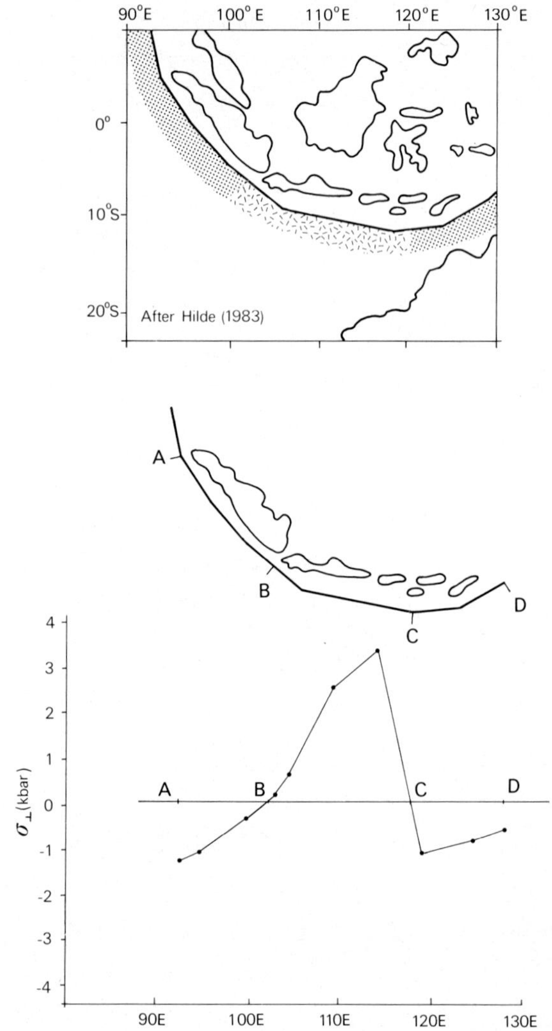

Fig. 10 - Top: Lateral variations in graben development in the seaward trench slope of the Indo-Australian plate subducting in the Indonesian trench system. Symbols and denote well-developed grabens in subducting plate and areas where graben development is not obvious (after Hilde, 1983). Bottom: Lateral variation of $\sigma_\perp$, the stress component normal to the trench axis. Sign convention: tension positive, compression negative.

mation of grabens. Regional compression has the opposite effect.

A (similar) analysis of the Indonesian subduction zone, based on the regional stress field calculated by Cloetingh and Wortel (1985, 1986), is in progress. Fig. 10 (bottom part) shows the stress normal to the trench axis ($\sigma_{\perp}$) for segments of the Indonesian zone. The top part of the figure, after Hilde (1983), indicates that grabens are well developed in the Java part of the trench system, whereas they are "not obvious" in the Sumatra part of the system. Again there is a close correspondence between the extent to which grabens are developed and the regional stress field, similar to the Peru-Chile Trench (see also Fig. 7).

It follows that the regional stress field plays an important role in controlling the topography of the seaward trench slope and, via the sediment trapping mechanism (Fig. 7), the process of of sediment accretion onto the continental margin. Taking into account the role of the regional lithospheric stress field in graben formation not only explains first-order latitudinal changes in accretion and subduction of sediments in the Peru-Chile Trench; it is also expected to contribute to understanding similar variations observed in other subduction zones.

TEMPORAL VARIATIONS IN SUBDUCTION DYNAMICS AND VERTICAL TECTONICS

In the two previous sections on fragmentation and accretion the emphasis was on spatial variations in the subduction related forces (as in Fig. 4). Through the stress field these spatial variations induce spatial variations in tectonic processes, such as accretion. It should be noted, however, that the forces associated with subduction may not only vary spatially but also in time. Underlying causes are changes in plate geometry and changes in the age of the lithosphere at convergent plate boundaries (see Wortel and Vlaar, 1978, and Wortel, 1984, for such changes in the South American subduction zone). The latter changes affect the geometry of the downgoing slab (see Figs. 2 and 3) and the most important driving forces, namely the ridge push and the slab pull. Temporal variations in subduction zone dynamics are reflected in the lithospheric stress field and may cause vertical movements.

Vertical motions and apparent sealevel fluctuations

Recently it was demonstrated (Cloetingh et al., 1985) that variations in regional stress fields are capable of producing rapid vertical movements of the lithosphere, or short-term apparent sealevel fluctuations, of a magnitude deduced from the stratigraphic record (Kominz, 1984). A transition from tension of 1 kbar to compression of the same magnitude produces a net uplift (or apparent sealevel fall) of up to about 40 metres at the edge of the sedimentary basin. The deflections of the lithosphere at other stress levels can be scaled in proportion to the magnitude of the applied horizontal stresses, with a stress field of a few hundred bars corresponding to a deflection of about 10 metres. The dependence of the vertical displacement on the age of the underlying lithosphere is demonstrated in Fig. 11. The differential uplift Δw, defined as the difference in deflection for the change in stress from tension to a compression of equal magnitude is computed for the edge of the basin as a function of variations in the in-plane stress. Fig. 11 demonstrates that variations in relative sealevel of at least ten metres can be caused by changes in the regional stress field of the order of a few hundred bars. If these stress changes occur on a timescale of 1 Ma, then the associated rates of sealevel change are on the order of at least 1 cm/1000 years. Therefore Cloetingh et al. (1985) were able to argue that glacial fluctuations (Pitman and Golovchenko, 1983) are not the only mechanism capable of producing apparent sealevel changes with a magnitude and rate inferred from the stratigraphic record.

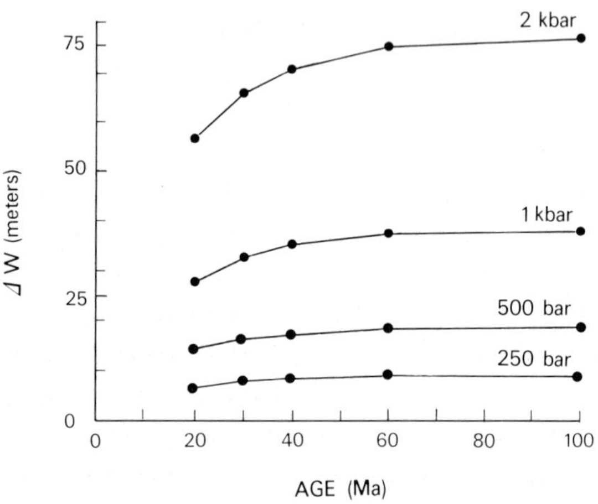

Fig. 11 - Differential uplift in metres at basin edge due to superposition of variations in intraplate stress field on flexure caused by sediment loading. The uplift is plotted as a function of the age of the underlying lithosphere. Results are given for a fixed stress transition from 2, 1, 0.5 and 0.25 kbar tension to 2, 1, 0.5 and 0.25 kbar compression, respectively.

Cloetingh et al.'s (1985) model also explains simultaneous fluctuations in apparent sealevel in neighbouring basin settings. Their tectonic mechanism for sealevel fluctuations explains the observed correlations (Bally, 1982) between the timing of changes in sealevel inferred from the seismostratigraphic record and tectonic events (see also Meulenkamp and Hilgen, 1986). In general short-term changes in apparent sealevel with magnitudes in excess of 50 metres must be related to major reorganizations at convergent plate boundaries, fragmentation of plates (Wortel and Cloetingh, 1983) and collision processes (Cloetingh and

Wortel, 1985). As such, the stratigraphic record provides a new source of information on paleostress fields (Cloetingh, 1985). Examination of individual basins in a wide range of tectonic settings, in connection with independent numerical modelling of paleostresses (Wortel and Cloeting, 1981) is required to fully exploit the potentials of this new approach in basins analysis.

Vertical motions associated with collision

A situation with significant temporal changes in subduction dynamics is encountered in the context of collision of plates (see Fig. 12). The resistive forces acting on a subducting slab mainly arise from frictional or shearing processes. They counteract the downdip component F_{sp} of the gravitational force F_{gr} acting on the slab (see Fig. 13). If the relative convergent motion of two plates comes to an end, for example in a situation depicted in Fig. 12, the resistive forces strongly decrease in magnitude or even vanish. Thus, upon cessation of plate convergence the gravitational force acting on a dipping slab effectively increases in magnitude.

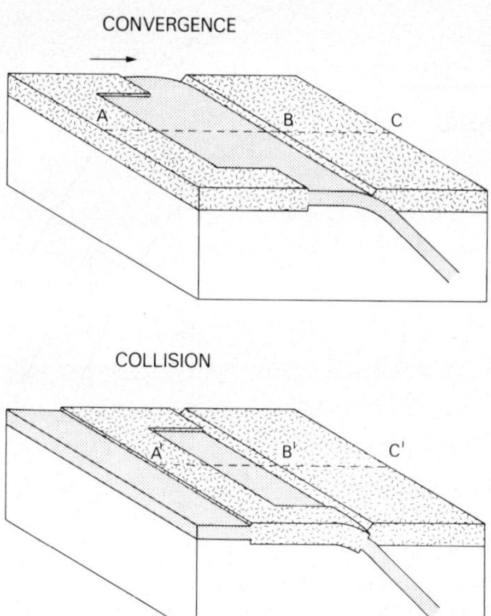

Fig. 12 - Schematic diagram of two convergent plates (top), with oceanic lithosphere being consumed. Collision (bottom) gives rise to a "land-locked" basin. No convergence takes place and the fate of the trapped oceanic lithosphere is controlled by its gravitational instability.

In a discussion of the dynamics of "land-locked" oceanic basins, such as shown schematically in Fig. 12, Le Pichon and Angelier (1981) and Le Pichon and

Alvarez (1984) emphasized the importance of roll-back, that is the vertical motion of the subducted slab under the action of gravitational forces (component F_{nm}), leading to an oceanward migration of the trench (to the left in Fig. 13). Le Pichon and co-workers focused on the back-arc region or the overriding plate (between B' and C' in Figs. 12 and 13). In this region roll-back induces extension, and hence subsidence.

Here we want to draw attention to implications of the new dynamic situation for the region near the plate contact (B' in Figs 12 and 13) and the area seaward of the trench (A'-B' in Figs. 12 and 13). Of interest here is that the flexure of the oceanic lithosphere not only migrates away from the former overriding plate but, in addition, changes in shape. The effective increase in gravitational force acting on the slab induces an increase in curvature of the plate, thereby producing a higher "bulge", seaward of the trench.

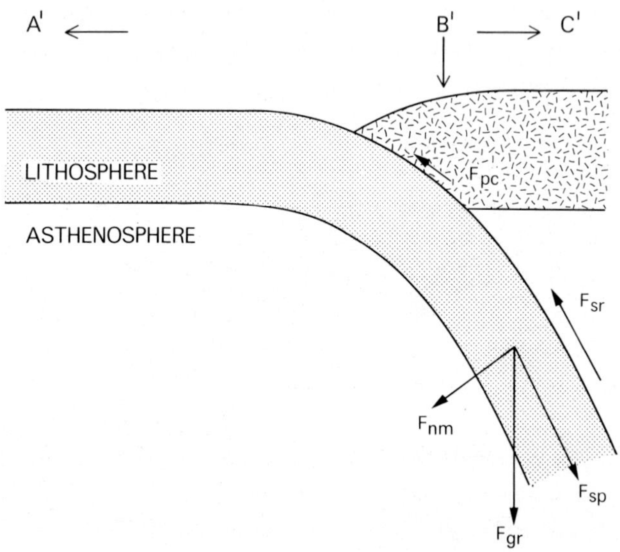

Fig. 13 - Cross section along line A'B'C' in Fig. 12. When convergence comes to an end, the velocity-dependent resistive forces F_{pc} and F_{sr} are reduced or vanish. Thus, the gravitational force on the slab effectively increases. This causes a change in shape of the bulge of the downbending lithosphere, seaward of the trench.

Preliminary calculations have shown that, depending on the age of the trapped oceanic lithosphere, vertical movements may be induced over distances of 100-200 km from the trench (between A' and B', Figs 12 and 13). Near the plate contact this may have serious tectonic effects (vertical motion, tilting), whereas farther away from the plate contact the magnitude may still be relevant in the context of sealevel fluctuations (see also Cloetingh, 1986).

Of special interest in this situation are the timescales involved. The induced vertical motion is originally controlled by the process of collision. The decay of the induced motion, however, is determined by the processes which govern the evolution of the subducted slab after termination of subduction. Primarily these processes are of thermal origin, which implies a timescale in the order of 10 Ma or several tens of million years. This is likely to be longer (but not necessarily so) than the timescale of the process which caused the cessation of subduction.

CONCLUSIONS

It is shown that the idea of a compensated slab pull force is not realistic. Partial or over-compensation of the slab pull force implies a net force transmitted from the subducted slab to the lithospheric plate to which the slab is attached. This net force, and in particular the way it varies along the convergent plate boundary, strongly affects the regional stress field in the lithosphere. This spatially varying stress field is subject to temporal variations, through the dependence of the main plate tectonic (driving) forces on the age of the lithosphere near the trench. Taking into account of the slab pull being uncompensated and its effect on the lithospheric stress field sheds light on several, hitherto unexplained spatial and temporal variatons in tectonic processes (accretion and subduction of sediments, vertical motions near active continental margins, plate fragmentation).

ACKNOWLEDGEMENT

We thank Professor N.J. Vlaar for stimulating cooperation and encouragement, and for significant contributions to the research described in this paper. Gerald Wisse (Delft University of Technology) is thanked for support with the finite element calculations.

REFERENCES

Backus, G., Park, J. and Garbasz, D., 1981. On the relative importance of the driving forces of plate motion. Geophys. J. R. Astron. Soc., 67: 415-435.

Bally, A.W., 1982. Musings over sedimentary basin evolution. Phil. Trans. R. Soc. London, A305: 325-338.

Chapple, W.M. and Tullis, T.E., 1977. Evaluation of the forces that drive the plates. J. Geophys. Res., 82: 1967-1984.

Cloetingh, S.A.P.L., 1982. Evolution of passive continental margins and initiation of subduction zones. PhD-Thesis, University of Utrecht, 111 pp.

Cloetingh, S., 1985. Intraplate stresses as a mechanism for sealevel variations. EOS, Trans. Am. Geophys. Un., 66: 1060.

Cloetingh, S., 1986. Tectonics of passive margins: implications for the stratigraphic record. Geol. Mijnbouw, in press.

Cloetingh, S., McQueen, H., and Lambeck, K., 1985. On a tectonic mechanism for regional sealevel variations. Earth Planet. Sci. Lett., 75: 157-166.

Cloetingh, S. and Wortel, R, 1985. Regional stress field of the Indian plate. Geophys. Res. Lett., 12: 77-80.

Cloetingh, S. and Wortel, R., 1986. Stress in the Indo-Australian plate. Tectonophysics, in press.

Elsasser, W.M., 1967. Convection and stress propagation in the upper mantle. Princeton Univ. Tech. Rep. 5, 65 pp.

England, P. and Wortel, R., 1980. Some consequences of the subduction of young lithosphere. Earth Planet. Sci. Lett., 47: 403-415.

Forsyth, D. and Uyeda, S., 1975. On the relative importance of the driving forces of plate motion. Geophys. J. R. Astron. Soc., 43: 163-200.

Goetze, C. and Evans, B., 1979. Stress and temperature in the bending lithosphere as constrained by experimental rock mechanics. Geophys. J. R. Astron. Soc., 59: 463- 478.

Hey, R., 1977. Tectonic evolution of the Cocos-Nazca spreading center. Geol. Soc. Am. Bull., 88: 1404-1420.

Hilde, T.W.C., 1983. Sediment subduction versus accretion around the Pacific. Tectonophysics, 99: 381-397.

Hilde, T.W.C. and Sharman, G.F., 1978. Fault structure of the descending plate and its influence on the subduction process. EOS, Trans. Am. Geophys. Union, 59: 1182.

Kominz, M., 1984. Oceanic ridge volumes and sea-level change - an error analysis. Am. Assoc. Petrol. Geol. Memoir, 36: 109-127.

Le Pichon, X. and Angelier, J., 1981. The Aegean Sea. Phil. Trans. R. Soc. London, A300: 357-372.

Le Pichon, X. and Alvarez, F., 1984. From stretching to subduction in back-arc regions: dynamic considerations. Tectonophysics, 102: 343-357.

Lister, C.R.B., 1975. Gravitational drive on oceanic plates caused by thermal contraction. Nature, 257: 663-665.

McKenzie, D.P., 1969. Speculations on the consequences and causes of plate motion. Geophys.J. R. Astron. Soc., 18: 1-32.

McKenzie, D.P., 1970. Temperature and potential temperature beneath island arcs. Tectonophysics, 10: 357-366.

Meulenkamp, J.E. and Hilgen, F.J., 1986. Event stratigraphy and tectonics of the Hellenic and Calabro-Sicilian arcs. In: F.C. Wezel (Editor), this volume.

Oxburgh, R. and Parmentier, E.M., 1977. Compositional and density stratification in oceanic lithosphere. J. Geol. Soc. London, 133: 343-355.

Pitman III, W.C. and Golovchenko, X., 1983. The effect of sealevel change on the shelfedge and slope of passive margins . Soc. Econ. Paleont. Mineral. Spec. Publ., 33: 41-58.

Ruff, L. and Kanamori, H., 1980. Seismicity and the subduction process. Phys. Earth Planet. Inter., 23: 240-252.

Schweller, W.J., Kulm, L.D. and Prince, R.A., 1981. Tectonics, structure and sedimentary framework of the Peru-Chile Trench. Geol. Soc. Am. Memoir, 154: 323-349.

Vlaar, N.J., 1975 The driving mechanism of plate tectonics. In: Borradaile, G.J. et al. (Eds.), Progress in Geodynamics. North-Holland, Amsterdam, pp. 234-245.

Vlaar, N.J. and Wortel, M.J.R., 1976. Lithospheric aging, instability and subduction. Tectonophysics, 32: 331-351.

Warsi, W.E.K., Hilde, T.W.C. and Searle, R.C., 1983. Convergence structures of the Peru Trench between 10°S and 14° S. Tectonophysics, 99: 313-329

Wortel, M.J.R., 1980. Age-dependent subduction of oceanic lithosphere. PhD-Thesis, University of Utrecht, 147 pp.

Wortel, R., 1982. Seismicity and the rheology of subducted slabs. Nature, 296: 553-556.

Wortel, M.J.R., 1984. Spatial and temporal variations in the Andean subduction zone. J. Geol. Soc. London, 141: 783-791.

Wortel, R., 1986. Deep earthquakes and the thermal assimilation of subducting lithosphere. Geophys. Res. Lett., 13: 34-37.

Wortel, R. and Cloetingh, S., 1981. On the origin of the Cocos-Nazca spreading center. Geology, 9: 425-430.

Wortel, R. and Cloetingh, S., 1983. A mechanism for the fragmentation of oceanic plates. AAPG Memoir, 34: 793-801.

Wortel, M.J.R. and Cloetingh, S.A.P.L., 1985. Accretion and lateral variations in tectonic structure along the Peru-Chile Trench. Tectonophysics, 112: 443-462.

Wortel, M.J.R. and Vlaar, N.J., 1978. Age-dependent subduction of oceanic lithosphere beneath western South America. Phys. Earth Planet. Inter., 17: 201-208.

PALEOMAGNETISM IN ARCUATE MOUNTAIN BELTS

W. LOWRIE and A. M. HIRT
Institut für Geophysik, ETH-Hönggerberg, 8093 Zürich, Switzerland

ABSTRACT

A selection of paleomagnetic investigations in different fold belts shows that the data can only be reliably used for evaluation of curvature of the mountain arc when (a) they have been determined by the most rigorous laboratory and analytical techniques and (b) there is detailed information about the local tectonic setting of each sampling site. Early paleomagnetic results from Japan must be discounted because the samples were not demagnetized; more recent studies appear to imply deformations within Japan that are not observed in the field. Application of paleomagnetism in remote regions with difficult access like the Himalaya and the Antarctic is handicapped by the tectonic complexity of the regions and sparseness of available data. In high latitudes, moreover, large declination uncertainties reduce the interpretative value of the declination data. When the paleomagnetic and structural geological conditions are fulfilled, linear regression analysis can be applied to paleomagnetic declinations and the corresponding fold axis trends. The slope of the regression line can be tested with a t-statistic against zero slope (no oroclinal bending) or any chosen value, such as unit slope (originally straight fold belt). Paleomagnetic data suggest that oroclinal bending may have formed the Iberian arc in northern Spain, but they do not support oroclinal bending as the reason for the observed curvature of the Swiss Jura mountains or the Umbrian-Marches Apennines in Italy.

INTRODUCTION

It is a matter of prime geological interest to know whether the observed curvature of a mountain arc is an original feature or the result of oroclinal bending, by which is understood the tectonically induced change of curvature of an originally straight or arcuate mountain belt (Carey, 1958). Paleomagnetic data provide a suitable basis for analysing such changes. The inclination of the stable component of the paleomagnetic vector can often be interpreted in terms of north-south translation of a region. The declination may be compared to a reference value and any significant difference taken as indication of tectonic rotation about a vertical axis. Obviously, if conclusions are to be made about tectonic deformations in which the basement is involved, the cover rocks which are investigated paleomagnetically must be autochthonous. Nevertheless, paleomagnetic data can also give useful clues to the tectonic history of para-autochthonous or allochthonous domains. For example, by establishing part of an apparent polar wander (APW) path in the southern Tethyan limestone sequences of the Umbrian Apennines VandenBerg et al. (1978) were able to demonstrate convincingly the African affinity of the probably allochthonous sedimentary formations in this part of central Italy.

Paleomagnetic data can provide the structural geologist with very useful information. The paleomagnetist, on the other hand, is dependent on the structural geologist for field geological control. The optimum results are to be expected when this symbiotic relationship is fully developed. Unfortunately, in some cases paleomagnetic directional data have been interpreted without adequate consideration of the structural geological implications. This has occasionally led to tectonic inferences that contradicted geological field observations. Consequently, some geologists have become sceptical of any use of paleomagnetic data. Yet, paleomagnetic directions which have been properly derived should not be ignored or lightly dismissed. They can constitute valuable geological information that support or contradict theories of the tectonic evolution of a region. This requires that certain fundamental conditions be fulfilled.

The first condition is that the local tectonics of the region of interest should be well enough known so that the established paleomagnetic vector can be corrected as necessary for change of the original horizontal position (tectonic tilt, or bedding correction), including correction for possible plunge of the fold axis. Various errors leading to false interpretation of the paleomagnetic signal can arise when the bedding correction is incomplete (MacDonald, 1980). The geological data should be measured directly in the field but it is often difficult to derive the plunge and trend of fold axes without extensive observations near each site. In well studied areas high quality, detailed tectonic maps often give these data adequately.

The second, important requirement is that the rocks used for a test of the oroclinal bending hypothesis must possess a stable remanent magnetization which was acquired prior to the tectonic event. This excludes rocks that have suffered extensive distortional internal deformation during tectonism which might have disturbed the primary remanence to a degree that can not be compensated. Distortional deformation can also alter the bedding attitude. In studies of paleomagnetism in strained rocks the restoration of bedding to its original horizontal attitude can not be achieved by the usual simple tilt correction (rigid-body rotation) alone, but requires additional corrections for the effects of strain on bedding and the remanent vector (Cogné and Perroud, 1985; Lowrie et al. 1986). Secondary remagnetizations are also excluded in most circumstances. The derivation of paleomagnetic results that are sufficiently well defined for use in tectonic interpretation places high standards on the methodologies used in the determination and analysis of the remanent magnetization. Most modern studies meet these requirements but some notable earlier studies did not.

A third requirement is that the ages of the formations investigated should be known as well as possible so that the directions can be compared to reference values. If the paleomagnetic declination varies along a mountain arc it is important to know how much of the variation is due to apparent polar wander during the time span represented by the formations in the study. This variation must be removed before the declinations can be interpreted in terms of subsequent tectonic rotations.

Declination data can only be interpreted meaningfully in this way when the corresponding inclinations are consistent. For example, Ries et al. (1980) deduced oroclinal bending of the Variscan orogenic belt in the Cantabrian region of northern Spain from the declination variation of paleomagnetic vectors. However, the associated inclination data are badly scattered, the stability of the remanent magnetization isolated by thermal cleaning is only minimally documented, and the adequacy of the remanence characteristics for tectonic interpretation is suspect. An interpretation of declinations where the inclinations are inconsistent loses credibility.

In this paper examples are given of paleomagnetic analyses of the arcuate shape of some folded mountain belts. In the first part, studies in three complex orogenic belts are described to illustrate problems that can arise when the paleomagnetic data are sparse and are not accompanied by detailed structural geological control. The results can only be interpreted in a broad sense as suggestive of the tectonic history. In the second part a method of statistical analysis is discussed which allows more rigorous interpretation of paleomagnetic declinations in terms of oroclinal bending when the fold axis azimuth is known at each site.

LARGE SCALE TECTONIC INTERPRETATIONS

Formation of the Japanese arc

Kawai et al. (1961) found that the directions of natural remanent magnetization (NRM) of rocks of pre-Tertiary age from southwest Japan differed systematically from those from northeast Japan. A declination difference of 40° suggested bending of the axis of the Japanese islands by this amount (Fig. 1a). The paleomagnetic data in this study were not subjected to the more rigorous demagnetization procedures and vector analysis techniques employed in modern investigations. More recently, using thermally demagnetized data from different rock types of Cretaceous to Middle Miocene age dated by the fission-track method, a clockwise rotation of 58° has been inferred for southwest Japan (Otofuji and Matsuda, 1983, 1984). An anticlockwise rotation of 47° was deduced for northeast Japan from igneous rocks of similar age, dated by fission-track

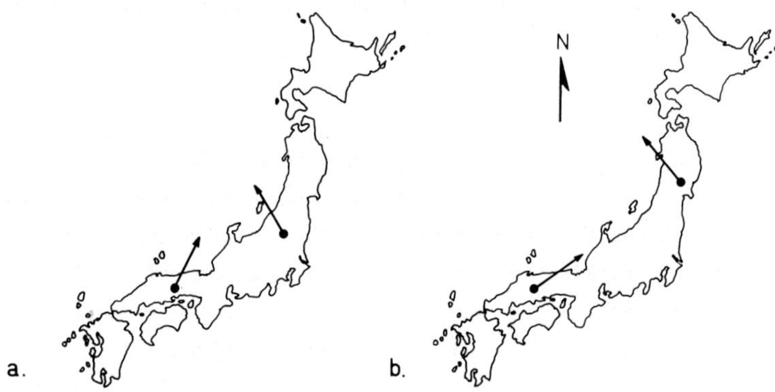

Fig. 1. Representative paleomagnetic declinations from northeast and southwest Japan; reported by (a) Kawai et al. (1961) and (b) Otofuji et al. (1985a).

and K-Ar methods (Otofuji et al., 1985b). In these studies pilot samples were subjected to detailed progressive demagnetization and vector analysis, but the major portion of the samples were then demagnetized at an 'optimum' temperature determined from the pilot studies. This is not as satisfactory as when progressive demagnetization and vector analysis are applied to each sample.

The newer paleomagnetic data imply a relative rotation of more than 100° between northeast and southwest Japan, much greater than the 40° implied by the earlier undemagnetized data (Fig. 1b). Otofuji et al. (1985a) formulated an interpretative model in which the 3,000 m isobath margins of the north and south segments of the Japanese arc were fitted against the Asian continental margin (Fig. 2a). This model claims that Japan achieved its present position and attitude by double fan-shaped spreading in the Sea of Japan between 21 and 11 Ma ago (Fig. 2b).

Lallemand and Jolivet (1986) were strongly critical of this model, of the continental margin fitting and the implications of the paleomagnetic data. Based on a synthesis of structural data enlarged by marine geological information from the Japan sea, they proposed separation of Japan from an earlier position against the Asian continent by a succession of pull-apart rifting and back-arc spreading (Fig. 2c). This did not involve as much internal rotation within Japan as implied by the paleomagnetic data. Field observations on land give no indication of the serious deformation that would result from the paleomagnetic model, and the agreement of structural directions in the north and south argues against major bending of Honshu (Lallemand and Jolivet, 1986).

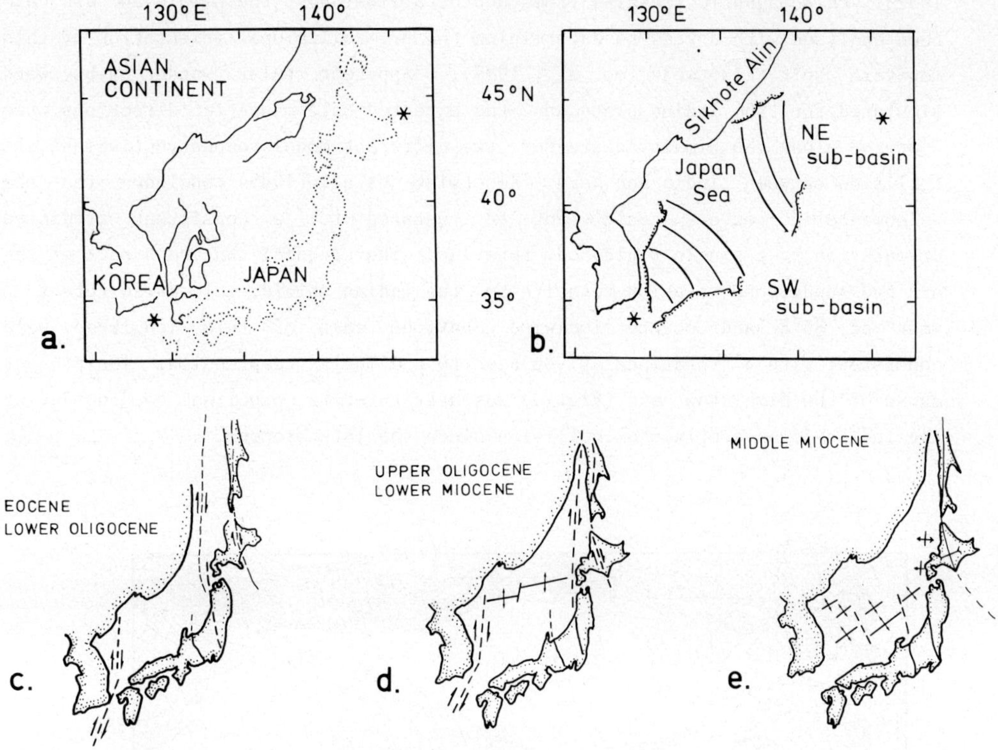

Fig. 2. Contrasting models proposed for the evolution of the Japanese arc: (a,b) double fan-shaped opening of the Japan Sea (Otofuji et al., 1985a) and (c-e) pull-apart rifting and back-arc spreading (Lallemand and Jolivet, 1986).

This disagreement between paleomagnetic and structural geological results can not be readily explained. The field geologists, therefore, dismiss the paleomagnetic results (Lallemand and Jolivet, 1986). This situation indicates the need for a concerted re-evaluation. In particular, it seems important that strict rejection criteria based on complete demagnetization of each sample be applied to the paleomagnetic data, and that detailed structural geological control be obtained at each site.

Curvature of the Himalayan Arc

The deformation of the Himalayan-Tibetan region has been modelled by the indentation of a hard, northward-moving Indian plate into a plastic Asian plate (Molnar and Tapponier, 1975). The subsequent evolution of southeast Asia was described by a 'propagating extrusion tectonics' model (Tapponnier et al.,

1982). Paleomagnetic results from southern Tibet and the Himalayan arc have been analysed with a view to determining the pre-collisional orientation of this mountain belt (Klootwijk et al., 1985). Apparent polar wander paths were simulated for the Indian plate and the expected paleomagnetic directions were computed along the Himalayan arc for two different model reconstructions of the collision between India and Asia. Klootwijk et al. (1985) concluded that the paleomagnetic vectors could not be restored to a consistent predicted orientation by a single rigid-body rotation. They deduced that each part of the arc had undergone rotation relative to the Indian shield; clockwise rotations west of 85°E and counterclockwise rotations east of this longitude were consistent with a rotational pivot near to the Turfan depression. The present curve of the Himalayan arc (Fig. 3) was attributed to oroclinal bending during the Indian-Asian continental collision since the late Miocene.

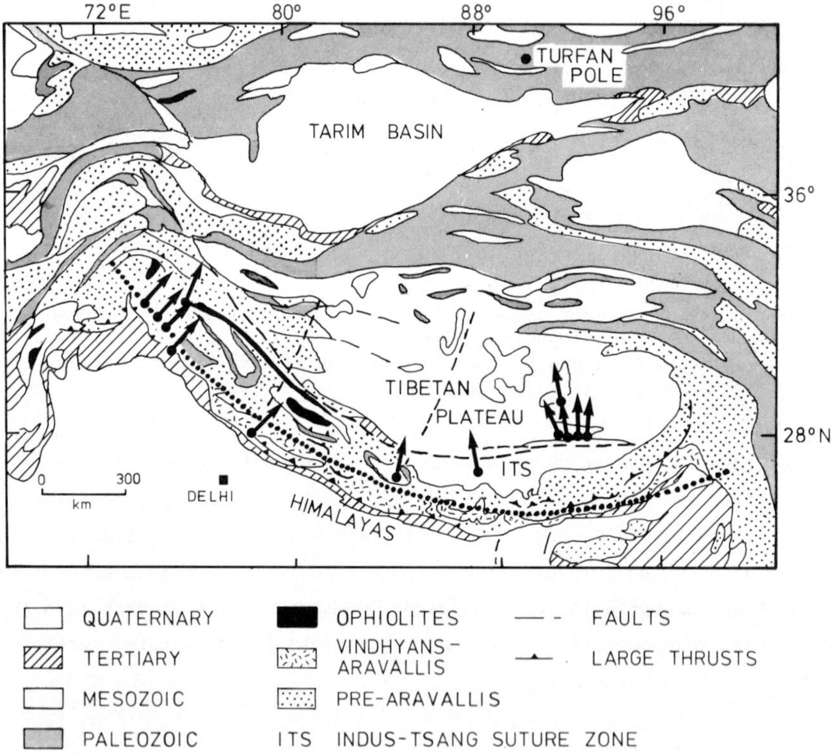

Fig. 3. Paleomagnetic declinations normal to the regional trend of the Himalayan arc (heavy dotted line) are consistent with oroclinal bending about a rotational pivot near the Turfan depression (after Klootwijk et al., 1985).

This tectonic interpretation is made in very broad terms and is based on comparatively few paleomagnetic data distributed thinly over a huge region. However, the agreement is promising and justifies additional investigations to see if the conclusions are borne out by more intensive regional coverage of the paleomagnetic and field geological data.

Southern Andes and Antarctic Peninsula

A similar situation where small paleomagnetic collections have been used to interpret the tectonic history of large regions is encountered in the southern Andes and the Antarctandes. Because of the difficult access to these remote regions, paleomagnetic and field geological data are very sparsely distributed. These problems are compounded by an intrinsic property of the earth's dipole magnetic field. When rocks are magnetized in high paleolatitudes, the inclination of the remanent magnetization is very steep and the declinations have large errors associated with them. If a paleomagnetic direction with inclination I has confidence circle with radius a_{95}, the declination error (dD_{95}) is given by:

$$\sin(dD_{95}) = \sin(a_{95})/\cos(I).$$

Demarest (1983) discussed the conventional methods of computing declination uncertainty and pointed out that this formula overestimates the error. However, it is adequate to illustrate the difficulty encountered in analyzing paleomagnetic data from the southern Andes and Antarctic peninsula. At these high latitudes, paleoinclinations are usually steeper than about 65°. For paleomagnetic data with a_{95} less than 15° the declination has an uncertainty range of about 60°.

Dalziel et al. (1973) found that the declinations measured in igneous intrusive rocks of late Mesozoic age varied around the 90° bend of the outer Magellan arc in the southern Andes by approximately the same amount as the change in structural trend. Burns et al. (1980) found evidence in Jurassic and Cretaceous igneous rocks from the inner Magellan arc for a much smaller bending of around 40° to 27°. From this they deduced that the Mesozoic arc was originally curved at the time of magnetization of the rocks. These southern Andean results show much between-site variability and the declinations have large uncertainties. Van der Voo and Channell (1980) concluded that (a) the paleomagnetic data neither supported nor contradicted oroclinal bending, and (b) the density of paleomagnetic coverage was inadequately low for the purpose of describing oroclinal bending in such a highly deformed region.

Paleomagnetic data from the Antarctic peninsula are also sparse and suffer from the same handicap of very steep inclinations. The interpretation of the data is less ambiguous, however. Consistent paleomagnetic declinations along the Antarctic peninsula give no support for oroclinal bending (Dalziel et al., 1973; Kellogg and Reynolds, 1978). The suggestion that paleomagnetic data from the Orville coast indicate oroclinal bending in the southern part of the S-shaped peninsula (Kellogg, 1980) has been subsequently discounted. Watts et al. (1984) attributed the difference in declinations between the Orville and Lassiter coasts to apparent polar wander and different magnetization ages of the poorly dated intrusives in these two parts of the peninsula. This emphasizes again the need for good age control in the rocks studied, and the importance of having satisfactory reference directions to take into account the effects of apparent polar wander.

DETAILED ANALYSIS OF CURVATURE IN MOUNTAIN BELTS

In the above examples, the relationships between paleomagnetic data and structural trends were evaluated in very broad terms. Sometimes, as in the Himalayan arc, southern Andes or Antarctic peninsula, the geology of the area is not yet well enough known to permit more detailed analysis. In contrast, when well-dated rocks are studied from a region that has good structural geological control for each site, more rigorous evaluation is possible. This is especially useful when the trend of the fold axis is known at each site (Fig. 4a).

Ries et al. (1980) illustrated several ways in which the paleomagnetic declination (D) could change with fold axis trend (F) along a curved mountain belt as a result of oroclinal bending. They tested the relationship in the Iberian arc by plotting the angle between fold axis and declination against fold axis trend for each of three sets of rocks of Cambrian, Silurian-Devonian and early Carboniferous age, respectively. Unfortunately, the analysis was made on undemagnetized data of dubious quality; no valid conclusions can be made from this data set.

Schwartz and Van der Voo (1983), in an evaluation of oroclinal bending in the Appalachians, analyzed the deviations in declination ($dD = D_r - D$) and fold axis trend ($dF = F_r - F$) from corresponding reference values (D_r, F_r), for which they used the appropriate regional means. This is acceptable for a study in which rocks of a single age are involved. However, if the samples represent an age span in which significant apparent polar wander may have occurred, the declination reference value (D_r) should be computed for each age from the appropriate APW path (Eldredge et al., 1985). The expected relationships between the declination and fold axis deviations are illustrated in Fig. 4(b-d).

If a mountain arc develops by bending of an originally straight folded range, the stable pre-bending declinations will change by the same amount as the fold

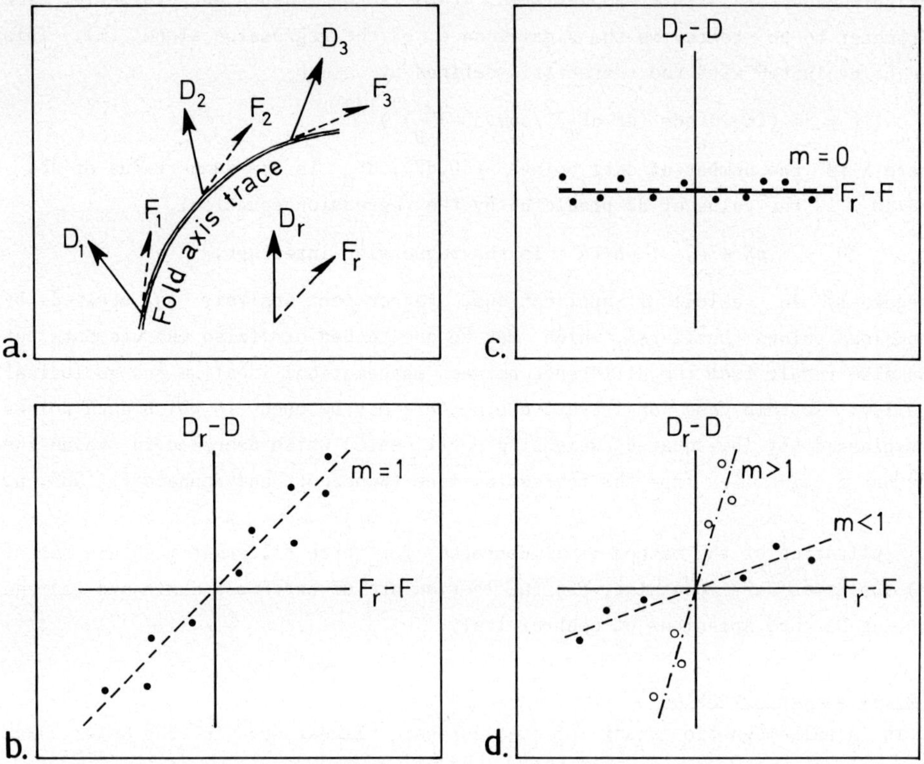

Fig. 4. (a) Arcuate mountain belt in which paleomagnetic declinations (D) are measured at sites where the fold axis trend (F) is known. D_r and F_r are the respective reference values.
(b-d) Plots of declination deviation (D_r-D) and fold axis deviation (F_r-F). The significance of the slope (m) indicates oroclinal bending in (b) and (d); lack of significance in (c) suggests the observed arc is a primary feature.

axis trend at each site; the deviations give a plot with unit slope (Fig. 4b). On the other hand, if the mountain arc originally had the same curvature as at present (or if the magnetizations post-date the bending) there will be no appreciable change of declination along the arc; the deviations give a plot with zero slope (Fig. 4c). It may be that the arc formed with an original curvature that was subsequently tightened or opened by later oroclinal bending. In these cases the gradient of the deviations plot is less than unity if the bend has been tightened and is greater than unity if the bend has been opened (Fig. 4d). Each situation described by a significantly non-zero gradient corresponds to some degree of oroclinal bending.

The occurrence of oroclinal bending can therefore be tested by linear regression analysis, in which declination deviation (dD) is the dependent variable and fold axis trend deviation (dF) is the independent variable. The parameter to be tested is the significance of the regression slope (m). This can be evaluated with the t-statistic defined by

$$t = m \left((N-2)\ \mathrm{sum}\ (dF-dF_m)^2/\mathrm{sum}(dD-dD_p)^2\right)^{1/2}$$

where N is the number of data points (dD,dF), dF_m is the mean value of dF, and dD_p is the value of dD predicted by the regression equation:

$dD_p = m\ dF + c$, where c is the regression intercept.

Frequently in geological applications, regression analysis is spoiled by anomalous points (outliers), which may be due to bad or false measurements but can also result from the difference between mathematical idealism and geological reality. In this case an iterative procedure may be used in which data points are biassed by 'bi-squared' weighting coefficents which decrease in value the further a point lies from the regression line (Wonnacott and Wonnacott, 1985, p. 327).

Application of the method is illustrated for three different mountain belts: (a) the Swiss Jura mountains, (b) the Iberian arc of northwest Spain and (c) the Umbrian-Marches Apennines of central Italy.

The Swiss Jura Mountains

In a paleomagnetic study of the Jurassic limestones in the Swiss Jura mountains Johnson et al. (1984) had difficulties with late secondary remagnetization components but found stable primary directions which gave a positive fold test and defined a European Jurassic pole position. Eldredge et al. (1985) used data from 24 sites distributed over half the length of the mountain belt in an analysis of the origin of the curvature. At each site the declinations were compared with the strike of bedding. The implicit assumption of cylindrical folds, although reasonable for long anticlinal structures, may not be adequate for the box-shaped folds found in the Swiss Jura (Trümpy, 1980). The correlation coefficient was found to be low and the slope of the regression line was much flatter than expected for oroclinal bending.

We have further evaluated the data used by Eldredge et al. (1980, Table 2) by extending the analysis to test the regression slope. The simple regression line has a slope of 0.159 and comparison with zero slope gives a t-statistic of 0.74, which is less than the test-value of 2.07 for the slope to be significant at the 95% level. Therefore, as deduced by Eldredge et al. (1980), the Jurassic limestone paleomagnetic declinations do not support oroclinal bending as the origin of curvature in the Swiss Jura mountains.

The Iberian Arc

Undemagnetized data from 32 sites around the Iberian arc in northwest Spain were interpreted by Ries et al. (1980) as evidence that the arc is a secondary feature, formed by bending of an originally straighter fold-belt during the Variscan orogeny. Although they also listed thermally demagnetized directions for 20 sites, they published little documentation of the magnetization stability. Demagnetization increased the within-site directional dispersion at 14 of the sites, and at more than half the sites the demagnetized directions were more than 10° different from the undemagnetized values. The interpretation of the undemagnetized data is unjustifiable. Moreover, Ries et al. (1980) neglected the mixed polarities indicated by positive and negative inclinations. They incorrectly selected the declination nearest to the fold axis regardless of the inclination of the magnetization.

The demagnetized data have been re-evaluated for this paper (Table 1). Large within-site confidence circles at many sites after demagnetization indicate that they can not be used; sites with radius of confidence larger than 15° are not considered further. Sites where the declination makes an angle of more than 90° with the fold axis were also rejected. Paleomagnetic data from only 10 sites are retained to test the orocline hypothesis (Fig. 5a). Only three sites are not Upper Silurian to Lower Devonian in age, so a single reference declination has been used. A t-test of the slope (m=0.696) of the regression line (Fig. 5b) compared to zero slope gives t = 5.18. This is greater than the null hypothesis value, t = 2.31, and suggests that the data from the 10 sites indicate some degree of oroclinal bending. A t-test against unit slope gives t = 2.25, which is marginally smaller than the reference value and indicates that the observed slope does not differ significantly from unity. Thus, the few accepted data suggest that the arcuate shape of the Iberian arc is the result of oroclinal bending of a perhaps originally straight fold belt. We have no quality control over the 10 points which are also too few to support the conclusions strongly. A new study is needed to verify the findings.

The Umbrian Apennines

The most extensively studied rocks from this region are the middle Cretaceous to Eocene Scaglia limestones. In magnetostratigraphic studies Lowrie and Alvarez (1977) noted a marked change of magnetic declination with stratigraphic level in the Scaglia Rossa limestone. Similar observations were interpreted by VandenBerg et al. (1978) to be the result of apparent polar wander during deposition. In a paleomagnetically based investigation of tectonics in Umbria, Channell et al. (1978) attempted to compensate for the declination variation

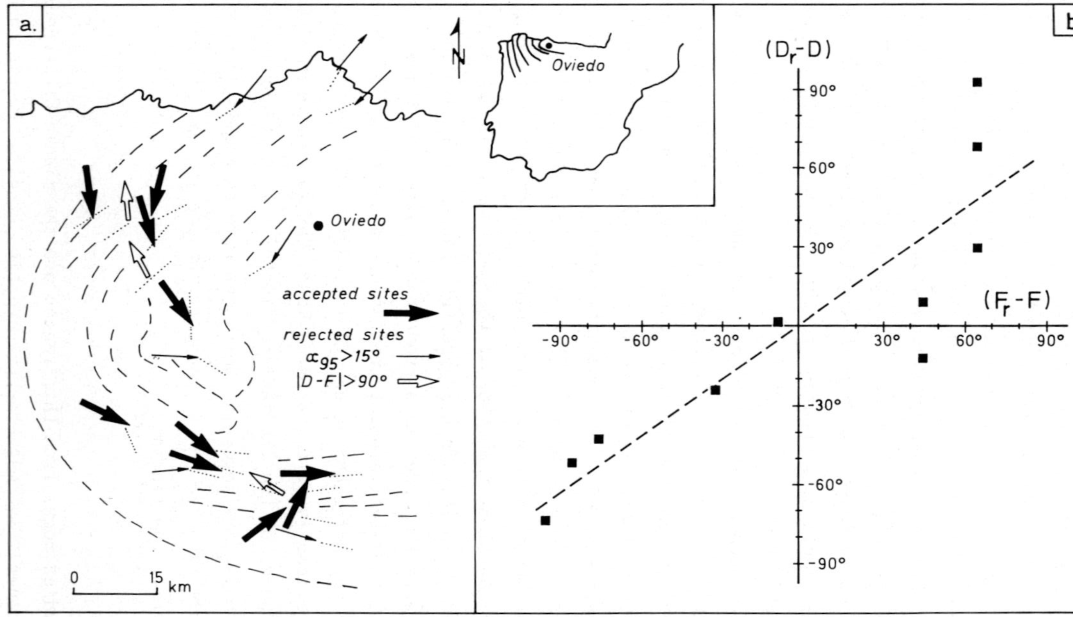

Fig. 5. (a) Paleomagnetic declination at 10 of the sites in the Iberian arc studied by Ries et al. (1980). Dotted lines are measured fold axes and dashed lines are schematic. (b) The regression slope is not significantly different from unity, suggesting oroclinal bending may have formed the arc.

TABLE 1

Statistical analysis of deviations of declinations and fold axis trends from regional mean values in the Iberian arc. Re-analysis of demagnetized data of Ries et al. (1980) using the method of Eldredge et al. (1985). Data were rejected because of: (a) a_{95} more than 15°, (b) (D-F) more than 90°.

Site	a_{95}	Paleomagnetic Direction Incl	Dec1	Fold Axis F	(F_r-F)	(D_r-D)**
Middle Cambrian						
1	17	rejected (a)				
4	14	-13	121 (301)	107	rejected (b)	
50	15	-39	150 (330)	228	rejected (b)	
58	14	32	169	231	-84	-52
60	11	39	141	178	-31	-24
61	20	rejected (a)				
Upper Silurian-Lower Devonian						
3	28	rejected (a)				
5	15	3	48	80	67	69
6	28	rejected (a)				
10	18	rejected (a)				
16	7	23	160	221	-74	-43
17	15	6	191	241	-94	-74
18	15	-41	173 (353)	234	rejected (b)	
23	13	43	129	100	47	-12
24	15	30	108	100	47	9
25	12	30	23	80	67	94
29	15	9	116	154	-7	1
Lower Carboniferous						
13	42	rejected (a)				
27	11	9	87	80	67	30
63	59	rejected (a)				

Regression line: slope = 0.696 Correlation coefficient = 0.878
intercept = -0.5

**Deviations based on regional mean reference values D_r = 117, F_r = 147

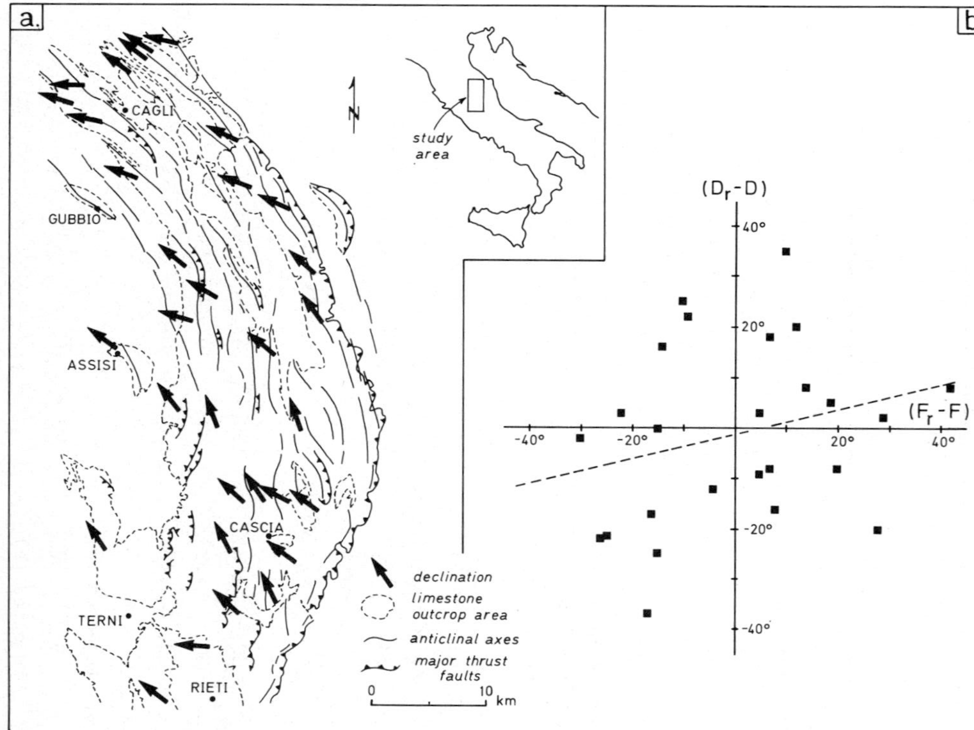

Fig. 6. (a) Paleomagnetic declinations of the Maiolica limestone in Umbria (Hirt and Lowrie, 1986). (b) The regression slope does not significantly differ from zero, suggesting the curvature of the arc is a primary feature.

with age by comparing paleontologically dated sites from north and south Umbria; they found significant differences. Van der Voo and Channell (1980) considered several possible explanations for the dependence of paleomagnetic declination on position along the fold belt and preferred one in which the arcuate shape results from differential décollement changing gradationally along the arc.

Eldredge et al. (1985) re-examined the Scaglia Rossa data in Channell et al. (1978) and found that the slope of the best fit line was close to unity. They concluded that the Scaglia Rossa data supported oroclinal bending, at least in the sedimentary cover. The paleomagnetic data in Channell et al. (1978) were obtained several years earlier, using thermal or alternating field demagnetization. The quality of the original paleomagnetic data was acceptable according to contemporary standards, but at that time stricter multicomponent analysis of the paleomagnetic vector was not customary, and many site mean directions were obtained by a minimum dispersion criterion. Also, the Scaglia Rossa is not a suitable formation for regression analysis, because of the apparent age dependence of declination. A different reference declination would be needed at each site. This detailed knowledge is not available; Eldredge et al. (1985) used a single reference declination for all the sites. Moreover, because the original publication did not give fold axes orientations at each site, they determined these geometrically from a sketchmap (Fig. 4 in Channell et al., 1978). However, the fold axes in this map are largely schematic and were not intended for this type of use. For these reasons, the conclusion of Eldredge et al. (1985) that Scaglia Rossa paleomagnetic data indicate oroclinal bending in Umbria is invalid.

Magnetostratigraphic studies in several sections of the Late Jurassic and Early Cretaceous Maiolica limestone in northern Umbria showed no significant trends of declination with stratigraphic position (Lowrie and Alvarez, 1984). Although weakly magnetized, the limestone possesses a stable component of magnetization which can be derived by vector analysis. This characteristic remanence (ChRM) gives a positive fold test, implying that it was acquired before the folding (Hirt and Lowrie, 1986). The declinations of ChRM at 25 Maiolica sites throughout the fold belt are fairly uniform (Fig. 6a). As evidenced by the data from magnetic stratigraphy sections, between-site variation can not be associated with apparent polar wander during deposition of the limestone. Comparison with coeval directions from autochthonous Gargano show that the Umbrian region is only slightly rotated (Lowrie, 1986). Therefore, the regional mean value of declination was used as reference. The trend of the fold axis at each site was obtained from field observations, 1:50,000 scale geological maps and the recent tectonic maps of the Northern

Apennines (C.N.R., 1982). Again, the regional mean was used as reference trend.

The plot of declination and fold axis trend deviations is shown in Fig. 6b. A t-test of the regression slope (m=0.259) against zero slope gives a t-value of 1.38, less than the 95% confidence level of 2.07. The statistical analysis shows that the regression slope is not significantly different from zero.

The data from the Maiolica limestone do not support the hypothesis of oroclinal bending in the Umbrian-Marches Apennines. The Umbrian limestone sequences are usually taken to be allochthonous on a Triassic evaporite décollement level. The uniformity of declinations throughout the fold belt implies that the sedimentary cover was emplaced without much internal disruption. The arcuate shape of the mountain range may be determined by the positions of overridden thrust ramps controlled by an earlier fault system.

DISCUSSION AND CONCLUSIONS

Paleomagnetic directions that have been determined by rigorous laboratory and analytical methods provide a useful geological parameter for evaluating possible oroclinal bending. Suitable reference directions must be available to permit correction for the effects of polar wander if the rocks investigated span a large time interval. The full potential of the data can only be realized when accompanied by detailed structural geological investigation of the tectonic setting of each site.

Regression analysis allows the relation between declinations and fold axes to be tested statistically. This method suggests that the curvature of the Variscan folds in the Iberian arc results from oroclinal bending, but the arcuate forms of the overthrust carbonate sequences in the Swiss Jura mountains and Umbrian-Marches Apennines are primary features. However, even when the statistical test shows no significant change of declination along the arc, this does not positively exclude a form of oroclinal bending. Knowles and Opdyke (1968), in a paleomagnetic test of the origin of curvature of the northern Appalachian fold belt, found no support for a secondary origin of the observed curvature, but pointed out that the curved shape could have been achieved by differential movement along 'en échelon' strike-slip faults. The curvature in both the Swiss Jura and the Umbrian Apennines may, therefore, result from differential amounts of décollement during emplacement of the overthrust sequences (Channell et al., 1978; Van der Voo and Channell, 1980).

ACKNOWLEDGEMENTS

We thank the Swiss National Science Foundation, grant 2.684-0.82, for financial support. Contribution 512, Institut für Geophysik, ETH-Zürich.

REFERENCES

Burns, K.L., Rickard, M.J., Belbin, L., and Chamalaun, F., 1980. Further palaeomagnetic confirmation of the Magellanes orocline. Tectonophysics, 63: 75-90.

Carey, S.W., 1958. A tectonic approach to continental drift. In: Symposium on Continental Drift. Univ. of Tasmania, Hobart, 177-355.

Channell, J.E.T., Lowrie, W., Medizza, F., and Alvarez, W., 1978. Palaeomagnetism and tectonics in Umbria, Italy. Earth Planet. Sci. Lett., 39: 199-210.

Cogné, J.-P., and Perroud, H., 1985. Strain removal applied to paleomagnetic directions in an orogenic belt: the Permian red slates of the Alpes Maritimes, France. Earth Planet. Sci. Lett., 72: 125-140.

Consiglio Nazionale di Ricerche, 1982. Carta strutturale dell' Appennino Settentrionale. M. Boccaletti and M. Coli (Editors), Publ. No. 49.

Dalziel, I.W.D., Kligfield, R., Lowrie, W., and Opdyke, N.D., 1973. Paleomagnetic data from the southernmost Andes and the Antarctandes. In: D.H. Tarling and S.K. Runcorn, Implications of Continental Drift to the Earth Sciences, Vol. 1. Academic Press, New York, 87-101.

Demarest, H.H., Jr., 1983. Error analysis for the determination of tectonic rotation from paleomagnetic data. J. Geophys. Res., 88: 4321-4328.

Eldredge, S., Bachtadse, V., and Van der Voo, R., 1985. Paleomagnetism and the orocline hypothesis. Tectonophysics, 119: 153-179.

Hirt, A.M., and Lowrie, W., 1986. Paleomagnetism of the Umbrian-Marches orogenic belt. Earth Planet. Sci. Lett., in press.

Kawai, N., Ito, H., and Kume, S., 1961. Deformation of the Japanese Islands as inferred from rock magnetism. Geophys. J. R. astr. Soc., 6: 124-130.

Kellogg, K.S., 1980. Paleomagnetic evidence for oroclinal bending of the southern Antarctic peninsula. Geol. Soc. Amer. Bull., 91: 414-420.

Kellogg, K.S., and Reynolds, R.L., 1978. Paleomagnetic results from the Lassiter coast, Antarctica, and a test for oroclinal bending of the Antarctic peninsula. J. Geophys. Res., 83: 2293-2299.

Klootwijk, C.T., Conaghan, P.J., and Powell, C. McA., 1985. The Himalayan arc: large-scale continental subduction, oroclinal bending and back-arc spreading. Earth Planet. Sci. Lett., 75: 167-183.

Knowles, R.R., and Opdyke, N.D., 1968. Paleomagnetic results from the Mauch Chunk formation: a test of the origin of curvature in the folded Appalachians of Pennsylvania. J. Geophys. Res., 73: 6515-6526.

Lallemand, S., and Jolivet, L., 1986. Japan Sea: a pull-apart basin? Earth Planet. Sci. Lett., 76: 375-389.

Lowrie, W., 1986. Paleomagnetism and the Adriatic promontory. Tectonics, in press.

Lowrie, W., and Alvarez, W., 1977. Late Cretaceous geomagnetic polarity sequence: detailed rock and palaeomagnetic studies of the Scaglia Rossa limestone at Gubbio, Italy. Geophys. J. R. astr. Soc., 51: 561-581.

Lowrie, W., and Alvarez, W., 1984. Lower Cretaceous magnetic stratigraphy in Umbrian pelagic limestone sections. Earth Planet. Sci. Lett., 71: 315-328.

Lowrie, W., Hirt, A.M., and Kligfield, R., 1986. Effects of tectonic deformation on the remanent magnetization of rocks. Tectonics, in press.

MacDonald, W.D., 1980. Net tectonic rotation, apparent tectonic rotation and structural tilt correction in paleomagnetic studies. J. Geophys. Res., 85: 3659-3669.

Molnar, P., and Tapponnier, P., 1975. Cenozoic tectonics of Asia: effects of a continental collision. Science, 189, 419-426.

Otofuji, Y., and Matsuda, T., 1983. Paleomagnetic evidence for the clockwise rotation of southwest Japan. Earth Planet. Sci. Lett., 62: 349-359.

Otofuji, Y., and Matsuda, T., 1984. Timing of rotational motion of southwest Japan inferred from paleomagnetism. Earth. Planet. Sci. Lett., 70: 373-382.

Otofuji, Y., Matsuda, T., and Nohda, S., 1985a. Opening mode of the Japan sea inferred from the paleomagnetism of the Japan arc. Nature, 317: 603-604.

Otofuji, Y., Matsuda, T., and Nohda, S., 1985b. Paleomagnetic evidence for the Miocene counter-clockwise rotation of northeast Japan - rifting process of the Japan arc. Earth Planet. Sci. Lett., 75: 265-277.

Ries, A.C., Richardson, A., and Shackleton, R.M., 1980. Rotation of the Iberian arc: palaeomagnetic results from north Spain. Earth Planet. Sci. Lett., 50: 301-310.

Schwartz, S.Y., and Van der Voo, R., 1983. Paleomagnetic evaluation of the orocline hypothesis in the central and southern Appalachians, Geophys. Res. Lett., 10: 505-508.

Tapponnier, P., Peltzer, G., Le Dain, A.Y., Armijo, R., and Cobbold, P., 1982. Propagating extrusion tectonics in Asia: new insights from simple experiments with plasticine. Geology, 10: 611-616.

Trümpy, R., 1980. The geology of Switzerland. Schweiz. Geol. Kommission, Wepf and Co., Basel, 104p.

VandenBerg, J., Klootwijk, C.T., and Wonders, A.A.H., 1978. The late Mesozoic and Cenozoic movements of the Italian peninsula: further paleomagnetic data from the Umbrian sequence. Geol. Soc. Amer. Bull., 89: 133-150.

Van der Voo, R., and Channell, J.E.T., 1980. Paleomagnetism in orogenic belts. Rev. Geophys. Space Phys., 18: 455-481.

Watts, D.R., Watts, G.C., and Bramall, A.M., 1984. Cretaceous and Early Tertiary paleomagnetic results from the Antarctic peninsula. Tectonics, 3: 333-346.

Wonnacott, R.J., and Wonnacott, T.H., 1985. Introductory Statistics, Fourth Edition. John Wiley and Sons, New York, 649p.

GEOCHEMICAL AND ISOTOPIC SYSTEMATICS OF EASTERN SUNDA ARC VOLCANICS : IMPLICATIONS FOR MANTLE SOURCES AND MANTLE MIXING PROCESSES.

R.VARNE[1] and J.D.FODEN[2]

[1]Geology Department, University of Tasmania, G.P.O. Box 252C Hobart, Tasmania 7001, Australia.
[2]Geology Department, University of Adelaide, G.P.O. Box 498 Adelaide, S.A. 5001, Australia.

ABSTRACT

Active volcanoes of the eastern Sunda arc, west of the arc/continent collision between the Sunda-Banda arc system and northwest Australia, are erupting mafic volcanics that range from calc-alkaline basalts, through shoshonitic trachybasalts, to leucite tephrites and olivine leucitites. All display typically arc-like geochemistry.

The eastern Sunda volcanic arc is located over a north-dipping Benioff-Wadati seismic zone, across a transition from Cretaceous and Tertiary arc lithosphere of near-continental thickness to younger arc lithosphere with near-oceanic thickness. To the south of the volcanic arc, Eastern Indian Ocean lithosphere gives way eastwards to northwest Australian lithosphere.

K, Rb, Sr, Ba, La, and Nb contents of the eastern Sunda mafic volcanics increase threefold along the arc towards the collision zone. The increase in K-group element contents is well correlated with increase in $^{87}Sr/^{86}Sr$ values from about 0.7040 to 0.7055, and decrease in $^{143}Nd/^{144}Nd$ values from about 0.51290 to 0.51257.

Systematic geochemical and Sr, Nd, Pb isotopic changes reflect the involvement in eastern Sunda arc magmatism of at least two sources, one that is yielding material poor in K-group elements with low $^{87}Sr/^{86}Sr$ values, and another that yields material rich in K-group elements with high $^{87}Sr/^{86}Sr$ values. There is no strong evidence that either is recently subducted upper crustal sialic material. Isotopic data suggest that subcontinental mantle may be the source of the K-rich arc component with high $^{87}Sr/^{86}Sr$ values, whereas suboceanic mantle may be the source of the K-poor component. The compositional variations in the arc volcanics may reflect the operation of large-scale tectonic processes that are mixing suboceanic and subcontinental mantles around the collision zone.

INTRODUCTION

The tectonic processes of orogenic zones might be better understood if the petrological, geochemical, and isotopic characteristics of arc magmas could be used to identify the composition, origin, and conditions of melting of their source materials.

The Sunda-Banda arc system of Indonesia is the site of the world's only active continent-arc collision, and its volcanic rocks are typically arc-like in their geochemistry (Whitford et al., 1979; Foden and Varne, 1980; Hutchison, 1982; Wheller et al., in press).

Occurring east of the continent-arc collision zone (Fig. 1), the Banda volcanics are widely cited as a prime example of the involvement of continent-derived sialic material in arc volcanism. West of the collision zone, the volcanic rocks of the eastern Sunda segment of the Sunda-Banda arc system display a compositional range from tholeiite to leucitite, wider than that of any other currently-active volcanic arc.

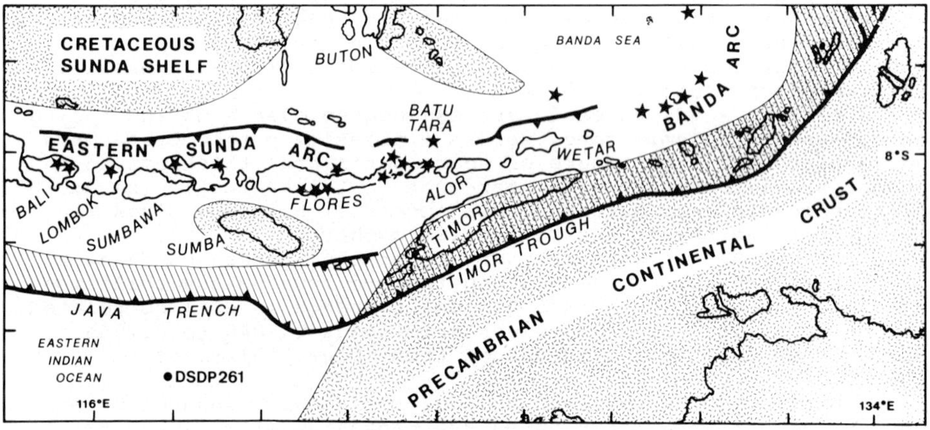

Fig. 1. Sketch map showing eastern Indonesia and western Australia. Distribution of continental crust after Hamilton (1979) and Pigram and Panggabean (1983); location of accretionary wedge from Silver et al., (1983); back-arc thrusts from Hamilton (1979) and Silver et al., (1983); active volcanoes shown by stars.

This arc system therefore offers a unique setting for empirically evaluating the effect of various source materials, including subcontinental lithosphere, on the composition of arc volcanics.

CRITICAL GEOCHEMICAL DIFFERENCES BETWEEN ARC AND NON-ARC MAFIC VOLCANICS

Although there is broad agreement among petrologists and geochemists that the tholeiitic and sodic alkaline volcanics of non-orogenic settings have probably been derived mainly by fractional crystallization of primary melts of mantle peridotite, there is no such concensus about the origins of the volcanic rocks of orogenic zones.

The correlation between the K_2O contents of arc volcanoes and the depth to the underlying Benioff-Wadati zone (Hatherton and Dickinson, 1969) was originally interpreted within the plate tectonic concept to imply that arc magmatism was simply the result of melting of the downgoing slab. This idea has now been largely discarded, essentially for geochemical reasons. Many arc volcanics seem too magnesian or too potassic to be partial melts of oceanic crust of basaltic composition, and experimental studies of the high-pressure melting of basaltic compositions have yielded melts that are compositionally unlike the mafic volcanics of arcs.

However, comparisons of the trace element compositions of mafic volcanics from non-orogenic settings with those of arcs have revealed marked and fairly systematic differences. These differences are difficult to explain if volcanics both of orogenic and non-orogenic settings are formed by melting of mantle peridotite alone.

In particular, mafic arc volcanics are generally poorer in Ti-group elements, particularly Nb, than are non-orogenic volcanics both from intra-continental settings, and from major ocean basins (Chayes, 1964; Pearce and Cann, 1973, and see Fig. 3A). The richer in K-group elements are the volcanics, the more marked the differences become.

In non-orogenic volcanics, enrichment in K-group elements tends to be accompanied by enrichment in Na, in Ti-group elements, and in light rare earth elements (LREE). Because the K-group, Ti-group, and LRE elements behave fairly coherently in these rocks, they have been viewed collectively as a group of trace elements which are all similarly 'incompatible' during partial melting of upper mantle (i.e. they partition strongly into melts in equilibrium with Mg-rich olivine, pyroxene, garnet and spinel).

In arc volcanics, by contrast, these elements no longer seem to behave as a coherent group. When compared with non-orogenic alkaline volcanics, arc volcanics that display similar amounts of enrichment in K-group elements are relatively less enriched in LREE and Na, and remain characteristically poor in Ti-group elements (Foden and Varne, 1980).

Special explanations have been advanced to account for these distinctive geochemical features of arc volcanics.

It has been suggested that if the mantle wedge beneath the arc is composed of peridotite that has already undergone magma extraction, perhaps at a mid-oceanic spreading centre, leading to its depletion in 'incompatible' elements, then the low Ti-group element concentrations of arc volcanics might merely reflect low abundances in their mantle source (Green, 1976). It has also been argued that their mantle source might not necessarily be poorer in Ti-group elements than the sources of non- orogenic volcanics, but that its mineralogy might include refractory phases which retain Ti-group elements during melting (e.g. Green, 1980).

Neither of these explanations of the low Ti-group element abundances of arc volcanics accounts for the wide variations in K-group element contents that also characterize arc volcanics.

Because K-group element concentrations of arc volcanics are poorly correlated with concentrations of Na and Ti, their behaviour also attracts special explanations.

First, it has seemed necessary to postulate a separate arc source of K-rich material, particularly if Ti-poor peridotite which has already undergone melt extraction is also believed to be involved in arc magmagenesis. It has been speculated that subducted sediment or altered oceanic crust might be a source of such K-rich material. Second, processes unique to the orogenic setting have been invoked to account for the separation of the K-group elements from other 'incompatible' trace elements. It has been suggested that a K-enriched fluid or melt derived from K-bearing sediments and altered volcanics of the subducting slab may rise into the mantle wedge and there enrich the sources of the arc magmas (Nicholls and Ringwood, 1973).

In this article, we first present a set of representative analyses of carefully selected Quaternary-Recent mafic volcanic rocks from the eastern Sunda arc, including much new major, trace and isotopic analytical data, and then interpret these data both according to current petrogenetic theories and in the light of our knowledge of their particular geological setting. In particular, we evaluate the hypothesis that K-group enrichment in the arc volcanics may result from recent subduction of upper crustal sialic material.

REGIONAL SETTING OF EASTERN SUNDA ARC VOLCANOES

The geological and tectonic history of the Indonesian region is complex. The events that led up to the appearance of the eastern Sunda arc in its present location are not yet well understood. In this brief discussion of the geological setting of the eastern Sunda arc, we focus on those geological and tectonic factors which might be influencing the composition of the rocks of the Quaternary-Recent eastern Sunda volcanoes.

In the Sunda-Banda arc-trench system, earthquake foci form a seismic zone about 500 km in width, extending along the length of the arc. Beneath the Javanese, eastern Sunda and Banda arc sectors, the earthquakes occur at depths ranging from <70 km up to 650-680 km, but beneath Sumatra the deeper earthquakes are absent (Fitch, 1970; Cardwell & Isacks, 1978). Most of the shallow earthquake foci occur in the forearc: together with the deeper foci, the earthquakes define a northward-dipping Benioff-Wadati zone, consistent with northward subduction of Eastern Indian Ocean lithosphere beneath the volcanic arc (Hamilton, 1979).

At depths greater than 100 km below the eastern Sunda arc, the Benioff-Wadati zone dips at 65°-70° (Hamilton, 1979). Few Benioff-Wadati zone earthquakes have been recorded from depths between 250 and 350 km beneath Java, Bali and Sumbawa (Fitch and Molnar, 1970; Cardwell and Isacks, 1978), possibly indicating the zone of magma production (cf. Sugimura and Uyeda, 1973, 177-182). If depths to the top of the Benioff-Wadati zone beneath the active eastern Sunda volcanoes are also taken as a guide to the depth of magma generation, these lie in general between about 150 to 250 km (Hutchison, 1982). It seems reasonable to speculate that Sunda arc magmatism may originate at pressures of at least 50 kb, and, possibly, as high as 80 kb.

Source materials of Sunda-Banda magmatism could include subducted Eastern Indian Ocean lithosphere, continental lithosphere involved in and around the collision zone (Varne, 1985, and Fig. 1), and subarc mantle, which beneath the eastern Sunda arc might also be Eastern Indian Ocean lithosphere, trapped when the arc developed in its present location (Hamilton, 1979).

In the easternmost, Banda, sector of the arc system (Fig. 1), the suturing of northwest Australia to the volcanic arc seems near-complete and the arc is now tightly bounded to its south by continental lithosphere (Hamilton, 1979). Much of the floor of the

Banda Sea enclosed by the arc of Banda volcanoes seems to be composed of Miocene continent-derived material (Silver et al., 1985), but the possibility remains that the Banda Sea may be partly floored by oceanic lithosphere (Pigram and Panggabean, 1983; Silver et al., 1985).

To the west of the Banda sector, north and west of Wetar, the collision between continent and arc still seems to be proceeding. Back-arc thrusts (Fig. 1) may be developing as a result of subduction polarity reversal following collision (Hamilton, 1979; Silver et al., 1983).

The regional structural trends of this sector of the Sunda-Banda arc system, where the continental fragment of Sumba, and parts of northern Timor, lie within the fore-arc of the active collision zone (Fig. 1), are less regular than those of the arc sectors to the west or east. To the south of the volcanic arc, the rifted continental lithosphere of northwest Australia gives way westward to oceanic lithosphere of the Eastern Indian Ocean, which bounds the eastern Sunda and Java-Sumatra arc sectors to their south.

The crust of the Eastern Indian Ocean south of the eastern Sunda arc includes igneous and sedimentary rocks of Late Jurassic and Early Cretaceous age, as well as Cainozoic sediments. Created during the Jurassic segmentation of Gondwanaland that formed the present rifted continental margin of northwest Australia (Veevers, 1977), it represents some of the oldest surviving ocean lithosphere.

This westerly transition, south of the arc, from continental to oceanic lithosphere, is accompanied north of the arc by transition westwards from the troughs of the Banda Sea and the Flores and Bali Basins to the shallow Sunda Shelf north of Java. Paralleling these changes in the crustal matrix of the volcanic arc, its lithosphere also thickens and becomes older westward.

In the eastern Sunda sector, seismic studies indicate crustal thicknesses near Bali of about 18 km (Curray et al., 1977; Ben-Avraham and Emery, 1973), and the oldest known arc crust is of Miocene age (van Bemmelen, 1949), whereas the arc crust of Java is up to 30 km thick (Curray et al., 1977; Ben-Avraham and Emery, 1973), is structurally more complex than the crust of the eastern Sunda islands (van Bemmelen, 1949), and includes Mesozoic material (Hamilton, 1979). West again, the lithosphere of Sumatra contains materials ranging back to Palaeozoic continental margin sediments.

The eastern Sunda segment of the Sunda-Banda arc system therefore seems to be sandwiched between two converging continental margins, and is located east of a transition from lithosphere with near-continental thicknesses into younger, thinner, lithosphere, possibly with oceanic affinities, and west of an arc sector where suturing of colliding continent to volcanic arc seems largely completed.

Beneath Java, it seems that magmas rise through thick arc crust, and that their source materials may involve subarc mantle which includes pre-Tertiary material. Beneath the eastern Sunda arc, the arc crust is thinner and younger, and the subarc mantle may be preMiocene Eastern Indian Ocean lithosphere, trapped during the mid-Tertiary initiation of the arc. It seems possible that the magmas feeding the volcanoes of Bali and Lombok, in the eastern Sunda arc (Fig. 1), are mainly derived from suboceanic mantle.

Within the collision zone, from Sumbawa to Wetar, the sources of the magmatism could also include continental crust and mantle, as well as Eastern Indian Ocean lithosphere and subarc mantle (Varne, 1985). In contrast, although the Banda volcanoes east of the active collision zone have developed over a subduction zone, there is no clear indication that Indian Ocean lithosphere is at present subducting beneath the arc. Judged by their tectonic settings alone, involvement of continental lithosphere seems more likely in Banda magmatism than in the magmatism of the eastern Sunda arc.

There is a tendency for the mafic volcanics of the active eastern Sunda volcanoes to become more K_2O-rich eastwards along the arc towards the collision zone (Foden and Varne, 1980), although this pattern of along-arc K_2O enrichment may not have been as clear in early Quaternary times and could be a transient feature. Wheller et al. (in press) have shown that similar along-arc variations occur elsewhere in the Sunda-Banda arc, apparently refecting different stages in the growth of the arc system. As in the eastern Sunda arc, these variations over-ride the simultaneous tendency for K_2O contents to increase with depth to Benioff-Wadati zone.

DISTRIBUTION, GEOCHEMISTRY AND ISOTOPIC CHARACTERISTICS OF EASTERN SUNDA VOLCANICS

Distribution

The active volcanoes of Bali (Fig. 1), Batur and Agung, are

erupting a basalt-andesite-dacite association low enough in K_2O to be transitional to arc tholeiite (Whitford and Nicholls, 1976; Wheller and Varne, in press). On Lombok island, east of Bali, the basalt-andesite-dacite association of the Rindjani volcano is clearly calcalkaline (Foden and Varne, 1980, 1981a; Foden, 1983).

On Sumbawa, east of Lombok, the active shoshonitic volcanoes of Tambora and Sangeang Api belong to a ne-trachybasalt-trachyandesite association (Foden and Varne, 1980, 1981b). Volcanics from the extinct Quaternary volcanoes of Soromundi and Sangenges, on Sumbawa, include rare calcalkaline rocks, as well as shoshonitic and leucititic rocks (Foden and Varne, 1980, 1981b).

The timing of its appearance, as well as the distribution of the K-rich volcanism in the eastern Sunda arc, suggest that the K-rich silica-unsaturated volcanics may have appeared in the arc after the late Neogene collision with northwest Australia had begun (Varne, 1985). Similar rocks are known from Miocene sequences in Sulawesi but have not yet been found among the pre-collisional calcalkaline early Neogene (middle Miocene?) volcanics on Lombok and Sumbawa (van Bemmelen, 1949; and our reconnaissance sampling).

Petrographic and geochemical variations

Petrographic characteristics of eastern Sunda rocks are described by Foden and Varne (1980, 1981a,b), and Foden (1983, in press). Brouwer (1943), Van Bemmelen (1949), Neumann van Padang (1951), and Hutchinson (1982) also provide petrographic information.

To summarize, each volcano contains one or more suites, each characterized by mafic volcanics and their associated differentiates (Foden and Varne, 1980; Foden and Varne, 1981b). Mineralogical characteristics of the mafic eastern Sunda volcanics are a reflection of their major element compositions, principally the variations in SiO_2 and K_2O contents (Fig. 2).

Mafic volcanics of the tholeiitic and calc-alkaline suites are mainly ol- and hy-normative, and their differentiates are Q-normative andesites and dacites (Foden, 1983; Wheller and Varne, in press; Table 1). Typically, tholeiitic and calc-alkaline basalts may be distinguished petrographically by the lack of groundmass olivine and the presence of orthopyroxene reaction rims on early- precipitated olivine phenocrysts. The lack of orthopyroxene or pigeonite, the presence of groundmass olivine and

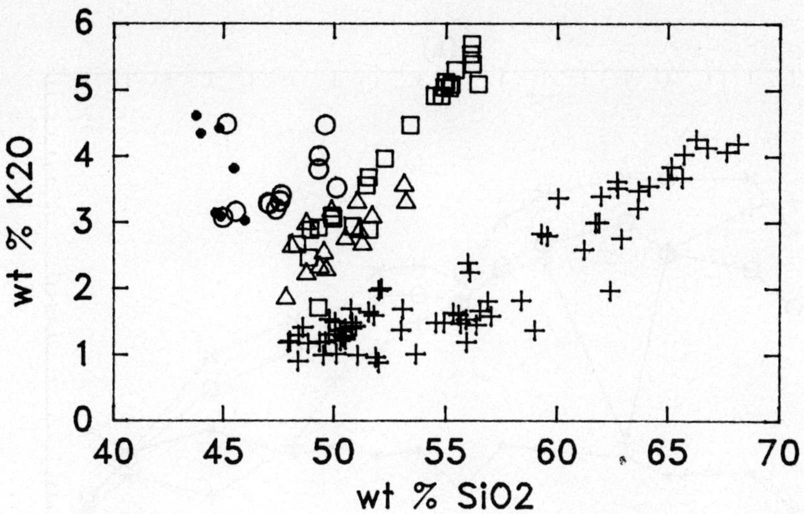

Fig. 2. K_2O-SiO_2 variation diagram for eastern Sunda volcanoes. Leucitite suites: G.Sangenges (spots) and Soromundi (circles); shoshonitic suites: Tambora (squares) and Sangeang Api (triangles); calc-alkaline suite: Rindjani (crosses).

alkali feldspar, and norms containing up to 10% ne, distinguish the shoshonitic trachybasalts, which are associated with ne-bearing differentiates (Foden, 1986; Table 2). Leucite tephrites and leucitites may both have leucite phenocrysts. Plagioclase and leucite both occur in tephrites whereas leucitites lack plagioclase. These rocks are rich in normative ne and di, and some contain normative lc (Foden and Varne, 1980, 1981b; our new unpublished data).

Tables 1, 2 and 3 contain major and trace element analyses of eastern Sunda rocks, particularly selected to be representative of the compositional range encountered in the mafic volcanics, although differentiated rocks are also included. Details of major and trace element analytical methods are given in the caption to Table 1. Sr and Nd isotopic data are also given for some of these rocks, from Jenner et al. (in prep.).

Chondrite-normalized comparison of mafic eastern Sunda minor and trace element compositions with those of typical non-arc volcanics (Fig. 3A) shows that the arc volcanics display the relative impoverishment of Nb, compared with K-group elements, that is a particular characteristic of arc geochemistry.

(A)

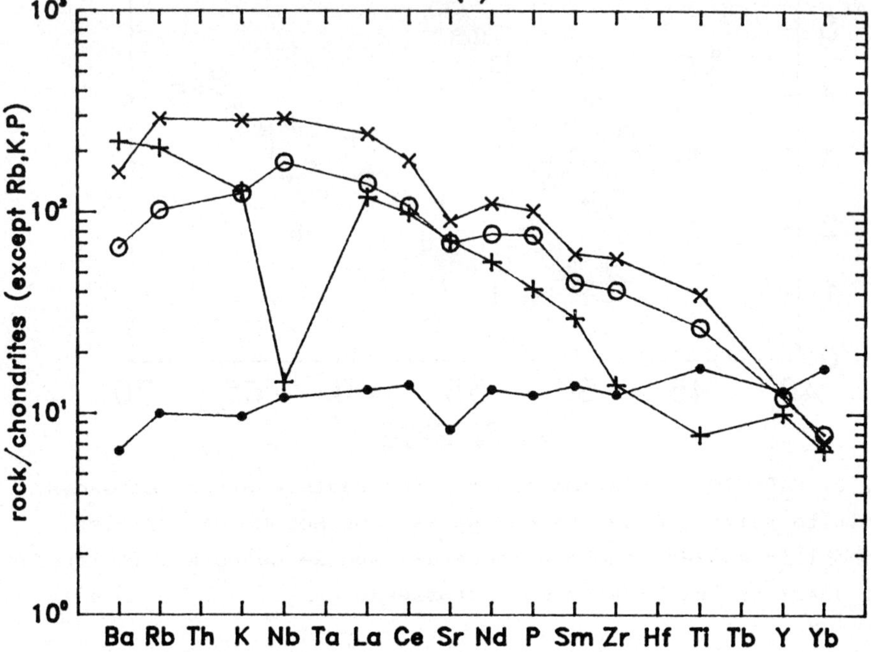
10³
10²
10¹
10⁰
rock/chondrites (except Rb,K,P)
Ba Rb Th K Nb Ta La Ce Sr Nd P Sm Zr Hf Ti Tb Y Yb

(B)

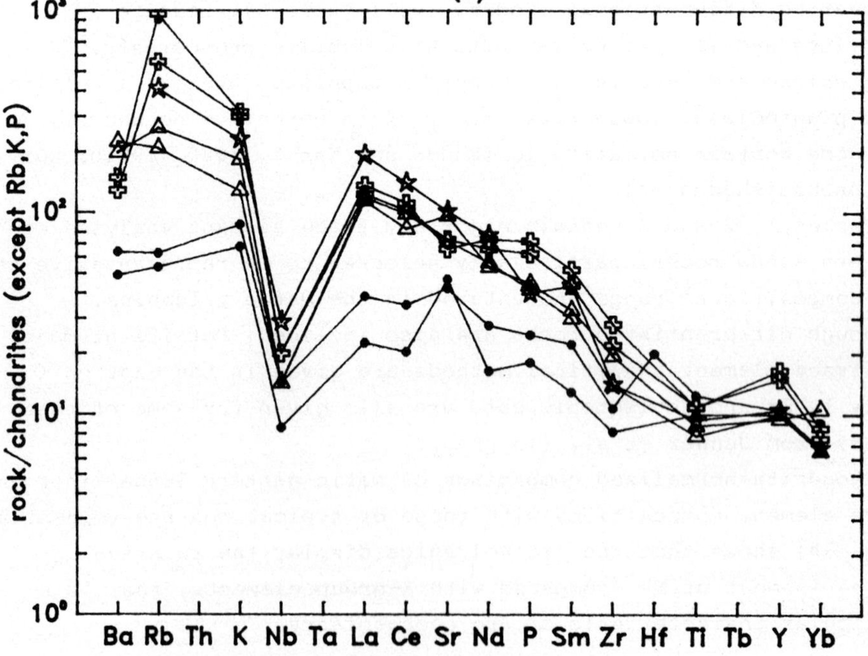
10³
10²
10¹
10⁰
rock/chondrites (except Rb,K,P)
Ba Rb Th K Nb Ta La Ce Sr Nd P Sm Zr Hf Ti Tb Y Yb

Foden and Varne (1980) commented that when basalts, trachybasalts, and leucitites (Tables 1-3) were compared at similar MgO contents and Mg/(Mg+Fe) values, the progression from hy- and Q-normative to ne- and lc-normative compositions is not marked by significant enrichment in TiO_2 and Na_2O (see also Wheller et al., in press), nor by significant change in K/Nb, despite a substantial increase in K-group elements and LREE, increases in Nb/Zr and Ba/Sr values, and a decrease in K/Rb values (Fig. 3B).

It is particularly notable that the eastern Sunda mafic volcanics retain their typically arc-like geochemical pattern (Fig. 3A), with its characteristic trough at Nb, when normalized against chondrite, throughout the compositional range from basalts poor in K-group and LRE elements, through shoshonites, to leucitites rich in K-group and LRE elements (Fig. 3B). This confirms that the eastern Sunda olivine leucitites are geochemically an extreme expression of the arc volcanicity (Foden and Varne, 1980; Wheller et al., in press).

These geochemical variations within the eastern Sunda mafic volcanics are only simple to explain if K, Rb, Nb and the LRE all come little-fractionated from the same source (Varne, 1985). This behaviour is evidence against the existence of a special enrichment mechanism, confined to the orogenic setting, that affects the K-group elements alone.

Fig. 3 (see facing page). A:Chondrite-normalized comparison diagram to display geochemical patterns typical of orogenic and non-orogenic volcanics, including ocean-floor basalt (Table 5, Analysis 1) from D.S.D.P. Site 260, Eastern Indian Ocean (.); basanite with lherzolite xenoliths (Frey et al., 1978, Tables 2 & 6, Sample 2679) from Victoria, eastern Australia (o); olivine leucitite (Cundari, 1973, Table II , Sample BYL 2; unpublished data from F. Frey; Nb calculated from Ta assuming Nb/Ta=16) from N.S.W., eastern Australia (x); phonolitic tephrite from Sangeang Api, eastern Sunda arc (Table 2, Analysis 6).

B: Chondrite-normalized comparison diagram for mafic eastern Sunda volcanics (data from Tables 1-3).

Normalizing factors are Ba: 6.9, Rb: 0.35, K: 120, Nb: 0.35, La: 0.328, Ce: 0.865, Sr: 11.8, Nd: 0.63, P:46, Sm: 0.203, Zr: 6.84, Ti: 620, Y: 2.0, Yb: 0.22 (Thompson, 1982).

TABLE 1
Chemical and isotopic analyses of representative Quaternary volcanics of the basalt-andesite-dacite series of the Rindjani volcano, Lombok Island, eastern Sunda arc.

	1	2	3	4	5	6	7	8
Major elements								
SiO_2	48.32	49.47	50.25	50.43	52.90	55.09	61.82	65.03
TiO_2	0.69	1.28	1.03	1.00	1.06	0.95	0.71	0.67
Al_2O_3	10.53	17.93	18.92	18.34	18.50	18.78	17.25	16.85
FeO	9.19	9.60	9.39	10.20	9.74	8.33	5.50	3.68
MnO	0.17	0.20	0.19	0.22	0.18	0.18	0.19	0.14
MgO	14.02	6.60	5.38	5.64	3.44	3.23	1.88	1.23
CaO	14.38	10.39	10.14	9.94	8.68	7.43	4.25	3.04
Na_2O	1.50	2.99	3.12	3.01	3.75	3.69	4.91	4.88
K_2O	0.90	1.26	0.98	1.00	1.37	1.59	3.00	3.81
P_2O_5	0.15	0.26	0.19	0.20	0.21	0.27	0.40	0.23
Tot.	99.85	99.98	99.59	99.98	99.83	99.54	99.91	99.56
LOI	2.33	0.83	0.58	1.33	1.27	2.13	0.35	1.04
Trace elements								
Rb	21	22	19	16	30	37	62	96
Cs	-	0.23	-	-	0.51	0.54	-	2.97
Sr	452	497	552	490	510	495	403	331
Ba	-	441	338	393	378	504	-	836
Hf	-	4.00	-	-	2.62	3.07	-	5.85
Zr	53	73	56	60	107	122	150	276
Nb	2	5	3	4	2	5	-	13
U	-	2.17	-	-	0.66	0.95	-	2.96
Th	-	6.83	-	-	2.54	3.17	-	10.90
Y	13	21	20	23	28	29	33	41
Sc	50	33	27	35	27	20	11	10
V	-	313	302	-	-	186	-	39
Cr	510	259	23	29	39		38	29
Ni	151	76	25	23	6		-	2
La	11	12.6	7.4	-	12.1	12.8	-	26.0
Ce	-	30.2	17.6	-	26.6	27.5	-	56.5
Pr	-	3.6	-	-	3.29	4.24	-	6.32
Nd	-	17.3	10.3	-	14.4	16.9	-	25.1
Sm	-	4.5	2.61	-	3.28	4.09	-	4.98
Eu	-	1.3	0.88	-	1.05	1.24	-	1.26
Gd	-	4.7	2.8	-	3.6	4.12	-	4.96
Dy	-	4.3	2.75	-	4.05	4.13	-	4.75
Er	-	2.5	1.65	-	2.48	2.65	-	3.22
Yb	-	2.0	1.49	-	2.31	2.69	-	3.23
$^{87}Sr/^{86}Sr$	0.70389	0.70395	0.70396	0.70399	0.70408	0.70401	0.70394	0.70398
$^{143}Nd/^{144}Nd$	0.512776	0.512883	0.512907	0.512890	0.512885	0.512884	0.512878	0.51289

Major and trace elements, except Hf, Cs, U, Th and REE, determined by XRF at the Universities of Adelaide and Tasmania. Hf, Cs, U and Th determined by spark-source mass spectrometry by Dr R.S.Taylor at RSES, ANU. REE determined by J.D.F. at the University of Adelaide using isotope dilution techniques. Sr and Nd isotope analyses by G.A.Jenner (Jenner et al., in prep.). Analysis 1: U of T 48001 ankaramite; analysis 2: 41632 high-Al basalt; analysis 3: 48007 high-Al basalt; analysis 4: 41676 high-Al basalt; analysis 5: 41634 low-Si andesite; analysis 6: 41622 low-Si andesite; analysis 7: 48026 high-K andesite; analysis 8: 41671 dacite.

TABLE 2
Chemical and isotopic analyses of representative Quaternary volcanics of the shoshonitic ne-trachybasalt-trachyandesite series of Tambora (analyses 1-4) and Sangeang Api (analyses 5-8) volcanoes, Sumbawa Island, eastern Sunda arc.

	1	2	3	4	5	6	7	8
Major elements								
SiO_2	48.78	51.49	55.09	56.10	47.77	48.02	50.43	53.05
TiO_2	0.96	0.87	0.69	0.61	0.81	0.91	0.84	0.69
Al_2O_3	17.93	17.15	19.66	19.49	13.05	18.88	19.00	19.21
FeO	9.96	9.02	5.96	5.26	9.41	9.60	8.74	7.33
MnO	0.21	0.18	0.18	0.18	0.18	0.12	0.22	0.22
MgO	5.56	4.65	2.71	1.85	8.65	4.33	3.65	2.69
CaO	10.44	9.04	5.47	4.50	13.90	10.59	9.71	8.05
Na_2O	3.14	4.16	4.56	5.79	3.79	4.27	4.12	4.69
K_2O	2.47	2.89	5.03	5.69	1.85	2.64	2.75	3.56
P_2O_5	0.38	0.40	0.54	0.45	0.44	0.47	0.40	0.40
Tot:	99.83	99.85	99.89	99.92	99.85	99.83	99.86	99.89
LOI	1.07	0.59	0.80	0.54	1.14	1.09	0.87	1.05
Trace elements								
Rb	206	93	136	142	73	92	84	112
Sr	1141	1031	1085	981	850	1140	943	1010
Ba	807	791	1209	1334	1550	1370	-	2250
Zr	82	104	145	153	107	122	150	134
Nb	4	4	7	7	5	5	4	4
Y	23	25	25	25	20	19	21	27
Sc	26	26	10	6	39	22	20	11
V	335	255	189	89	-	186	-	39
Cr	13	7	4	4	228	38	29	38
Ni	20	15	4	5	48	10	8	5
La	-	-	-	38.7	38.9	37.5	-	40.9
Ce	55	64	81	79.7	85.4	71.7	-	85.0
Pr	-	-	-	8.01	8.84	7.85	-	7.77
Nd	27	26	29	30.6	35.7	33.9	-	33.1
Sm	-	-	-	5.23	6.06	6.70	-	5.94
Eu	-	-	-	1.56	1.69	1.93	-	1.78
Gd	-	-	-	4.23	4.77	6.08	-	5.17
Dy	-	-	-	3.84	3.93	5.11	-	-
Er	-	-	-	2.16	1.83	2.42	-	2.77
Yb	-	-	-	2.26	1.44	2.33	-	2.84
$^{87}Sr/^{86}Sr$	0.70389	0.70400	0.70392	0.70394	0.70502	0.70487	0.70474	0.70485
$^{143}Nd/^{144}Nd$	0.512821	0.512806	0.512830	0.512806	0.512650	0.512667	0.512743	0.512706

For analytical methods, see Caption, Table 1. Analysis 1: U of T 48040 ne-trachybasalt; analysis 2: 48048 ne-trachybasalt; analysis 3: 48054 ne-trachyandesite; analysis 4: 48062 ne-trachyandesite; analysis 5: 48067 phonolitic tephrite; analysis 6: 48073 phonolitic tephrite; analysis 7: 4807 ne-trachyandesite; analysis 8: 48081 ne-trachyandesite.

TABLE 3
Chemical and isotopic analyses of representative Quaternary volcanics of the leucititic series of Sangenges (analyses 1-3) and Soromundi (analyses 4-7) volcanoes, Sumbawa Island, eastern Sunda arc.

	1	2	3	4	5	6	7
Major elements							
SiO_2	44.63	45.17	50.55	47.25	47.78	48.09	49.24
TiO_2	1.12	1.13	0.98	0.97	0.96	0.93	0.84
Al_2O_3	11.30	13.55	18.51	14.10	15.91	16.47	17.68
FeO	12.01	11.93	9.07	10.32	10.68	10.30	9.13
MnO	0.20	0.20	0.18	0.22	0.25	0.23	0.22
MgO	8.50	7.28	4.45	7.75	5.86	5.58	3.38
CaO	15.41	13.99	9.67	13.19	12.24	11.45	10.05
Na_2O	1.75	1.57	3.77	2.46	2.56	3.08	4.68
K_2O	4.31	4.52	2.21	3.32	3.31	3.44	4.00
P_2O_5	0.76	0.66	0.46	0.42	0.43	0.42	0.62
Total:	99.99	100.00	99.85	100.00	99.98	99.99	99.84
LOI	1.40	1.34	0.90	1.32	2.15	1.27	1.33
Trace elements							
Rb	195	337	61	143	112	107	144
Sr	779	865	724	1190	1520	1580	2400
Ba	892	1030	-	-	2640	2680	-
Zr	191	150	138	96	98	103	119
Nb	7	7	3	10	12	12	15
Y	33	30	32	21	22	26	23
Sc	54	51	26	39	35	29	16
V	442	402	-	-	319	314	-
Cr	35	212	48	209	87	61	39
Ni	46	29	8	28	22	15	4
La	44.7	40.3	-	63.0	70.0	-	-
Ce	96.0	88.2	-	121.0	147	149	-
Nd	46.4	41.6	-	48.4	51.0	50.0	-
Sm	10.6	8.87	-	8.82	-	-	-
Eu	3.11	2.52	-	2.39	-	-	-
Gd	9.46	7.05	-	6.09	-	-	-
Dy	6.84	5.50	-	4.11	-	-	-
Er	2.68	2.26	-	1.49	-	-	-
Yb	1.88	1.67	-	1.49	-	-	-
$^{87}Sr/^{86}Sr$	0.70550	0.70534	0.70415	0.70485	0.70479	0.70480	0.70527
$^{143}Nd/^{144}Nd$	0.512571	0.512571	0.512818	0.512660	0.512706	0.512660	0.512612

For analytical methods, see Caption, Table 1. Analysis 1: 48115 olivine leucitite; analysis 2: 48119 olivine leucitite; analysis 3: 48112 ne-trachyandesite; analysis 4: 48135 leucite tephrite; analysis 5: 48139 leucite tephrite; analysis 6: 48136 leucite tephrite; analysis 7: 48143 leucite tephrite.

Sr, Nd, Pb, O and Be isotopic characteristics

$^{87}Sr/^{86}Sr$ values of eastern Sunda volcanics range from about 0.7039 to 0.7055, and are negatively correlated with their $^{143}Nd/^{144}Nd$ values, which range from about 0.512907 (ε Nd = +5.3) to 0.51257 (ε Nd = -1.37), with a trend following the "mantle array" (Tables 1,2 and 3; Fig. 4).

Increasing K_2O in volcanics with SiO_2 < 55 wt% correlates with increasing $^{87}Sr/^{86}Sr$ (Fig. 5) and therefore with increasing Nd/Sm and Rb/Sr values (Tables 1, 2 and 3, and see Figure 3B), and decreasing $^{143}Nd/^{144}Nd$ values (Fig. 4).

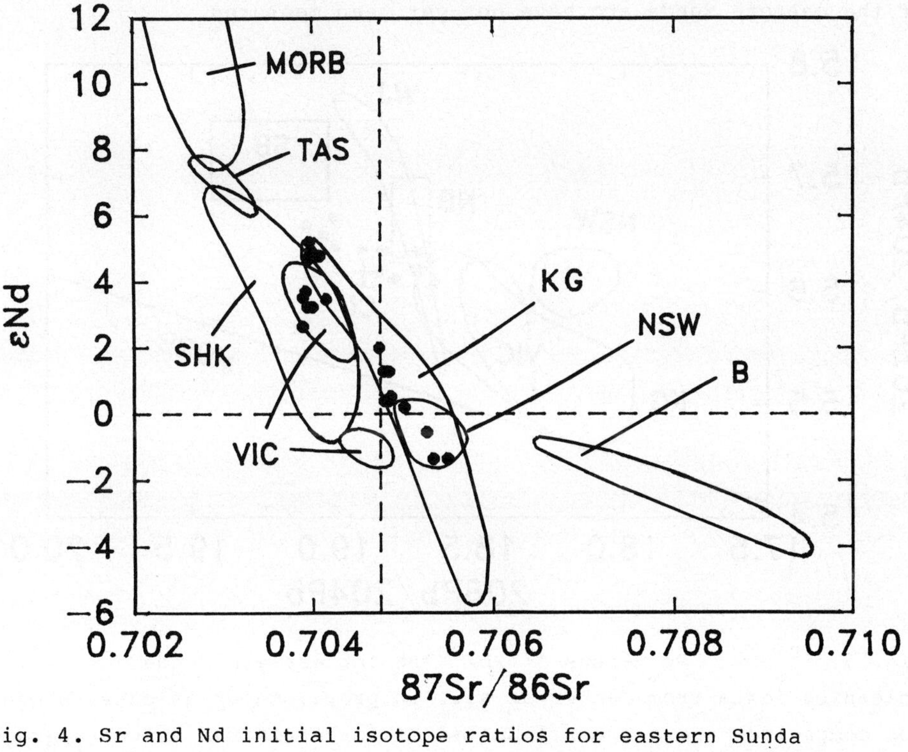

Fig. 4. Sr and Nd initial isotope ratios for eastern Sunda volcanics (spots). For comparison, Field B is for Banda volcanics (Whitford et al. 1981); Field Kg is for Kerguelen ocean-island volcanics (Dosso et al. 1979; White and Hofmann, 1982); eastern Australian sodic alkaline volcanics shown as Field SHK for Southern Highlands and Kiama, N.S.W. (Menzies and Wass, 1983), VIC for Victoria and TAS for Tasmania (McDonaugh et al. 1984); NSW denotes the field for eastern Australian leucitites (Nelson et al. 1985).

Increasing K_2O also correlates, although less well, with increasing $^{206}Pb/^{204}Pb$ from 18.65 to 19.08, $^{207}Pb/^{204}Pb$ from 15.61 to 15.67, and $^{208}Pb/^{204}Pb$ from 38.67 to 39.23 (Jenner et al., in prep.; Fig. 5). $\delta^{18}O$ relative to SMOW ranges from about +7.0 to about +5.5, with a tendency for oxygen to become lighter in the K_2O-rich rocks (Varne and Wheller, unpublished data; Jenner et al., in prep.).

Preliminary studies of three historical lavas from Batur volcano on Bali, and one from Ebulobu volcano on Flores, show their ^{10}Be contents to be very low (Tera et al., in press), providing no evidence supporting the involvement in the arc magmatism of young sedimentary material. However, ^{10}Be contents of K-rich volcanics of the eastern Sunda arc have not yet been measured.

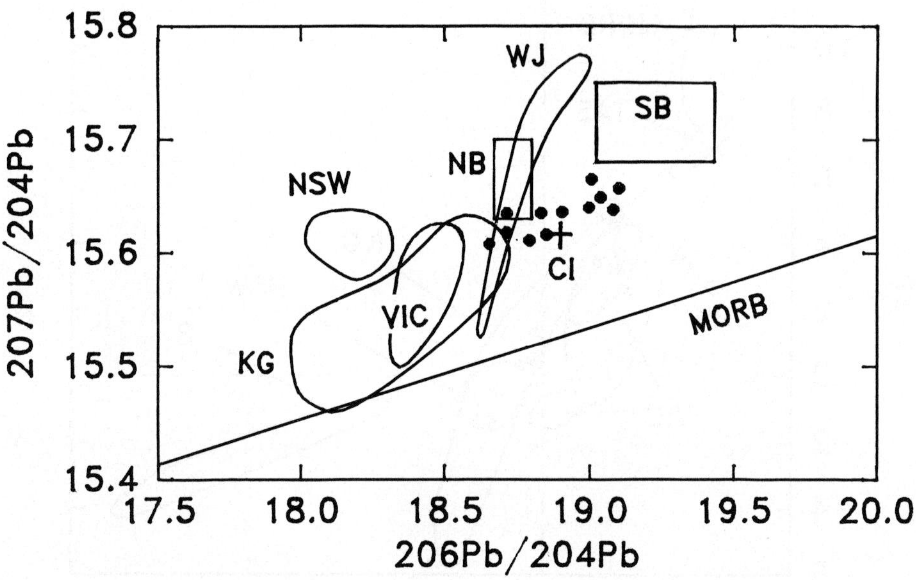

Fig. 5. $^{207}Pb/^{204}Pb$ versus $^{206}Pb/^{204}Pb$ for eastern Sunda volcanics (data from Jenner et al., in prep.; shown as dots. Shown for comparison are a cross labelled CI for Christmas Island average (Hart, 1984); boxes for north Banda islands (NB) and south Banda islands (SB) from Morris and Hart (1980); fields for Kerguelen (KG) from Dosso et al. (1979) and White and Hoffman (1982), for West Java (WJ) from Whitford (1975b), for basanites (VIC) from Victoria and leucitites (NSW) from N.S.W., eastern Australia, from Cooper and Green (1969) and Nelson et al. (in press). Also shown is a reference MORB-line from White (1985).

GEOCHEMICAL AND ISOTOPIC CONSTRAINTS ON EASTERN SUNDA MAGMA SOURCES: IS SUBDUCTED SEDIMENT INVOLVED?

The wide but correlated and seriate geochemical and isotopic variations displayed by the eastern Sunda arc volcanics suggest that their source materials either include two, or more, components with contrasted geochemical and isotopic characteristics which mix together, possibly during melting (Varne, 1985), or that these seriate variations characterize the source materials themselves.

It seems necessary to postulate that, at one extreme, a source exists for the material poor in K-group elements, with relatively low $^{87}Sr/^{86}Sr$ =< 0.704, high $^{143}Nd/^{144}Nd$ values, relatively less radiogenic $^{207}Pb/^{204}Pb$ values, and with $\delta^{18}O$ of about +7.0. This material is most prominent in the basaltic volcanics of the Bali-Lombok sector, farthest removed from the arc-continent collision zone.

Another source seems necessary for the material rich in K-group elements, with $^{87}Sr/^{86}Sr > 0.7055$, low $^{143}Nd/^{144}Nd$ values, relatively radiogenic $^{207}Pb/^{204}Pb$ values, and slightly lighter oxygen, whose presence is most apparent in the olivine leucitites erupted near the zone of the currently-active collision between the volcanic arc and continental lithosphere of northwest Australia.

Is sialic continental crust the main source of this K-rich material?

We assess this possibility in three ways. First, we inspect the Sr, Nd, Pb and O isotopic ratios of the eastern Sunda volcanics for the effects that in theory may betray the input of sialic continental crust; next, we review the compositions of sediments from the floor of the Eastern Indian Ocean, south of the eastern Sunda islands, as representative of the sedimentary material that may be subducting beneath the eastern Sunda arc, and compare their compositional characteristics with those of the arc volcanics; and, finally, we compare the eastern Sunda volcanics with the Banda volcanics from the arc segment east of the collision zone, which are widely accepted as the prime example of incorporation of sialic-sediment-derived material in arc magmatism.

Possible isotopic effects of sialic involvement in arc magmatism.

Most geochemists would expect that input of sialic continental crust into arc magmatism would be identifiable by its effect on Sr, Nd, Pb, and O isotopic ratios.

Whitford and Jezek (1982) argued for involvement of continental crust material in west Java volcanic rocks, either in the magma source regions or during ascent to the surface. They pointed out that the west Java rocks define a steep trend on the $^{207}Pb/^{204}Pb$ versus $^{206}Pb/^{204}Pb$ diagram (Fig. 5), a trend which is also followed by north Banda arc volcanics (Fig. 5; Morris and Hart, 1980), but is clearly unlike the trend followed by the eastern Sunda volcanics from Lombok and Sumbawa, which is similar to trends followed by ocean-island basalts of the Indian Ocean (Fig. 5).

The generally high $^{87}Sr/^{86}Sr$ values in arc volcanics, when compared with those of North Atlantic and Pacific ocean-floor basalts, have been ascribed to the presence of material that originated in sialic continental crust where high $^{87}Sr/^{86}Sr$ values have been evolved. This material is believed to contribute strontium to the oceanic lithosphere, either directly in continent-derived sediments, or indirectly, via seawater involved in the alteration of ocean-floor basalts. When this lithosphere is subducted, it is believed to release hydrous fluids or melts enriched in K-group elements, including Sr. These fluids or melts

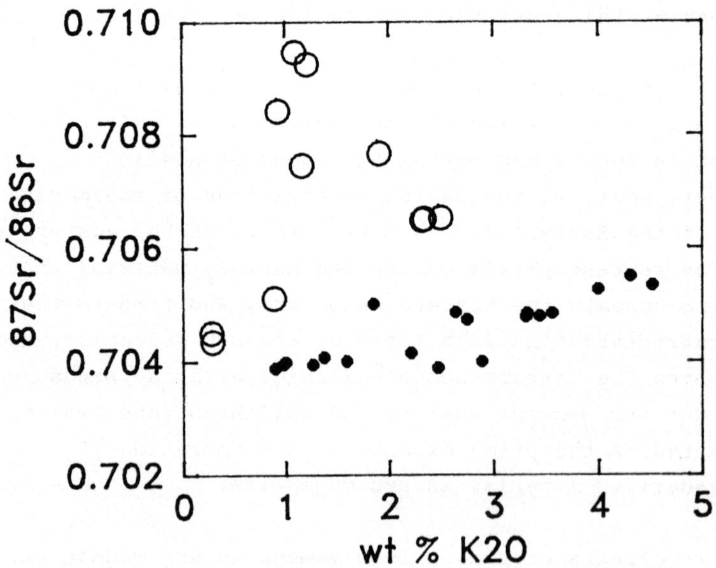

Fig. 6. K_2O versus $^{87}Sr/^{86}Sr$ for eastern Sunda volcanics with SiO_2 < 55 wt % (spots; data of Tables 1-3) and Banda volcanics with SiO_2 < 60 wt % (open circles; data of Magaritz et al. 1978).

are thought to rise into the mantle overlying the subducting slab, and there introduce the bulk of the radiogenic Sr and other K-group elements that later appear in the arc volcanics.

At first sight, such a process might apparently explain the good correlation between eastern Sunda K_2O contents and $^{87}Sr/^{86}Sr$ values (Fig. 6). However, the rationale of the process is its postulated ability to separate and concentrate K-group elements alone. Therefore, it will not account for the geochemical and isotopic correlations displayed by the mafic eastern Sunda volcanics, where increase in K-group elements is also accompanied by increase in Nb, in La, Nd, and La/Yb (Fig. 3B), and in $^{143}Nd/^{144}Nd$ values (Fig. 4).

Sr and Nd isotopic characteristics of eastern Sunda volcanics also define a trend within the "mantle array" (Fig. 4), ranging from positive values of εNd in the K-poor basalts with relatively low $^{87}Sr/^{86}Sr$ values < 0.7040 to negative values of εNd in the K-rich leucitites with relatively high $^{87}Sr/^{86}Sr$ values $>$ 0.7055. This trend is not consistent with selective addition of radiogenic Sr to arc magma sources. Alteration of ocean-floor basalts affects their Sr isotopic ratios, but leaves their Nd isotopic ratios little-disturbed. Any material contributed by altered basaltic crust of the Eastern Indian Ocean is unlikely to lie in the Sr and Nd isotopic "mantle array".

Like the Pb isotopic data (Fig. 5), the Sr and Nd isotope data for eastern Sunda volcanics resemble fields defined by mafic volcanics of eastern Australia and of Indian Ocean islands (Fig. 4), where no recent sedimentary contamination is suspected.

Evaluation of Eastern Indian Ocean sediments as a possible source of the K-rich component in eastern Sunda magmatism.

The Eastern Indian Ocean crust south of the Sunda arc includes igneous and sedimentary rocks of Mesozoic and Cainozoic age.

D.S.D.P. Site 261 (12°57'S, 117°54'E) lies south of Sumbawa (Fig. 1), about 200 km from the Sunda Trench on the northeast side of the Argo Abyssal Plain, and Site 260 (16°9'S, 110°18'E) lies south of central Java in the Gascoyne Abyssal Plain, about 700 km from the Sunda Trench.

Thick sediments at Sites 260 and 261 fall into two main types, clays which are mainly Mesozoic in age and calcareous oozes which are mainly Cainozoic in age (Cook, 1974). Clastic material from

TABLE 4
Averaged chemical analyses of Mesozoic (analyses 1-4) and Cainozoic (analyses 4-8) abyssal plain sediments of the Eastern Indian Ocean south of the eastern Sunda arc (Robinson et al., 1974, Tables 2 & 4).

	1	2	3	4	5	6	7	8
Major elements								
SiO_2	56.6	63.9	77.0	56.2	15.8	11.8	48.1	56.0
TiO_2	0.62	0.54	0.26	0.60	0.20	0.11	0.60	0.55
Al_2O_3	11.7	11.1	4.5	12.9	4.8	2.2	13.7	14.1
Fe_2O_3	6.8	6.5	3.4	5.8	2.1	1.3	6.8	4.3
MnO	0.41	0.25	0.32	0.05	0.12	0.02	0.80	1.2
MgO	2.5	2.2	1.6	2.4	1.2	1.4	3.0	2.3
CaO	2.1	1.6	1.8	3.7	36.7	42.5	4.9	1.1
Na_2O	3.0	1.9	2.0	2.3	1.9	1.6	2.8	4.8
K_2O	3.0	2.1	1.7	2.4	0.8	0.5	2.7	2.2
P_2O_5	0.13	0.16	0.08	0.14	0.07	0.09	0.40	0.14
LOI	13.0	9.8	7.3	13.6	36.4	38.5	16.2	13.0
Total	99.86	100.05	99.96	100.09	100.89	100.02	100.00	99.69
Trace elements								
Sr	140	164	90	100	1100	1500	170	130
Ba	950	936	250	200	470	260	600	1300

Analysis 1: zeolite clay; 2: siliceous clay; 3: radiolarian ooze; 4: nanno clay; 5: nanno ooze; 6: foram nanno ooze; 7: zeolitic clay; 8: radiolarian clay.

TABLE 5
Chemical and isotopic analyses of basalts from the Eastern Indian Ocean, south of the eastern Sunda arc.

	1	2	3		1	2	3
Major elements				Trace elements			
SiO_2	51.86	49.32	51.48	Rb	3.5	8.5	-
TiO_2	1.75	3.40	1.30	Sr	98	97	97
Al_2O_3	13.84	15.68	15.01	Ba	45	35	7
FeO	12.25	10.98	9.09	Zr	85	150	88
MnO	0.20	0.29	0.18	Nb	4.2	7.1	0.84
MgO	5.84	7.33	8.52	Y	26	42	46
CaO	10.25	6.72	10.80	La	4.3	6.0	2.34
Na_2O	2.72	3.48	2.85	Ce	12	18	11
K_2O	0.14	0.88	0.54	Nd	8.3	15	7.1
P_2O_5	0.13	0.31	-	Sm	2.8	5.3	3.11
Total	98.98	98.39	99.77	Yb	3.7	5.1	3.4
LOI	1.15	2.68	2.68				
$(^{87}Sr/^{86}Sr)_m$					0.70433	0.70394	-
$(^{87}Sr/^{86}Sr)_i$					0.70414	0.70348	-

Analyses 1 and 2: major and trace element data from Robinson and Whitford (1974), Sr isotope data from Whitford (1975b). Analysis 3 from Frey et al. (1977), except Nb from Robinson and Whitford (1974).

$(^{87}Sr/^{86}Sr)_m$) is the measured isotopic composition. $(^{87}Sr/^{86}Sr)_i$ is the calculated initial value using an estimated age of 140 Ma (Heirtzler et al., 1973). 1: D.S.D.P. Site 260 basalt (also see Figure 3A). 2: D.S.D.P. Site 261 average Unit B basalt. 3: D.S.D.P. Site 261 Unit A basalt (34-1:75-77).

northwest Australia is scarce (Hamilton, 1979). Representative analytical data for Eastern Indian Ocean sediments are listed in Table 4, taken from Robinson et al. (1974), showing them to vary particularly widely in their contents of SiO_2, Al_2O_3, CaO, Na_2O and K_2O. Their contents of Sr and Ba also show little tendency to behave coherently (Table 4). The addition to K-poor mantle peridotite of any of these sediments seems unlikely in itself to provide the regular major and K-group trace element variations displayed on a regional scale by the eastern Sunda volcanic rocks.

It has been argued that, because K-group elements are more soluble in aqueous fluids than are Ti-group elements, a fluid evolved from the slab during subduction might preferentially extract the K-group elements from the sedimentary component of the slab, and transport them into the source regions of the arc magmatism. Alternatively, melts of the subducted sediments might rise into the mantle above the downgoing slab, similarly enriching it in K-group elements (Nicholls and Ringwood, 1973).

Typically, these sediments are as rich in Na_2O as they are in K_2O (Table 4). They seem as likely to provide Na to an aqueous solution or melt as they are to provide K-group elements. Increase in K-group elements in the volcanic rocks is not accompanied by increase in Na.

Important differences persist when comparisons between Sunda arc volcanics and these Eastern Indian Ocean sediments are restricted to their K, Ba, and Sr contents. Sediments and volcanics differ both in their absolute and in their relative concentrations of K, Ba, and Sr (Fig. 7, and compare data of Tables 1-4).

Thorough mixing of these sediments, with their variable K, Ba, and Sr contents, would be required to produce the relatively coherent behaviour of the K, Ba, and Sr contents of the arc volcanics (Fig. 7).

^{10}Be, apparently derived from subducted sediments involved in arc magmatism, has been detected in volcanics in several arcs elsewhere (Brown et al., 1982; Tera et al., in press). Although no measurements of ^{10}Be seem to have been made on Eastern Indian Ocean sediments, much of the sedimentary section is composed of Quaternary and Late Tertiary material in which ^{10}Be is probably present.

The lack of ^{10}Be in Sunda volcanics (Tera et al., in press)

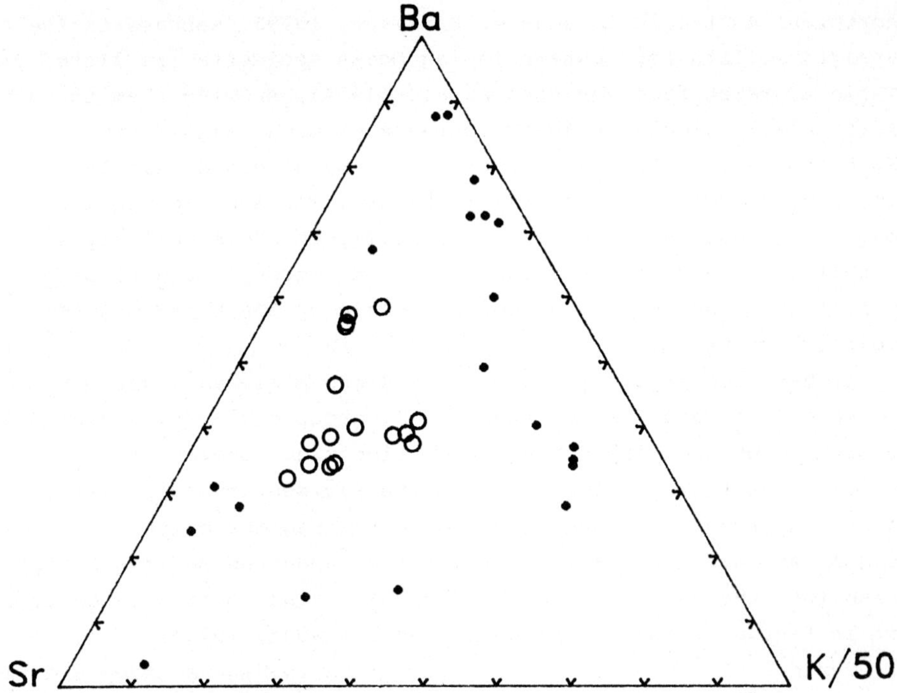

Fig. 7. Ba-Sr-K/50 compositional triangle to compare Eastern Indian Ocean sediments (spots; data of Cook, 1974) with eastern Sunda volcanics (open circles; data of Tables 1-3).

suggests, either, that sedimentary material is not present in the Sunda volcanic rocks in significant amounts, or that if it is present, the mixing process that must have operated to homogenize the subducted sediment to produce the arc-like behaviour of K, Ba and Sr also excluded the younger parts of the sedimentary column.

We conclude that there is no positive evidence, as yet, that material similar to these Eastern Indian Ocean D.S.D.P. sediments is the source of the K-group-element-enriched component in the eastern Sunda magmatism.

Comparison of eastern Sunda and Banda volcanics.

Combined oxygen and strontium isotope analysis of volcanic rocks also offers a means of identifying sediment involvement in magmas, because sedimentary and metasedimentary rocks are commonly enriched in ^{18}O as well as in ^{87}Sr when compared with mantle-derived rocks (James, 1981).

The Banda arc volcanics, from the arc segment immediately east of the Sunda arc (Fig. 1), are frequently cited as an excellent example of the incorporation of sialic sediment-derived material in arc magmatism. We have already pointed out that, judging by their tectonic setting alone, involvement of continental lithosphere seems more likely in Banda volcanoes than in the volcanoes of the eastern Sunda arc.

Magaritz et al. (1978) showed that $^{87}Sr/^{86}Sr$ values of Banda arc lavas correlate crudely with their $\delta^{18}O$ values, and range from mantle-like values to values believed due to the involvement of subducted sialic continental crust.

If the K-group element contents of the eastern Sunda arc volcanics are largely derived from subducted sediment, then it would be predicted that increasing K_2O might be accompanied both by increasing $^{87}Sr/^{86}Sr$ and by increasing $\delta^{18}O$ values.

Oxygen isotope ratios of eastern Sunda mafic volcanics are inconsistent with this expectation. The $\delta^{18}O$ values of mafic eastern Sunda volcanics fall in the range +5.5 to +7.0 per mil (Varne and Wheller, unpublished data; Jenner et al., in prep.).

In contrast to the eastern Sunda volcanics, the Banda volcanics, which are predominantly orthopyroxene-bearing Q-normative andesites and dacites, have Sr and O isotopic characteristics that range from mantle-like values of $^{87}Sr/^{86}Sr \sim 0.704$ and $\delta^{18}O \sim$ +7 permil, to values of $^{87}Sr/^{86}Sr \sim 0.709$ and $\delta^{18}O \sim +9$ permil. This trend is interpreted by Magaritz et al. (1978) to be a result of involvement in the Banda arc magmatism of sialic continental crust with $^{87}Sr/^{86}Sr \sim 0.716$ and $\delta^{18}O \sim +15$ permil, and it is notable that rare cordierite- bearing silica-rich "ambonite" volcanics with these crust-like Sr and O isotopic characteristics are known from the Banda arc (Magaritz et al. 1978).

The Banda volcanics therefore exhibit mineralogical, as well as isotopic, characteristics consistent with the involvement of sialic continental crustal material in their genesis. When compared with the eastern Sunda volcanics, they also lack the relatively simple correlation of Sr, Nd and O isotopic ratios with their major- and trace-element compositions, and in particular, with their K_2O contents (Fig. 6), suggesting that incorporation of sialic material produces irregular geochemical changes.

We conclude that the related geochemical and isotopic variations of the eastern Sunda volcanics seem theoretically and empirically inconsistent with the hypothesis that potassium enrichment in the arc volcanism is mainly due to the recent subduction of upper crustal sialic material beneath the arc.

Comparison of mafic volcanics of the Eastern Indian Ocean and the eastern Sunda arc.

All of the Mesozoic basaltic rocks of the Eastern Indian Ocean crust that were sampled at D.S.D.P. Sites 260 and 261 have undergone some low-temperature alteration after emplacement (Robinson and Whitford, 1974). All display near-chondritic normalized REE patterns (Robinson and Whitford, 1974), and are tholeiitic (Robinson and Whitford, 1974; Frey et al. 1977). Most are Q- or ol- and hy-normative, and the single ne-normative sample may have acquired this characteristic as a result of alteration (Frey et al. 1977).

Some basaltic rocks from Sites 260 and 261 (Table 5; Fig. 3A) resemble modern N-MORB-type ocean-floor basalts (Robinson and Whitford, 1974; Frey et al., 1977). Tholeiites with compositions more like those of E-MORB or oceanic island tholeiites (Table 5) also occur as a TiO_2-rich unit at Site 261 and a sill intruding sediments above oceanic basement at Site 260.

K_2O contents of the rocks range from 0.01 to 0.92 wt %. Alteration processes do not seem to have greatly enriched the rocks in K_2O (Table 5), and their Ba, Zr and Nb contents are relatively low, even in the most fractionated rocks.

$^{87}Sr/^{86}Sr$ values (Table 5) for Site 260 and 261 basalts lie, as Whitford (1975b) noted, at the lower end of the range of $^{87}Sr/^{86}Sr$ values of Sunda arc lavas. Whitford (1975b) also suggested alteration of Site 260 and 261 rocks might account for their relatively high $^{87}Sr/^{86}Sr$ values. However, similar values occur in oceanic island basalts of the Indian Ocean (Fig. 4).

The geochemical and Sr isotopic characteristics of the basalts at Sites 260 and 261 seem to point to the existence of Indian Ocean oceanic crust and mantle that could be related to the source of the material with $^{87}Sr/^{86}Sr < 0.704$ that dominates the compositions of the arc volcanoes on Bali and Lombok.

At Site 211, south of Sumatra, amphibole-bearing olivine basalts and dolerite sills were recovered (Hekinian, 1974). Although they are altered, the richness of these rocks in TiO_2, Na_2O, K_2O, and P_2O_5 (Hekinian, 1974) and their high La/Yb values (Frey al., 1977), suggest that they form part of a province of alkaline volcanism, including that of Christmas Island, related to the formation of shoal areas trending east from 95°E to the Sunda trench. Christmas Island possesses Pb isotopic characteristics (Hart, 1984) that are similar to those of the eastern Sunda arc volcanoes (Fig. 5).

GEOCHEMICAL AND ISOTOPIC SIMILARITIES BETWEEN SUBCONTINENTAL AND SUBOCEANIC MANTLES AND EASTERN SUNDA ARC VOLCANISM

We have shown that source materials of the eastern Sunda arc volcanics include two, or more, components with contrasted geochemical and isotopic characteristics. Neither of these materials seems likely to be derived from recently-subducted sialic crust or sediments and altered basalt from Indian Ocean crust.

One eastern Sunda source is providing material poor in K-group elements, with relatively low $^{87}Sr/^{86}Sr$ =< 0.7040, high $^{143}Nd/^{144}Nd$ values, relatively less radiogenic $^{207}Pb/^{204}Pb$ values, and with $\delta^{18}O$ of about +7.0. This material is most prominent in the basaltic volcanics of the Bali-Lombok sector, farthest removed from the arc-continent collision zone.

Another source is providing material rich in K-group elements, with $^{87}Sr/^{86}Sr > 0.7055$, low $^{143}Nd/^{144}Nd$ values, relatively radiogenic $^{207}Pb/^{204}Pb$ values, and relatively light oxygen, whose presence is most apparent in volcanic rocks near the collision zone.

There are at least two possible mantle sources which are both involved in the collision zone where these materials could be originating : suboceanic mantle of the Eastern Indian Ocean region, and "Gondwanaland" subcontinental mantle associated with Palaeozoic and Proterozoic continental lithosphere (Varne, 1985).

Other geochemists have already pointed out that the mantle sources of oceanic island basalts may satisfy some of the geochemical and isotopic requirements of volcanic sources in the Marianas (Stern and Ito, 1983), the Aleutians (Morris and Hart, 1983), and in Fiji (Gill, 1984).

Geological and tectonic arguments (Hamilton, 1979) suggest that the eastern Sunda arc may be built across pre-Miocene Indian Ocean lithosphere, and Indian Ocean lithosphere probably makes up the subducting slab beneath the Sunda arc.

Pb, Sr and Nd isotopic similarities (Fig. 4 and 5), although based on sketchy information, suggest that the volcanism of Indian Ocean islands and eastern Sunda arc volcanics could share an isotopically similar source, and the geochemical and Sr isotopic characteristics of D.S.D.P. Sites 260 and 261 tholeiites (Table 5) imply the existence of Indian Ocean mantle south of the eastern Sunda arc which could be related to the source of the material with $^{87}Sr/^{86}Sr$ =< 0.704 that dominates the compositions of the arc volcanoes on Bali and Lombok.

Some southeast Australian sodic alkaline volcanics (Menzies and Wass, 1983; McDonaugh et al., 1985) fall within Sr and Nd isotopic ranges that overlap those of Indian Ocean volcanics (Fig. 4), showing that Indian Ocean mantle Sr and Nd isotopic characteristics may be shared in part by Australian subcontinental mantle. However, the eastern Sunda K-rich volcanics have relatively high $^{87}Sr/^{86}Sr$ values (Fig. 6), and ultrapotassic volcanics are unknown from the Indian Ocean basin, implying that Indian Ocean mantle may not be the source of the K-rich material in the arc volcanics.

The isotopic and geochemical changes along the arc towards the collision zone between arc and continent, and scanty stratigraphic and tectonic evidence suggesting that the markedly K-rich arc volcanics made their appearance only after the collision had begun, could be interpreted to suggest that the arrival of the K-rich material followed the collision of the arc with northwest Australian continental lithosphere.

Varne (1985) inferred that subcontinental mantle may therefore be the source of the K-rich material appearing in the arc volcanoes, and that K-poor suboceanic and K-rich subcontinental mantle materials may be mixing around the collision zone to provide the sources for the eastern Sunda volcanics.

As well as providing sodic alkaline volcanics, Australian subcontinental mantle is known to be the source of continental ultrapotassic magmatism which displays the Ba/La and Ba/Nb values that were formerly believed to be characteristic of arc volcanism alone, and Varne (1985) has speculated that the distinctive gechemical features both of arc volcanics and of continental ultrapotassic suites may be due to melting of perovskite-bearing mantle, possibly at high pressures (Varne, in prep.).

Northwest Australian lamproites (Jaques et al., 1984) and

eastern Australian ultrapotassic olivine leucitites (Cundari, 1973) were both erupted in Miocene times through Palaeozoic continental lithosphere. The lamproites have unusual Pb isotopic compositions (Fraser et al., 1985; Nelson and others, in press), unlike those of the arc volcanics.

The leucitites have Sr, Nd, and Pb isotopic compositions which could be due (Nelson et al., in press) to mixing of the "normal" southeast Australian mantle which yielded the Cainozoic sodic alkaline volcanics (McDonaugh et al., 1985), with varying amounts of a mantle component with $^{87}Sr/^{87}Sr > 0.7055$, $^{143}Nd/^{144}Nd < 0.51163$, and $^{206}Pb/^{204}Pb$ values ranging up to 19.04 and $^{207}Pb/^{204}Pb$ values ranging up to 15.64 (Fig. 4 and 5). The isotopic characteristics of this component resemble those required of the K-rich component involved in the eastern Sunda rocks.

Our information is inadequate at present either to distinguish properly between the isotopic characteristics of Australian and Indian Ocean mantles, or to make proper comparison between them and the inferred source characteristics of the eastern Sunda volcanics. All three seem to have isotopic characteristics broadly related to those of Hart's (1984) Southern Hemisphere "Dupal" mantle domain, a globe-encircling isotopic mantle anomaly, centred on latitude 30°S, which is characterized by $^{87}Sr/^{86}Sr$ values falling in the range 0.703-0.705, and relatively high $^{207}Pb/^{204}Pb$, and $^{208}Pb/^{204}Pb$ values. However, some young Australian sodic volcanics have Pb isotopic compositions (Cooper and Green, 1969) that define an array which lies athwart the Indian Ocean and Sunda trends (Fig. 5), possibly ruling out Australian subcontinental mantle as a candidate for the source of the low $^{87}Sr/^{86}Sr$ material in the eastern Sunda arc.

We conclude that recently-subducted upper crustal sialic material is unlikely to be the cause of the systematic geochemical and isotopic variations in the eastern Sunda volcanics (Foden and Varne, 1980; Wheller et al., in press). These variations seem to reflect the existence of tectonic processes operating on a regional scale that are mixing suboceanic and subcontinental mantles around the collision zone (Varne, 1985).

ACKNOWLEDGEMENTS

Our field work in the Lesser Sunda Islands has been funded by the Australian Research Grants Committee and the University of

Tasmania, arranged by the Foreign Scholars Division of the Indonesian Institute of Science, and sponsored by the University of Gadjah Mada. Major and trace element analytical data were obtained at the Geology Departments of the Universities of Tasmania and Adelaide, and at the Research School of Earth Sciences, Australian National University, and Sr and Nd isotope analytical data at Abteilung Geochemie, Max-Planck Institute fur Chemie, Mainz.

REFERENCES

Ben-Avraham, Z. and Emery, K.O., 1973. Structural framework of Sunda Shelf. Amer. Assoc. Petrol. Geol. Bull., 57: 2323-2366.

Brouwer, H.A., 1943. Leuciethoudende en leucietvrije gesteen van den Soromandi of het eiland Soembawa. Ned. Akad. van Wet., Afd. Nat. 52: 303-397.

Brown, L., Klein, J., Middleton, R., Sacks,I.S, and Tera, F., 1982. ^{10}Be in island-arc volcanoes and implications for subduction. Nature, 299: 718-720.

Cardwell, R.K. and Isacks, B.L., 1978. Geometry of the subducted lithosphere beneath the Banda Sea in eastern Indonesia from seismicity and fault plane solutions. J. Geophys. Res., 83: 2825-2838.

Chayes, F., 1964. A petrographic distinction between Cenozoic volcanics in and around the open oceans. J. Geophys. Res., 69: 1573-1588.

Cook, P.J., 1974. Major and trace element geochemistry of sediments from Deep Sea Drilling Project, Leg 27, Sites 259-263, Eastern Indian Ocean. In: Veevers,J.J., Heirtzler,J.R., et al.(Editors), Initial Reports of the Deep Sea Drilling Project, Volume 27, U.S.Government Printing Office, Washington, pp. 418-498.

Cooper, J.A. and Green, D.H., 1969. Lead isotope measurements on lherzolite inclusions and host basanites from western Victoria, Australia. Earth Planet. Sci. Lett., 6: 69-76.

Cundari, A., 1973. Petrology of the leucite-bearing lavas in New South Wales. J. Geol. Soc. Australia, 20: 465-492.

Curray, J.R., Shor Jr, G.G., Raitt, R.W. and Henry, M., 1977. Seismic refraction and reflection studies of crustal structure of the eastern Sunda and western Banda arcs. J. Geophys. Res., 82: 2479-2489.

Dosso, L., Vidal, P., Cantagrel, J.M., Lameyre, J., Marot, A. and Zimine, S., 1979. "Kerguelen: continental fragment or oceanic island?": petrology and isotopic geochemistry evidence. Earth Planet Sci. Lett. 43: 46-60.

Fitch, T.J., 1970. Earthquake mechanisms and island arc tectonics in the Indonesian-Philippine region. Bull. Seismol. Soc. Amer., 60: 565-591.

Fitch, T.J. and Molnar, P., 1970. Focal mechanisms along inclined earthquake zones in the Indonesian-Philippine region. J. Geophys. Res., 75: 1431-1444.

Foden, J.D., 1983. The petrology of the calcalkaline lavas of Rindjani volcano, east Sunda arc: a model for island arc petrogenesis. J. Petrol., 24: 98-130.

Foden, J.D., 1986. The petrology of Tambora volcano, Indonesia: a model for the 1815 eruption. J. Volcanol. Geotherm. Res., 27, 1-41.

Foden, J.D. and Varne, R., 1980. The petrology and tectonic setting of Quaternary-Recent volcanic centres of Lombok and Sumbawa,

Sunda arc. Chem. Geol., 30: 201-226.
Foden, J.D. and Varne, R., 1981a. The geochemistry and petrology of the basalt-andesite-dacite suite from Rinjani volcano, Lombok: implications for the petrogenesis of island arc, calc-alkaline magmas. In: Barber,A.J. and Wiryosujono,S. (Editors), The Geology and Tectonics of Eastern Indonesia. Geol. Res. Develop. Centre Spec. Publ., 2, Bandung, pp. 115-134.
Foden, J.D. and Varne, R., 1981b. Petrogenetic and tectonic implications of near-coeval calcalkaline to highly alkaline volcanism on Lombok and Sumbawa islands in the eastern Sunda arc. In: Barber,A.J. and Wiryosujono,S. (Editors), The Geology and Tectonics of Eastern Indonesia, Geol. Res. Develop. Centre Spec. Publ., 2, Bandung, pp. 135-152.
Fraser, K.J., Hawkesworth, C.J., Erlank, A.J., Mitchell, R.H. and Scott-Smith, B.H., 1985. Sr, Nd and Pb isotope and minor element geochemistry of lamproites and kimberlites. Earth Planet. Sci. Lett. 76: 57-70.
Frey, F.A., Dickey, J.S., Thompson, G. and Bryan, W.B., 1977. Eastern Indian Ocean DSDP Sites: correlations between petrography, geochemistry and tectonic setting. In: Heirtzler,J.R. et al. (Editors), Indian Ocean geology and biostratigraphy, A.G.U. Washington, pp. 189-256.
Frey, F.A., Green, D.H. and Roy S.D., 1978. Integrated models of basalt petrogenesis: a study of quartz tholeiites to olivine melilitites from south eastern Australia utilizing geochemical and experimental petrological data. J. Petrol., 19: 463-513.
Gill, J.B., 1984. Sr-Pb-Nd isotopic evidence that both MORB and OIB sources contribute to oceanic island arc magmas in Fiji. Earth Planet. Sci. Lett., 68: 443-458.
Green, D.H., 1976. Experimental testing of "equilibrium" partial melting of peridotite under water-saturated, high-pressure conditions. Can. Mineral., 14: 255-268.
Green, T.H., 1980. Island arc and continent-building magmatism- a review of key geochemical parameters and genetic processes. Tectonophysics, 63: 367-387.
Hamilton, W., 1979. Tectonics of the Indonesian region. United States Geol. Surv. Prof. Pap., 1078, 345 pp.
Hart, S.R., 1984. A large-scale isotope anomaly in the southern hemisphere mantle. Nature, 309: 753-757.
Hatherton, T. and Dickinson, W.R., 1969. The relationship between andesite volcanism and seismicity in Indonesia, the Lesser Antilles, and other island arcs. J. Geophys. Res., 74: 5301-5310.
Heirtzler, J.R., Veevers, J.J. et al., 1973. Age of the floor of the Eastern Indian Ocean. Science, 180: 952-954.
Hekinian, R., 1974. Petrology of igneous rocks from Leg 22 in the northeastern Indian Ocean. In: von der Borch,C.J., Sclater,J.G. et al. (Editors), Initial Reports of the Deep Sea Drilling Project, Volume 22. U.S. Government Printing Office, Washington, pp. 413-447.
Hutchison, C.S., 1982. Indonesia. In: Thorpe,R.S. (Editor), Andesites: Orogenic Andesites and related rocks. John Wiley & Sons, Chichester, England, pp. 207-224.
Jaques, A.L., Lewis, J.D., Smith, C.B., Gregory,G.P., Ferguson, J., Chappell, B.W. and McCulloch, M.T., 1984. The diamond-bearing ultrapotassic (lamproitic) rocks of the West Kimberley region, western Australia. In: Kornprobst,J. (Editor), Kimberlites. I: Kimberlites and related rocks. Elsevier, Amstrdam, pp. 222-254.
James, D.E., 1981. The combined use of oxygen and radiogenic isotopes as indicators of crustal contamination. Ann. Rev.

Earth. Planet. Sci., 9: 311-344.
Jenner, G.A., White, W.M., Dupre, B.R., Foden, J.D. and Kerrich, R. (in prep.). Geochemistry of Quaternary volcanics from the east Sunda arc, Indonesia: isotopic and incompatible element constraints on the nature of their source region.
McDonaugh, W.F., McCulloch, M.T. and Sun,S.S., 1985. Isotopic and geochemical systematics in Tertiary-Recent basalts from southeastern Australia and implications for the evolution of the sub-continental lithosphere. Geochimica et Cosmochimica Acta, 49: 2051-2067.
Magaritz, M., Whitford, D.J. and James, D.E., 1978. Oxygen isotopes and the origin of high-$^{87}Sr/^{86}Sr$ andesites. Earth Planet. Sci. Lett., 40: 220-230.
Morris, J.D. and Hart, S.R., 1980. Lead isotope geochemistry of the Banda arc. Eos, 61: 1157.
Morris, J.D. and Hart, S.R., 1983. Isotopic and incompatible element constraints on the genesis of island arc volcanics from Cold Bay and Amak Island, Aleutians, and implications for mantle structure. Geochimica et Cosmochimica Acta, 47: 2015-2030.
Menzies, M.A. and Wass, S.Y., 1983. CO_2- and LREE-rich mantle below eastern Australia: a REE and isotopic study of alkaline magmas and apatite-rich mantle xenoliths from the Southern Highlands Province, Australia. Earth Planet. Sci. Lett., 65: 287-302.
Nelson, D., McCulloch, M.T. and Sun, S.S., in press. The origins of ultrapotassic rocks as inferred from Sr, Nd and Pb isotopes. Geochimica et Cosmochimica Acta.
Neumann van Padang, M., 1951. Catalogue of the Active Volcanoes of the World Including Solfatara Fields, Part 1. International Volcanological Association, Napoli, 271 pp.
Nicholls, I.A. and Ringwood, A.E., 1973. Effect of water on olivine stability in tholeiites and the production of silica-saturated magmas in the island arc environment. J. Geol. 81, 285-300.
Nicholls, I.A. and Whitford, D.J., 1976. Primary magmas associated with Quaternary volcanism in the western Sunda arc, Indonesia. In: R.W. Johnson (Editor), Volcanism in Australasia, Elsevier, Amsterdam, pp. 77-90.
Pearce, J.A. and Cann, J.R., 1973. Tectonic setting of basic volcanic rocks determined usingtrace element analyses. Earth Planet. Sci. Lett., 19: 290-300.
Pigram, C.J. and Panggabean, H., 1983. Age of the Banda Sea, eastern Indonesia. Nature, 301: 231-234.
Robinson, P.T. and Whitford, D.J., 1974. Basalts from the Eastern Indian Ocean, DSDP Leg 27. In: Veevers,J.J., Heirtzler,J.R., et al. (Editors), Initial Reports of the Deep Sea Drilling Project, Volume 27. U.S. Government Printing Office, Washington, pp. 551-560.
Robinson, P.T., Thayer, P.A., Cook, P.J. and McKnight, B.K., 1974. Lithology of Mesozoic and Cenozoic sediments of the Eastern Indian Ocean, Leg 27, Deep Sea Drilling Project. In: Veevers, J.J., Heirtzler, J.R. et al. (Editors), Initial Reports of the Deep Sea Drilling Project, Volume 27. U.S. Government Printing Office, Washington, pp. 1001-1048.
Silver, E.A., Gill, J.B., Schwartz, D., Prasetyo, H. and Duncan, R.A., 1985. Evidence for a submerged and displaced continental borderland, north Banda sea, Indonesia. Geology, 13: 687-691.
Silver, E.A., Reed, D., McCaffrey, R. and Joyodiwiryo, Y., 1983. Back arc thrusting in the eastern Sunda arc, Indonesia: a consequence of arc-continent collision. J. Geophys. Res., 88: 7429-7448.

Stern, R.J. and Ito, E., 1983. Trace-element and isotopic constraints on the source of magmas in the active Volcano and Mariana island arcs, western Pacific. J. Volcanol. Geotherm. Res., 18: 461-482.

Sugimura, A. and Uyeda, S., 1973. Island arcs: Japan and its environs. Elsevier, Amsterdam, 247 pp.

Tera, F., Brown, L., Morris, J., Sacks, I.S., Klein, J. and Middleton, R., in press. Sediment incorporation in island-arc magmas: inferences from ^{10}Be. Geochimica et Cosmochimica Acta.

Thompson, R.N., 1982. Magmatism of the British Tertiary Volcanic Province. Scott. J. Geol., 18: 49-107.

Van Bemmelen, R.W., 1949. The geology of Indonesia, Vol 1A. Government Printing Office, The Hague, 732 pp.

Varne, R., 1985. Ancient subcontinental mantle: a source for K-rich orogenic volcanics. Geology, 13: 405-408.

Veevers, J.J., 1977. Models of the evolution of the Eastern Indian Ocean. In: Heirtzler, J.R. et al., Indian Ocean geology and biostratigraphy. A.G.U., Washington, pp. 151-163.

Wheller, G.E. and Varne, R., in press. Genesis of dacitic magmatism at Batur volcano, Bali, Indonesia: implications for the origins of strato-volcano calderas. J. Volcanol. Geotherm. Res.

Wheller, G.E., Varne, R., Foden, J.D. and Abbott, M.J., in press. Geochemistry of Quaternary volcanism in the Sunda-Banda arc, Indonesia, and three-component genesis of island arc basaltic magmas. J. Volcanol. Geotherm. Res.

White,W.M., 1985. Sources of oceanic basalts: Radiogenic isotopic evidence. Geology, 13: 115-118.

White,W.M. and Hofmann.A.W.,1982. Sr and Nd isotope geochemistry of oceanic basalts and mantle evolution. Nature, 296: 821-825.

Whitford, D.J., 1975a. Strontium isotopic studies of the volcanic rocks of the Sunda arc, Indonesia, and their petrogenetic implications. Geochim. Cosmochim. Acta, 39: 1287-1302.

Whitford, D.J., 1975b. Geochemistry and petrology of volcanic rocks from the Sunda arc, Indonesia. unpubl. Ph.D. thesis, Australian National University.

Whitford, D.J., Foden, J.D. and Varne, R., 1978. Sr isotope geochemistry of calcalkaline and alkaline lavas lavas from the Sunda arc in Lombok and Sumbawa, Indonesia. Carnegie Institution of Washington Yearbook, 77: 613-620.

Whitford, D.J. and Jezek, P.A., 1979. Origin of Late Cenozoic lavas from the Banda Sea, Indonesia: trace element and Sr isotope evidence. Contrib. Mineral. Petrol., 68: 141-150.

Whitford, D.J. and Jezek, P.A., 1982. Isotopic constraints on the role of subducted sialic material in Indonesian island-arc magmatism. Geol. Soc. Amer. Bull., 93: 504-513.

Whitford, D.J. and Nicholls, I.A., 1976. Potassium variation in lavas across the Sunda arc in Java and Bali. In: R.W. Johnson (Editor), Volcanism in Australasia. Elsevier, Amsterdam, pp. 63-75.

Whitford, D.J., Nicholls, I.A. and Taylor, S.R., 1979. Spatial variations in the geochemistry of Quaternary lavas across the Sunda arc in Java and Bali. Contrib. Mineral. Petrol., 70: 341-356.

Whitford, D.J., White, W.M. and Jezek, P.A., 1981. Neodymium isotopic composition of Quaternary island arc lavas from Indonesia. Geochimica et Cosmochimica Acta, 45: 989-995.

III. THE ALPINE-MEDITERRANEAN REGION

MECHANISM OF FORMATION OF FOLD BELTS: THE ALPINE-CARPATHIAN REGION

EUGENE V. ARTYUSHKOV[1] and MICHAEL A. BAER[2]
[1]Institute of Physics of the Earth, Moscow (USSR)
[2]Ministry of Geology of the USSR, Moscow (USSR)

ABSTRACT

The development history of the Alpine-Carpathian segment of the Alpine belt is considered. Since the Triassic numerous deep basins on continental crust have been formed there by rapid subsidence without significant stretching or thrust loading. They originated from the destruction of the lower crust under the asthenospheric upwelling. Two oceanic basins were produced by sea-floor spreading in the Jurassic.

Deep basins on oceanic and attenuated continental crust were underlain by a thin lithosphere. From the Late Jurassic and until the Pliocene time they were intensely compressed, which formed the Alps, Carpathians and Dinarides. Compression began in the deepest basins and then covered the basins of a moderate depth. No compression occurred in adjacent cratonic regions with a thick lithosphere. The present fold belts follow the boundary between the past deep basins and the surrounding stable regions.

Compression of the crust in each region ceased after the crustal surface emerged by several hundreds of meters above sea level. This permits estimating the force that produces the compression. High mountains in the folded regions were formed long after the termination of compression. They could have been produced by upwelling of a hot asthenosphere and phase transformations in the lower crust.

INTRODUCTION

The development of fold belts includes three basic phenomena: 1) the formation of deep basins on oceanic and continental crust, 2) an intense compression of the crust and 3) the formation of high mountain ranges (Aubouin, 1965; Dewey et al., 1973; Khain and and Mikhailov, 1985). According to popular ideas these phenomena are associated with the following processes. Oceanic basins form by sea-floor spreading. Deep basins on continental crust form due to lithospheric stretching (McKenzie, 1978; Sclater et al., 1980; Le Pichon and Angelier, 1981, and others), thrust loading (Beaumont

1981), or thermal relaxation in previously stretched regions (Sleep, 1971). Convergent plate motions produce subduction of oceanic lithosphere and obduction of ophiolitic blocks (Plate Tectonics, 1980). High mountains arise from an intense compression of deep basins on the attenuated continental crust or underthrusting of one plate of continental crust of normal thickness under another.

The above scheme explains both horizontal movements (sea-floor spreading and compression of the crust) and vertical movements (formation of deep basins and high mountains) in fold belts in terms of horizontal plate motions. Our analysis of the structure of fold belts (the Urals, Appalachians, Scandinavian Caledonides, the Alpine and Verkhoyansk belts) has revealed that their development took place in a considerably different way (Artyushkov and Baer, 1983, 1984, 1986). Oceanic basins were indeed produced by sea-floor spreading. Most deep basins on continental crust, however, formed without significant stretching or thrust loading. They were produced by a very rapid subsidence of the duration $\sim$1-10 Ma or by a slow subsidence of the duration $\sim 10^3$ Ma. An intense compression and thickening of the crust took place only in oceanic basins or in deep basins on continental crust formed by rapid subsidence. None of cratonic blocks was intensely shortened.

It has been earlier shown on some examples that folding and mountain building represent independent phenomena (Artyushkov et al., 1982). The crustal shortening commonly produced an uplift of the crustal surface only of several hundreds of meters above sea level. A high relief in each region formed after the termination of folding in this region. According to these results the scheme of the development of fold belts appears to be more complicated than is commonly believed. Sea-floor spreading and compression of the crust in these regions are indeed a result of horizontal plate motions. The formations of deep basins on continental crust and high mountains are, however, independent of plate motions and associated with almost purely vertical crustal movements.

Our preceding analysis was mostly oriented on the formation of

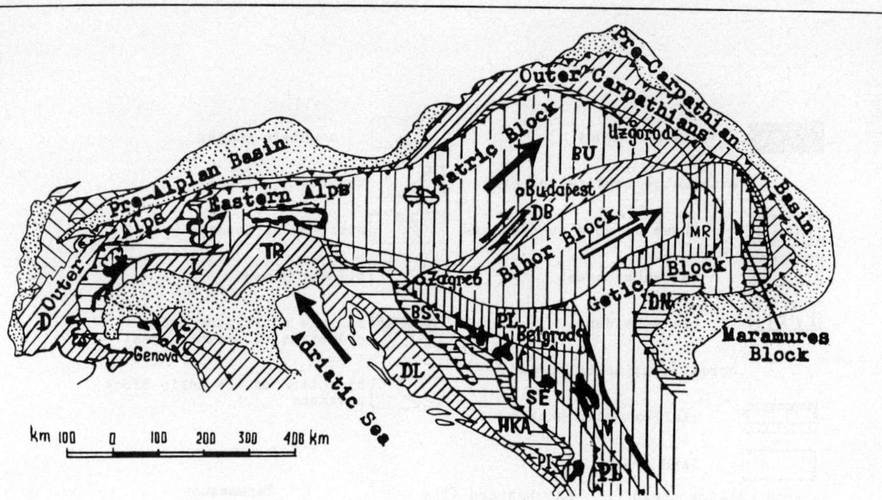

Fig. 1

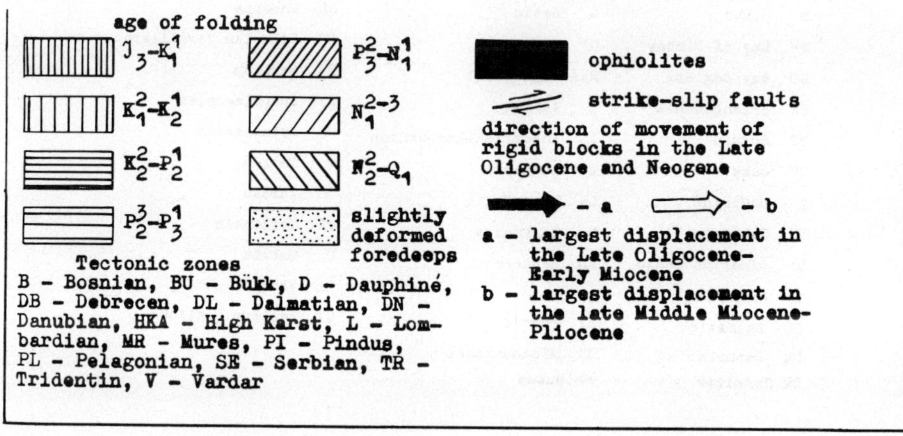

J_3	= Late Jurassic	P_2^1	= early Eocene	N_1^1	= early Miocene
K_1^1	= early Early Cretaceous	P_2^3	= late Eocene	N_1^2	= middle Miocene
K_1^2	= late Early Cretaceous	P_3^1	= early Oligocene	N_1^3	= late Miocene
K_2^1	= early Late Cretaceous	P_3^2	= late Oligocene	N_2^2	= mid. late Pliocene
K_2^2	= late Late Cretaceous			Q_1	= Pliocene

Fig. 1. Schematic tectonic map of the Alpine-Carpathian segment of the Alpine belt (after Aubouin et al., 1970; Cousin, 1970; Debelmas, 1975; Sandulescu, 1975; Debelmas et al., 1980; Baer, 1983, and others).

oceanic crust

attenuated continental crust

three or more subsidences

two subsidences

one subsidence

normal continental crust

shallow-water shelf

land areas

mountain ranges

thrusts and nappes

strike-slip faults

olistostromes

regions of folding and/or obduction of oceanic crust

direction of the Adria Block movement

Ac	Alpine chain	EA	Eastern Alps	P	Parnassus
AN	Antalya	EM	East Mediterranean	PI	Pindus
AP	Apulia	EP	Eastern Pontides	PL	Pelagonian
B	Bihor	G	Getic	RH	Rhodope
BB	Bay of Biscay	GV	Gavrovo	SC	Silesian Cordillera
BD	Bey Daglari	HKA	High Karst	SE	Serbian
BR	Briançonnais	I	Ionian	SP	Subpelagonian
BS	Bosnia	IL	Insubrian-Lombardian	SS	Subsilesian
BU	Bükk	L	Lombard	T	Tatric
D	Dauphiné	LN	Lago Negro	TC	Tarcau
DB	Dobragea	M	Maramures	TR	Tridentin
Dc	Dinarian chain	MDR	Menderes	U	Umbria
DD	Domuz Dag	MG	Magura	VL	Valais
DL	Dalmatia	MI	Moesia	WP	Western Pontides
DN	Danubian	MM	Midmountains		
DR	Durmitor	MR	Mures		

Fig. 2. Paleotectonic reconstructions of the western part of the Alpine belt (a) Late Paleozoic-Early Triassic, (b) Late Triassic (after Biju-Duval et al., 1977; Belov, 1981; Kovacs, 1982; Baer, 1983, and others).

In Fig. 2b a similar facies change is observed in the NS direction in the following blocks: Eastern Alps (EA), Tatric (T), Bihor (B) and Maramures (M). The Carpathian Keuper facies zone is represented by terrestrial and lagoonal deposits, the Main Dolomite and Dachstein limestone facies zone is represented by shallow-water and barrier reef deposits, the Hallstatt limestone facies zone is represented by deep-water sediments. This indicates that the above blocks were located along the northern margin of the Dinarian deep-water basin as shown in Fig. 2b. Dashed line shows the zone of continental break-up which produced the Neo-Tethys.

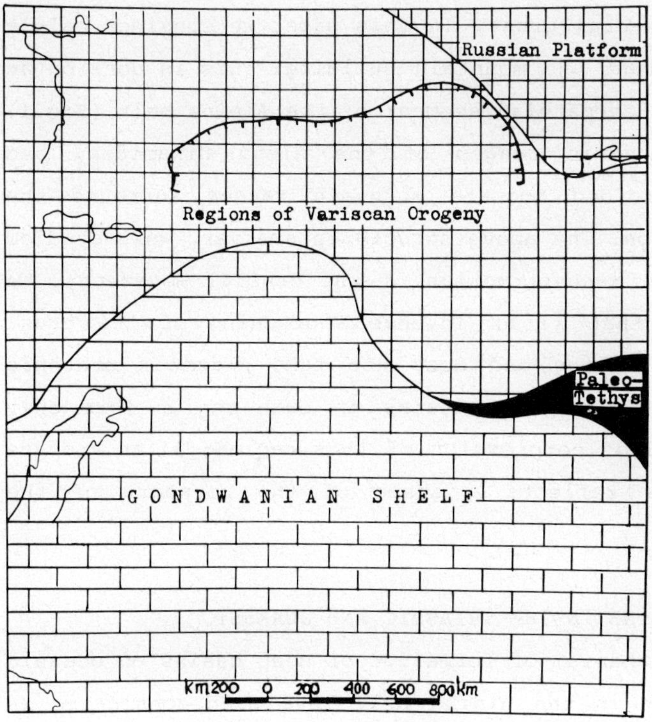
Russian Platform
Regions of Variscan Orogeny
Paleo-
Tethys
G O N D W A N I A N   S H E L F
km 200 0 200 400 600 800 km

– Fig. 2a

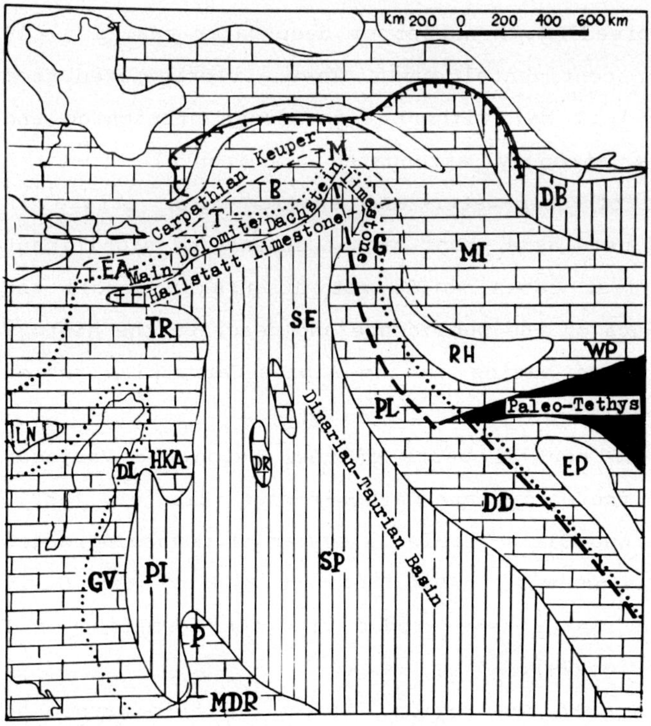
km 200 0 200 400 600 km
Carpathian Keuper
M
B
T
Main Dolomite Dachstein limestone
Hallstatt limestone
EA
G
MI
DB
TR
SE
RH
WP
Paleo-Tethys
PL
Dinarian-Taurian Basin
LN
DL
HKA
DR
EP
DD
GV
PI
SP
P
MDR

– Fig. 2b

deep basins on continental crust. In this paper we consider mainly compression of the crust and mountain building. This is done on an example of the Alpine-Carpathian segment of the Alpine belt (Fig.1). It includes the mountain ranges of the Alps, Dinarides, and Carpathians. These folded arcuate mountain ranges surround the Intra-Carpathian region. The above structures are very well studied wich permits reliable reconstruction of the crustal movements. We are trying to show that 1) an intense shortening of the crust occurs only in deep basins and does not take place in cratonic continental regions; 2) high mountains in each region form only after the termination of compression of this region; 3) an arcuate shape of fold belts reflects a shape of the boundary of the compressed basins.

FORMATION OF DEEP BASINS IN THE TRIASSIC AND JURASSIC

The history and mechanism of formation of deep basins on oceanic and continental crust in the Alpine belt have been considered in Artyushkov and Baer (1984). Several oceanic basins were formed there by sea-floor spreading. A number of deep-water basins (~ 1--3 km) were formed on continental crust. They all originated from rapid subsidence (~ 1-10 Ma) without significant stretching and were underlain by a strongly attenuated continental crust. A conformable position of deep-water and shallow-water strata in numerous sections was used as a proof of the absence of significant stretching. Here in most cases we do not discuss in detail the original geological data on the subsidence and describe the history of the formation of deep basins in the Alpine-Carpathian region only briefly.

In the northern and central parts of the region the Variscan folding terminated in the early Late Carboniferous (Ziegler, 1982). In the Late Carboniferous-Early Permian a system of mountain ranges and intramountain basins was formed. By the end of the Early Triassic the mountain ranges were eroded with the formation of a shallow-water shelf (Fig. 2a).

In the southern part of the region, on the northern margin of

Gondwana, cratonic conditions existed from the Early Paleozoic. In the Early Triassic this also represented a shallow-water shelf. In the mid-Middle Triassic a deep-water Dinarian Basin (Fig. 2b) of a depth 1.5 km $\lesssim h_o \lesssim$ 2.5 km was formed without significant stretching in a few million years. Its deposits are widespread in the Halstatt Nappe in the Eastern Alps, Inner Dinarides and innermost zones of the West and East Carpathians.

The crustal uplift took place on the shelf north of the Dinarian Basin at the transition between the Triassic and Jurassic. Sea-floor spreading occurred there with the formation of the Vardar oceanic basin, ~ 200-500 km wide (Fig. 3a), which represented the northern part of the Neo-Tethys.

The second subsidence of continental crust took place in the late Early Jurassic (Fig. 3a). In 3-10 Ma the basins, ~1-1.5 km deep, were formed on a shelf in the Alps and Carpathians. At the same time the floor of the Dinarian Basin subsided to $\gtrsim$ 2.5 km.

Opening of the Central Atlantics began in the late Middle Jurassic, which produced a sinistral displacement of Africa with respect to Europe (Lancelot, 1980; Dercourt et al., 1985). In the Alpine belt this displacement caused sea-floor spreading in the Apennines, Western and Northern Alps and West Carpathians with the formation of the Ligurian-Piedmontian oceanic basin, 200-600 km wide (Fig. 3b). The southwestern end of this basin was connected by a transform fault with an opening oceanic basin in the Central Atlantics. In the southeastern part of the Alpine-Carpathian region relative movement of Africa and Europe produced compression of the Vardar oceanic basin with a strike normal to the trend of the Ligurian-Piedmontian Basin (see next section).

The third rapid subsidence of continental crust occurred in the Late Jurassic which resulted in a widening of deep basins (~ 1.5-3 km) in the Alps and Carpathians (Fig. 3b). Additional deepening probably occurred in the Dinarian Basin.

By the end of the Jurassic a large deep-water basin on oceanic and attenuated continental crust was formed (Fig. 3b). Its boundary with a shallow-water European shelf on the north was close to the present boundary of the Alps and Carpathians.

oceanic crust

attenuated continental crust

three or more subsidences

two subsidences

one subsidence

normal continental crust

shallow-water shelf

land areas

mountain ranges

thrusts and nappes

strike-slip faults

olistostromes

regions of folding and/or obduction of oceanic crust

direction of the Adria Block movement

Ac	Alpine chain	EA	Eastern Alps	P	Parnassus
AN	Antalya	EM	East Mediterranean	PI	Pindus
AP	Apulia	EP	Eastern Pontides	PL	Pelagonian
B	Bihor	G	Getic	RH	Rhodope
BB	Bay of Biscay	GV	Gavrovo	SC	Silesian Cordillera
BD	Bey Daglari	HKA	High Karst	SE	Serbian
BR	Briançonnais	I	Ionian	SP	Subpelagonian
BS	Bosnia	IL	Insubrian-Lombardian	SS	Subsilesian
BU	Bükk	L	Lombard	T	Tatric
D	Dauphiné	LN	Lago Negro	TC	Tarcau
DB	Dobragea	M	Maramures	TR	Tridentin
Dc	Dinarian chain	MDR	Menderes	U	Umbria
DD	Domuz Dag	MG	Magura	VL	Valais
DL	Dalmatia	MI	Moesia	WP	Western Pontides
DN	Danubian	MM	Midmountains		
DR	Durmitor	MR	Mures		

Fig. 3. Paleotectonic reconstructions of the western part of the Alpine belt (a) late Early Jurassic, (b) late Late Jurassic-early Early Cretaceous. Position of Africa and Eurasia and lines of continental break-up here and in Figs. 4 and 5 are shown according to Dercourt et al., 1985.

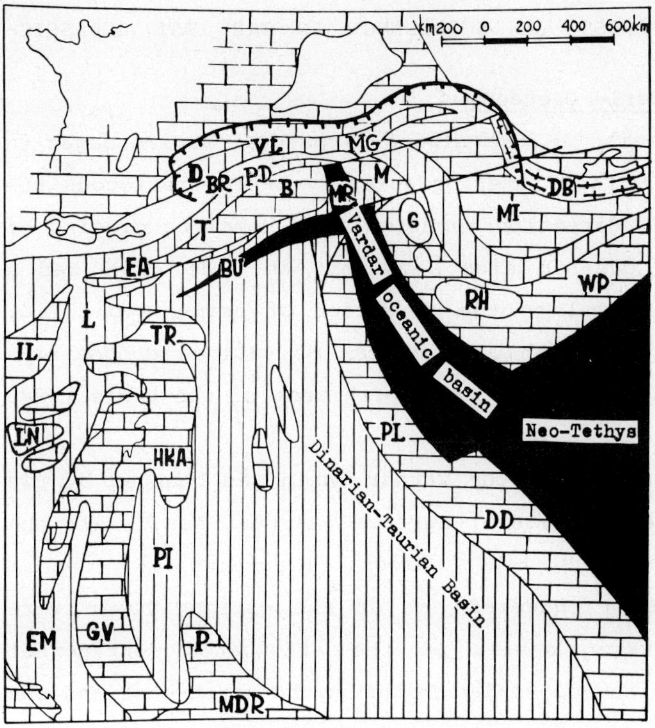
km 200 0 200 400 600km
VL
MG
D
BR
PD
B
M
MR
DB
MI
G
T
Vardar
oceanic
basin
EA
BU
WP
RH
L
TR
IL
Neo-Tethys
IN
HKA
PL
Dinarian-Taurian Basin
DD
PI
EM
GV
P
MDR

Fig. 3a

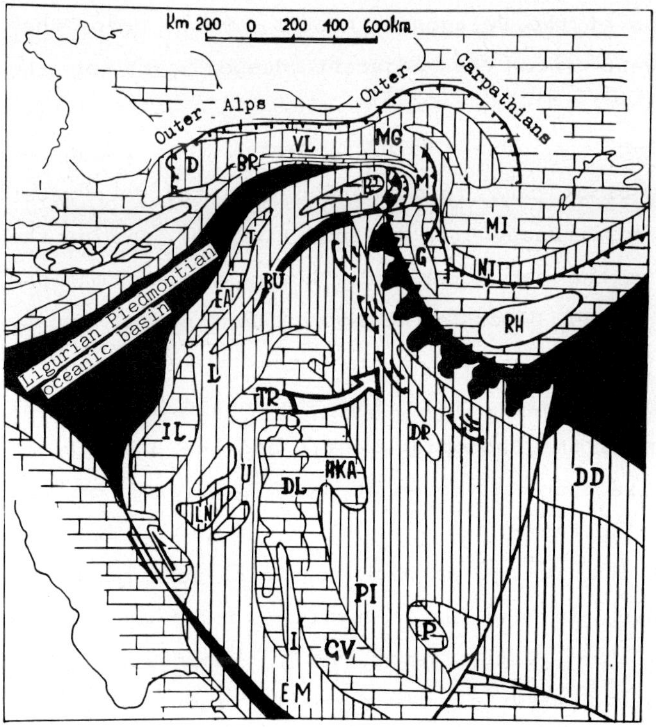
Km 200 0 200 400 600km
Outer Alps
Outer Carpathians
VL
MG
D
BR
B
M
MI
G
NT
T
BU
EA
Ligurian Piedmontian oceanic basin
RH
L
TR
IL
DR
DD
U
HKA
DL
LN
PI
P
I
GV
EM

Fig. 3b

COMPRESSION OF THE CRUST AND SUBSIDENCE IN THE LATE JURASSIC--NEOGENE

Late Jurassic-early Early Cretaceous

An intense compression in the Alpine-Carpathian region began in the Late Jurassic. In the Kimmeridgian-early Early Cretaceous the compression in the NE direction resulted in a complete closing of the Vardar Basin on oceanic crust, 200-500 km wide, that had a trend approximately normal to the direction of compression (Fig. 3b). This was associated with a volcanism of various composition (andesitic, basaltic, acidic) to the north-west of the basin (Savu et al., 1978).

On the east and north-west the Vardar Basin was bounded by two elevated blocks where the crustal surface was near sea level: the Rhodopian Massif and the Bihor Block. Fragments of oceanic lithosphere (ophiolites) squeezed out of the closing basin were juxtaposed on deep-water basins on the attenuated continental crust: the Transilvanian-Bucovinian Basin on the north-east and the submerged Pelagonian Massif on the Dinarian Basin on the south--west. From the region of the Pelagonian Massif, ~1 km deep, they slid to the south-west into the adjacent deeper part of the Dinarian Basin (~3 km).

After the closing of the Vardar Basin the Rhodopian Massif and the Bihor Block came into contact with the deep-water Dinarian and Transilvanian-Bucovinian basins (Fig. 3b). During the following collision in the early Early Cretaceous no deformations occurred in the two former blocks with normal continental crust. Under their pressure shortening by 2-3 times took place in the two latter basins (100-300 km wide) with the attenuated continental crust. The Getic Zone of the Rhodopian Massif and the Bihor Block with the opposite direction of movement were separated by the Transdanubian strike-slip fault.

Despite an intense shortening, no high mountains were formed in the compressed regions in the Late Jurassic and early Early Cretaceous (Aubouin, 1965; Dimitrijevič, 1974; Rampnoux, 1970). Almost no coarse clastic material was deposited and the crustal surface emerged above sea level only by about several hundreds of meters.

Late Early Cretaceous - early Late Cretaceous

In this epoch compression occurred mainly in the NW direction and was normal to the compression in the Late Jurassic - early Early Cretaceous . In the Late Aptian-Albian it resulted in a shortening of a basin on oceanic crust that was approximately normal to the direction of compression (Fig. 4a). This basin, 200-600 km wide, was located in the Ligurian Zone of the Apennines, Inner Piedmont Zone of the Western and Northern Alps and Inner Pieninian Zone of the West Carpathians. In the first zone compression mainly occurred from a counter-clockwise rotation of the Iberian Block (Dercourt et al., 1985). In the second and third zones it was produced by the NW movement with a counter-clockwise rotation of the Adria Block. This movement is supposed to have resulted from sea-floor spreading in the Eastern Mediterranean.

The above oceanic basin was almost completely closed in the Alps and West Carpathians. After an intense shortening the crustal surface (or more exactly the surface of a melange of blocks of oceanic lithosphere) slightly emerged above sea level (Oberhauser, 1968; Elter and Trevisan, 1973). This region was a source for flysch deposits in the adjacent basins on the attenuated continental crust. No andesitic volcanism took place. This probably indicates that the compressed oceanic lithosphere was not subducted deeply into the mantle and remained near the surface under the basin.

Between the Albian and Cenomanian an intense compression also occurred in the axial zone of the Dinarian Basin on the attenuated continental crust (Fig. 4a). This basin was formed by three rapid subsidences and had a great depth. The region, 200 km wide, was shortened by 2-3 times.

Between the Cenomanian and Turonian a new rapid subsidence took place on continental crust. A basin, 100-200 km wide, was formed on the European shelf in the Danubian Zone of the South Carpathians.

(Fig. 4a) (Sandulescu, 1975; Nachev, 1980). Rapid subsidence also occurred on rigid microcontinental blocks (Briançonnais, East Alpine, Bihor and Tatric). The depth of the basins was of ~1 km.

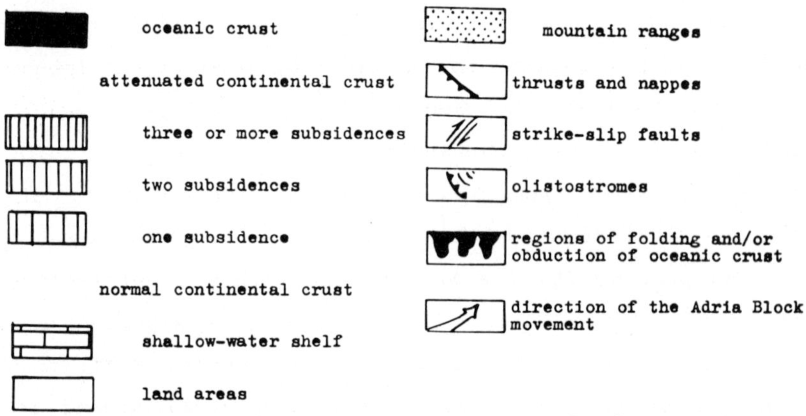

Ac	Alpine chain	EA	Eastern Alps	P	Parnassus
AN	Antalya	EM	East Mediterranean	PI	Pindus
AP	Apulia	EP	Eastern Pontides	PL	Pelagonian
B	Bihor	G	Getic	RH	Rhodope
BB	Bay of Biscay	GV	Gavrovo	SC	Silesian Cordillera
BD	Bey Daglari	HKA	High Karst	SE	Serbian
BR	Briançonnais	I	Ionian	SP	Subpelagonian
BS	Bosnia	IL	Insubrian-Lombardian	SS	Subsilesian
BU	Bükk	L	Lombard	T	Tatric
D	Dauphiné	LN	Lago Negro	TC	Tarcau
DB	Dobragea	M	Maramures	TR	Tridentin
Dc	Dinarian chain	MDR	Menderes	U	Umbria
DD	Domuz Dag	MG	Magura	VL	Valais
DL	Dalmatia	MI	Moesia	WP	Western Pontides
DN	Danubian	MM	Midmountains		
DR	Durmitor	MR	Mures		

Fig. 4. Paleotectonic reconstructions of the western part of the Alpine belt (a) late Early-early Late Cretaceous, (b) late Late Cretaceous- Early Eocene.

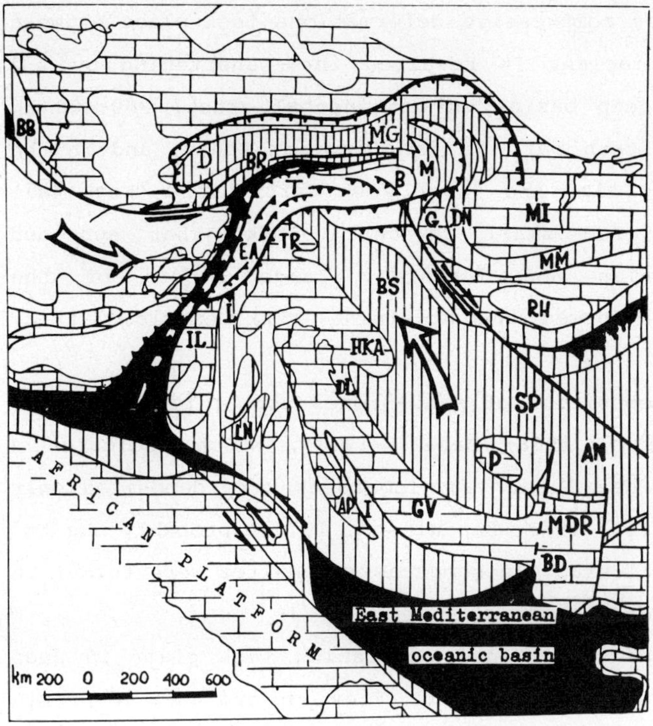
BB
D
BR
MG
B
M
T
EA
TR
G
DN
MI
MM
BS
RH
I
IL
HKA
DL
SP
LN
P
AN
AP
I
GV
MDR
BD
AFRICAN PLATFORM
East Mediterranean
oceanic basin
km 200 0 200 400 600

Fig. 4a

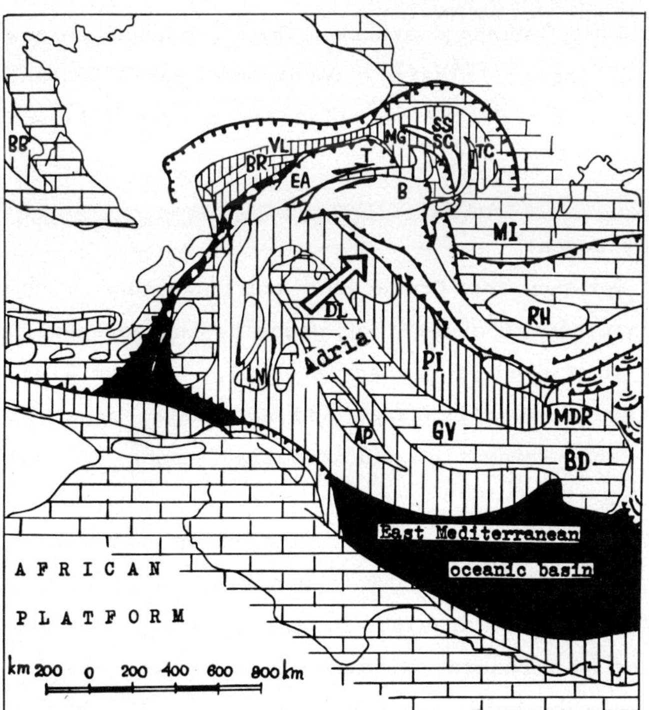
BB
VL
BR
EA
T
MG
SS
SC
TC
B
MI
DL
RH
Adria
PI
LN
AP
GV
MDR
BD
East Mediterranean
oceanic basin
AFRICAN
PLATFORM
km 200 0 200 400 600 800 km

Fig. 4b

The next episode of compressive deformations took place between the Turonian and Coniacian. It resulted in a shortening by 2-3 times of a band of deep basins on continental crust, 300-400 km wide, in the northwestern part of the Dinarian Basin, and in the basins of the Eastern Alps and Inner West Carpathians (Fig. 4a). Compression also covered small adjacent blocks that subsided between the Cenomanian and Turonian. Large nappes of the sedimentary cover were formed: Austro-Alpine, Bükk, Gemeric, Tatric, and others.

In most regions deposition of sand-stones took place after the compression (Oberhauser, 1968; Mahel, 1974), which indicates a small altitude above sea level. Conglomerates were deposited only locally (Andrusov et al., 1973; Sandulescu, 1975), probably due to erosion of small cordilleras. In some cases bauxites were formed in the absence of an intense erosion.

Until the Paleogene no intense shortening took place in deep flysch basins in the Outer Carpathians (Figs. 3b and 4a). Compression resulted in displacements of the attenuated crust along thrust faults for a comparatively short distance. This produced narrow chains of islands-cordilleras (Silesian, Maramures, Sub-Silesian, Kumanian). They were a source of the material for a coarse flysch in the adjacent basins.

Late Cretaceous - Early Eocene

At the transition between the Cretaceous and Paleocene a new rapid subsidence of continental crust took place to the south-west of the Dinarian Basin in a band of the Adria shelf, ~200 km wide. This formed deep basins in the Pre-Karst and High-Karst zones of the Outer Dinarides and Tridentin and Friulian zones at the Southern Alps (Fig. 4b).

In the Paleocene-Early Eocene an intense compression occurred in the Alpine Belt in the NE direction. This was associated with a rapid collision of Africa and Eurasia. In the Alpine-Carpathian region a broad band of the remnant Dinarian Basin - the Bosnian Zone, ~300 km wide, was shortened by about two times. The lithosphere of this basin was underthrust to the north-east under the Inner Dinarides. Thin plates of the sedimentary cover were detached from the basement, juxtaposed on each other and emplaced to the south-west on the Pre-Karst Zone.

During the preceding epoch of compression the Tatric Block was displaced by ~100 km to the north-west (Fig. 4a). In the Paleocene the granitic and sedimentary layers of this block were displaced by ~300 km to the north-east into the Inner Pieninian Zone (Fig. 4b). As a result the Tatric Block that was originally located to the south-west of the Bihor Block (see Fig. 4a) appeared to be located to the north-west of it.

Under the pressure of the Tatric Block the sedimentary cover of the Pieninian Zone was detached from the basement and strongly shortened with the formation of a folded arc. The NE displacement of the Bihor Block produced compression by~100 km in the adjacent inner part of the flysch basin of the Outer Carpathians (Fig. 4b). This increased the curvature of the fold belt near the junction of the East and South Carpathians.

After the compression the crustal surface in all the basins emerged above sea level; however, conglomerates were deposited only locally (Andrusov et al., 1973; Mahel, 1974). This again indicates that no high mountains were formed.

Compression during the Late Jurassic-Early Eocene formed a general folded structure of the Inner Alps, Inner Carpathians and the Intra-Carpathian region. No intense compression took place there later.

Late Eocene - Early Oligocene

A new rapid subsidence in the Alpine-Carpathian region took place in shelf regions in the early Late Eocene (Fig. 5). In the Dalmatian Zone of the Outer Dinarides this produced a deeper-water basin, ~150-200 km wide. In the Outer Alps deep basins, ~100-150 km wide were formed in the Dauphine and Helvetic zones. These were the last large basins on continental crust that formed in the Alps and Dinarides by rapid subsidence.

Between the Eocene and Oligocene an intense compression took place all along the Alpine belt and was associated with the collision of Africa and Europe (Dercourt et al., 1985). In the Alpine-Carpathian region compression occurred mostly in deep basins on continental crust in NS and NE directions. In the Outer Dinarides this produced shortening by 2-3 times of deep basins, 300-400 km wide, that formed in the Middle Triassic, Late Cretaceous and Paleogene (Fig. 5). The compression formed the Outer Dinarides.

Compression of the same intensity took place in the inner part of the Outer Alps in a band of basins, 100-200 km wide, that formed by rapid subsidence in the Early Jurassic, Late Jurassic and Early Late Cretaceous. After this epoch of compression all deep basins in the Alps that had formed prior to the Late Eocene became intensely shortened. As a result the margin of the Alpine folded arc reached an arcuate boundary of the above basins. For the first time the Alpine fold belt acquired an arcuate shape.

No intense compression took place in the Carpathians in the Late Eocene-Early Oligocene. By the end of the Eocene the flysch basin, 200-300 km wide, in the Outer Carpathians was filled with sediments up to a considerable extent. Between the Eocene and Oligocene a new rapid subsidence took place there which considerably increased the depth (Fig. 5) (Baer and Shcherba, 1984).

There are no Early Oligocene conglomerates in the Dauphine--Helvetic Basin in the Alps and the Ionian Basin in the Dinarides (Debelmas, 1979; Savoyat, 1977). Hence the adjacent regions of the Alps and Dinarides just after the termination of the Pyrenean Orogeny were not highly elevated above sea level.

In the Alps and Dinarides the mountains, ~ 1 km high, were formed only in the Late Oligocene, i.e., 5-10 Ma after the compression between the Eocene and Oligocene (Aubouin, 1965; Trümpy 1973). The mountain ranges were gradually destroyed by erosion by the end of the Miocene.

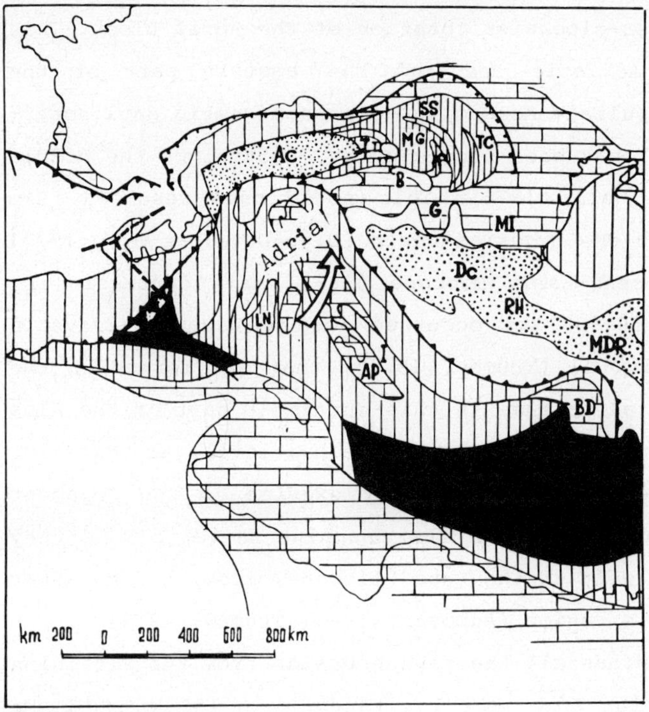

Fig. 5. Paleotectonic reconstruction of the western part of the Alpine belt for the Late Eocene-Early Oligocene (see legends in Figs. 2 - 4). Mountain ranges are shown for Late Oligocene-Early Miocene time.

Late Oligocene - Pliocene

In the Late Oligocene and Early Miocene the collision of Africa and Europe produced compression of the Alpine-Carpathian region in the NW direction (Dercourt et al., 1985). In this epoch an intense crustal shortening, however, took place not only along the western and northern margins of the region, i.e., in the Alps, and West Carpathians. It also occurred in the East Carpathians that were approximately parallel to the direction of collision. This required considerable displacements of comparatively rigid blocks between Africa and Europe in the direction normal to that of the continental collision. It is most probable that this compression occurred from a counter-clockwise rotation of the Adria Block. This block includes the Adriatic Sea and the eastern part of the Apennine Peninsula (Apulia). According to paleomagnetic data rotation of the Adria Block has been suggested by $\sim 25°$ since the Middle Eocene (Vanderberg et al., 1978). This could have resulted from sea-floor spreading in the Ligurian Basin (Biju-Duval et al., 1977).

In the Late Oligocene-Early Miocene (30-17 Ma) compression by about two times (50-100 km) occurred in the Dauphine-Helvetic Basin of the Alps (Fig. 1) (Trümpy, 1973), that was formed in the Late Eocene. The crustal surface in this outermost part of the Alps emerged by several hundreds of meters above sea level. At that time the mountains of moderate height already existed in the adjacent inner part of the Outer Alps. In the Dauphine-Helvetic Zone they were, however, formed only in the Late Pliocene, i.e., $\sim$15 Ma after the compression of this region (Aubouin, 1965; Trümpy, 1973).

In the West Carpathians all the flysch basins from the Pieninian Zone to the Sub-Silesian Zone were shortened by $\sim$150 km 30-17 Ma ago (Andrusov et al., 1973; Mahel, 1974). This was associated with the NNE displacement of the Tatric Block. In the East Carpathians during the same epoch compression of only 50-70 km took place from the displacement of the Bihor Block (Sandulescu, 1975; Baer and Shcherba, 1984). These blocks with a different magnitude of the displacement were separated by the Zagreb-Zemplin dextral strike--slip fault (see Fig. 1).

Since the Middle Miocene a collision of Africa and Europe again occurred in the NE direction (Dercourt et al., 1985). In the late Middle-early Late Miocene (13-11 Ma) an intense compression occurred in the East Carpathians. Several nappes of the sedimentary cover of the outer part of the flysch basins were emplaced on each other with a shortening about 100 km (Fig. 1). Then this pile of nappes was emplaced on the adjacent margin of the Russian Platform. The visible amplitude of this displacement is of $\sim$50 km. Hence in the above epoch the total shortening in the East Carpathians was $\gtrsim$150 km. This probably occurred from the ENE movement of the Bihor Block. The ENE displacement of the Tatric Block was considerably smaller. It was separated by a strike-slip fault from the West Carpathians (Royden et al., 1982).

The last compression took place 2-3 Ma ago in the Foredeep of the Carpathians. This occurred in a region $\sim$200 km long near the junction of the East and South Carpathians (Baer and Shcherba, 1984). The compression had a moderate magnitude up to several tens of kilometers.

After the compression in the Oligocene-early Late Miocene the crustal surface slightly emerged above sea level. Deposition of conglomerates in the Trans-Carpathian Basin and the Foredeep of the Carpathians began only in the mid-Late Miocene. This indicates the formation of an elevated relief ($\sim$ 0.5 - 1 km) in the West and East Carpathians 11 - 10 Ma ago.

In the East Carpathians this took place just after the termination of an intense compression (13-11 Ma). No new crustal shortening however, occurred during the uplift. In the West Carpathians the uplift took place 6 Ma after the termination of folding.

The main uplift in the Carpathians that formed their present relief took place only recently - in the Late Pliocene - Early Pleistocene (Hofstein, 1964; Feldbarg, 1985). This occurred long after the termination of the last epoch of an intense folding in the West (15 Ma), East (8 Ma) and South (65 Ma) Carpathians. This intense uplift in the Carpathians was concomitant with the formation of mountain ranges all over the Alps and Dinarides.

In contrast to the Alps, Dinarides and the other segments of the Alpine belt, no exposures of the basement of the flysch basins are known in the Carpathians. All the consolidated crust was underthrust during compression. Basic volcanic rocks of the tholeiitic type were found in some places in the lower part of the flysch deposits (Sandulescu, 1975; Lomize, 1983). This permitted the suggestion that the inner part of these basins was underlain by the Late Jurassic oceanic crust (Herz and Savu, 1974; Dercourt et al.,

1985). The composition of the tholeites does not, however, permit to relate them reliably to the oceanic type (Lomize, 1983). In contrast to many other folded regions, no ophiolitic blocks with the structure typical of oceanic crust were found in the flysch basins of the West and East Carpathians. Hence, this problem still remains unsolved.

Subduction of oceanic lithosphere in the Carpathians in Neogene time is usually supposed, according to the calc-alkaline volcanism, to have occurred 18-3 Ma ago (Royden et al., 1982, and others). Such a volcanism is indeed typical of island arcs; however, it took also place in many other regions that were very far from the subduction zones. For instance, calc-alkaline volcanism was typical of the Dinarian-Taurian Basin in the Middle Triassic (Celet et al., 1977). No subduction occurred in the Alpine belt at that time. The Paleozoic oceanic crust had already been subducted in the Carboniferous, while the formation of a new oceanic crust began only between the Triassic and Jurassic.

FORMATION OF THE INTRA-CARPATHIAN BASINS IN THE MIOCENE-PLIOCENE

A system of basins of a considerable depth was formed in the Intra-Carpathian region in this epoch (Fig. 6). The crust is considerably thinner there (22-25 km, Stegena et al., 1975) than under the adjacent mountain belts. Sclater et al. (1980) suggested that the subsidence and thinning of the crust occurred from lithospheric stretching by about two times that took place 13 Ma ago. No deep basins were formed immediately after the supposed stretching because of the concomitant upwelling of a hot asthenosphere to the crust. The following gradual subsidence mainly resulted from cooling of the mantle. In the above paper an intense stretching before the subsidence was not, however, proved. The analysis of the sedimentary cover structure in this basin revealed no intense stretching just before the subsidence (Artyushkov and Baer, 1984).

Royden et al.,(1982) suggested that stretching by 75-100 km had occurred earlier (16-13.5 Ma ago) and produced a system of minor

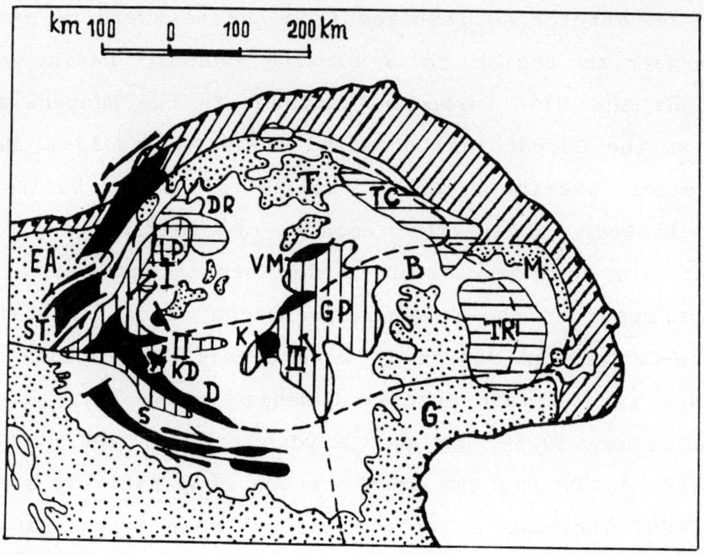

Fig. 6

region of a slight Neogene-Pleistocene subsidence

regions of an intense Pannonian subsidence

regions of an intense Badenian subsidence
a - with significant stretching
b - without significant stretching

regions of Pre-Neogene folding

regions of Neogene folding

boundaries of rigid blocks

	Pannonian basins		
GP	Great Plain	TC	Trans-Carpathian
LP	Little Plain	TR	Transylvanian
	Badenian basins	V	Vienna
D	Drava	VM	Vatta-Maklar
DR	Danube-Raba		Rigid blocks
K	Kiskun	EA	Eastern Alps
KD	Kadarkut	T	Tatric
S	Sava	B	Bihor
ST	Styrian	M	Maramures
		G	Getic

Fig. 6. Scheme of location of the Neogene deep basins in the Intra--Carpathian region.
(after Royden et al., 1982; Horvath and Royden, 1983; Balla, 1985; Pogacsas, 1985; Pospisil and Vass, 1985; Nikolaev, 1986).

basins. The stretching resulted from the eastward movement of the Intra-Carpathian region to a closing oceanic basin between the Carpathians and East European Platform. In the Neogene an intense folding in the Carpathians occurred 30-17 Ma and 13-11 Ma ago (see the previous section). The formation of minor basins occurred exactly between these two epochs of folding and could not accomodate for the compression in the Carpathians.

Let us consider the Neogene subsidence of continental crust in the Intra-Carpathian region in more detail. An intense subsidence took place in two main epochs: Badenian time, 16.5-13 Ma ago and Pannonian time, 10.5-2 Ma ago (Royden et al., 1982; Pospisil and Vass, 1985). During the former epoch comparatively small basins were formed. They can be related to two main types. Larger peripheral basins (Danube-Raba, Transylvanian, Trans-Carpathian, see Fig. 6) formed by rapid subsidence without significant stretching. Terrestrial and shallow-water deposits were abruptly overlain by deeper-water marls and turbidites (Ciupagea et al., 1970; Geology and Hydrocarbons of the Ukrainian Carpathians, 1971). The transition is several meters thick. The depth of the basins was of 0.5 km.

The strata formed before the subsidence were not significantly disrupted within the basins. This indicates no significant extension during the subsidence. Consider, e.g., a profile across the Danube-Raba Basin (Fig. 7a). In the central part it includes a half-graben AC. An absolute extension in this structure is approximately equal to a horizontal projection AB=1.8 km of the normal fault that bounds the half-graben on the south-west. The initial width of the half-graben approximately equals the segment BC = 13.6 km. The stretching factor β is AC/BC = 1.13, which corresponds to an extension of 13%. This half-graben is, however, filled by the Cretaceous deposits 2 and 3. Hence the main extension occurred in the Cretaceous, i.e., long before the Badenian subsidence. The stratum 4 (Helvetic) that deposited just before this subsidence is practically continuous. It is only slightly disrupted by some normal faults; however, they ensure the total extension along the profile of Fig. 7a of only several per cent.

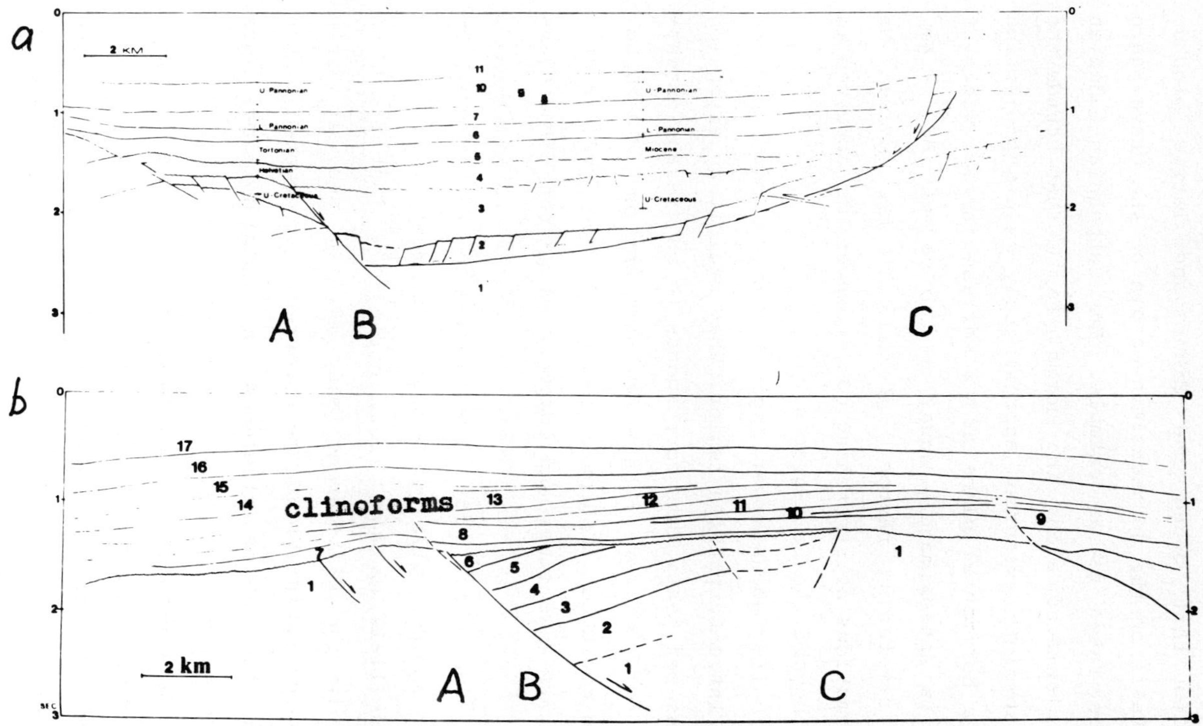

Fig. 7. Seismic time sections through two Badenian basins in the Intra-Carpathian region (Pogacsas, 1985).

a. Southwestern part of the Danube-Raba Basin (see profile I in Fig. 6).

b. The Kadarkut trough (see profile II in Fig. 6). 1=Pre-Tertiary formations, 2-6=Carpathian-Badenian formations, 7-17=Upper Badenian-Sarmatian-Pannonian-Pleistocene formations. Formations filling the trough (2-6) are overlain by younger sediments with erosional unconformity. Among the Pannonian formations seismic sub-units 10-14 represent prograding delta facies (clinoforms).

The half-grabens formed in Badenian time (Vienna, Styrian, Kadarkut, Kishkun, Vatta-Maklar and others, see Fig. 6) can be related to the second basin type. They are located near large strike-slip faults that bound the Bihor and Tatric blocks.

The Kadarkut Basin on the northeastern margin of the Drava Basin (Fig. 7b) can be taken as an example. The Carpathian-Badenian strata 3-6 are displaced along the normal fault that bounds the half-graben. An absolute extension approximately equals projection of the fault on a horizontal plane AB $\approx$ 2 km. The initial width of the half-graben is approximately equal to the segment BC 7 km. This gives the stretching factor $\beta \sim 1.3$ in the graben. No significant deformations of stretching can, however, be observed in the same profile beyond the half-graben.

The Badenian half-grabens cover together only a small portion ($\sim$5-10%) of the Intra-Carpathian region. They exhibit an extension up to several tens per cent. This corresponds to an average extension of the region of no more than a few per cent or 10-20 km. Such an extension could not apparently produce the observed crustal attenuation by about two times and compression in the Carpathians by 150-200 km.

In a few million years most Badenian basins were filled by sediments. Then a slight uplift took place over most of the Intra-Carpathian region 11-12 Ma ago, i.e., during the last epoch of intense folding in the East-Carpathians. 10.5 Ma ago, in Early Pannonian time, a new rapid subsidence occurred that covered a much wider area than the basins stretched in Badenian time. This produced two deep basins: the Great Plain and Little Plain basins (Fig. 6) which form together the Pannonian Basin (Pogacsas, 1985).

After the rapid subsidence terrestrial and shallow-water deposits were abruptly replaced by turbidites. Then in $\sim$3 Ma the basins were filled by clinoforms (Fig. 8a), up to 700-800 m high. This approximately corresponds to the initial water-depth in the basins. In the second half of the Late Miocene and Pliocene the clinoforms were overlain by 1-3 km of terrestrial deposits.

Pannonian strata that deposited after the rapid subsidence rest

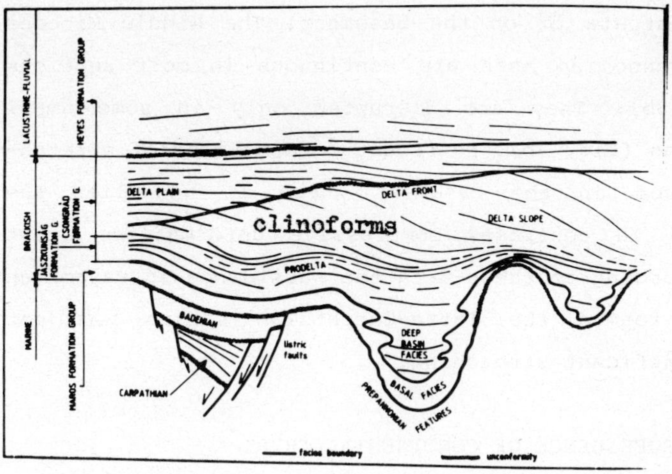

Fig. 8a

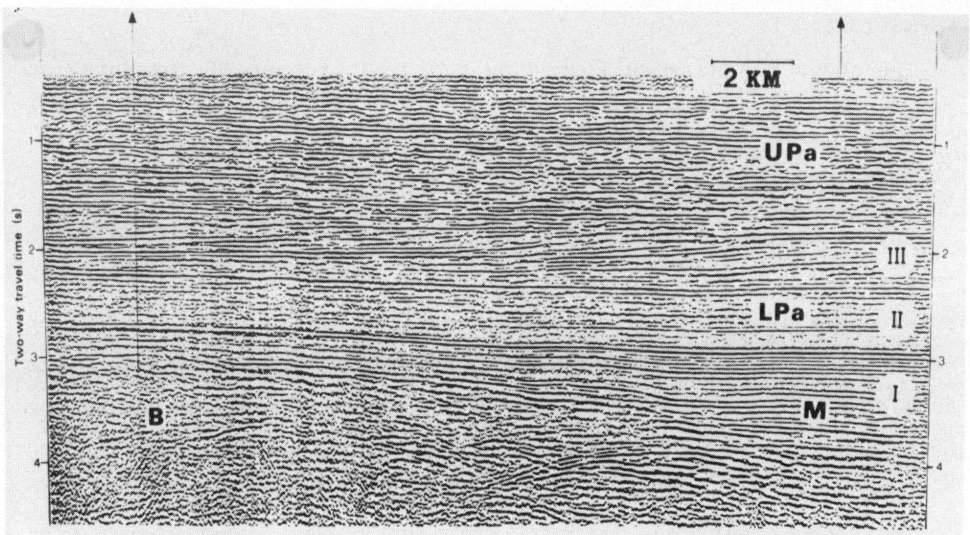

Fig. 8b

Fig. 8. The Pannonian basins of the Intra-Carpathian region. a. General scheme of Neogene basins evolution and sediment filling of the Pannonian Basin (after Pogacsas, 1985).
b. Seismic time section through the Mako depression of the Great Plain of the Pannonian Basin (after Varga and Pogacsas, 1981) (see profile III in Fig. 6).
B = Basement, M = Miocene layers, LPa = Lower Pannonian, UPa = Upper Pannonian sediments.

on Middle Miocene strata or on the basement. The Middle Miocene strata under the Pannonian ones are continuous in most sections (see, e.g., Fig. 8b). They are disrupted only in some small Badenian half-grabens (Fig. 7b). No recent deformations of stretching can be observed in the basement where it underlies the Pannonian deposits. In contrast, compressive deformations exist in some places. According to these data the subsidence in Pannonian time that mainly formed the Intra-Carpathian basins was not associated with significant stretching.

MECHANISM OF RAPID SUBSIDENCE OF CONTINENTAL CRUST

Deep basins on continental crust in the Alpine-Carpathian region were formed without significant stretching (see Artyushkov and Baer, 1984 and previous section). Let us discuss possible causes of their formation.

The above basins were formed by rapid subsidence of the duration ~1-10 Ma. This precludes a significant role of thermal relaxation which has a much larger characteristic time of ~50 Ma (Sleep, 1971).

Rapid subsidence of continental crust and emplacement of large nappes commonly took place in different epochs. For instance, the subsidences in the Middle Triassic and late Early Jurassic occurred before the onset of compression in the Late Middle Jurassic. The subsidence in a broad shelf region in the Outer Dinarides in the Late Eocene took place between two epochs of compression: in the Paleocene and Early Oligocene.

In some cases an intense subsidence was concomitant with compression; however, these processes occurred far one from another. For example, the subsidence in a large flysch basin in the East Carpathians in the Early Oligocene took place when compression occurred in the Alps and Dinarides, i.e., 300-400 km from the East Carpathians. The above time and space relationships preclude thrust loading as a mechanism of the subsidences.

According to the condition of isostatic balance the crust under deep basins formed without thrust loading should be strongly attenuated. The same conclusion follows from the relation between the intensity of shortening and the crustal thickness. Most folded arcs in the Alpine-Carpathian region were formed by the shortening of deep basins on continental crust by several times and are now underlain by crust 30-55 km thick (3-5 km of rocks could have been removed by erosion). Hence, before the compression the crust in these basins was several times thinner. Consider some typical examples.

In the early Early Cretaceous the Transylvanian-Bucovinian Basin and the inner part of the Dinarian Basin were shortened by 2-3 times, which produced the Inner East Carpathians and Inner Dinarides. No intense compression occurred in these regions after the Early Cretaceous. The present crustal thickness is $h_c \sim 40\text{-}45$ km (Sollogub et al., 1980). In the early Late Cretaceous shortening of the same intensity in the northwestern part of the Dinarian Basin and the adjacent regions formed the Eastern Alps and the Inner West Carpathians. The crustal thickness is $\sim 40\text{-}55$ km and ~ 35 km, respectively (Mueller and Lowrie, 1980). Compression by 2-3 times of the Dauphine and Piedmont Basins took place in the Early Miocene and Early Oligocene respectively. This produced the Western Alps underlain by the crust 40-50 km thick (Menard and Thouvenot, 1984). Hence the crustal thickness in all these basins before the compression was $h_c \sim 15\text{-}25$ km.

Most deep basins on continental crust formed on a shallow-water shelf. The crustal thickness in shelf regions is $h_c \sim 35\text{-}40$ km (Belyaevsky, 1974, 1981). Rapid subsidence and thinning of the crust without significant stretching at the surface should be associated with the distruction of the lower crust. The removal of the lower crust is apparently possible only under upwelling of the asthenosphere to the base of the crust.

Two main situations can be considered. Subcrustal erosion produced by convective flows in the mantle (Gilluly, 1963) is associated with the removal of the material of the lower crust from beneath the basin to the adjacent regions. This should produce the crustal uplift in these regions concomitantly with the subsidence.

Furthermore, rapid erosion seems to be possible only under a very high temperature of the lower crust which strongly reduces the viscosity. Rapid subsidence occurred mostly on a cool shelf and was associated with no intense uplift in the adjacent regions. This precludes subcrustal erosion as a mechanism.

In our previous papers we suggested a gabbro-eclogite transformation as a possible cause of rapid subsidence of continental crust. This occurs under upwelling of hydrous asthenosphere (anomalous mantle) at moderate temperature T $\sim$ 800-900° C. Under the contact with anomalous mantle basaltic layer (where it exists) can rapidly transform into eclogite with a high density $\rho \sim 3.5$-3.6 g/cm^2. The dense eclogite tears off the crust and sinks into the anomalous mantle of low viscosity. This can destroy all the basaltic layer or a considerable part of it. The attenuated crust subsides with the formation of a deep basin.

The mechanism of rapid gabbro-eclogite transformation in the lower crust meets many difficulties (Artyushkov and Baer, 1986). Other known mechanisms, however, meet even more difficulties. This is why we are using the above mechanism as a working hypothesis.

All the rapid subsidences in the Alpine-Carpathian region were preceded by slight crustal uplift. Slight volcanism preceded some subsidences or took place during them. For instance, a regional crustal uplift and slight basic volcanism occurred 8 Ma before the subsidence in the Middle Triassic that formed the Dinarian Basin (Ziegler, 1982). Slight calc-alkaline volcanism took place during the subsidence (Celet et al., 1977). Similarly, the subsidence between the Cenomanian and Turonian was preceded by a slight crustal uplift (Andrusov et al., 1973; Sandulescu, 1975). Slight uplift and volcanism could be associated with upwelling of anomalous mantle at moderate temperature that caused the subsidences.

CONDITIONS FOR AN INTENSE SHORTENING

We have considered the formation of three folded ranges: the Alps, Carpathians and Dinarides. They were formed by an intense compression of deep basins on oceanic and continental crust. All the basins formed by sea-floor spreading were completely closed. The total volume of the oceanic lithosphere blocks emplaced on the adjacent regions is very small. As in the other belts the folded structures of the Alpine-Carpathian region are mainly built up of a strongly shortened crust of deep basins on continental crust.

Compression started in the oceanic basins in the axial part of the region. Then it spread to the deepest adjacent basins on continental crust that were formed by two or three rapid subsidences. At the latest stages compression occurred in the marginal basins of a comparatively small depth. This is a well known phenomenon of migration of folding from the axis of the geo-syncline belt to the adjacent platforms (frame) (Aubouin, 1965). It indicates that compression of deeper basins requires a smaller outer force. This can be explained as follows.

Thickening of the crust from compression increases its potential energy proportionally to the square of the crustal thickness h_c^2. Neglecting a finite strength of the lithosphere the force necessary for compression of the crust up to the thickness h_c also increases as h_c^2 (Artyushkov, 1973, 1986; Artyushkov et al., 1982):

$$\Sigma = [(\rho_c - \rho_o)(\rho_m - \rho_c)/(\rho_m - \rho_o)]\, g h_c^2/2 \qquad (I)$$

Here is the density of the crust, ρ_o - the density of water, ρ_m the density of the mantle and g - the gravity. In the regions where the crustal surface is above sea level ρ_o should be taken as zero.

Compression of the crust should start where the force Σ necessary for compression takes minimum values. According to (I) this occurs where the crustal thickness h_c has the smallest values, i.e., inthe basins on oceanic crust and deepest basins on continental crust.

As in the other fold belts no cratonic block was shortened in the Alpine-Carpathian region. This means that cratonic lithosphere of a large thickness $d \gtrsim 100$ km has a high strength and cannot be compressed by the forces that drive plate motions. An intense shortening of only deep basins indicates that they were underlain not only by a thin crust but by a thin lithosphere too.

Oceanic basins form by sea-floor spreading above upwellings of the asthenosphere at high temperature T ~ 1300°C (Dmitriev et al., 1984). Deep basins on the attenuated continental crust form by rapid subsidence above upwellings of the asthenosphere at moderate temperature T ~ 800-900°C (Artyushkov and Baer, 1983, 1984, 1986). In both cases the initial upwelling produces a long living thermal inhomogeneity (~ 50 Ma) which decreases the lithospheric thickness.

In tectonically active regions upwelling of the asthenosphere repeats several times with the intervals of several tens of million years (ibid). Migration of new portions of the asthenospheric anomalous mantle into the regions with reduced lithospheric thickness (traps) rejuvenates the thermal inhomogeneity and decreases again the lithospheric thickness (Yanshin et al., 1977; Artyushkov, 1983). Then an intense shortening of thin crust becomes possible in the basins that are approximately normal to the direction of compression. For example, the Vardar Basin of the NW strike was closed in the Late Jurassic by compression in a NE direction. At that time the Ligurian-Piedmontian Basin of the NE strike remained comparatively stable. It was shortened in the late Early Cretaceous when compression took place in the NW direction.

Collision of lithospheric plates produces shortening of deep basins located between the plates. The front of folding gradually migrates to the adjacent cratons. Folding stops after this front reaches the boundary of deep basins on continental crust with the adjacent shelf. The boundary of such basins is commonly of an arcuate shape. Their shortening produces arcuate fold belts that follow the outer boundary of the basins. For example, the boundary of the present deep-water Black Sea Basin with the Russian Platform has an arcuate shape. Shortening of the Black Sea in the NS direction would produce a fold belt along this boundary.

THICKENING OF THE CRUST AFTER THE COMPRESSION AND THE FORCE NECESSARY FOR IT

Compression of the crust commonly ceases after the crustal surface reaches several hundreds of meters above sea level. This equals the altitude of high continental platforms where the crustal thickness is $_c = h_c^o \sim$ 40-45 km (Belyaevsky, 1974, 1981). An intense compression occurs when the crust is underlain by a heated anomalous mantle with a decreased density. In such regions the crustal surface should emerge to the level of continental platforms after it reaches the thickness $h_c < h_c^o$

Deep basins on continental crust arise due to the destruction of the lower crust that can originally be denser than the upper crust. The crust formed by the compression of deep basins where the basaltic layer has been completely destroyed will consist of only sialic rocks of an average density $\rho_{sl} \sim$ 2.65-2.70 g/cm^3.

Let us designate the density and thickness of the anomalous mantle layer by ρ_{am} , $_{am}$ and the density of the crust of continental platforms by ρ_c^o . Then it follows from the condition of isostatic balance that the maximum thickness of a sialic crust that can be reached after the compression is:

$$h_{sl}=\left[(\rho_m - \rho_c^o)\, h_c^o-(\rho_m - \rho_{am})\, h_{am}\right]/(\rho_m - \rho_{sl}) \qquad (2)$$

Let us take ρ_m = 3.35 g/cm^3, ρ_{am} = 3.30 g/cm^3, h_{am} = 20 km, ρ_c^o = 2.8 g/cm^3, ρ_{sl} = 2.65 g/cm^3. Then for h_c^o = 40-45 km we obtain:

$$h_{sl} = 30\text{-}34 \text{ km} \qquad (3)$$

In the Alpine-Carpathian region the crust of such a thickness exists in the West Carpathians and Innermost Dinarides (Sollogub et al., 1980).

According to (I) the force necessary to produce the crust of this thickness by compression is:

$$\Sigma = (2.5\text{-}3.2)\ 10^9 \text{ bar cm} \qquad (4)$$

During the compression the crust is underlain by the anomalous mantle, which increases the force Σ as compared with the value (4). In this case using relation (47) in Artyushkov (1973) Σ can be represented as

$$\Sigma = \left[\rho_{sl}\ (\rho_m - \rho_{sl})\ g h_{sl}^2 + \rho_{am}\ (\rho_m - \rho_{am})\ g h_{am}^2 + 2\rho_{sl}\ (\rho_m - \rho_{am})\ g h_{sl}\ h_{am}\right]/(2\rho_m) \qquad (5)$$

Taking g = 10^3 g/cm^3 and the above values of the other parameters we obtain:

$$\Sigma = (2.8\text{-}3.6)\ 10^9 \text{ bar cm} \qquad (6)$$

The main forces that drive plate motions: the ridge push Σ_{RD}

and subduction pull Σ_{SD} are smaller than (6): $\Sigma_{RD} \sim 2\cdot10^9$ bar cm, $\Sigma_{SD} \sim 0.6\cdot10^9$ bar cm (Artyushkov, 1983). Their concentration on fold belts is necessary to produce the observed compression (Artyushkov, 1986).

Under many mountain ranges the crustal thickness is considerably higher than the values of (3): $h_c \gtrsim h_c$ =40-45 km. For example, in the Outer Dinarides $h_c \sim$ 40-45 km (Sollogub et al., 1980). In the Outer East Carpathians $h_c \sim$ 55-60 km. In the South Carpathians $h_c \sim$ 48-52 km (Dobrev and Shchukin, 1974). In the Alps h_c reaches 50-55 km (Mueller and Lowrie, 1980). If the density of the crust just after the compression was the same as in present time high mountains would have arisen by the end of folding. A thick crust underlaid by the anomalous mantle could be located at the same level as high platforms only if it was denser than in the platform region.

If rapid subsidence of continental crust really occurs from gabbro-eclogite transformation, then only rocks that are denser than the underlying anomalous mantle tear off the crust and sink into the mantle. A layer of intermediate composition may exist in many regions between the granitic and basaltic layers. It can include blocks of both granite and gabbro. After the transformation of gabbro to eclogite the average density of this layer ρ_i may be still lower than that of the underlying anomalous mantle: $\rho_i > \rho_{am}$. Then this intermediate layer will be preserved in the lower crust after the subsidence.

In the process of compression light sialic rocks mainly accumulate in the upper crust, while denser rocks concentrate in the lower crust (Artyushkov, 1986). Then it may be expected that in folded regions dense rocks of an intermediate composition are mainly located in the lower crust.

A high density layer of this type probably now exists under the Foredeep of the Carpathians. The crustal thickness $h_c \sim$ 65 km (Sollogub et al., 1980) is higher there than under the adjacent Russian Platform (40-45 km) and the Outer East Carpathians (55-60 km). It can be supposed that dense rocks underthrust the crust of the Russian Platform during the compression of the East Carpathians.

Let us suppose that after the compression the crust includes an intermediate layer of the density ρ_i and thickness h_i. Then according to the condition of isostatic balance the maximum crustal thickness that can be reached after the compression is

$$h_c = h_{sl} + \left[(\rho_i - \rho_{sl})/(\rho_m - \rho_{sl})\right] h_i \quad (7)$$

where h_{sl} is determined by (2). Let us take $\rho_i = 3.25$ g/cm^3, $h_i = 25$ km, $\rho_{sl} = 2.65$ g/cm^3, $\rho_m = 3.35$ g/cm^3. Then we find from (7), (3)

$$h_c = 51\text{-}55 \text{ km} \quad (8)$$

Let us assume for the sake of simplicity that the density of the intermediate layer equals that of the anomalous mantle under the crust: $\rho_i = \rho_{am}$. Then the anomalous mantle and the intermediate layer can be considered as a layer of the density ρ_{am} and the total thickness $h_i + h_{am}$. Assuming, e.g., that after the compression the crust includes the sialic layer with $\rho_{sl} = 2.65$ g/cm^3, $h_{sl} = 30\text{-}34$ km, and the intermediate layer with $\rho_i = 3.30$ g/cm^3, $h_i = 20$ km. Let the crust be underlain by the anomalous mantle layer with $\rho_{am} = 3.30$ g/cm^3, $h_{am} = 20$ km. Then taking $h_{am} = 40$ km in (5) we obtain:

$$\Sigma = (3.4\text{-}4.1)\ 10^9 \text{ bar cm} \quad (9)$$

This force is only by ~ 20% larger than the force (6) necessary for compression of the sialic crust.

POSSIBLE MECHANISMS OF MOUNTAIN BUILDING

At least eight episodes of intense compression occurred in the Alpine-Carpathian region in the Mesozoic and Cenozoic. They produced the fold belts of the Alps, Carpathians and Dinarides.

In the Inner Carpathians an intense folding terminated 85 Ma ago. In the Outer Carpathians it terminated 60 Ma ago in the South Carpathians, 17 Ma ago in the West Carpathians and 11 Ma ago in the East Carpathians. The mountains of moderate height (~0.5-1 km) were formed 10-11 Ma ago; however, high mountains in all the Carpathians originated only 2-3 Ma ago.

In the Dinarides an intense folding terminated 30 Ma ago. The mountains of moderate height were formed mainly in the Inner Dinarides 20-28 Ma ago. The formation of the present mountain range began 3 Ma ago.

In the Inner Alps an intense compression terminated 85 Ma ago; in the inner zones of the Outer Alps it terminated 33 Ma ago. 5-10 Ma after mountains of a moderate height were formed in both these regions. During this uplift folding without formation of high mountains occurred in the outer zones of the Outer Alps. It terminated 17 Ma ago. Then 3 Ma ago a new intense uplift covered all the Alps which formed the present high relief.

As was mentioned above in the Variscan epoch mountain building in the Alpine-Carpathian region began in the late Late Carboniferous, while an intense shortening terminated in the early Late Carboniferous.

These data indicate that mountain building in the Alpine--Carpathian region was not directly associated with compression of the crust and represented an independent phenomenon. This resembles the formation of high intraplate plateaus that arise without significant compression after a long period of cratonic development. For instance, in the Northern Tien-Shan an intense folding terminated 220 Ma ago (Tectonic Map of Eurasia, 1966). Then cratonic conditions were typical of the region in the Mesozoic--Eocene. The crustal uplift that formed the present high relief began only in the Oligocene 35 Ma ago and was not associated with a new compression (Makarov and Spiridonov, 1983).

An intense crustal uplift without compression requires a considerable density decrease in the crust or mantle or thickening of the crust by some other process. Let us consider the main possibilities.

Tectonically active regions are characterized by repeated up-wellings of the anomalous mantle. The thickness and temperature (and density) of the anomalous mantle that fills the trap under a folded structure can be different in different epochs. Let us designate the density and thickness of the layer of the anomalous mantle under the crust during compression by ϱ^{0}_{am}, h^{0}_{am}. Let the anomalous mantle layer of the density ϱ^{1}_{am}, h^{1}_{am} underlie the crust in the other epoch. Then the additional uplift of the crustal surface is:

$$\Delta \zeta_1 \left[(\varrho - \varrho^{1}_{am})\, h^{1}_{am} - (\varrho_m - \varrho^{0}_{am})\, h^{0}_{am}\right] / \varrho_m \tag{10}$$

Let us take $\varrho^{0}_{am} = 3.30\ g/cm^3$, $h^{0}_{am} = 20$ km, $\varrho^{1}_{am} = 3.25\ g/cm^3$, $h^{1}_{am} = 30\text{-}40$ km. Then

$$\Delta \zeta_1 = 0.6\text{-}0.9 \text{ km} \tag{11}$$

Compensation of the relief mainly by the anomalous mantle should be typical of the mountain ranges with a thin crust. For example, in the West Carpathians $h_c \sim 30\text{-}35$ km. According to (3) without a strong decrease in the density of the mantle even a purely sialic crust of this thickness should be several hundreds of meters above sea level. The average altitude of the West Carpathians is $\zeta \sim 1$ km. Hence this mountain range is compensated mostly by a decrease in the density in the mantle. The same situation can be supposed for the Inner Dinarides ($h_c \sim 30\text{-}40$ km, $\zeta \sim 1.5$ km) and Inner Carpathians ($h_c \sim 30\text{-}40$ km, $\zeta \sim 1.3$ km). These mountain regions of a moderate height are underlain by the crust that is thinner than that of high continental platforms. The anomalous mantle that compensates the uplift may be located just under the crust or it can be separated from the crust by a layer of the mantle lithosphere, several tens of kilometers thick.

Segregation of the basaltic magmas takes place from a high temperature anomalous mantle ($T \sim 1300°$) on the mid-oceanic ridges. This produces a layer of basalts, 4-5 km thick, in oceanic crust. Similar phenomenon can occur under fold belts on the continents. Under compressive stresses typical of such regions basaltic magmas are unable to reach the surface and will freeze under the crust.

The formation of a new layer of gabbro of the density ϱ_g and a thickness h_g in the lower crust produces an uplift by

$$\Delta \zeta_2 = \left[(\varrho_m - \varrho_g) / \varrho_m\right] h_g \tag{12}$$

Taking $\varrho_g = 2.95\ g/cm^3$, $h_g = 5$ km we have:

$$\Delta \zeta_2 = 0.6 \text{ km} \tag{13}$$

Several repeated upwellings of anomalous mantle at a high

temperature can produce a new basaltic layer under the sialic crust.

As was discussed in the previous section the lower crust of fold belts can include large volumes of rocks of intermediate composition. After the termination of compression the basic material in these rocks exists mainly in a form of dense eclogite, which is stable under low or moderate temperature T ~ 800-900°C. Upwelling of anomalous mantle at high temperature to the base of the crust can produce an inverse transformation of eclogite into a lighter gabbro. This should result in the additional crustal uplift by

$$\Delta \zeta_3 = (\Delta \rho_i / \rho_i) \; h_i^1 \qquad (14)$$

Here $\Delta\rho_i$ is a decrease in the density in the intermediate layer, ρ_i^o - the initial density of this layer before the uplift, h_i^1 - the thickness of the intermediate layer after the uplift.

Assuming, e.g., that the intermediate layer before the uplift includes forty volumetric per cent of granite with $\rho = 2.7$ g/cm^3 and 60% of eclogite with $\rho = 3.55$ g/cm^3. Then its average density is $\rho_i^1 = 3.21$ g/cm^3. Let us suppose that after heating all the eclogite was transformed into gabbro with $\rho = 2.95$ g/cm^3. Then the average density of the intermediate layer becomes $\rho_i^1 = 2.86$ g/cm^3, $\Delta\rho = 0.35$ g/cm^3. Let the transformation eclogite-gabbro occurs in the lowermost layer of the crust of the final thickness $h_i^1 = 10$-15 km. Then we find from (15) that

$$\Delta\zeta_3 = 1.1\text{-}1.6 \text{ km} \qquad (15)$$

The additional uplift $\Delta\zeta_3$ from eclogite-gabbro transformation occurs concomitantly with the additional uplift $\Delta\zeta_1$ from upwelling of a hot anomalous mantle. According to (2), (15) the combination of these effects can produce the crustal uplift by

$$\Delta\zeta = 1.7\text{-}2.5 \text{ km} \qquad (16)$$

Segregation of the basaltic magmas from the anomalous mantle can increase the magnitude of the uplift.

Most mountain ranges in the Alpine-Carpathian region have a small or moderate average height ζ, although some peaks reach a high altitude. Thus ζ is of ~ 1 km in the West Carpathians, $\zeta \sim 1.2$ km in the East Carpathians, $\zeta \sim 1.5$ km in the South

Carpathians, Dinarides and Outer Alps. Only in the Inner Alps the average altitude of the relief is $\zeta \sim 2.5$ km. The formation of these mountain ranges can, in principle, be explained by the above three mechanisms.

The available seismic and gravity data could be used for revealing a relative role of the above mechanisms in the formation of the Alps, Dinarides and Carpathians. This problem, however, requires a special paper.

The traps at the base of the lithosphere arise after the first asthenospheric upwelling that forms deep basins. They continue to exist during the stage of folding. Filling of these traps by a hot anomalous mantle after thickening of the crust from compression eventually produces high mountains. No upwelling and no uplift occurs in the adjacent cratonic regions, since they are underlain by a thick lithosphere.

CONCLUSIONS

As follows from the above consideration the development of the Alpine-Carpathian segment of the Alpine belt occurred in a rather simple way. In the Triassic-Middle Jurassic several deep basins were formed on oceanic and continental crust. Oceanic basins were produced by sea-floor spreading. Deep basins on continental crust originated from rapid subsidence without significant stretching or thrust loading. The subsidence occurred due to the destruction of the lower crust under upwelling of the anomalous mantle at moderate temperature.

Compression of the crust began in the Late Jurassic and proceded until the Pliocene. All deep basins formed before the beginning of compression were shortened several times. A number of new deep basins on continental crust were formed during the above epoch of compression. They were also shortened except the Intra-Carpathian basins that formed quite recently.

Compression began in the oceanic basins and deepest basins on continental crust (2-3 km). Then it covered the basins on continental crust of a moderate depth ($\sim$ 1 km). The intensity of compression in the latter basins was considerably lower. Compression ceased after deep basins were shortened and the front of folding reached their boundary with stable cratonic regions. The present arcuate fold belt of the Alps, Carpathians and Dinarides follows this boundary.

Deep basins on a thin crust were underlain by the anomalous mantle. This strongly decreased the lithospheric thickness and permitted an intense shortening under the convergent plate motions. No cratonic block with a thick lithosphere was intensely compressed. This indicates that the forces driving plate motions are insufficient for compression of cratonic lithosphere.

After all the epochs of compression the crustal surface in the compressed regions emerged only up to several hundreds of meters above sea level. This permits us to estimate the force that compresses the crust in fold belts as $\Sigma \sim (3\text{-}4)\ 10^9$ bar cm.

The crustal compression began $\sim$ 150 Ma ago. The first mountains of moderate height in the Alps, Dinarides and Carpathians were formed only 28-10 Ma ago. During this uplift no compression occurred in the uplifting regions. The present high relief originated 2-3 Ma ago, i.e., 8 Ma after the termination of the last intense compression in the Alpine-Carpathian region. This means that the formation of high mountains was not associated with thickening of the crust from compression. Mountain building can be caused by upwelling of a hot asthenosphere and/or decrease in the density of the lower crust from phase transformations.

Thus, the basic processes that ensure the development of fold belts are rapid subsidence of continental crust and sea-floor spreading. They produce deep basins underlain by thin crust and thin lithosphere. Such weak regions become easily compressed under the convergent plate motions. This increases again the crustal thickness, and the crustal surface slightly emerges above sea level. The following upwelling of a hot mantle into the traps under the compressed regions eventually results in the formation of high mountains.

REFERENCES

Andrusov, D., Bystritsky, Ya. and Fusan, O.,1973. Geological structures of the West Carpathians. Guidebook for excursions X Congress Carp. - Balk. Geol. Assoc. Geol. Inst. D. Stur., Bratislava, 50 pp.

Artyushkov, E.V., 1973. Stresses in the lithosphere caused by crustal thickness inhomogeneities. J. Geophys. Res., 78: 7675--7708.

Artyushkov, E.V., 1983. Geodynamics. Elsevier, Amsterdam, 312 pp.

Artyushkov, E.V., 1986. The forces driving plate motions and compression of the crust in fold belts. Mid-Term Report of the Interunion Commission on the Lithosphere, Symposium 106. Am. Geophys. Union, Washington, D.C..

Artyushkov, E.V. and Baer, M.A., 1983. Mechanism of continental crust subsidence in fold belts: the Urals, Appalachians and Scandinavian Caledonides. Tectonophysics, 100: 5-42.

Artyushkov, E.V. and Baer, M.A., 1984. Mechanism of continental crust subsidence in the Alpine belt. Tectonophysics, 108: 193--228.

Artyushkov, E.V. and Baer, M.A., 1986. Mechanisms of formation of deep basins on continental crust in the Verkhoyansk fold belt; miogeosynclines and cratonic basins. Tectonophysics, 122: 217-245

Artyushkov, E.V., Baer, M.A., Sobolev, S.V. and Yanshin, A.L., 1982. Mechanism of formation of fold belts. Sovetskaya Geologia, 9: 22-36 (in Russian).

Aubouin, J., 1965. Geosynclines. Amsterdam, Elsevier, 335 pp.

Aubouin, J., Blanchet, R., Cadet, J.-P., Celet, P., Charvet, J., Chorowicz, J., Cousin, M., Rampnoux, J.-P., 1970. Essai sur la géologie des Dinarides. Bull. Soc. Géol. Fr., 7 Sér., XII (6): 1060-1095.

Baer, M.A., 1983. Carpathians and Dinarides in the Mesozoic. Geotektonika, No.2: 58-70.

Baer, M.A. and Shcherba, I.G., 1984. Late Alpine history of the evolution of the East Carpathians. Byull. Mosk. O-va Ispyt. Prir., Otd. Geol. LIX(4): 47-62 (in Russian).

Balla, Z., 1985. The Carpathian loop and the Pannonian basin: a kinematic analysis. Geophysic. transactions, 30(4): 313-353.

Beaumont, C., 1981. Foreland basins. Geophys. J. of Roy. Astron. Soc. 65: 291-329.

Belov, A.A., 1981. Tectonic evolution of the Alpine Fold Region. Nauka, Moscow, 212 pp. (in Russian).

Belyaevsky, N.A., 1974. The Earth Crust of the territory of the USSR. Nedra, Moscow, 280 pp. (in Russian).

Belyaevsky, N.A., 1981. The structure of the Earth Crust of the continents according to the geological and geophysical data. Nedra, Moscow, 432 pp. (in Russian).

Biju-Duval, B., Dercourt, J. and Le Pichon, X., 1977. From the Tethys ocean to the Mediterranean seas: a plate tectonic model of the evolution of the Western Alpine system. In: Structural history of the Mediterranean basins, Editions Technip., Paris: pp. 143-164.

Celet, P., Cadet, J.P., Charvet, J. and Ferriere, J., 1977. Volcano-sedimentary and volcano-detritic phenomena of Mesozoic age in dinarid and hellenic ranges: a comparison. In: Structural history of the Mediterranean basins. Edition Technip., Paris, pp. 35-46.

Ciupagea, D., Pauca, M. and Ichem, Tr., 1970. Geologia Depresiunii Transilvaniei, Editura Academiei Republicii Socialiste Romania, Bucuresti, 256 pp.

Cousin, M., 1970. Esquisse géologique des confins italo-yougoslaves: leur place dans les Dinarides et les Alps méridionales. Bull. Soc. Géol. Fr., 12(6): 1034-1047.

Debelmas, J., 1975. Réflexions et hypothèses sur la paléogéographie crétacée des confins alpino-apenniniques. Bull. Soc. Géol. Fr., 7 Sér., XVII(6): 1002-1012.

Debelmas, J., 1979. Alpes Savoic et Dauphiné. Guides géologiques régionaux. Masson, Paris-New York-Barcelone-Milan, 215 pp.

Debelmas, J., Oberhauser, R., Sandulescu, M. and Trümpy, R., 1980. L'arc alpino-carpathique. In: J. Aubouin, J. Debelmas and M. Latreille (Editors), Geology of the Alpine Chains born of the Tethys. Bureau de Recherches Géologiques et Minieres, Paris, pp. 86-96.

Dercourt, J., Zonenshain, L.P., Ricou, L.-E., Kazmin, V.G., Le Pichon, X., Knipper, A.L., Grandjaequet, C., Sborshchikov, J.M., Boulin, J., Sorokhtin, O., Geyssant, J., Lepvrier, C., Biju-Duval, B., Sibuet, J.-C., Savostin, L.A., Westphal, M. and Lauer, J.-P., 1985. Présentation de 9 cartes paléogéographiques au 1/20.000.000 s'étendant de l'Atlantique au Pamir pour la période du Lias à l'Actuel. Bull. Soc. Géol. Fr., 8 Sér., I(15): 637-652.

Dewey, J.E., Pitman, W.C. III, Ryan, W.B.F., Bonnin, J., 1973. Plate Tectonics and the evolution of the Alpine system. Bull. Geol. Soc. Am., 84(10): 3137-3180.

Dimitrijevic, M.D., 1974. The Dinarides: a model based on the new global tectonics. In: S. Jankovic (Editor), Metallogeny and concepts of the geotectonic development of Yugoslavia, Belgrade, pp. 141-178.

Dmitriev, L.V., Sobolev, A.V., Ukhanov, A.V., Malysheva, T.V. and Melson, W.J., 1984. Primary differences in oxigene fugacity and depth of melting in the mantle source regions for oceanic basalts. Phys. Earth. Planet. Inter., 70: 303-310.

Dobrev, T.B. and Shchukin, Yu.K., 1974. Geophysical Fields and Seismicity of the Eastern Part of the Carpathian-Balkanian Region. Nauka, Moscow, 171 pp. (in Russian).

Elter, P. and Trevisan, L., 1973. Olistostromes in the tectonic evolution of the Northern Apennines. In: K.A. de Jong and R. Scholten (Editors), Gravity and Tectonics. Wiley, New York, pp. 183-196.

Feldbarg, N.E., 1985. On the main features of the orogenic stage of the evolution of the East Carpathians. In: Proceeding reports XIII Congress Carp.-Balk. Geol. Assoc., p. I, sect. 2, Tectonics. Publ. Geol. Inst., Poland, Cracow, pp. 350-353.

Geology and Hydrocarbons of the Ukrainian Carpathians, 1971. Nedra,

Moscow, 329 pp. (in Russian).
Gilluly, J., 1963. The tectonic evolution of the western United States, Q. J. Geol. Soc. London, 119:133-174.
Herz, N. and Savu, H., 1974. Plate tectonics history of Romania. Bull. Geol. Soc. Am., 85: 1429-1440.
Hofstein, I.D., 1964. Neotectonics of the Carpathians. Izv. Akad. Nauk. Ukr.SSR, Kiev, 192 pp.
Horvath, F., and Royden, L., 1983. Mechanical model for the formation of the Intra-Carpathian basins. Anuarul. Inst. Geol. Geofiz., LX: 65-73.
Khain, V.E. and Mikhailov, A.E., 1985. General Geotectonics. Nedra, Moscow, 326 pp. (in Russian)
Kovacs, S., 1982. Problems of the "Pannonian Median Massif" and the plate tectonic concept. Contributions based on the distribution of Late Paleozoic-Early Mesozoic isopic zones. Geol. Rundschau, 71(2): 617-640.
Lancelot, Y., 1980. Birth and evolution of the "Atlantic Tethys" (Central North Atlantic). In: J. Aubouin, J. Debelmas and M. Latreille (Editors), Geology of the Alpine Chains born of the Tethys. Bureau de Recherches Géologiques et Minières, Paris, pp. 215-223.
Le Pichon, X. and Angelier, J., 1981. The Aegean Sea. Trans. R. Soc. London, Ser. A., 300: 357-372.
Lomize, M.G., 1983. Tectonic settings of geosynclinal volcanism. Nedra, Moscow, 194 pp. (in Russian).
Mahel, M., 1974. Tectonics of the Carpathian-Balkan regions. Explanations of the Tectonic Map of the Carpathian-Balkan regions and their Foreland, Geol. Inst. D. Stur., Bratislava, 455 pp.
Makarov, V.I. and Spiridonov, H.B., 1983. Neotectonics of the Rhodope Massif and the adjacent regions. Article II. Geotektonika, N 5: 23-46.
McKenzie, D., 1978. Some remarks on the development of sedimentary basins. Earth Planet. Sci. Lett., 40(1): 25-32.
Menard, G. and Thouvenot, F., 1984. Ecaillage de la lithosphére europénne sous les Alpes Occidentales: arguments gravimétriques et seismiques liés à l'anomalie d'Ivrea. Bull. Soc. Géol. Fr., 7. Sér., XXVI(5): 875-884.
Mueller, St. and Lowrie, W., 1980. Die geodinamische Entviklung des Westlihen Mittelmeerraumus und der Alpen. Fermesung Fotogrammetric Kultur. Technik., 78(12): 470-495.
Nachev, J., Sultanov, A. and Angelov, V., 1980. The flysch in Bulgaria. Guide book. Bulgarian Acad. Sci., Probl. Comm. IX, Sofia, 120 pp.
Nikolaev, V.G., 1986. Pannonian Basin. Nauka, Moscow, 480 pp.
Oberhauser, R., 1968. Beiträge zur Kenntnis der Tektonik und der Paläogeographie während der Oberkreide und dem Paläogen im Ostalpenraum. Jahrb. Geol. Bundesanstalt, 111: 115-145.
Plate tectonics, 2nd edition, 1980. Edited by I.M.Bird. Am. Geophys. Un. 992 pp.
Pogacsas, G., 1985. Seismic stratigraphic features of Neogene sediments in the Pannonian Basin. Geophys. Transactions Eötvös

Lorand Geophys. Institute of Hungary, Budapest, 30(4): 373-410.
Pospisil, L. and Vass, D., 1985. Influence of the structure of the lithosphere on the formation and development of intramountain and back molasse basins of the carpathian mountains. Geophys. Transactions Eötvös Lorand Geophys. Institute of Hungary, Budapest, 30(4): 355-371.
Rampnoux, J.-P., 1970. Regards sur les Dinarides internes yougoslaves (Serbie-Monténégro oriental): stratigraphie, évolution paléogéographique, magmatisme. Bull. Soc. Géol. Fr., 7 Sér., XII(6): 948-966.
Royden, L.H., Horvath, F. and Burchfiel, B.C., 1982. Transform faulting, extension and subduction in the Carpathian Pannonian region. Bull. Geol. Soc. Am., 93(8): 717-723.
Sandulescu, M., 1975. Essai de synthèse structurale des Carpathes. Bull. Géol. Soc. Fr., 7 Sér., XVII(3): 299-358.
Savoyat, E., 1977. La zone ionienne. In: Réunion Extraordinaire de la Société Géologique de France in Grèce. Bull. Soc. Géol. Fr., 7 Sér., XIX(1): 10-20.
Savu, U., Nastaseanu, S., Lupu, M., Nicolae, J., 1978. Ophiolites and sedimentary formations in South Apuseni and Southern Carpathians. Guidebook Field Works 21/2.2 Groups, Bucharest, 43 pp.
Sclater, J.G., Royden, L., Horvath, F., Burchfiel, B.C., Semken, S. and Stegena, L., 1980. The formation of Intra-Carpathian basins as determined from subsidence data. Earth Planet. Sci. Lett., 51: 139-162.
Sleep, N.K., 1971. Thermal effects of the formation of Atlantic continental margins by continental break up. Geophys. J.R. Astron. Soc., 24: 325-350.
Sollogub, V.B., Guterh, A. and Prosen, D., 1980. Structures of the Earth Crust of the Central and East Europe according to geophysical investigations. Naukova Dumka, Kiev, 206 pp.
Stegena, L., Geczy, B. and Horvath, F., 1975. Late Cenozoic evolution of the Pannonian Basin. Tectonophysics, 26 (12): 71-90.
Tectonic Map of Eurasia, 1966. Nauka, Moscow.
Trümpy, R., 1973. The sequence of orogenic events in the Central Alps. In: K.A. de Jong and R. Scholten (Editors), Gravity and Tectonics. Wiley, New York, pp. 233-256.
Vandenberg, J., Klootwijk, C., Wonders, A., 1978. Late Mesozoic and Cenozoic movements of the Italian Peninsula: further paleomagnetic data from the Umbrian sequence. Bull. Geol. Soc. Am., 89: 133-150.
Varga, I. and Pogacsas, G., 1981. Reflection seismic investigations in the Hungarian part of the Pannonian Basin. Earth Evol. Sci., 3/4: 232-239.
Yanshin, A.L., Artyushkov, E.V. and Schlesinger, A.E., 1977. Main types of large structures of the lithospheric plates and possible mechanisms of their formation. Dokl. Akad. Nauk. SSSR, 234(5): 1175-1179.
Ziegler, P.A., 1982. Geological Atlas of Western and Central Europe. Elsevier, Amsterdam.

Reprinted from *The Origin of Arcs*, by F.-C. Wezel (Editor), pp. 233—251 Elsevier Science Publishers B.V., Amsterdam © 1986

THE WESTERN ALPINE ARC : NEW DATA AND HYPOTHESIS

Jacques DEBELMAS
Geology Dept. University, F 38031, Grenoble, France.

ABSTRACT

The Western Alpine arc is, at the same time, inherited and induced. It is inherited (i) from Variscan and Late Variscan dispositions which were rejuvenated during the Jurassic? and (ii) from the paleogeographical pattern built by the oceanic Liguria-Piemont spreading. It is induced (or more exactly exagerated) by the Insubric impingement and the later Apenninic folding.

RESUME

L'arc des Alpes occidentales est à la fois hérité et induit. Il est hérité de structures hercyniennes et tardihercyniennes qui ont rejoué au cours du Jurassique, ainsi que de dispositions paléogéographiques résultant de l'ouverture de l'océan liguro-piémontais. Mais il est aussi induit (ou plus exactement exagéré) par le poinçonnement du promontoire insubrien et par le plissement tardif de l'Apennin ligure.

Many geologists have dealt with the geological interpretation of the Western Alpine arc. The first of these, E. Argand (1916) considered this feature to be a result of the compression of a geosynclinal belt between a curved offset of the Africa margin and a symetrical inset of the European margin, both probably of Variscan origin. The resulting folds fit in between the two opposite structures.

On the other hand, J. Goguel (1963) thought that the Western Alpine arc did not reveal an arcuated paleogeography but the lateral rebound of the northward displacement of the Central and Eastern Alps. The Mesozoic paleogeography would not have been arcuated in its shape but roughly E-W.

Thus, at that time, two opposed conceptions attempted to explain the curvature of the Alpine arc. But, as early as 1970, the introduction of Plate tectonics allowed a new interpretation of the Alps as a collision range, the collision itself being occasionally made more complex by the occurence of intermediate microplates between the two major ones (Africa and Eurasia).

Among others, Laubscher's interpretation (1971) was a good example of this new approach to the problem : the Western Alpine arc would be the result of the westward motion of an "Insubric microplate" between two lateral shears, the northern one between Mt Blanc and Aar, the southern one between Pelvoux and Argentera (Fig.I). Unfortunately, whilst the first may be (perhaps) observed in

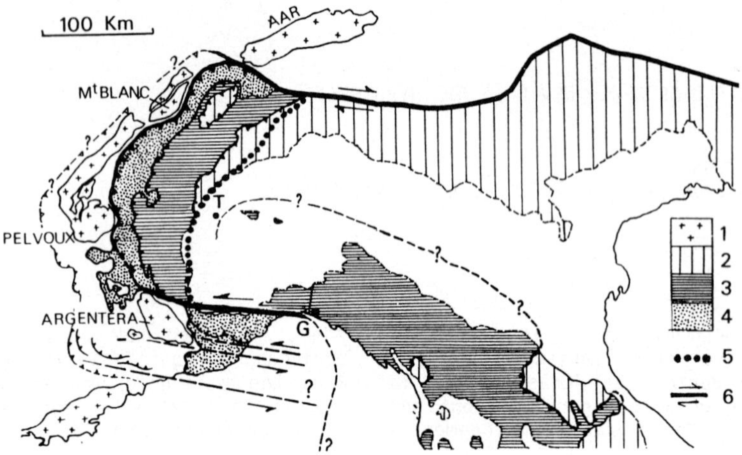

Fig.1. Structural scheme of the Alpine-Apenninic arc after Laubscher, 1971 (simplified ,Debelmas,1972).
1. External Crystalline massifs ; 2. South-Alpine zone ; 3. Alpine and Apenninic oceanic zones ; 4. Briançonnais zone and Embrunais flysch nappes ; 5. Ivrea gravity anomaly ; 6. Assumed shear zones ; G. Genoa ; T. Torino.

the field, the second faulted zone is very dubious. Besides, Laubscher's interpretation assumes that, predating this E-W Oligocene or Miocene motion, a Paleogene N-S motion had occured, as in Goguel's hypothesis.

M. Gidon (1974) put forward, for the first time, a reorientation of the tectonic collisional stress by the accidents limiting the main Alpine structural zones. Such accidents would rework in dextral shears along a curved pattern outlining the rotation of an intermediate Insubric microplate inside the sinistral Mesogean transform zone. Perhaps due to an improper comparison with meteorological processes (where the Coriolis strength occurs), Gidon's hypothesis did not receive sufficient attention.

Finally, L.E. Ricou (1984), linking and modernizing Goguel's and Laubscher's conceptions, assumes again a rectilinear and E-W trend to the early Alpine paleogeography, on one hand, and, on the other, assumes a sinistral northward strike-slipping of the Internal Alpine zones, in connection with the displacement of the Austroalpine nappes. This meridian translation would have been followed by the westward motion of the Internal zones between two lateral shear zones, inducing the arcuate shape of the Western Alps. L.E. Ricou agrees with M. Gidon on the importance of the strike-slipping motions between the main structural zones but considers their curvature as a recent feature.

The present paper intends to take another look at the problem with the help of data from the Variscan structures, the Mesozoic paleogeography and new observations on Alpine tectonics and seismicity.

A. DATA FROM THE VARISCAN STRUCTURES

Although the general frame of the Variscan chain is quite well known in France (Autran, 1980 ; Autran et Cogné, 1980), the available data are relatively vague in the Alpine area owing to the scarcity of reference land-marks.

The External Crystalline massifs belong to the Variscan axial zone, early folded from the Devonian to the Early Carboniferous. Paleomagnetic data from the Belledonne massif (Westphal, 1976) clearly show that the Late Variscan structures ("synclinaux houillers") were not wrenched by the Alpine folding : the Variscan magnetic field is always parallel to the stable Europe Permian trend. This is also true for the Argentera massif (Bogdanoff et al., 1977), but the Late Variscan structures are there roughly perpendicular to the Belledonne trends. Both sketch a capital letter L for this Late Variscan pattern.

Between Belledonne and Argentera, the Pelvoux massif is not so reliable, because significant Late-Variscan structures are lacking and its eastern part moved through about 45° in an anticlockwise rotation during the first Alpine folding event.

The " zone houillère briançonnaise" shows a pre-Namurian folding and a thick sedimentary sequence from the Namurian to the Autunian. In this respect, it is very similar to the Sarre-Lorraine basin : both are symetrically disposed and mark the N and S boundaries of the Variscan axial zone. Paleomagnetic data are scarce but they show an anticlockwise rotation of 45°. However the samples come from the Pelvoux latitude, i.e. from a zone which has lately been wrenched. Hence, these results cannot be considered as significative.

In other respects, there are sufficient indications that, in Carboniferous time, the Briançonnais sediments were laid down upon a thin crust, this because the Briançonnais basement is invaded by numerous tholeitic sills, with even some calc-alkaline influences (Parison, 1984). This feature has played a major role during the Alpine folding, making cleavages and décollements easier.

The South-Alpine domain s.l. was, for a long time, interpreted as a dependance of the Gondwana land, laying out of the Variscan orogenic zone proper. This conclusion is now out of date (Castellarin & Vai, 1980) : the platform where the Variscan folding weakens and where the Variscan metamorphism disappears begins with the Carnic Alps and extends to the Dinarides. The largest part of the South-Alpine domain shows sequences which have been folded and metamorphized in the Late Carboniferous. So they belong to the Southern Variscan External zone which appears broader here than in the previous interpretations.

This South-Alpine Variscan pattern was overprinted by Late-Variscan faults (e.g. the Judicaria fault or the Insubric line). But these faults have to be replaced in their palinspastic position since paleomagnetic data show a post-

Triassic anticlockwise rotation of about 60° of the whole domain with regard to the stable Europe (Vandenberg & al., 1976 ; Westphal, 1976). Replaced in their primitive positions, the above mentioned great faults become E-W (for the Judicaria fault) and NNW-SSE (for the Insubric and Gail lines) (Fig.2). The

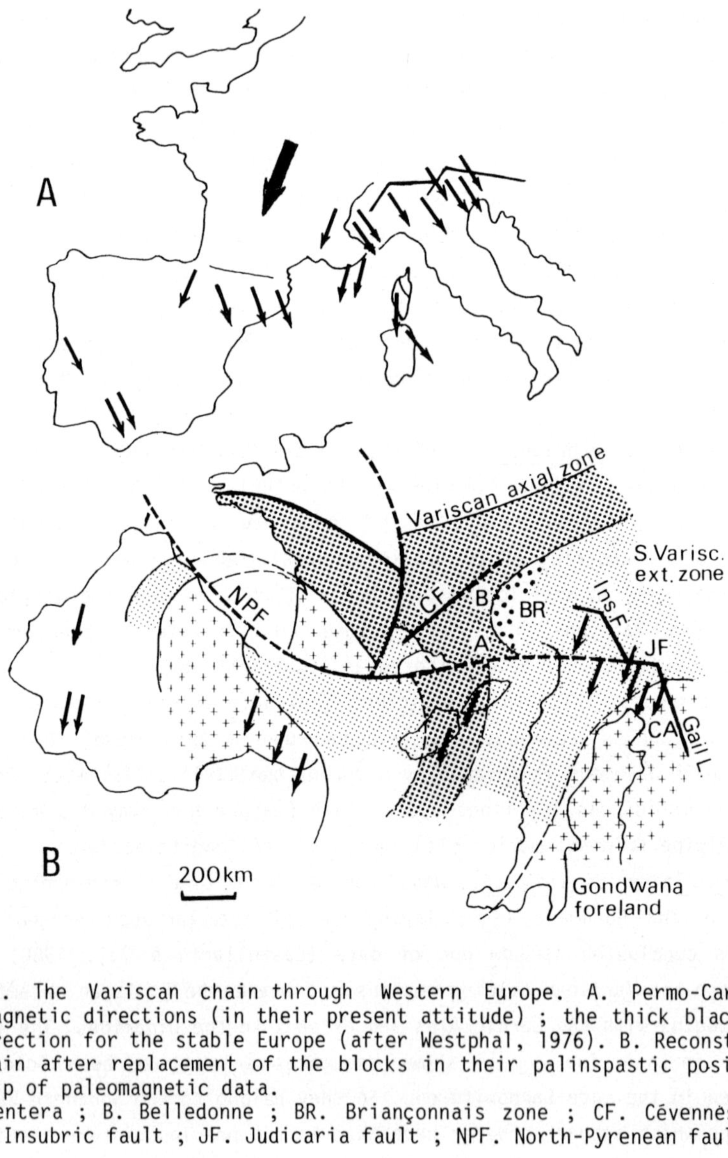

Fig. 2. The Variscan chain through Western Europe. A. Permo-Carboniferous paleomagnetic directions (in their present attitude) ; the thick black arrow is the direction for the stable Europe (after Westphal, 1976). B. Reconstitution of the chain after replacement of the blocks in their palinspastic positions with the help of paleomagnetic data.
A. Argentera ; B. Belledonne ; BR. Briançonnais zone ; CF. Cévennes faults ; Ins.F. Insubric fault ; JF. Judicaria fault ; NPF. North-Pyrenean fault.

first may be the prolongation of a classic and important Late Variscan accident, well-known in the SW of Europe, the North-Pyrenean fault (Arthaud & Matte, 1975). The second is more difficult to explain, but is parallel to the Late

Paleozoic compression (probably the so-called Saalic phase) which is believed to have induced the Late Variscan structures of the SW Europe (Arthaud & Matte, op.cit.).

However, the exact primitive position of this South-Alpine domain is difficult to fit, for want of precisions about the length and the direction of its displacement in Mesozoic times. Fig.2 is only a possibility among many others, built on the assumption of the prolongation of the N-Pyrenean fault in the Judicaria line.

Finally, the Maures massif, Corsica and Sardinia belong to the Variscan axial zone. The latter two are to be considered as being in their palinspastic position close to the Languedoc-Provence coast. Such a reconstitution implies a quasi-meridian direction for this Mediterranean branch of the Variscan axial zone.

From all the preceding data, it is possible to build the pattern of Fig.2, which is of course a very hypothetical one. From this figure, it appears that the roughly E-W branch of the Central European Variscan chain, on one hand, and the roughly N-S Corsica-Sardinia axis on the other hand, were arranged in a curved shape (in good agreement with the paleomagnetic data provided by the External Crystalline massifs) at the place of the future Alpine arc.

It is important to notice also the position of the Briançonnais Carboniferous basin inside the curve of the Variscan chain. Perhaps this basin induced the late curvature of the Variscan arc exactly as the Jura around the Tertiary molassic basin. Such an hypothesis implies a Late Paleozoic phase, probably the above mentionned Saalic phase (Middle Permian). This phase has been underestimated in the SE of France : it folded the Gard and St Etienne Carboniferous basins on the Massif Central border ; it folded the Stephanian synclines of the External Crystalline massifs ; and, in a part of the Briançonnais zone (Vanoise), it induced the thrust of the internal side of the Carboniferous basin with emplacement of crystalline basement slices (Sappey gneiss). This phase trends along a NW-SE axis, which is consistent with the conclusions of Arthaud et Matte (1975) from other data. But the NW-SE direction of the Late Variscan features in the Argentera massif implies the reorientation of the stress along the pre-existing curvature of the Variscan chain.

B. DATA FROM THE ALPINE MESOZOIC PALEOGEOGRAPHY

The Alpine belt, like all collision chains, formed along an oceanic domain, the so-called Liguria or Liguria-Piemont realm, and its two continental margins (European m. to the NW, South-Alpine m. to the SE) (Bernouilli & Lemoine, 1980 ; Lemoine, 1984, 1985). This oceanic realm is now classically interpreted as a NE-SW trending pull-apart basin, the spreading of which implies two transform faults, a Northern-Penninic fault and a South-Tethyan fault which is beyond the scope of the present paper (Fig.3).

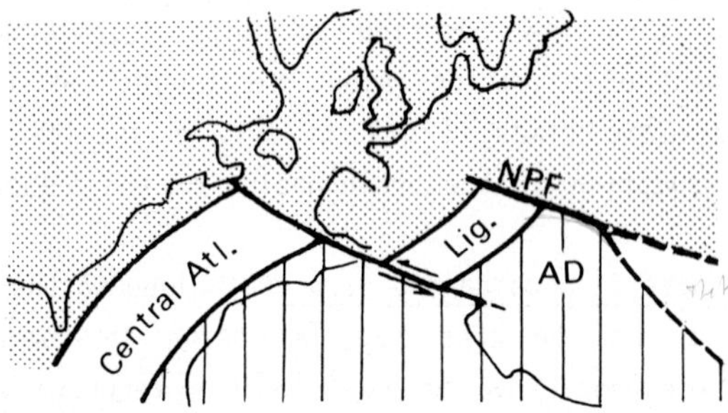

Fig. 3 The Liguria-Piemont oceanic domain in the framework of the Atlantic spreading (after Lemoine, 1984).
AD. Adriatic promontory ; NPF. North-Penninic fault.

But, if the overall pattern seems simple enough, the detailed reconstruction of its different parts appears more complicated.

On both sides of the oceanic area, the continental margins were certainly thinned and fragmented into blocks tilted along normal faults. It is reasonable to think that this topography of more or less parallel belts has been preserved in the less deformed parts of the Alpine area, i.e. on the edge of the French Massif Central. There, such blocks effectively exist and trend NE-SW, parallel to Late Variscan faults, the so-called "Cévennes faults", which controlled the Mesozoic paleogeography.

I. Further to the East, in the External Alpine domain, such features may be observed in the Crystalline Belledonne massif, with the same N50 direction, here again confirmed by paleomagnetism. The connected faults turn to N10 in the South of the massif where their synsedimentary (Lias-Dogger) activity has been well documented (see for example, Barféty & Gidon, 1984 ; Lemoine & al., 1981), but this N10 orientation is not fairly supported by paleomagnetic data (the calculated directions are too dispersed). In the Argentera massif, the direction of the blocks is N140 (supported by paleomagnetism) and controls, in the same way, the Jurassic paleogeography. This latter is outlined by NW-SE swells and troughs (Dardeau, 1983) (Fig.8) which are also inherited from earlier Variscan features.

Therefore, it is clear that, at the emplacement of the future Alpine arc, the European margin was already characterized, from Early Mesozoic times, by a curved pattern or, at least, a pattern largely opened to the NE (Fig.4).

2. In the Internal French-Italian Alps, the Mesozoic paleogeographical structures have not been preserved. However, their existence is classically

assumed and influences the interpretation of the series (Lemoine, 1984), but there is no paleomagnetic support of the primitive orientation of the blocks. They are assumed to be more or less parallel to the general trend of the oceanic opening, i.e. NE-SW. In any case, an arcuated disposition can be shown.

3. The transverse features. The Western Alpine domain was limited to the N and S by transverse structureswhich require explanation (Fig.4).

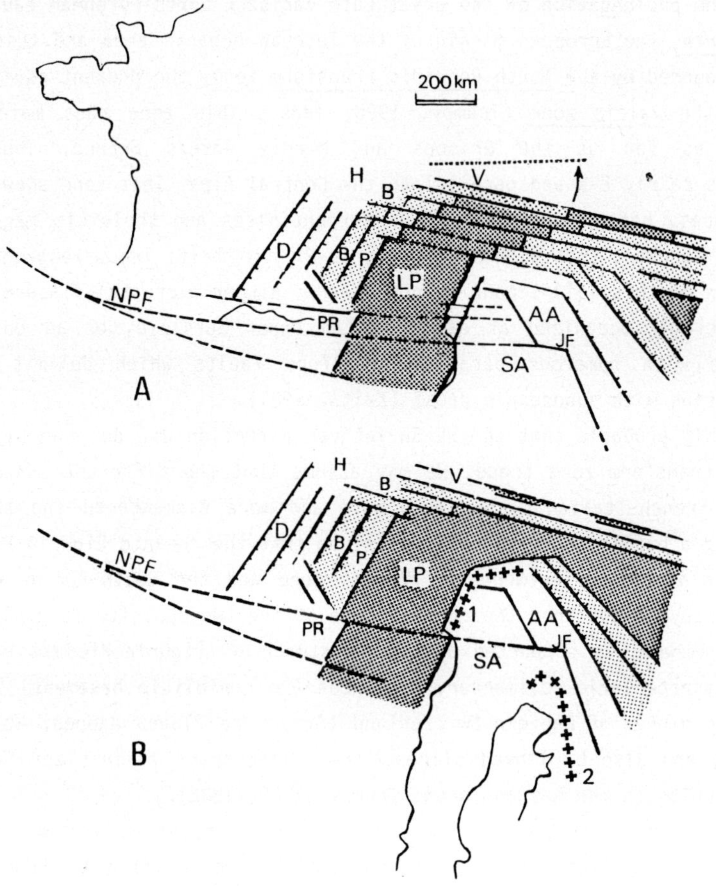

Fig.4. Palinspastic reconstitution of the Western Alpine area in the Early Cretaceous.
AA. Austroalpine zone ; B. Briançonnais zone ; D. Dauphiné zone ; H. Helvetic zone ; J.F. Judicaria fault ; LP. Liguria-Piemont zone ; NPF. North-Pyrenean fault ; P. Piemont zone ; PR. Provence platform ; SA. South-Alpine zone ; V. Valais zone.
Little cross lines : paleogeographical features (I. Insubria-Lombardia ridge, Debelmas, 1972 - 2. Belluno trough, Aubouin et al., 1970).
Light stippled : continental thinned crust ; dark stippled : oceanic crust.

To the South, the External domain, at least, butts against the Provence platform. This E-W feature is difficult to replace in the Late Variscan framework of SW Europe, owing to its probable northward displacement in Tertiary times and the probable W-E displacement of the Iberian microplate and Corsica-Sardinia axis along the North-Pyrenean fault. However, it is probable that the Provence E-W ridge marked the northern limit of this microplate in its palinspastic position. In such a hypothesis, the N-Provence flexure could possibly be the prolongation of the great Late Variscan North-Pyrenean fault.

To the North, the European margin of the Tethyan oceanic area and this area itself were bounded by the North-Penninic transform zone, the present equivalent of which is the Valais zone (Trümpy, 1980, 1985). This zone runs mainly in Switzerland, as far as the Grisons and hardly enters France. Thus its development is fairly E-W and parallel to the Central Alps. This zone shows some oceanic features, notably some slices of serpentinites and tholeitic magmatism but much less abundant that within the Liguria-Piedmont belt. The paleogeographic interpretation of this Valais zone is not clear : either a crustal fissure, with a bottom partially oceanized as early as the Upper Jurassic, or an entirely ensialic zone with numerous parallel transform faults which delimit small pull-apart basins with suboceanic crust (Kelts, 1981).

It is highly probable that the NE-SW Tethyan direction did not sharply turn into the E-W transform zone trend. We may assume that the different structural zones of the French-Italian margin were more and more dismembered and shifted eastward along a bundle of transform faults, so that the oceanic Liguria-Piedmont material could insert in between the Valais zone and the South-Alpine margin (Fig.4) ; whereby it outlined the curved shape of the two opposite margins.

Such a framework is supported by the existence of Liguria-Piedmont remants (with their caracteristic sedimentary sequences and ophiolitic basement) in the South-Penninic nappes of Eastern Switzerland (Arosa and Platta nappes, Weissert & al., 1985), and also by linked slices with Briançonnais affinities (Falknis and Sulzfluh units in the Grisons area, Trümpy, 1980, 1985).

To conclude this development, the part of the European margin which gave birth to the Western Alpine arc is well bounded to the S and N by transverse features. The Western Alpine arc fits exactly in such limits and overprints Variscan and early Mesozoic inherited structures having an arcuated or L shape. Such coincidences cannot be the result of pure hazard.

4. The South-Alpine margin. In its present state, it is made up, to the South, by the South-Alpine zone proper, relatively weakly folded and where paleostructures may be preserved, and, to the North, by the thrust and sliced Austroalpine slabs where the paleostructures probably no longer exist.

Most authors have emphasized the "Cévennes" (NE-SW) direction of the main faults of the South-Alpine zone, as, for instance, the Judicaria fault which is a late Variscan fracture, with a well documented Mesozoic synsedimentary activity (Castellarin, 1982) (Fig.5). An E-W section of this margin gives a set

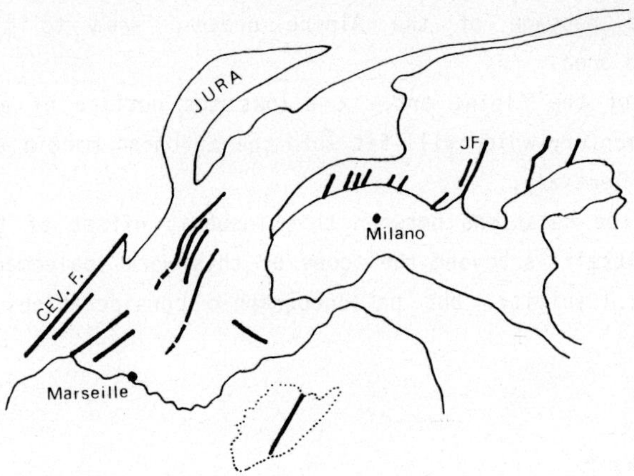

Fig.5. Synsedimentary Jurassic faults in the Alpine area, in their present orientation (after Lemoine, 1984). Cev.F : Cévennes faults ; JF : Judicaria fault.

of more or less tilted blocks similar to those within the European margin. This apparent symetrical disposition on both its margins tends to support the assumption of the "Cévennes" direction of the oceanic stripe. However a difficulty arises from the paleomagnetic data which appears to turn the Austroalpine and South-Alpine zones by 60°, in an anticlockwise rotation, to replace them in their palinspastic position. By this means, all apparent "Cévennes" faults become E-W (Fig.4). But other features become more or less parallel to the oceanic area (N40) : these are the classic South-Alpine paleogeographical zones, i.e. the Trente and Frioul ridges, on both sides of the Belluno trough (Aubouin et al., 1970). The Insubria-Lombardia ridge, deduced from sedimentological studies in the Lombardia Cretaceous flysch, easily finds its place in this scheme (Fig.4B).

In this new frame, the Insubric and Gail fault-line is more enigmatic, just as the domain which extended at this time between this fault-line and the North-Penninic fault.

This domain could be ensialic but strongly broken by the double system of the marginal faults of the oceanic area, on one hand, and the faults of the N-Penninic transform zone, on the other hand (Fig.4A).

However, this domain could also be oceanic, with a NW-SE trend, opening to the SE, between lateral normal fault escarpments on both sides (Fig.4B). The Insubric fault would be one of the Southern bundle. This second assumption offers a triple advantage :

1. It makes easier the closing up of this domain and the northwards wandering of the South-Alpine and Austroalpine block during the Late Cretaceous and Early Tertiary.
2. It explains the passage of the Alpine oceanic area to the Eastern Mediterranean oceanic area.
3. For the problem of the Alpine arc, it allows the outline of an acute or arcuate Insubric promontory which will fit into the European margin as early as the beginning of the Tertiary.

The question of the relations between this Insubric offset of the African plate and the plate itself is beyond the scope of this work (paleomagnetic data suggest a joint and liability, but paleogeographic considerations suggest a possible independance).

C. STRUCTURAL DATA

C.1. The collisional pattern. Most authors agree on the arcuate shape of the Western Alps as a result of the Eurasia-Africa collision process. As early as 50 Ma, this collision occured along a N-S to NNW-SSE direction, which was imposed by the Atlantic spreading and the connected motion of the Africa and Eurasia plates (Biju-Duval & al., 1976 ; Patriat & al., 1982) (Fig.6).

Such a collision direction is in good agreement with the E-W to ENE-WSW trend of the first Provence, Alpine and Austrian Upper or Middle Cretaceous folds, but not with the apparent westward displacement of the Insubric promontory invoked by most authors at the end of the deformation process to explain the curvature of the Alpine chain.

C.2. The recent structural field works. These works havebeen undertaken since 1980 in various areas of the Alps in an attempt to understand the mechanics of the Alpine deformation for a modern appraisal. They reveal a variable scheme which has been well documented in the Southern Vanoise massif (North French Alps, Platt & Lister, 1984) and in the Briançon area (Tricart, 1984).

C.2.1. In the Southern Vanoise massif, the first recognizable Alpine deformation is characterized by a NW trending stretching lineation associated with a flat-lying foliation. This lineament which is oblique to the transverse of the present axis of the Alps, is considered to be probably connected with the major nappe displacement and involved a NW directed subhorizontal shear.

In the second deformation, always linked with foliation, an E-W stretching direction occurs, with N-verging folds. The deformation was a simple dextral shear. Such a phase ended with a metamorphic peak of greenschist facies, dated at 38 Ma (Late Eocene). Of course, these two first phases occured below a thick overburden and reveal a ductile-type deformation.

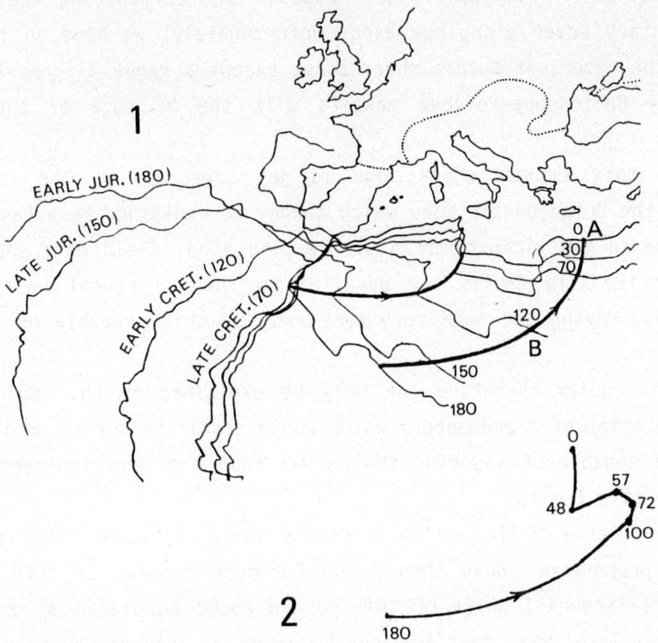

Fig.6. Relative motions of Africa with respect to Europe from Jurassic to Present. 1. After Patriat & al., 1982 - 2. After Biju-Duval & al., 1976. In both reconstructions, the N-S motion trend is the same since the last 50 Ma.

In the third phase, backthrusting and backfolding directed S or SE, took place under low-grade conditions and was followed by differential uplifts along steep faults and by strike-slip displacements.

It is interesting to notice that a similar evolution was described in the Central Alps (Steck, 1984) : so, there was, from the Central Alps to the N French Alps, an homogeneous tectonic stress-field which is consistent with the NNW displacement of the South-Alpine block. In any case, there is indication of a westward translation of the Insubric block.

C.2.2. The Briançon area, i.e. the area where the Alpine arc has the best shape, does not reveal such originally transverse features in regard to the general Alpine trend. All the deformation phases (including backthrusting) seem more or less homoaxial, with an always outward directed stretching lineation (Malavieille & al., 1984).

But the question is whether such a structural homogeneity hides or not a more complex history in so far as most authors assume the existence of a sinistral shear zone, parallel to the general NNW-SSE stress-field (Goguel, 1963 ; Ricou, 1984). On the west side of this shear zone, the nappe displacement would not have occured to the NNW, as in Vanoise, but to the W or SW, as the lateral "wash" of the main movement. In other words, the NNW motion would induce sinistral rotations of basement blocks capable of reorganizing the folding of their sedimentary cover along new axes. Unfortunately, we have no indications, at least in the cover, of such a shear zone, except a vague alignment of the SW ending of the Belledonne-Pelvoux massifs with the NE edge of the Argentera massif.

Moreover, this assumed shear zone is not supported by the very regular curvature of the Briançonnais zone which cannot be explained by a basement-block reorganization in the vicinity of a strike-slip zone. Field data show that the observable strike-slip faults are parallel to the structural units (Gidon & Barféty, 1975), having the same curvature and, by this, capable of sliding one against another.

Such a so regular curvature can only be explained by the impingement and plastic indentation of a promontory which would itself be curved or in rotation, and therefore capable of inducing the normal faults of the impinged margin to act as strike-slip faults.

Furthermore, each of these listric faults would represent ramps along which the Insubric promontory could thrust the European margin. By this means, the most eastern (=internal) units of this margin would be placed at a relatively deep structural level with ductile-type deformation. This disposition makes easy the adaptation of the impinged margin to the shape of the impinging promontory.

C.3 But is such a mecanism supported by the paleogeographic pattern ? As early as the Aptian, the Insubric promontory (=Adriatic platform) drove northward with an anticlockwise rotation of about 60° (Fig.6, curve between B and A). Such a movement is only possible through the disppearance of the domain lying immediately N of the Insubric platform (Fig.4). If the domain was oceanic, it could have disappeared easily by subduction. If it was continental, its disappearance by flaking off is more difficult to understand, except if its crust was sufficiently thinned.

At the end of the Late Cretaceous northwards wandering, the Insubric block arrived against the WNW-ESE North Penninic zone which was folded and flaked off, with N vergence. The Insubric block was deviated to the NW, in a way which was always made easy by the subduction of the Tethyan oceanic crust. At the extremity of this strike-slipping, the collision between the Insubric block and the European margin occured along a SW-NE direction (first phase in Vanoise) (Fig.7).The Insubric block acted as an impinging bulge and the resulting stress-field spread out as an 180°-opened fan, more especially as the normal faults of the European margin were ramps arranged in a kind of hemicycle. This is in good agreement with the stretching lineations map in the Western Alps (Malavieille & al., 1984).

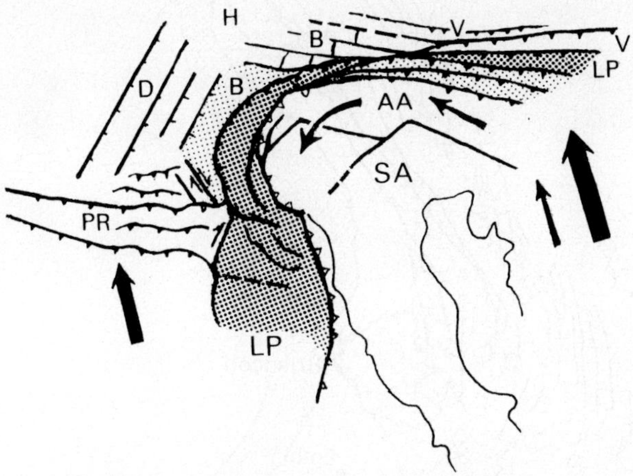

Fig.7. The collision between Europe and the Adriatic promontory in the Early Eocene. For the abreviations, see fig.4.
Thick black arrows : dominant compressive stress-field.
Thin black arrows : successive motions of the Adriatic block.

In the ductile Penninic domain, these normal faults were reactivated not only as thrust faults but also as strike-slipping planes, which is the expression of a reorientation of the tectonic stress all along these faults and perhaps also of the last rotation of the Insubric promontory.

In the less ductile External domain, the reorientation of the tectonic stress around the impinging bulge has been locally disturbed by the "Cévennes faults" or their satellites. The resulting arc shows several secondary virgations (Vialon, 1974) (Fig.8).

C.4. Geophysic data. Such a reorientation of the tectonic stress all along the Alpine arc is supported by the seismic study of the Western Alps.

According to G. Perrier & P. Vialon (1980), the Western Alps still show tectonic activity. One of the main seismic lineaments is localized along the front of the North External Crystalline massifs, exactly as if these massifs, from Belledonne to Aar, were being pushed northward, which implies a potential shear line one the SW side of the moving slab. This line does not appear in the cover structures but could be present deeper, in the basement, where it would correspond to the well-know Briançon-Imperia seismic lineament (Fréchet, 1978).

Along this lineament, the calculated axes of the maximum regional tectonic stress are always perpendicular to the Alpine arc (Fig.9). Thus, there is a rotation of the local stress-field all along the Alpine arc. With only the data of the Internal zones, the resulting scheme is a regular fan. With the observed anticlockwise deviations of the maximum stress axis in the External zones, the

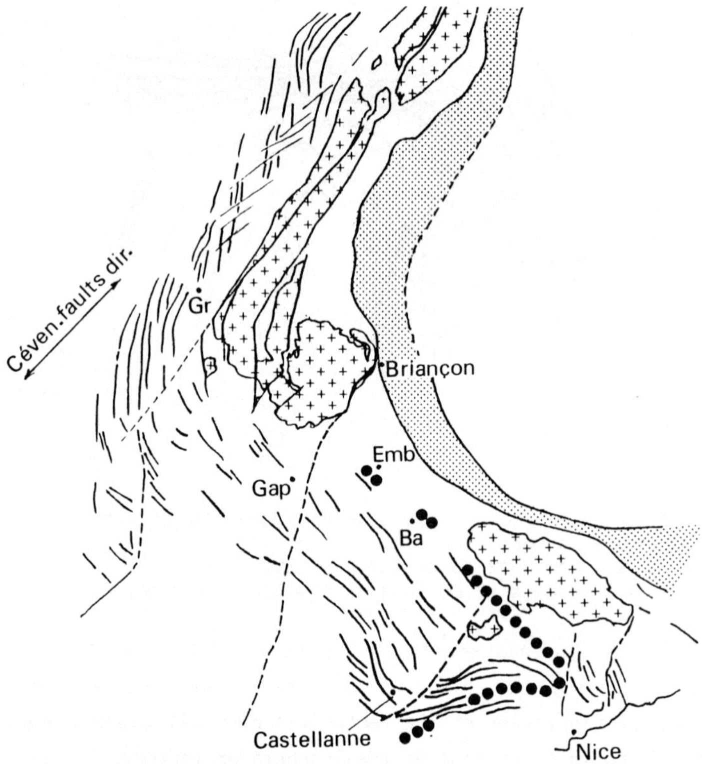

Fig.8 The cover folds pattern of the Western Alpine external zone.
Dashed lines : Late Variscan inherited faults
Black point lines : Jurassic paleoridges (after Dardeau, 1983)
Stippled : Briançonnais zone.
Ba. Barcelonnette ; Emb. Embrun ; Gr. Grenoble

resulting scheme shows a fan with radius in a spiral arrangement (Fig.9). Such a disposition recalls the scheme of Gidon (1974) and is in good agreement with the idea of the reorientation of the tectonic stress-field by the Late Variscan NE-SW, NW-SE and E-W features.

C.5. To end this structural discussion on the genesis of the Alpine arc, it is necessary to discuss the increasing curvature of the chain in the Maritime Alps. It is not at all evident that the approach of the E-W Provence structures is sufficient to explain this extreme incurvation in the genesis of which the Apenninic folding certainly took a part as Boccaletti & Guazzone (1970) assumed it twenty five years ago.

Beginning in the Late Oligocene, this folding process contributed to disturb the Paleogene Alpine arc. The latter was characterized by a SW vergence, the former by a NE vergence, which implies an intermediate transform fault. For a long time, this fault was assumed to be the Sestri-Voltaggio fault (Fig. 10),

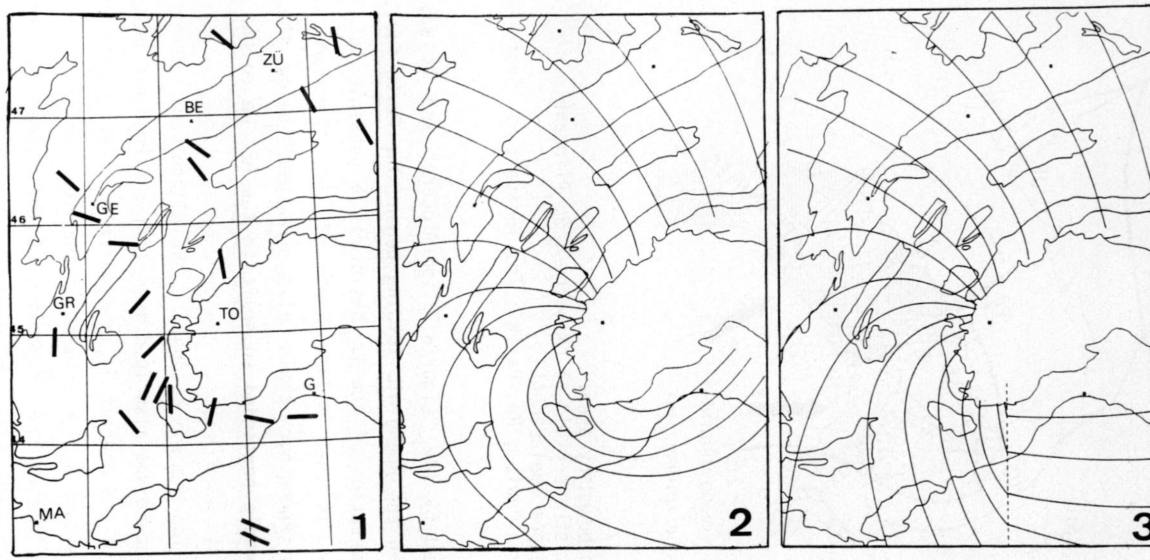

Fig.9. Present tectonic stree-field of the Western Alps (after Fréchet, 1978).
1. Focal solutions for the present seismic activity (calculated axis of the present maximum regional tectonic stress-field)
2 & 3. Assumed present regional stress-field (two variants).

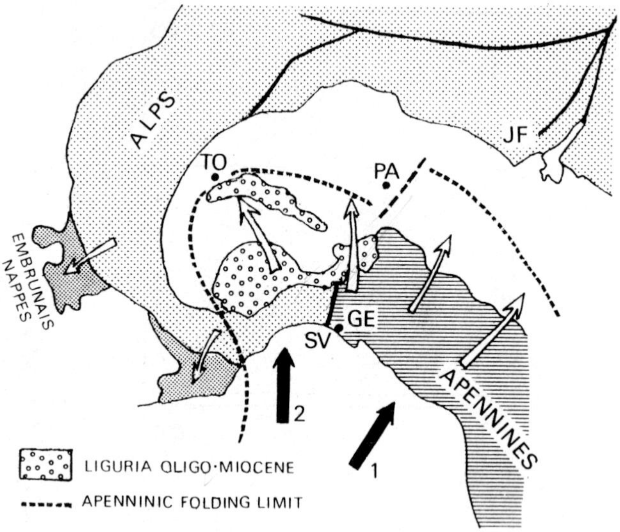

Fig.10. Alpine-Apenninic boundary area.
Black arrows : Apenninic stress-field (1. Miocene ; 2. Pliocene)
White arrows : gravity nappe emplacement
GE. Genoa ; J.F. Judicaria fault ; Pa. Pavia ; To. Torino ; SV. Sestri-Voltaggio fault.

but this fault is sealed by the "Liguria Oligo-Miocene" and it was already inactive at the time when the Apenninic folding began. The Pavia fault, hidden below the Neogene sedimentation of the Po basin, is a possible but improved solution. More important is that to the South of this transform zone, the Liguria nappes of the Apenninic chain also disappear below the Liguria Oligo-Miocene, but the whole continued to shift to the NE, up to the Pliocene included.

So, the Liguria Oligo-Miocene belongs to the allochthonous Apennines. But the Liguria Oligo-Miocene was also laid down upon the Penninic units of the Maritime Italian Alps and the contact is everywhere a stratigraphic one. From this, we are obliged to assume a displacement of the Maritime Alps units to the NE. Such a displacement implies the increasing of the Paleogene curvature of these Alps to the NE, by the means of probable but unknown intra-crustal cleavages and décollements.

D. SUMMARY

A Variscan orogenic arc, with a W or NW convexity, was present in Carboniferous time. Late Variscan block-faulting emplaces a structure opened to the E or NE, with an L-shape, which, for a large part, controlled fracturing of the European margin at the time of the Tethyan spreading, in the area comprised between the Leman lake and the Provence chain.

The Tethyan spreading gave birth to the Alpine oceanic area inside a complex pattern where an Insubric, more or less triangular, promontory progressively appears. At the end of the Africa-Europe rapprochement, this promontory impinged on the European margin, with a fan-shape reorientation of the corresponding stress-field.

In the Alpine External zone, this reorientation occured along SW-NE, NW-SE or E-W ramps. It gave birth to an external arc which was made up of successive minor virgations controlled by the most prominent NE-SW faults.

The Internal zones, the tectonic style of which is more ductile and the basement thinner as early as the Carboniferous, were deformed in a more souple style where a part of the deformation became lateral sliding along the old listric faults of the European margin. This reorientation of the tectonic stress-field along an almost perfect curve, is still supported by seismic data.

Small thrusts appeared in each compartment of these Internal zones, but they are minor structures compared to the large Austroalpine slabs and are not a problem for the replacement of their material in the core of the arc (objection raised by J. Goguel as early as 1963).

To conclude, the Western Alpine arc is, at the same time, inherited and induced. It is inherited from Variscan and Late Variscan dispositions, and induced (or more exactly exagerated) by the Insubric impingement and the later Apenninic folding.

Acknowledgements : I am gratefully indebted to M. Gidon and M. Lemoine for critical comments of this work and to R. Oliver for correcting my English.

REFERENCES

Argand, E., 1916. Sur l'arc des Alpes occidentales. Ecl.Geol.Helv., 14 : 145-191.

Arthaud, F. et Matte, Ph., 1975. Les décrochements tardihercyniens du SW de l'Europe. Tectonophysics, 25 : 139-171.

Aubouin, J., Sigal, J., Berland, J.P., Blanchet, R., Bonneau, M., Cadet, J.P., Guillot, P.L., Lacour, A., Piat, B., Vicente, J.Cl., 1970. Sur un bassin de flysch : stratigraphie et paléogéographie des flyschs crétacés de la Lombardie (versant sud des Alpes orientales, Italie). Bull.Soc.Géol.Fr. : 612-658.

Autran, A., 1980. Evolution structurale de la France du Protérozoïque aux distensions post-hercyniennes. C.R. 26ê C.G.I., coll. C7 : 10-17.

Autran, A., et Cogné, J., 1980. La zone interne de l'orogenèse varisque dans l'Ouest de la France et sa place dans le développement de la chaine hercynienne. C.R. 26ê C.G.I., Coll. C6 : 90-111.

Barféty, J.Cl. et Gidon, M., 1975. La place des failles longitudinales dans la structure du Briançonnais (Alpes occidentales, France). C.R.Ác.Sc.Paris, 281 : 1677-1680.

Barféty, J.Cl. et Gidon, M., 1984. Un exemple de sédimentation sur un abrupt de marge fossile : le Lias du versant Est du massif du Taillefer (zone dauphinoise, Alpes occidentales). Rev.Géogr. phys.Géol.dyn., 25 : 267-276.

Bernouilli, D. et Lemoine, M., 1980. Birth and evolution of the Tethys : the overall situation. B.R.G.M., mém. 115 : 168-179.

Biju-Duval, B., Dercourt, J. et Le Pichon, X., 1976. La genèse de la Méditerranée. La Recherche, 71 : 811-822.

Boccaletti, M. and Guazonne, G., 1970. La migrazione terziaria dei bacini toscani e la rotazione dell'Apennino settentrionale in une "zone di torsione" per deriva continentale. Mem.Soc.Geol.Ital., IX : 177-195.

Bogdanoff, S. et Schott, J.J., 1977. Etude paléomagnétique et analyse structurale dans les Schistes rouges permiens du Sud de l'Argentera. Bull.Soc.Géol.Fr. : 909-916.

Castellarin, A., 1982. Lineamenti ancestrali sudalpini. Guida alla geolog.Sud-Alpina, Bologna : 41-55.

Castellarin, A. & Vai, G.B., 1981. Importance of Hercynian tectonics within the framework of the Southern Alps. J.Struct.Geol., 3 : 477-486.

Dardeau, G., 1983. Le Jurassique des Alpes Maritimes (France), paléogéographie, contrôle tectonique, déformation ultérieure. C.R.Ac.Sc.Paris, 296 : 661-664.

Debelmas, J., 1972. A propos de quelques hypothèses récentes sur la genèse de l'arc alpino-apenninique. Rev.Géogr.Phys.Géol.Dyn., 14 : 229-244.

Fréchet, J., 1978. Sismicité du SE de la France et une nouvelle méthode de zonage sismique. Thèse Grenoble, 159 p.

Gidon, M., 1974. L'arc alpin a-t-il une origine tourbillonnaire ? C.R.Ac.Sc.Paris, 278 : 21-24.

Goguel, J., 1963. L'interprétation de l'arc des Alpes occidentales. Bull.Soc.Géol.Fr. : 20-29.

Kelts,K., 1981. A comparison of some aspects of sedimentation and translational tectonics from the Gulf of California and the Mesozoic Tethys, Northern Penninic margin. Ecl.Geol.Helv., 74 : 317-338.

Laubscher, H., 1971. The large-scale kinematics of the Western Alps and the Northern Apennines and its palinspastic implications. Am.J.Sc., 271 : 193-226.

Lemoine, M., 1984. La marge ouest de la Téthys ligure.In : G. Boillot, Les marges continentales actuelles et fossiles autour de la France. Masson, Paris: 157-248.

Lemoine, M., 1985. Structuration jurassique des Alpes occidentales et palinspastique de la Téthys ligure. Bull.Soc.Géol.Fr. : 126-137.

Lemoine, M., Gidon, M. et Barféty, J.Cl., 1981. Les masifs cristallins externes des Alpes occidentales : d'anciens blocs basculés nés au Lias lors du rifting téthysien. C.R.Ac.Sc.Paris, 292 : 917-920.

Malavielle, J., Lacassin, R. et Mattauer, M., 1984. Signification tectonique des linéations d'allongement dans les Alpes méridionales. Bull.Soc.Géol.Fr. : 895-906.

Parison, M.D., 1984. Problèmes pétrochimiques et structuraux en Vanoise septentrionale (Savoie). Thèse Paris-Sud, 358 p.

Patriat, Ph., Segoufin, J., Schlich, J., Goslin, J., Auzende, J.M., Breuzart, P., Bonnin, J., Olivet, J.L., 1982. Les mouvements relatifs de l'Inde, de l'Afrique et de l'Eurasie. Bull.Soc.Géol.Fr. : 363-373.

Perrier, G. et Vialon, P., 1980. Les connaissances géophysiques sur le SE de la France. Implications géodynamiques. In : Géologie des Pays Européens (France, Belgique et Luxembourg), Dunod, Paris : 269-276.

Plat, J.P. and Lister, G., 1984. Structural evolution of a nappe complex, Southern Vanoise massif, French Penninic Alps. J.Str.Geol., 7 : 19-36.

Ricou, L.E., 1984. Les Alpes occidentales : chaîne de décrochement. Bull.Soc.Géol.Fr. : 861-874.

Steck, A., 1984. Structures de déformation tertiaire dans les Alpes Centrales. Ecl.Geol.Helv., 77 : 55-10.

Tricart, P., 1984. From passive margin to continental collision : a tectonic scenario for the Western Alps. Am.J.Sc., 284 : 97-120.

Trümpy, R., 1980. An outline of the Geology of Switzerland. C.R.26è C.G.I. Guide-book Exc.G 10, 102 p.
Trümpy,R., 1985. Die Plattentektonik und die Entstehung der Alpen. Neujahrsblatt natf.Ges.Zürich, 187, 47 p.
Vandenberg, J. and Wonders, A., 1976. Paleomagnetic evidence of large fault displacement around the Pô Basin. Tectonophysics, 33 : 301-320.
Vialon, P., 1974. Les déformations synschisteuses superposées en Dauphiné. Bull.Suisse Min.Pétr., 54 : 663-690.
Weissert, H.J. and Bernoulli,D., 1985. A transform margin in the Mesozoic Tethys: evidence from the Swiss Alps. Geol.Rdschau, 74 : 665-679.
Westphal, M., 1976. Contribution du paléomagnétisme à l'étude des dérives continentales autour de la Méditerranée occidentale. Thèse Strasbourg, 304 p.
Westphal, M., Edel, J.B., Schott, J.J., 1978. Essai de synthèse des données paléomagnétiques et de la tectonique de plaque en Méditerranée occidentale. C.R.Som.Soc.Géol.Fr.,12-15.

SOUTHALPINE VERSUS PO PLAIN APENNINIC ARCS

A. CASTELLARIN and G.B. VAI

Dip. Scienze della Terra, V. Zamboni 67, I-40127 Bologna (Italy)

ABSTRACT

Previously known Apenninic arcs, buried under the Po Plain are compared with recently recognized arcuate systems within the Southern Alps.

The three main arcuate systems of the Po Plain are all of Neogene age (Upper Miocene to Upper Pliocene) and are roughly concentric with the older, more internal Apennine arcs. They originated by overthrusting of the Apennines onto the Po Plain foredeep. The arcuate pattern could be related to the rheologic buttresses represented by the two more rigid Mantua and Pavia crustal blocks, characterized by positive magnetic anomalies(basic intrusions) and by repeated Palaeozoic(?), Triassic and Neogene volcanic bodies; such blocks have prevented lateral propagation of décollement surfaces giving rise an overall arcuate pattern.

The three main arcuate systems of the Southern Alps are of different ages, show a remarkable time-space polarity to the SE and are not concentric. The western, innermost arc is of Palaeogene age; the intermediate arc, partly buried beneath the Lombardy Plain, is of Upper Miocene age and is sharply bent around the Adamello batholith; the eastern, outermost arc has the same Upper Miocene to Upper Pliocene age as the buried Apenninic arcs and is only developed beyond the front of the first two Southalpine arcs. Such an asymmetric setting can be related to the buttressing effect provided by the Venetian (Lessini to Dolomites) Platform acting as a rigid block, owing to its peculiar magmatic and depositional history since the Upper Palaeozoic (Hercynian granitoid intrusions, Permian volcanism, mid-Triassic volcanism and tectogenesis, Palaeogene volcanism).

The very close location of the Venetian Platform to the Mantua block strengthened a relevant part of the Padania crust accounting for the major time-space asymmetric distribution of the adjoining arc systems.

The epidermal, roughly symmetric opposite vergence of the two arc systems does not reflect symmetrical kinematics. In fact, from the above data, the Neogene Po Plain Apenninic arcs are better explained as overthrusts upon the Padania-Adriatic foredeep. Conversely, the coeval eastermost Southalpine arc originated by underthrusting of the Adriatic-Padania crust towards the NNW below the Southern Alps.

True, active A subduction is, therefore, suggested at the eastern Southalpine front (consistently with the Africa-Adria drift towards the N), whereas only relative, apparent or passive A subduction can be admitted for the Northapennine front.

Major arcuate segments in orogenic belts, such as the Western Alps arc, the Carpathian arc, the Gibraltar arc or the Hellenic arc have fascinated earth scientists for a long time.

Intermediate scale or less pronounced arcs, like the Jura Mountains arc or the Northern Apennine arc have also been treated and referred to (see Tectonic Map of Europe 1979, IGC, Moscow).

This paper intends to focus upon the importance of the smaller-scale arcs which are found at both sides of the Po Plain in the Northern Apennines and the Southern Alps and to discuss their possible origin and significance.

A minor arcuate internal structure within the Northern Apennine was evidenced first by the structural interpretation of the subsurface seismic survey of the Po Plain by AGIP in the 1950's. Three to four main folded arc systems (called "archi rampanti" by Gortani) were distinguished and named, and a diachronism of deformation from W to E was recognized (Gortani, 1956; Rocco and Jaboli, 1956). The formal, updated definition and description of the three main folded arcs (Monferrato F.A., Emilia F.A., Ferrara-Romagna F.A.) is due to Pieri and Groppi (1981).

A minor arcuate structural pattern within the Southern Alps, totally neglected previously, was first recognized only very recently by Castellarin (1984).

A brief description of these arcuate patterns shall precede a comparative analysis and a tentative regional interpretation of their kinematics, initially neglecting , possible, larger-scale dynamic implications.

THE BURIED PO PLAIN ARCS

Description

According to Pieri and Groppi (1981) and older AGIP works (Gortani, 1956; Rocco and Jaboli, 1956) the present orographic boundary of the Northern Apennines does not coincide with their structural front, which extends for more than 50 km into the Po Plain beneath the Quaternary cover.

Three main systems of broadly E-W trending, arc-shaped thrust belts have been named by Pieri and Groppi (1981):

1) The Monferrato Folded Arc is the western and innermost arc showing a Late Pliocene age of the latest diastrophic phase known.

2) The Emilia Folded Arc is the central arcuate system thrust onto the Pedealpine Homocline and widely colliding with the opposite facing, older, buried Southalpine front. The latest and most impor-

phase affecting this arc is also Late Pliocene in age.

3) *The Ferrara-Romagna Folded Arc* is the eastern composite system of arcs representing the more complex, major structural feature of the Po Plain. The Romagna folds, the more internal structure of the arc, are characterized by the structural disharmony of the clastic Tertiary formations with respect to the Mesozoic carbonates.

The arcuate pattern of these three systems is consistent with similar arcuate structures found in the peninsular Apennine chain.

Shortening in the buried arcs appears to increase eastward as the number of individual arcs increases. This is consistent with an anticlockwise rotation of the Apennines around a western pole (Pieri and Groppi, 1981).

A first quantitative estimate of shortening in the buried Apennine thrusts was attempted by Gasperi in Bartolini *et al.*(1983). At the same time, Pieri (1983) published excellent seismic profiles (with geologic interpretation) across the Po Plain thrusts, stimulating quantitative shortening estimates based on balanced sections (Bally *et al.*, 1985; Castellarin *et al.*, 1986) The resulting shortening figures range from about 5 km for the central Emilia arc and about 10 km for the eastern Ferrara-Bologna arc (Gasperi in Bartolini, 1983) to 20% (about 10 km) and 30% (about 20 km) for the same two arcs respectively (Castellarin *et al.*, 1986).

Careful analysis of two balanced seismic sections across the Emilia and Ferrara-Romagna arcs compared with the structural elements of the adjoining exposed Apennine margin (fig. 6), published by Castellarin *et al.*, 1986, has provided a more accurate picture of this polyarcuate, buried belt.

(1) The Po Plain arcs are severe compressional structures of Neogene (mainly Messinian to Late Pliocene) age originated from the coeval foredeep troughs of the Apennine chain. Thus, the buried Apennine front is largely overthrust onto the Po Plain and Adriatic foreland suggesting an A subduction (Bally and Snelson, 1980; Bally *et al.*, 1985) of relative or passive type (see later).

(2) Large, syntectonic, clastic sedimentary wedges characterize the main pulses of thrusting in the Po Plain during the Early Messinian -Late Pliocene time interval. Northward directed tectonic transport was still active in Middle to Upper Pliocene time as documented by less prominent, but still typical, syntectonic sedimentary wedges of this age. Conversely, the Quaternary depositional bodies display flat, lenticular shapes and are more readily interpreted as due to the filling of large, residual depressions unaffected by tectonic trans-

port and deformation along previous thrust planes.

(3) The imbricated thrust structure (Dahlstrom, 1970; Butler, 1982; Boyer and Elliott, 1982) is dominated by flats, with main décollement levels located at the base of Messinian (euxinic shales), at the Mesozoic-Tertiary boundary (Scaglia cinerea) or at the base of the Mesozoic (Burano evaporites and/or Verrucano pelites). The ramps usually cut across the Tertiary clastics. Frontal flat-ramp folds are narrow to absent, whereas almost each thrust front displays a prominent asymmetric anticline at its rear. Duplex-structures have not yet been detected. As a rule, the thrust planes are increasigly steeper from N to S.

(4) The time-space distribution of thrust planes appears to be quite irregular (Fig. 6). In fact the thrust systems advanced progressively toward the axis of the Po Plain between Upper Miocene and topmost Lower Pliocene, according to a piggy-back propagation of thrusts (Dahlstrom, 1970) as recognized by Ori and Friend (1984). Starting with the Middle Pliocene, the outermost frontal structures of the Po Plain arcs were sealed and deformation continued only at intermediate, southerly located fronts. A further, major, Quaternary tectonic reactivation affected mainly the previous structures along the present, exposed Apennine margin (the Pedeapennine Lineament Castellarin et al., 1986) according to an overstep propagation of thrusts (Dahlstrom, 1970). The reactivation of the Pedeapennine Lineament since the Middle Pliocene is related to the strong, opposite vertical movement of the Apennine versus the Padania sector. Owing to more than 5 km of vertical offset since the middle Pliocene, mechanical disconnection between buried and outcropping imbricated system is likely to have occurred along the Pedeapennine Lineament. Also the confining conditions increasing with time (mainly due to the load of sediments and of displaced units) prevented tectonic reactivation of buried thrusts at the end of Pliocene, except those located on less confined structural highs (Monestirolo, Ferrara, Longastrino). Conversely, at the Pedeapennine margin the steep dipping thrust planes were subject to decreasing geostatic loading by erosional and gravitational denudation controlled by uplift. Therefore, compressional reactivation was favoured. In this way, large tectonic transport of the Liguride Nappe, inside the Padan margin during Upper Pliocene to Lower Pleistocene time, and severe deformation of sediments of the same or even younger age at the Bologna-Modena margin, can be easily

explained. It appears that a northward propagation of the Po Plain imbrication system was prevented by the more rigid behaviour of undeformed crustal blocks and by the already achieved collisional stage between the opposite Southalpine and Apennine fronts at the outermost border of the Emilia arc. A possible explanation for both the overstep-propagating thrusts in the Apennine-Po Plain system and the origin of the Po Plain arcs was suggested by Castellarin et al. (1986) and are discussed in the next section.

Interpretation

The multiple-arc-shape of the buried Apennine front is a structure that has not yet been convincingly explained.

Neither a cojugate system of NE-SW and NW-SE trending strike-slip faults with major sinistral components along the WNW-ESE direction (the Gafsa trend according to Boccaletti et al., 1982, fig. 8) nor a NE-SW trending sinistral strike-slip displacement assumed along the Voghera-Tortona lineament (Ogniben, 1969) at the western side of the Emilia arc, are suitable to resolve the peculiar multiarcuate front. A recent hypothesis suggests that the arcs resulted from concurrent compressions and rotations: sinistral and dextral planar rotation of the structures would have produced the present outline of the front (Patacca and Scandone, in press).

On the other hand, significant strike-slip movement along the Apennine border, admitted to explain mostly the Mesozoic palaeogeography (i.e. the Emilia Fault proposed by Bosellini, 1981), is not consistent either with the subsurface geological Neogene reconstructions (Pieri and Groppi, 1981; Pieri, 1983, 1984; Rizzini and Dondi, 1978; Dondi et al., 1982; Mattavelli et al., 1983) or with Neogene-Quaternary stratigraphical and sedimentological data which display a lateral N to S continuity between the deltaic to alluvial deposits of the Po Plain border and the pedemontane distributory system of Apennine rivers, located in the rear (Francavilla and Colombetti, 1980; Ricci lucchi et al., 1981).

Modern models of thrust systems show oblique slip movements along the lateral sides of the arcs (Butler, 1982; Boyer and Elliott, 1982). Nevertheless, pre-kinematic reconstructions of stratigraphic-structural units and models of their possible differential shortening across the arcs are still unrealized. A possible exception might be the largely general and qualitative classical solution of fold junction and "virgation" of Alps and Southern Alps proposed in the past (see f.i. Argand, 1924; Dal Piaz, 1942). An attempt to give explanation to at least some of the basic geometrical problems by means of very simple kinematical hypotheses was made by Castellarin et al. (1986). It is summarized below.

1) Overthrusting of the Apennine chain toward the N and relative underthrusting of the adjoining undeformed Po Plain. The décollement area is confined by the linear boundary of the Po Plain foreland to the N and by the irregular protrusions of the Tortonian Apennine front to the S (Fig. 3). Thus, the décollement area has a narrower or larger width according to the convexity or concavity of its southern boundary. In this model the arcuate front of decoupled units depends upon the original width of the space subjected to décol

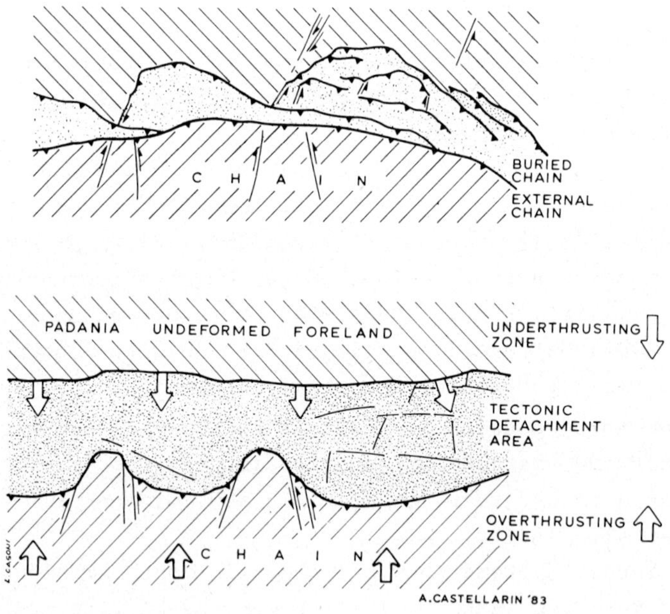

Fig. 3. Genetic model of the Po Plain buried arcs.
A: original stage (Upper Miocene): the décollement area is enclosed between irregular protrusions of the Tortonian Apennine front to the S and the linear margin of the undeformed Padania foreland. B: final stage (Upper Pliocene) (after Castellarin et al., 1986).

lement. When the decoupled area is completely consumed and the N and S boundaries come into contact or even overlap, the resulting front appears more advanced where the décollement area is larger, and more withdrawn where the area is narrower. Moreover, on the southern border dextral and sinistral strike-slip faults would be needed to allow differential advancement of more versus less-displaced parts of the arcs.

2) Overthrusting and underthrusting as in the previous model. In this hypothesis, however, the décollement area is confined by a linear boundary to the S, whereas the outline at the northern limit is

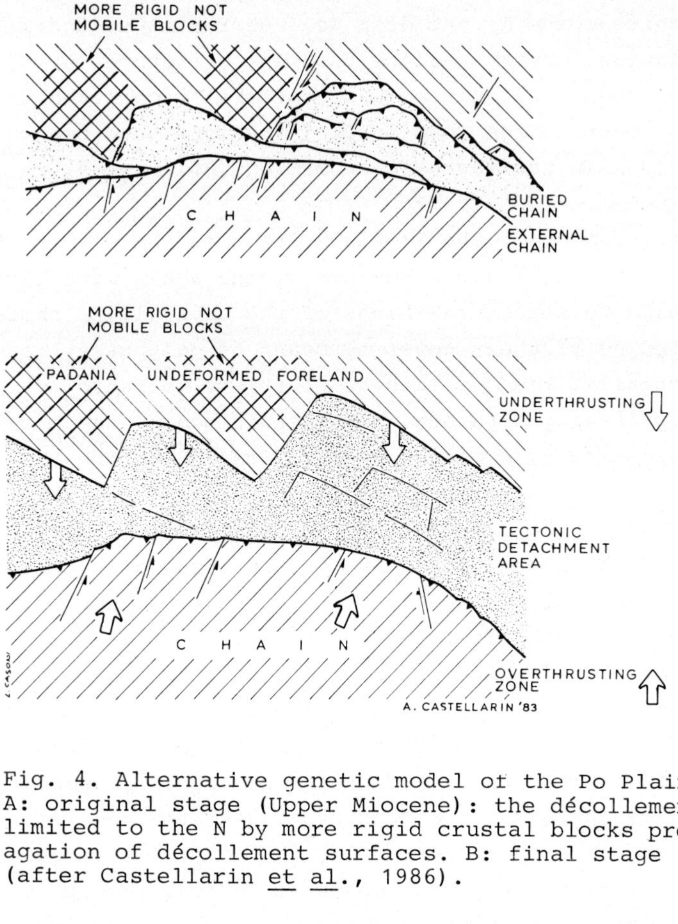

Fig. 4. Alternative genetic model of the Po Plain buried arcs. A: original stage (Upper Miocene): the décollement area is sinuously limited to the N by more rigid crustal blocks preventing lateral propagation of décollement surfaces. B: final stage (Upper Pliocene) (after Castellarin *et al*., 1986).

sinuously curved around more-rigid Padania crustal blocks preventing lateral propagation of décollement surfaces. This model does not entail systematic strike-slip displacements to provide complete mobilization of the décollement source area, whose former outline closely prefigured the arc-shaped resulting front. Strike-slip movements along oblique or transverse faults, however, can be easily admitted, as transfer faults, located between differentially shortened (or advanced) sectors of the arcuate thrust systems (Fig. 4).

We favour the second model which fits better the geological constraints of the Po Plain arcs. In fact the areas embraced by the adjoining Padania arcs correspond to more rigid crustal blocks. Such two areas, S of Pavia and Mantua, which can be called the Mantua and Pavia blocks, (Fig. 5), display remarkable aeromagnetic anomalies (Bolis et al., 1981; Arisi Rota and Fichera, 1985). These pattern are consistent with subsurface data (Pieri and Groppi, 1981): i) thick Triassic volcanics cored in the deep well near Mantua (Rodigo) and ii) Permian volcanics, or crystalline basement structural high (Battuda) and large Neogene volcanic bodies (Mortara) recognized in western Lombardy. Moreover, large deep basic intrusions have been assumed in order to explain the magnetic anomalies (Bolis et al., 1981).

The arc-shaped arrangement of the Po Plain front can be simply explained by admitting a more rigid behaviour of the zones with high positive anomaly linked to structural highs of the crystalline basement and to strengthening of their cover by means of polyphase, magmatic events. Particularly, the Mesozoic and Tertiary volcanic cycles, with their deep feeding apparata, subvolcanic bodies and possible shallow plutonic masses, provided a consolidation of the Po Plain upper crust, welding the Permo-Mesozoic and part of the Tertiary sedimentary cover to the crystalline basement. In this way, tectonic detachment of the cover would have been prevented in the intervening areas between the arcs and further advancement of the thrust system towards the N effectively blocked.

The consolidated crustal Po Plain blocks located between the arcs reacted to the frontal compressions mainly by severe downwarping and tilting, thus providing large space to sedimentation. In this way, the Parma-Reggio area with its upto 7 km thick sedimentary wedge of Plio-Pleistocene age appears to be consistent with the general framework. In this case, pure tectonic loading (Walcott, 1970; Beaumont, 1981; Quinlan and Beaumont, 1984) seems to be less important for increasing subsidence with respect to crustal downwarping linked

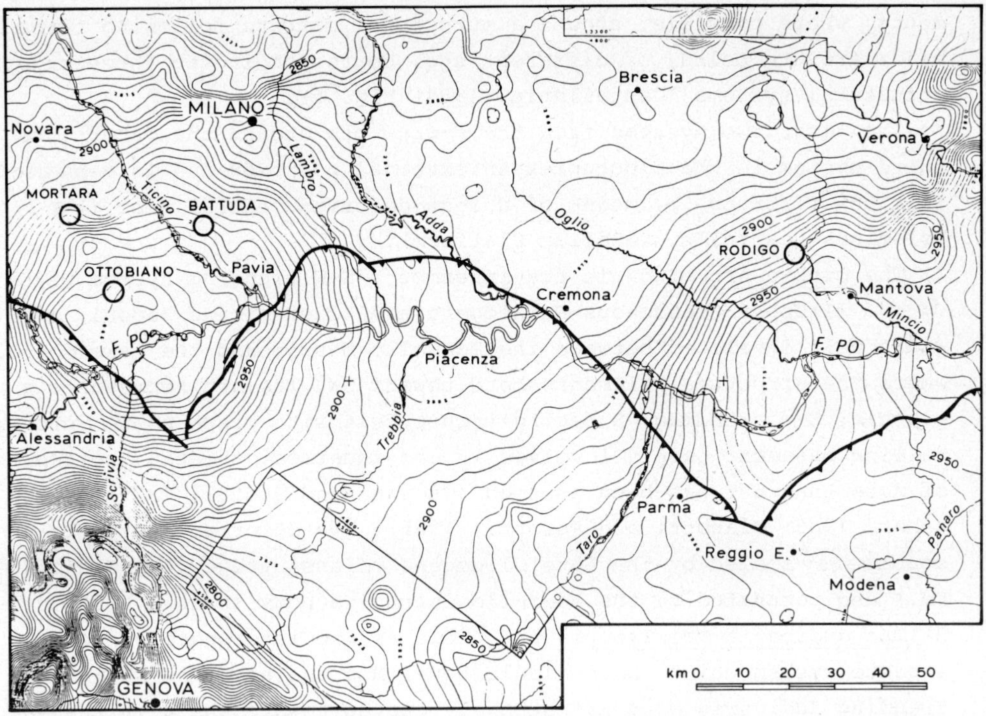

Fig. 5. Outline of the Po Plain buried arcs versus aeromagnetic anom aly. The two major positive anomalies fit well with the Mantua and Pavia blocks which separate the individual Po Plain arcs (data by AGIP,1981 and Pieri and Groppi, 1981; after Castellarin et al.,1986).

by tangential stress. The principal advantage of this hypothesis is the sinuous margin of the decoupling area around the former rigid blocks (Fig. 4)

This means that the postkinematic arc-shaped arrangement cannot be related to significant dextral and sinistral planar rotations of the front. This possibility is moreover contradicted by AGIP structural maps and profiles (Pieri and Groppi,1981) which show similar tectonic styles and shortening across the differently advanced portions of each arc.

THE SOUTHALPINE ARCS

Description

The quite different shortening values assumed in the western (more than 50%), central (up to 20%) and eastern (more than 65%) sectors generated many doubts about a simple, general internal structure of the Southern Alps. Careful dating of the individual structural elements, where possible, showed a structural pattern following three major arcuate belts, of different age, corresponding to three major structural systems (Castellarin, 1984)(Fig. 6).

The first two systems from the W grew one around the other; the third system in the E occupies an external position and is separated from the first two by means of a less deformed central area (the Dolomites-Lessini or Venetian Platform).

1) The Orobic Arc. This is the innermost, E-W trending system which is of (?)Upper Cretaceous-Palaeogene age (Gaetani and Jadoul, 1979); and is well developed between the Adamello Massif and the Lake's area, where the structural elements bend toward SW. The arc represents a classic, E-W trending, S facing thrust belt with pre-Westphalian crystalline basement deeply involved in the innermost sheets. Shortening of more than 50% has been assumed (Gaetani and Jadoul, 1979; Laubscher, 1985; Gianotti and Perotti, 1986). The deformed system is unconformably sealed by the Late Oligocene molasse stage (the Gonfolite Fm.) and postdated by the Adamello intrusive body (Brack, 1981).

2) The Judicaria-Val Trompia Arc. This is the intermediate, more arcuate system which is of Middle-Upper Miocene age. The NNE-SSW trending Judicaria axes are strongly curved toward the W in the Val Trompia and Brescia area, plunging below the Po Plain where they form wide, buried structures (Pieri and Groppi, 1981), representing the western half of this arc. The exposed eastern half is typically developed E of the Adamello Massif up to the Adige Valley. It is built by a system of E to S facing thrust sheets mainly detached from the Mesozoic cover except for the innermost, basement involving sheets. Active compressional tectonics (Castellarin and Sartori,1982, 1986; Castellarin, 1984) is largely favoured in explaining all the field data in alternative to wrench (Trevisan, 1938; Laubscher,1973) or pure gravitative (De Jong, 1967) tectonics. The age of deformation is mainly Middle to Upper Miocene (Castellarin and Sartori, 1986). The NNE-SSW trending Judicaria axes are sharply deflected to E-W south of the Garda Lake and buried beneath the Lombardy Plain, where they show the same deformation age (Pieri and Groppi, 1981). Comparable deflection is exposed south of the Adamello Massif.

The innermost and the intermediate systems together represent the entire western sector of the Southern Alps.

3) The Friuli-Julian Arc. This is the outer system, representing most of the eastern sector of the Southern Alps. It has a complex, concave external front of Plio-Quaternary age (the area of repeated Friuli earthquakes!). Older compressional phases, certainly active in the

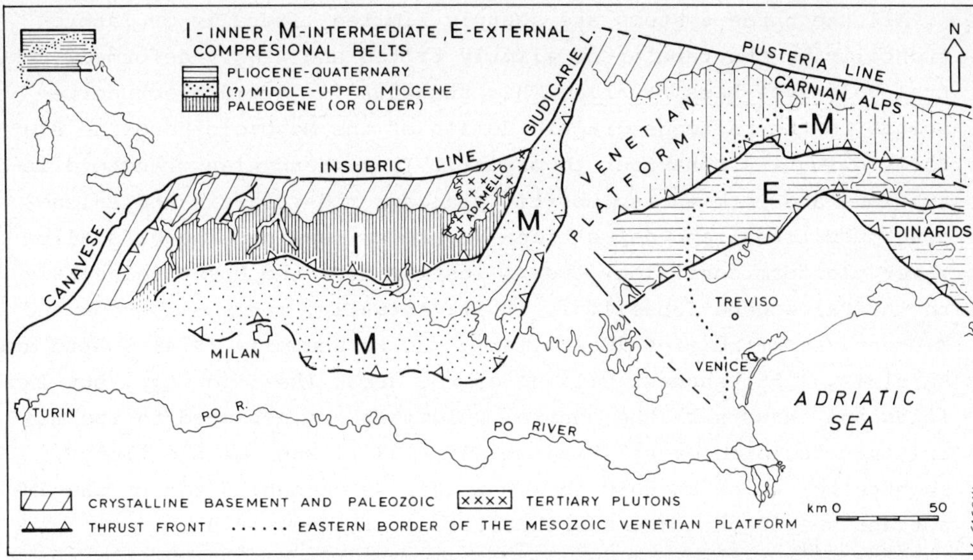

Fig. 6. General, chronologic classification of compressional fronts inside the Southern Alps (modified after Castellarin, 1984).

past, are not easily distinguishable from the severe younger deformation of the front. This is the most deformed part of the Southern Alps characterized by a mainly S facing, very tightly imbricated thrust-sheet belt largely involving the Upper Palaeozoic and, partly, also the Lower, Hecynian deformed Palaeozoic (Frascari _et al._, 1979, 1980; Castellarin _et al._, 1980b; Castellarin and Vai, 1981, 1982; Cavallin, 1981; Cavallin and Martinis, 1982; Cavallin and Giorgetti, 1983). The severe shortening, both in the Mesozoic cover and in the Palaeozoic basement, implies reduction to one third of the pre-Cretaceous depositional N-S extent (Castellarin, 1979; Vai, 1979). This maximum shortening is reached in the Friuli concavity of the front, where the thrust belt, limited to the N by the straight-linear chain of the Alps _s.s._, shows its narrowest width. This was taken as evidence of major, differential underthrusting (A subduction) of the Adriatic foreland toward the NNW (Castellarin _et al._, 1980b; Castellarin and Vai, 1982). The age of deformation of this belt ranges from Uppermost Cretaceous to Plio-Pleistocene (see f.i. the Eocene and lowermost Miocene unconformities: Selli, 1953, 1963; Ceretti, 1965; Carulli _et al._, 1983; Casati, 1969; Castellarin _et al._, 1980b, Sarti, 1980; Cavallin and Pirini, 1980; etc.). However, the Pliocene phase, which continues until present (Zanferrari _et al._, 1982), strongly overprints the entire belt, the frontal part of which appears to be the most recent within the Southern Alps.

All the three systems are sharply limited laterally or latero-frontally by the tabular or sligtly tilted and weakly deformed, central Dolomites-Lessini area. This structurally rigid block corresponds to the palaeogeographic domain of the Mesozoic Venetian Platform, owing its behaviour to magmatic (late Hercynian granitoid intrusions and Permian ignimbritic plateau; mid-Triassic and Palaeogene magmatites) and depositional (more or less persistent shallow water platform condition) events lasting from the Upper Palaeozoic to the Palaeogene (Bosellini, 1965). Actually, minor Oligocene and Miocene unconformities (Castellarin and Farabegoli, 1974; Conato and Martinis, 1955) suggest tilting of the area. The important, but local folds and thrusts of the central Dolomites are assigned to the mid-Triassic tectonic event (Pisa et al., 1980). Much of the Alpine shortening, which amounts to about 20%, is accounted for by the important Val Sugana basement thrust (Castellarin, 1979).

Interpretation

The structural relationship between inner and intermediate arcs (western sector) is easily explained in terms of classic orogenic time-space polarity. On the other hand, the mutual relationships of the two previous arcs and the third outer arc to the Venetian (Trento) Platform are much more difficult to explain. The major field evidences are: i) severe compressional structures developed in areas of ancient basinal environment; ii) mild to absent deformation in areas of large, brittle platform; iii) lack of significant strike-slip movement along the Judicaria Lineament (characterized by compressional structures).

Moreover, the highly arcuate outline of the intermediate arc, consistent with a fan-like arrangement of strain axes, is hard to explain in terms of rigid-block models.

It is tempting, therefore, to suggest a visco-elastic behaviour of different rock systems during deformation, as recently proposed by Castellarin and Sartori (1986), whose interpretation is summarized as follows (Fig. 7). An attenuated crustal block, corresponding to the eastern and southern still undeformed parts of the Lombardy Basin, limited to the W by the Adamello Massif plus the Orobic Chain and to the SE by the Venetian-Padania platform, both representing normal, rigid crust is assumed. Unidirectional NW-SE trending compressional stress, driven by the Po Plain-Adriatic block, would produce a rising flow of the underlying, ductile rock material against the lateral rigid blocks, whereas the overlying brittle rocks (upper crys-

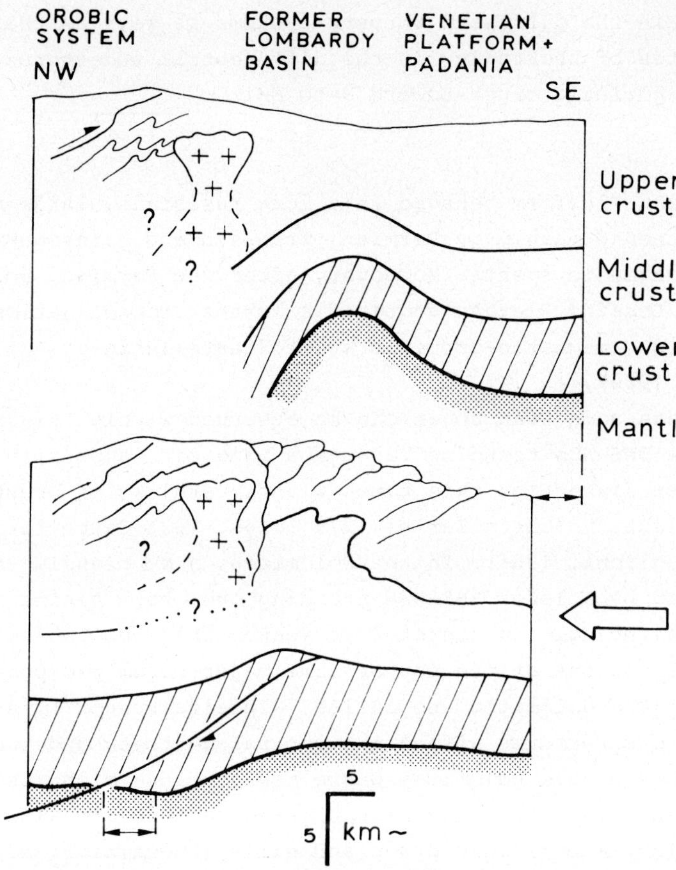

Fig. 7. Genetic model of the Judicaria-Val Trompia Southalpine arc. Crosses: Adamello batholith. Scale is approximate.

talline basement and sedimentary cover) would be squeezed out as imbricated sheets facing toward the E to S, less confined directions. This assumption is based on a crustal structure (Fig. 7) with ductile or "attenuated", low velocity middle crust (Giese and Morelli, 1975), implying a plastic versus rigid behaviour of the crustal system (Tapponier, 1977).

A major role of gravity tectonics in generating the Southalpine thrusts, previously overestimated, is being dismissed by recent authors; lack of denudated source areas, areal distribution of the three different thrust systems, their geometric relationships and their distinct age are the main reasons for supporting an active tectonic interpretation. Gravity tectonics may be responsible for only local, minor and epidermal effects. The main process possibly able to explain the different thrusts systems as well as the sharply different rates of shortening is the differential underthrusting of the Po Plain-Adriatic crust toward N to NW.

DISCUSSION

The Venetian Platform behaved as a more resistant, large crustal block strengthened mainly by Permian, Triassic and Palaeogene (Lessini Plateau) magmatic events. Moreover, after the Permian, Dolomites and northern Lessini Plateau acquired a further consolidation through the mid-Triassic magmatic-orogenic event (Castellarin et al., 1980a; Pisa et al., 1980).

Nevertheless some weak zones can be observed within this more rigid block. The ENE-WSW trending Valsugana Line corresponds in fact to a major thrust dissecting this block with several km of crustal shortening and uplift (Castellarin, 1979, figs. 1-3; Castellarin and Vai, 1982; Doglioni, 1986). In the Dolomites, a N-S trending thrust system crossed by a later ENE-WSW trending one, both Alpine in age, have been convincingly interpreted as responsible for the 'apical' tectonic duplications at the top of some major Dolomitic peaks (Doglioni, 1984) (the enigmatic, so-called "Gipfelfaltungen"). Anyway, these tectonic structures affect the central-northern and the westernmost sectors of the Dolomites where the Permian volcanics are thin or vanishing, the Triassic volcanic centers are almost absent and the Tertiary subvolcanic dykes and sills (Lucchini et al.,1983) are also very rare. To the S, toward the plain, the more rigid behaviour of this block is consistent with the subsurface and geophysical data characterized by high magnetic anomalies along the Adige River (AGIP, 1981), coupled with a gravimetric high, centered within the southern border of the Lessini Plateau (Giese and Morelli, 1975). To the SW, similar conditions, although less continuous, are largely represented in the Mantua area and widely in the Pavia region to the W (see above).

The main effect of this crustal setting was the sharp curvature (up to 90°) of the Giudicaria-Val Trompia arc. Its front overrides onto the Venetian Platform to the E and the Mantua and Pavia rigid blocks to the S.

The Venetian Platform was not greatly affected by compressive deformation on its western side; however, the adjacent basinal sectors to the E (Belluno and Iulian Basins: Winterer and Bosellini, 1981) underwent deformation produced by the NNW directed movements of the Adriatic microplate against the previously settled Alpine structures

(I-M of fig. 6). This was responsible for the greatest shortening in the Friuli-Carnia sector (Castellarin, 1979; Vai, 1979). The underthrusting gradually decreased to the W up to the Schio-Vicenza Line, a NW-SE trending, rejuvenated Mesozoic Lineament, located within an area strongly strengthened by middle Triassic volcanism and shallow plutonic intrusions (Recoaro), and by emplacement of large Palaeogene volcanic and subvolcanic bodies. The tectonic Southalpine front is, in fact, only slightly offset by the Schio-Vicenza Line. This is additional evidence for the weak deformation at this border of the outermost Southalpine arc. Its concave outline was originated by the differential underthrusting of the Adriatic block and possibly by the convex, angular shape of its northern border.

As for the Apennine arcs, the driving force responsible for the Messinian-Pliocene thrusting was related by some authors (e.g. Giglia, 1974) to the extension within the Tyrrhenian area and the internal Tuscan sector of the Apennines, where a system of tectonic depressions ("Graben") formed during the same time interval (Bartolini et al., 1983) (cf. gravitational spreading, Locardi, 1985; Wezel, 1985).

Nevertheless, the amount of extension in that area is inadequate to explain the amplitude, distribution and location of the Po Plain arcs, in addition to the older Apennine deformation fronts. In fact the Monferrato and partly the Emilia arcs appear to lack an adequate source area to their rear. The Oligo-Miocene Ligurian Basin cannot be considered for this purpose. Its opening may have originated older arcuate structures within the Apennine chain such as the Falda Toscana and the Cervarola-Falterona tectonic units (Boccaletti and Guazzone, 1974; Boccaletti et al., 1977; Carmignani et al., 1978). Therefore, a further meridian trending compression is needed to explain the Monferrato and Emilia arcs. Such compression can be easily assumed for the area between the converging African and European plates (Argand, 1924; McKenzie, 1972; Dewey et al., 1973; Boccaletti et al., 1982; Livermore and Smith, 1985). In this scenario the Tyrrhenian extensions, more than the cause of the Apennine thrustings, would appear as a consequence of the anticlockwise rotation of the Apennines pushed by the northward drift of Africa and thrusting over the decoupled Adriatic block.

Consequently, N-S directed, compressional Pliocene deformation of the Monferrato and Emilia arcs, as well as reactivation of previous western Alpine units can be inferred. The prominent torsion of the Alps around the Turin zone, where the Monferrato front faces the Padania corridor, close to the Western and Southern Alps; the long col lisional band of the Emilia arc front over the older, Southalpine, intermediate arc; and the small western Padania space unaffected by deformation are all good arguments in favour of a similar hypothesis. Therefore the Pliocene displacement of the Monferrato and Emilia arcs, the Southalpine front being inactive, could achieve the whole Pliocene space reduction by shortening of the Padania crust only on the Northapennine front. Conversely, in the eastern, Ferrara-Romagna arc, the area subjected to most rotational effects due to larger crustal extension in the rear (Tyrrhenian Sea and Tuscany), the northward directed, African drift-induced, compressional stress was reinforced and the Padania-Venetian crust underwent shortening on both fronts. To the N by means of underthrusting of the northern border of the Adriatic block, to the S with the overriding of the Apennines onto the Padania foredeep.

At this point it is worthwhile to compare and contrast the two distinct polyarcuate systems. Similarities and differences can be equally listed.

1) The two systems display epidermal opposite vergence, having a common foreland, i.e. the undeformed part of the Po Plain, which at first sight would appears as an homogeneous element. A closer inspection, however, suggests a morphologic differentiation of the Plain into three parts (fig. 10) and a structural subdivision into two parts, corresponding to the Pedealpine Homocline (Po Plain) and the Adriatic Homocline (Venetian Plain) of Pieri and Groppi (1981).

2) Also the threefold internal distinction of both arc systems is misleading both in terms of time and space (figs. 9 and 10). Spatially, the Po Plain arcs are broadly concentric with the older, inner Apennine arcs. The Southalpine arcs, on the contrary, sharply deviate from a simple concentric structure (fig. 10). Chronologically, the Po Plain arcs are of mainly Messinian to Pliocene age (disregarding a minor time polarity toward the E). Conversely, the Southalpine arcs show a major time polarity as the youngest arc is only developed in the eastern sector. This reprents the major asymmetrical element in comparing the geologic evolution of the two arc systems (fig. 10). It reflects a major structural difference between the Po Plain and the Southalpine arcs (see next point).

3) The epidermal, rougly symmetrical opposite vergence portrayed in the map (fig. 10) does not reflect the deeper kinematics as shown in cross section (fig. 8). The Po Plain Apenninic arcs owe their origin to major differential overthrusting of the Apennine chain onto the Padania-Adriatic foredeep. Conversely, the Southalpine, outer,

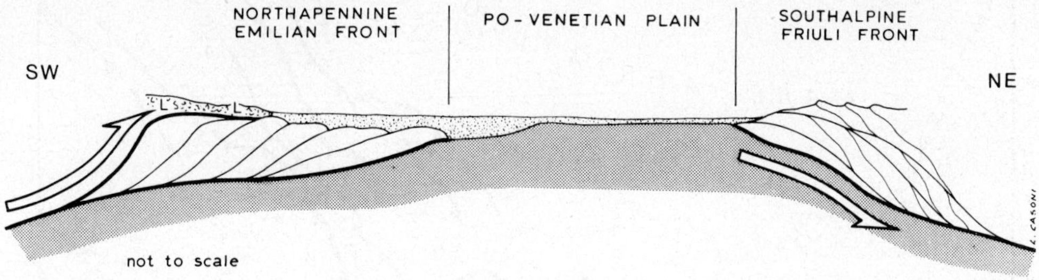

Fig. 8. Simplified kinematic S-N section across the eastern sectors of the Po Plain and the adjoining buried Apenninic and Southalpine fronts.

Friuli-Julian arc was generated at the same time by underthrusting of the Adriatic-Padania crust toward the NNW beneath the Southern Alps. Therefore, if true, active A subduction (Bally and Snelson, 1980) is suggested for the Southalpine front (in agreement with the Africa drift towards the N), only relative or passive A subduction can be admitted for the Northapenninic front. To allow coeval, active A subduction on both sides of the Po Plain a spreading axis would be needed, which seems disproved by the crustal and structural setting of the Po Plain.

4) Both arcuate systems are remarkably limited by more rigid, almost undeformed crustal blocks, whose location and size strongly influenced the development of the arc systems controlling their axial trends and the crossing of subsequently growing arc branches. Noteworthy in this context is the Venetian Platform, plus the Mantua and Pavia blocks. Of particular structural interest is also the contiguity, if not continuity, of the Venetian Platform to the Mantua block, what can explain many of the geometric and kinematic asymmetries of fig. 10.

The relevance of such blocks, acting as buttresses preventing the further lateral extension of the individual thrusts, appears critical for the origin of the arcuate patterns.

CONCLUSION

Time-space evolution of the Southalpine arcs is characterized by lateral accretion accomplished by addition of younger branches in an ESE (SE) direction. Accretion occurred by décollement of crustal elements (incorporated within the deformed belts) linked to an active,

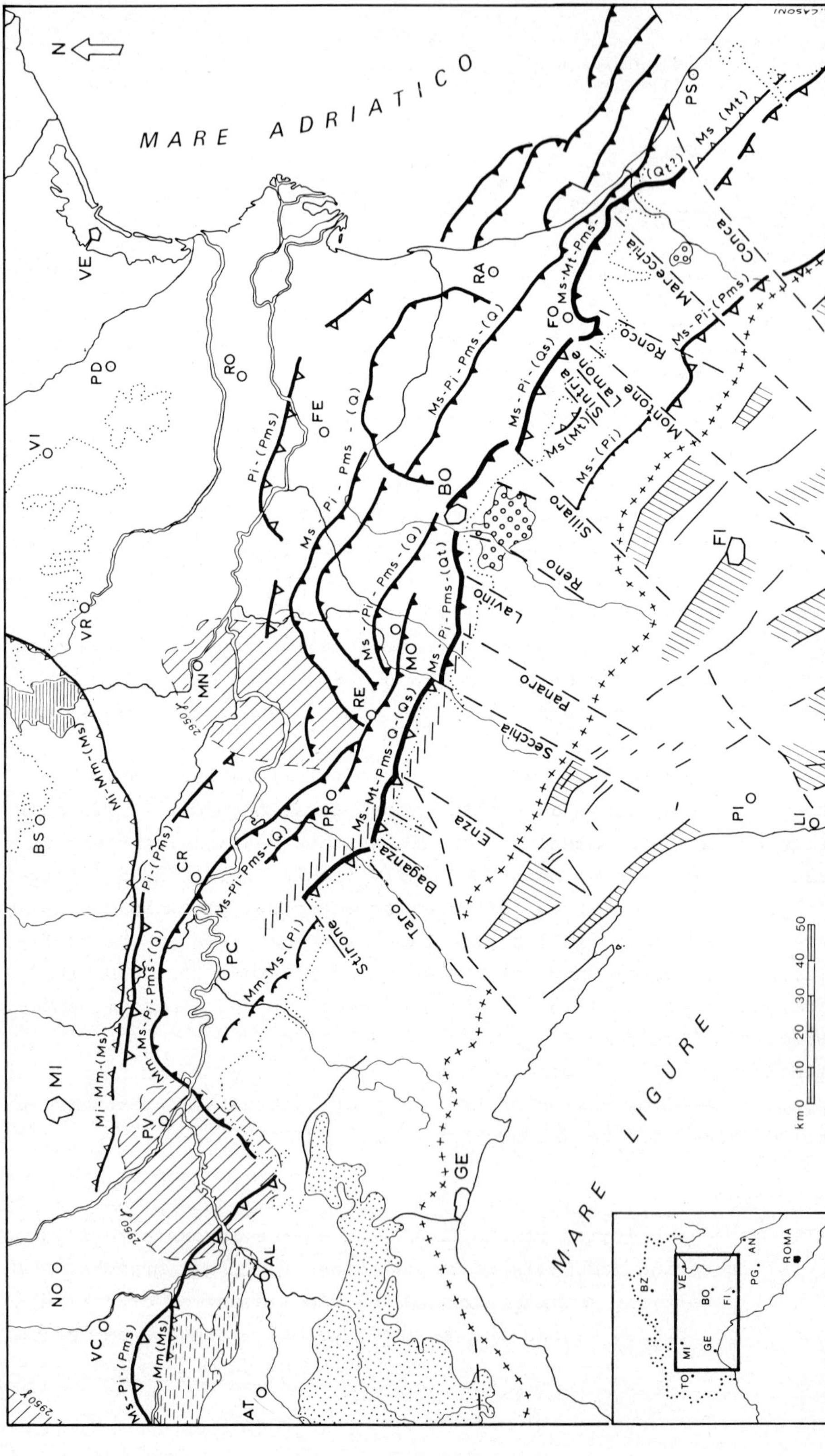
MARE ADRIATICO
MARE LIGURE
N
VE
PD
RO
VI
FE
VR
MN
BS
MI
CR
PC
PV
NO
VC
AL
AT
GE
PR
RE
MO
BO
FO
RA
PS
FI
PI
LI
Conca
Marecchia
Ronco
Montone
Lamone
Sintria
Sillaro
Reno
Lavino
Panaro
Secchia
Enza
Baganza
Taro
Stirone
Ms-Pi-Pms-(Q)
Ms-Mt-Pms-
Ms-Pi-(Qs)
Ms-(Mt)
Ms-(Pi)
Ms (Mt)
(Qt?)
Ms-Pi-(Pms)
Pi-(Pms)
Ms-Pi-Pms-(Qt)
Ms-Mt-Pms-Q-(Qs)
Mm-Ms-(Pi)
Mi-Mm-(Ms)
Mm(Ms)
km 0 10 20 30 40 50
L.CASONI
ROMA
AN
PG
BZ
TO

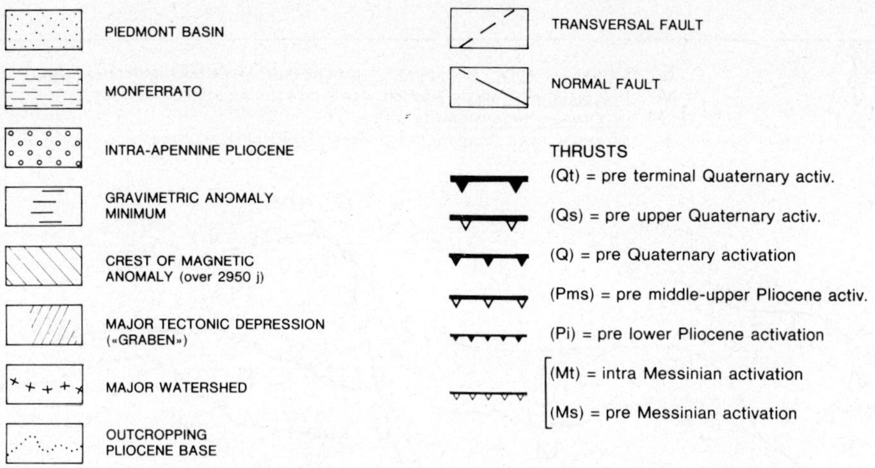

Fig. 9. Tentative chronologic classification of compressional fronts in the outer and in the buried Apennine chains. Major transverse faults connecting the different front segments are named. Captions refer to discrete time intervals of compressional synsedimentary phases. Blind and outcropping fronts are not graphically distinguished. Captions under brackets indicate the age of undeformed cover sealing the related lineament; in this way buried lineaments are differentiated and the age of last folding and thrusting is defined.

Key to captions: Mi: Lower Miocene; Mm: Middle Miocene; Ms: Upper Miocene; Mt: uppermost Miocene; Pi: Lower Pliocene; Pms: Middle to Upper Pliocene; Q: Quaternary undifferentiated; Qs: part Middle to part Upper Pleistocene; Qt: part Upper Pleistocene to Holocene.
Key to the last folding and thrusting: (Ms): pre-Messinian; (Mt): intra-Messinian; (Pi): pre-Lower Pliocene; (Pms): pre-Middle to Upper Pliocene; (Q): pre-Quaternary undifferentiated; (Qs): pre-part Middle to part Upper Pleistocene; (Qt): pre-part Upper Pleistocene to Holocene (after Castellarin *et al.*, 1986).

NNW-directed A-type subduction (Bally and Snelson, 1981) of the Adriatic microplate. Similar kinematics would imply dextral strike-slip components along the Periadriatic Lineament, as previously postulated (Laubscher, 1971, 1973, 1985,1986; Doglioni, 1986). Ancestral, more rigid crustal structures, however, were not appreciably rearranged by the assumed oblique-slip movements. Thrust belts ramped on and around more rigid blocks also where evidence of coeval strike-slip movement is lacking, as for the prominent, NNE trending, Neogene thrusts along the Judicaria Line (Castellarin and Sartori,1982).

The Po Plain arcs were partly controlled by the Upper Miocene to Pliocene extension in the N-Tyrrhenian and internal Apennine (Tuscany) areas. This extension produced rotational effects toward the NE which are thought to have enhanced the compressional regime induced

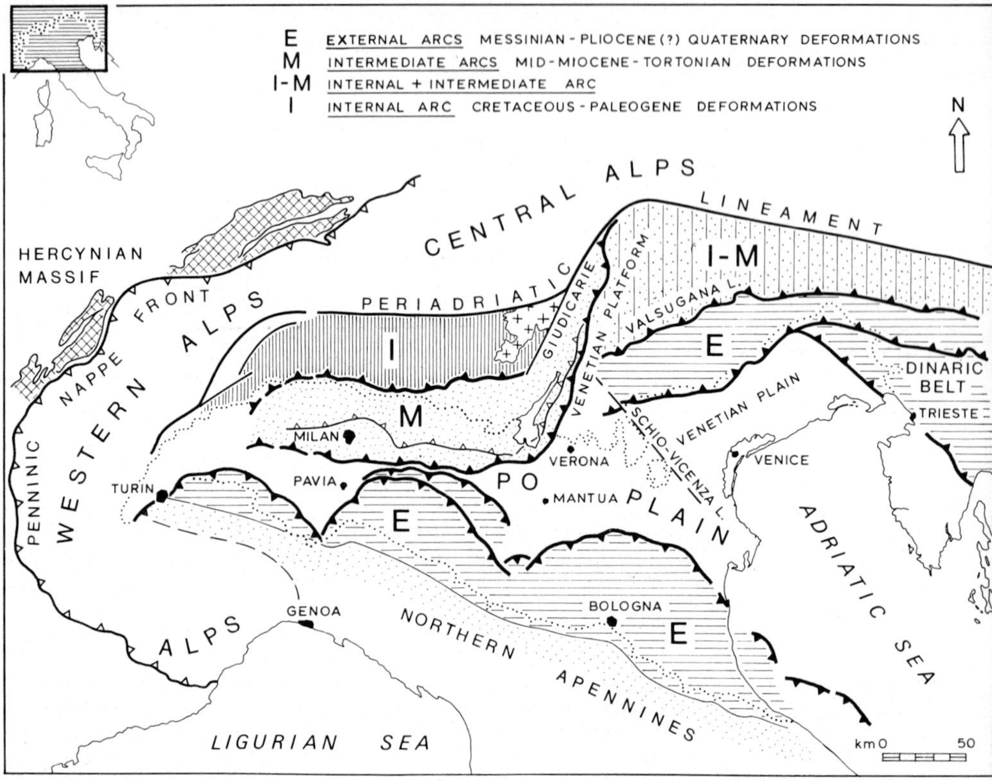

Fig. 10. Simplified structural map of the Po Plain and adjoining Northapenninic and Southalpine arcs devoided of the Quaternary undeformed cover.

by the African drift. The composite stress field acted mainly across the eastern Padania-northern Adriatic area, particularly at the still active, outer Southalpine Friuli-Julian front.

A critical point in our reconstruction of the internal Southalpine and Po Plain buried structures is the role played by the Verona-Vicenza homocline, which gently dips beneath the Pliocene-Quaternary units of the Po Plain (Pieri and Groppi, 1981, sez. 9). No significant duplication processes can be admitted along this border, where stratigraphical units seem welded to the underlying crust. Other possibilities, however, have been recently proposed. For instance, by means of blind thrusts producing extensive shortening of the upper crust in the Lessini Plateau, underneath the Po Plain (Roeder, 1985). Lack of any syntectonic lithosome, such as foredeep sedimentary wedges, along the assumed "blind" front and the regular blanket of Plio-Quaternary deposits across the Plain (Pieri and Groppi, 1981, sez. 9) are the principal arguments against important displacements on that part of the Southalpine margin during the Neogene.

Conversely, in the S part of the Plain, the foredeep is impressively developed (Ricci Lucchi, 1986),mainly in the Ferrara-Romagna arc.

To sum up, if the Lessini Plateau was not involved in the shortening, only the Ferrara-Romagna front could accommodate the differential deformation existing between the Lessini Plateau border and the eastern Southalpine sector (Friuli-Julian arc).

As to the basic conditions common to the Southalpine and the Apennine-Po Plain arcs, they are typically characterized by more rigid crustal blocks, whose size, location and shape severely influenced the arc system development, controlling their axial trends and their geometric relationships.

The Padania domain reached its present outline as a result of Upper Miocene-Pliocene heterochronous and quite asymmetric bilateral compressions, applied in different ways to its N and S borders. They caused a progressive reduction of the undeformed Po Plain space and its gradual narrowing. The Plio-Pleistocene high sedimentation rate within the tightened Padania corridor is related to the general downwarping of the Padania depocenter, possibly triggered by low-velocity mantle-channel centrifugal flow towards both the Alpine and the Apennine roots which, at the same time, were severely uplifted (see f.i. Panza et al., 1980). This subsiding trend of the Po Plain was enhanced by the Neogene compressions, as well as by the loads implied by displaced tectonic units and recently deposited sediments.

REFERENCES

AGIP, 1981. Italia, Carta Magnetica, scala 1:500.000, Foglio B. Milano.

Argand, E., 1924. La tectonique de l'Asie. Con. Géol. Int., XIII Belgique 1922, 1, 171-372.

Arisi-Rota, F. and Fichera, R., 1985. Magnetic interpretation connected to "geomagnetic provinces": the Italian case history. AGIP Expl. Ser., 29 pp. (pre-print).

Bally, A.W., Catalano, R. and Oldow, J.S., 1985. Elementi di tettonica regionale. Pitagora Ed., Bologna, 276 pp.

Bally, A.W. and Snelson, S., 1980. Realms of subsidence. In: A.D. Miall (Editor), Facts and principles of world petroleum occurrence. Canad. Soc. Petroleum Geol. Memoir, 6, 9-94.

Bartolini, C. et al., 1983. Carta neotettonica dell'Appennino Settentrionale. Note illustrative. Boll. Soc. Geol. It.,101,523-549.

Boccaletti, M. and Guazzone G., 1974. Remnant arcs and marginal basins in the Cenozoic development of the Mediterranean. Nature, 252, 5478.

Boccaletti, M., Guazzone, G. and Manetti, P., 1977. Evoluzione paleogeografica e geodinamica del Mediterraneo: i bacini marginali. Memorie Soc. Geol. It., 13/2, 162-199.
Boccaletti, M., Conedera, C., Dainelli P. and Gocev P., 1982. The recent (Miocene-Quaternary) regmatic system of the Western Mediterranean region. J. Petroleum Geology, 5, 1, 31-49.
Bolis, G., Cappelli, V. and Marinelli, M., 1981. Aeromagnetic data of the Italian area: instrumental to a better comprehension of the basement main characteristics in Italy. AGIP, 26 pp.(preprint)
Bosellini, A., 1965. Lineamenti strutturali delle Alpi Meridionali durante il Permo-Trias. Mem. Mus. St. Nat. Ven. Trid.,15,3, 1-72.
Bosellini, A., 1981. The Emilia Fault: a Jurassic fracture zone that evolved into a Cretaceous-Paleocene sinistral wrench fault. Bollettino Soc. Geol. It., 100, 161-169.
Boyer, S. and Elliott, D., 1982. Thrust systems. A.A.P.G. Bull., 66, 1196-1230.
Brack, P., 1981. Structures in the southwestern border of the Adamello intrusion (Alpi Bresciane, Italy). Schweiz. Min. Petr. Mitt., 61, 37-50.
Butler, R.W., 1982. The terminology of structures in thrust belts. J. Structural Geology, 4, 3, 239-245.
Carmignani, L., Giglia, G. and Kligfield, R., 1978. Structural evolution of the Apuane Alps: an example of continental margin deformation in the Northern Apennines, Italy. J. Geol.,86,487-504.
Carulli, G., Zucchi Stolfa, M.L. and Pirini Raddrizzani, C., 1983. L'Eocene di Monte Forcella (Gruppo del Monte Amariana - Carnia orientale). Memorie Soc. Geol. It., 24, 65-70.
Casati, P., 1969. Le fasi orogenetiche nelle Alpi Meridionali. Arti Grafiche Artigianelli Pavoniani Monza, 49 pp.
Castellarin, A., 1979. Il problema dei raccorciamenti crostali nel Sudalpino. Rendiconti Soc. Geol. It., 1, 21-23.
Castellarin, A., 1984. Schema delle deformazioni tettoniche sudalpine. Boll. Oceanologia Teor. Appl., 2, 105-114.
Castellarin, A. and Farabegoli, E., 1974. Cicli sedimentari di spiaggia nell'Oligocene di Cavalo. Giorn. Geologia, 39,2, 393-420.
Castellarin, A. and Vai, G.B., 1981. Importance of Hercynian tectonics within the framework of the Southern Alps. J. Structural Geology, 3, 4, 477-486.
Castellarin, A. and Vai, G.B., 1982. Introduzione alla geologia strutturale del Sudalpino.In: A. Castellarin and G.B. Vai (Eds.) Guida alla geologia del Sudalpino centro-orientale. Guide geol. reg. Soc. Geol. It., Bologna, 1-22.
Castellarin, A. and Sartori, R., 1982. Geologia della Linea delle Giudicarie Sud (tra Pieve di Bomo e Tione). Ibid., 103-114.
Castellarin, A. and Sartori, R., 1986. Il sistema tettonico delle Giudicarie, della Val Trompia e del sottosuolo dell'alta pianura lombarda. Memorie Soc. Geol. It., 26, 31-37.
Castellarin, A., Lucchini, F., Rossi, P.L., Simboli, G., Bosellini, A. and Sommavilla, E., 1980a. Middle Triassic Magmatism in Southern alps. II: a geodynamic model. Riv.It.Pal.Str.,85, 1111-1124.
Castellarin, A., Frascari, F. and Vai, G.B., 1980b. Problemi di interpretazione geologica profonda del Sudalpino orientale. Rendiconti Soc. Geol. It., 2, 55-60.
Cavallin, A., 1981. Fogli 24 Maniago e 39 Pordenone. In: A. Castellarin (Ed.), Carta tettonica delle Alpi Meridionali. Pubbl. n. 441, P.F.G., CNR, 46-50.
Cavallin, A. and Giorgetti, F., 1983. Elementi geologici e geofisici per un modello geodinamico dell'Italia nord-orientale e suoi rapporti con l'Appennino. Memorie Soc. Geol. It., 24, 79-87.
Cavallin, A. and Martinis,B., 1982. Gli scorrimenti del margine settentrionale della piattaforma carbonatica adriatica. In:Castellar-

in and G.B. Vai (Eds.), Guida alla geologia del Sudalpino centro-orientale, Guide geol. reg. Soc. Geol. It., Bologna, 349-359.

Cavallin, A. and Pirini Radrizzani, C., 1980. Il Miocene trasgressivo di Pian Cavallo (Prealpi Carniche): margine settentrionale della piattaforma adriatica. Riv. It. Pal. Strat., 86, 399-416.

Ceretti, E., 1965. La geologia del Gruppo del M. Plauris. Giorn. Geologia, 33, 1, 1-50.

Dahlstrom, C.D.A., 1970. Structural geology in the eastern margin of the Canadian Rocky Mountains. Can.Soc.Petr.Geol.,Bull.18,332-406.

Dal Piaz, Gb., 1942. Geologia della bassa Val d'Ultimo e del massiccio granitico di M. Croce. Mem. Mus.St.Nat.Ven.Trid.,5, 177-357.

De Jong, K.A., 1967. Tettonica gravitativa e raccorciamento crostale nelle Alpi Meridionali. Boll. Soc.Geol.It., 86, 4, 749-776.

Dewey, J.F., Pitman, W.C.III, Ryan, W.B.F. and Bonnin, J., 1973. Plate tectonics and the evolution of the Alpine system. Geol. Soc. Am. Bull., 84, 3137-3180.

Doglioni, C., 1984. I sovrascorrimenti nelle Dolomiti: sistemi di ramp-flat.Tecnoprint Bologna, 22 pp.

Doglioni, C., 1986. Tectonics of the Dolomites (Southern Alps, Northern Italy). J. Struct. Geol. (in press).

Dondi, L., Mostardini, F. and Rizzini, A., 1982. Evoluzione sedimentaria e paleogeografica nella Pianura Padana. In: G. Cremonini and F. Ricci Lucchi (Eds.), Guida alla geologia del margine appenninico-padano, Guide geol.reg.Soc.Geol.It., 47-48.

Francavilla, F. and Colombetti A., 1980. Lineamenti idrogeologici della pianura della Provincia di Bologna. Quad.IRSA-CNR,51,119-43.

Frascari, F., Vai, G.B. and Zanferrari, A., 1979. Profilo Carnico Centrale. Nota illustrativa sommaria. Rend.Soc.Geol.It.,1, 15-17.

Frascari, F., Semenza, E., Spalletta, C., Vai, G.B. and Venturini, C., 1980. Profilo Carnico B: Gailtal-M.S. Simeone. Nota illustrativa sommaria. Rend. Soc. Geol. It., 2, 17-20.

Gaetani, M. and Jadoul, F., 1979. The structure of the Bergamasc Alps. Ac. Naz. Lincei, Rend.Cl.Sc.F.M.N.,Ser.8, 66, 411-416.

Gianotti, R. and Perotti, C., 1986. Inquadramento strutturale. In: Guida all'escursione,Convegno di Geologia Lariana, 37-56,Pavia.

Giese, P. and Morelli, C., 1975. Crustal structure in Italy. In: L. Ogniben et al.(Eds.), Structural model of Italy.Quaderni Ricerca Scientifica, CNR, 90, 453-489.

Giglia, G., 1974. L'insieme Corsica-Sardegna e i suoi rapporti con l'Appennino settentrionale: rassegna di dati cronologici e strutturali. Rend.Sem.Fac.Scienze, Cagliari,suppl.v.13, 245-275.

Gortani, M., 1956. Recenti progressi nella conoscenza strutturale dell'Italia. In: F.Lotze (Ed.), Geotektonisches Symposium, Enke Stuttgart, 143-176.

Laubscher, H.P., 1971. Das Alpen-Dinariden Problem und die Palinspastik der südlichen Tethys. Geol.Rundschau, 60, 813-833.

Laubscher, H.P., 1973. Alpen und Plattentektonik. Das Problem der Bewegungsdiffusion und kompressiven Plattengrengen. Z. Deutsch. Geol.Ges., 124, 295-308.

Laubscher, H.P., 1985. The tectonics of the Western and Southern Alps:correlation between surface observations and deep structure. Second EGT Workshop, Venice 1985, ESF Strasbourg, 93-101.

Laubscher, H.P., 1986. The late Alpine (Periadriatic) intrusions and the Insubric Line. Mem. Soc. Geol. It., 26, 21-30.

Livermore,A. and Smith, A.G., 1985. Some boundary conditions for the evolution of the Mediterranean region. In: D.J. Stanley and F.-C. Wezel (Eds.), Geological evolution of the Mediterranean Basin. Springer-Verlag New York, 83-98.

Locardi, E., 1985. Neogene and Quaternary Mediterranean volcanism: the Tyrrhenian example. In: D.J. Stanley and F.-C. Wezel (Eds.), Geological evolution of the Mediterranean Basin. Springer-Verlag New York, 273-291.

Lucchini, F. et al., 1983. Petrochimica e dati radiometrici K/Ar e Rb/Sr dei "filoni basici" di Corvara in Badia e dei lamprofiri di Val Fiscalina (Dolomiti orientali). Min.Petr.Acta, 27, 233-249.

Mattavelli, L. et al., 1983. Geochemistry and habitat of natural gases in Po Basin,Northern Italy. A.A.P.G.Bull.,67,2239-2254.

McKenzie,D.P., 1972. Active tectonics of the Mediterranean region. Geoph.J.R. Astr. Soc., 30, 109-185.

Ori G.G. and Friend P.F., 1984. Sedimentary basins formed and carried piggyback on active thrust sheets. Geology, 122, 475-478.

Panza,G.F. et al., 1980. Struttura profonda dell'area mediterranea. Le Scienze, 141, 60-69.

Patacca, E. and Scandone P., 1986. Struttura geologica dell'Appennino emiliano-romagnolo: ipotesi sismotettoniche.

Pieri, M., 1983. Three seismic profiles through the Po Plain. A.A. P.G. Studies in Geology, Series 15, 3, 3.4.1./8-26.

Pieri, M., 1984. Storia delle ricerche nel sottosuolo padano fino alle ricostruzioni attuali. In: Cento anni di geologia italiana. Vol. giub. I Centenario S.G.I., Bologna Pitagora-Tecnoprint,155-176.

Pieri, M. and Groppi, G., 1981. Subsurface geological structure of the Po Plain. Pubbl. 414 P.F. Geodinamica, CNR, 23 pp.

Pisa, G. et al., 1980. Middle Triassic magmatism in Southern Alps. I: a review of general data in the Dolomites.Riv.It.Pal.Str.85,1093-1110.

Ricci Lucchi, F.,Colella, A., Ori, G.G. and Ogliani F., 1981. Pliocene fan-deltas of Intra-apenninic Basin, Bologna. In: F. Ricci Lucchi (Ed.), Excursion Guidebook,2nd Eur.Reg.Meet.Bologna, 79-162.

Ricci Lucchi, F., 1986. The foreland basin system of Northern Apennines and related clastic wedges: a preliminary outline. Giornale di Geologia, 47, 3, (in stampa).

Rizzini, A. and Dondi, L., 1978. Erosional surface of Messinian age in the subsurface of the Lombardian plain (Italy).Mar.Geol.,27,303-325.

Rocco, T and Jaboli, D., 1956. Geology and Hydrocarbons in the Po Basin. In:L.G. Weeks (Ed.),Habitat of oil, A.A.P.G., Tulsa, 1153-1167.

Sarti, M., 1980. Il Paleogene della Val Tremugna (Prealpi Carniche). Boll.Soc.Geol.It., 98, 87-108.

Selli, R., 1953. La geologia dell'alto bacino dell'Isonzo. Giornale di Geologia, 19, 1-153.

Selli, R., 1963. Schema geologico delle Alpi Carniche e Giulie occidentali. Giornale di Geologia, 30, 1-136.

Tapponier, P., 1977. Evolution téctonique du système alpin en Méditerranée: poinçonnement et écrasement rigido-plastique. Bull.Soc. Géol. France, s.7, 19, 437-460.

Trevisan, L., 1939. Il gruppo di Brenta. Mem.Ist.Geol.Padova,13,1-128.

Vai, G.B., 1979. Una palinspastica permiana della Catena Paleocarnica.Rend. Soc.Geol.It., 1 (1978), 25-27.

Wezel, F.-C., 1985. Structural features and basin tectonics of the Tyrrhenian Sea. In: D.J. Stanley and F.-C. Wezel (Eds.), Geological evolution of the Mediterranean Basin. Springer-Verlag New York, 153-194.

Winterer, E.J. and Bosellini, A., 1981. Subsidence and sedimentation on a Jurassic passive continental margin (Southern Alps, Italy). A.A.P.G. Bull., 65, 394-421.

BETIC-RIFIAN AND TYRRHENIAN ARCS : DISTINCTIVE FEATURES, GENESIS AND DEVELOPMENT STAGES

J.-P. BOUILLIN, M. DURAND-DELGA and Ph. OLIVIER

Laboratoire de Géologie méditerranéenne associé au C.N.R.S. n°145, Université Paul-Sabatier, 38 rue des Trente-Six-Ponts, 31400 Toulouse (France).

ABSTRACT

The Betic-Rifian and the Tyrrhenian Arcs bound both ends of the Maghrebide Chain. Their external zones originated from the South-Iberian,African and Apulian margins while the internal zones, with large basement nappes, originated from the Alboran, Kabylian, Peloritan and Calabrian ("ALKAPECA") South-European margins. During the Alpine convergence of the European and African plates, ALKAPECA was cut into blocks. The ejection along strike-slip faults of the blocks issued from the Western and Eastern extremities of ALKAPECA gave rise to the Arc of Gibraltar and to the Tyrrhenian Arc. The Western block (Alboran) was deformed as it was introduced into the area, reducing continuously throughout time, between Europe and Africa and as it collided with more and more external zones. The Eastern block (Aeolian) has bent the initially continuous African and Apulian margins. The Tyrrhenian Arc could be predeterminated during Jurassic distensional events : it corresponds to the paleogeographic angle made by the Ligurian basin (NE-SW trending) and the Mograbin basin (E-W trending).

INTRODUCTION

The South-Western part of the Western Mediterranean coast is edged over about 3000 km with an Alpine chain (fig.1) comprising the Betic Cordillera (Balearic Islands, South of Spain) and the Maghrebides (North of the Magrab, Sicily and

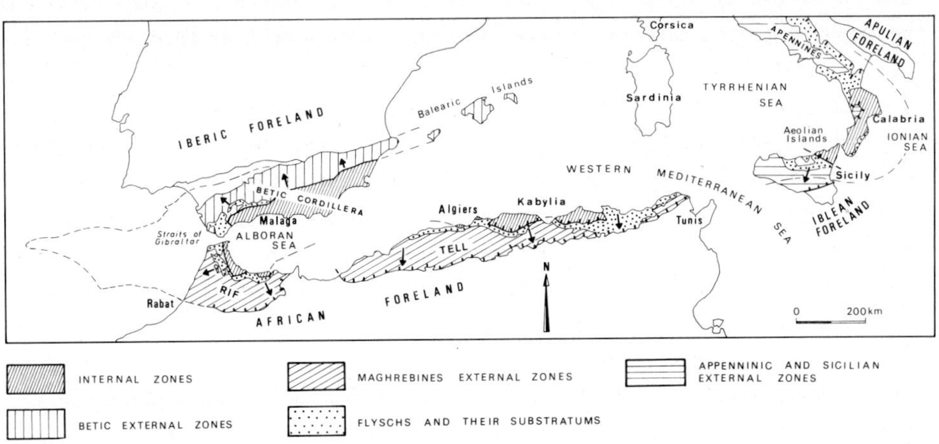

Fig.1. - The Alpine Chain of the Western Mediterranean area.

Calabria in the South of Italy).

At each of its ends this chain suffers an important curvature. Westward it is the Arc of Gibraltar which convexity faces the Atlantic Ocean and eastward the Tyrrhenian (or Siculo-Calabrian) Arc which convexity faces the Ionian Sea.

1 THE ARC OF GIBRALTAR

The Arc of Gibraltar (fig.2) comprises the Rifian and Western Betic parts of the Chain. The curvature is well defined, especially by units located at the boundary between Internal Zones and External Zones. Following this boundary from the Eastern Rif we can see that structural directions first trend roughly E-W in the Bokoya region (El Hoceima) and all along the Jebha fault. Then, from the area of Chaouen, they first bend to SE-NW and become more or less N-S until the Straits of Gibraltar. North of the Straits, the directions vary significantly around the N-S and after Gaucin trend SW-NE and finally roughly E-W to the North of Malaga and in the area of Granada.

1.1 THE UNITS INVOLVED IN THE ARC

The internal zones

From the deeper to the uppermost units we find :

- The Alpujarrides (Betic Cordillera) - Sebtides (Rif). It is a complex group of nappes with a basement and a sedimentary cover. The cover, consisting mainly of pelitic and carbonated triassic rocks, has been affected by a metamorphism of clearly Alpine age. The basement corresponds to deep origin rocks (mantellic Peridotites, Gneisses of the Granulite facies, Micaschists ...). The age of the emplacement of the Peridotites and the age of the high grade metamorphism are debated. For recent authors (Michard et al.,1983) these phenomena are linked and of Alpine age. However we think that ante-alpine structural and metamorphic phases have had a great intensity though most of them are now hidden and badly known.
- The Malaguides (Betic Cordillera) - Ghomarides (Rif) are constituted by nappes of dominant Paleozoic deposits, partly covered by thin Mesozoic and Eocene formations. The whole is generally devoid of important metamorphism.
- The Dorsale Calcaire and the Predorsalian Zone are a group of generally non metamorphic units comprising Paleogene and mainly Mesozoic rocks (where Trias and Lias are dominant). These carbonated deposits represent the remnants of the margin of the Malaguide-Ghomaride domain. Different parts of the continental shelf and of the slope (Predorsalian formations) could be found there. The Predorsalian shows flysch-like deposits at several times. A paleogeographic transition between the internal domain and the Flysch basin could be assumed by these flysch-like deposits.

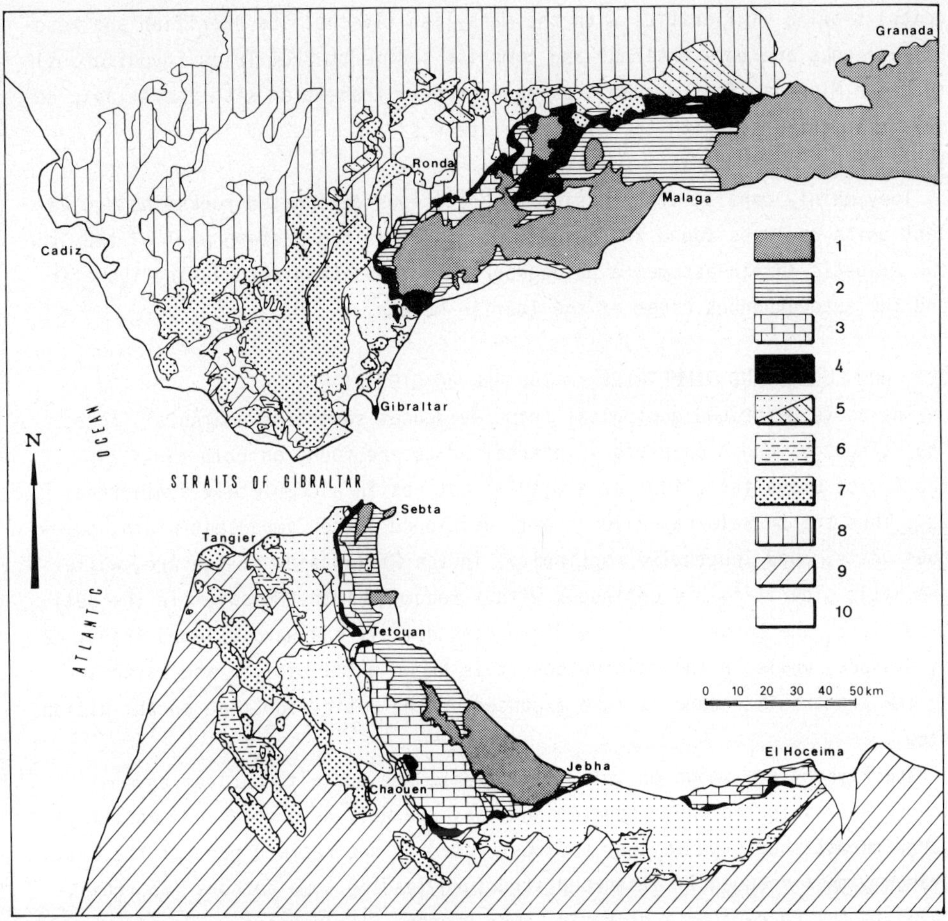

Fig. 2.- The Arc of Gibraltar. (1) Alpujarrides-Sebtides ; (2) Malaguides-Ghomarides ; (3) Dorsale calcaire ; (4) Pre-Dorsalian zone ; (5) Mauretanian flyschs and their possible substratum ; (6) Massylian flyschs ; (7) Numidian nappe ; (8) Betic external zones ; (9) Rifian external zones ; (10) Post-orogenic formations.

The Flyschs

Three groups are recognized: - The Mauretanian Flyschs which are the most internal, comprise frequently arkosic sediments dated from Neocomian to upper Oligocene ; - The Massylian Flyschs, which are the most external, comprise evolute sediments of Aptian-Albian age ; - The Numidian series dated from Oligocene to Middle Burdigalian, is mainly constituted by a thick sandy flysch of Aquitanian age. This coarse grained formation was deposited in a basin perhaps greatly superimposed to the previous Massylian zone.

The External Rifian zones

The Intrarifian zone comprises mainly Cretaceous formations (Ketama-Tangier units) showing similarities with the Massylian flyschs. The Prerifian and Mesorifian zones are more external and comprise sediments from Trias (gypsiferous) to Upper Miocene. They assume a paleogeographic transition with the Atlasic domain that is to say with the African plate s.s.

The External Betic zones

They mainly consist of Jurassic and Cretaceous carbonated rocks. No equivalent units could be found in the Rif. Moreover, in the Eastern part of the Chain, the Prebetic domain assumes a paleogeographic transition between the Subbetic and the autochthonous cover of the Iberian Meseta.

1.2 HOMOLOGIES AND SYMMETRIES IN THE ARC OF GIBRALTAR

The above described geological frame evidences several fundamental facts. The Malaguides and Alpujarrides internal units are found on both sides of the Arc of Gibraltar (Didon et al.,1973) but not in a completely symmetrical way. Thus the Dorsale calcaire is well developed (often important width, numerous units, good lengthwise continuity) in the Rifian side of the Arc, while in the Betic side it is discontinuous with a reduced width. Moreover in the Betic Cordillera, the Dorsale calcaire disappears quickly eastward (Sierra Arana, NE of Granada) while in the Maghrebides it is known until Sicilia and Calabria.

The Predorsalian zone is more expanded on the Betic side than on the Rifian side.

The Flyschs are known on both sides of the Arc but with a clear asymmetry. In the Rif as in the whole Maghrebides they are well developed and constitute large thrust-sheets while they are restricted to the western part of the Betic branch with a frequently disturbed structure leading some authors (Bourgois, 1978) to the concept of a gigantic Lower Miocene olistostrome.

The Betic and Rifian external zones are not directly related from a paleogeographic point of view despite the Almarchal unit, equivalent of the rifian unit of Tangier, being present in the southern part of the Campo of Gibraltar (Betic Cordillera). From the Internal zones (sedimentary cover of the Malaguides-Ghomarides) to the African foreland, crossing the Flyschs domain and the Rifian external zones, one can recognize, especially during the Mesozoic, a consistant paleogeographic evolution crosswise the Rif Chain. On the contrary the External and Internal Betic zones show a brutal discontinuity. Thus the Betic and Rifian Internal zones, with the Flyschs paleogeographically linked at their margin, build a single comprehensive geological domain which draws the Arc of Gibraltar despite some asymmetries of structure and despite the breaking of the Straits of Gibraltar open since the Pliocene. On the other hand, the Betic and Rifian External zones, despite the fact that they fit roughly the curvature of the Arc, constitute two separate domains, lately lumped together. It is useful, lastly, to notice that the "Homologies and symmetries" here introduced are not accepted by all the authors (Bourgois,1978 ; Wildi,1983).

1.3 STRUCTURE OF THE UNITS INVOLVED IN THE ARC OF GIBRALTAR

The main steps of structuration of the studied domain could be summarized as follows :

- The "Eo-alpine" phases, especially during Cretaceous and Eocene are poorly known;
- The first structuration in thrust-sheets of the Malaguides-Ghomarides domain can be dated from the Oligocene time (post Priabonian - ante Uppermost Oligocene). One can consider that these thrustings were moving from South or South-West towards North or North-East, considering the fact that during this event the Malaguides-Ghomarides overthrust the Alpujarrides-Sebtides which paleogeographic location was more Northern.
- The development of the Dorsalian slices and thrust-sheets can be dated from the Uppermost Oligocene. The displacements are directed towards the External domains. During the Uppermost Oligocene and/or the Lowermost Aquitanian, began the sedimentation of the "Oligo-Miocene" detritic formation, on the Betic-Rifian Internal zones

  In the Predorsalian domain and in the Flysch basin, from Upper Oligocene to Middle Burdigalian (included) there is a continuous accumulation of detritic deposits comprising a material of external origin (Numidian flysch).

  We think that during Middle and/or Late Burdigalian times a new important tectonic phase occurred. The Internal zones reacting as a whole, overthrust first the Predorsalian zone, then the Flysch units. Backthrusting of Predorsalian material, occurred in the Internal zones at the end of this period.
- The Middle Miocene (Langhian-Serravallian) corresponds to a period of comparative rest for tectonics. It is at the beginning of the Late Miocene that the development of thrust-sheets began again in the External domain and that the formation of the olistostromes of the Prerif, of the Guadalquivir and of the Horse-shoe Plain (Atlantic abyssal plain, westward of Gibraltar - Lajat et al., 1975) started. These olistostromes reproduce the Arc of Gibraltar, but this shape seems partly caused by the previous topography (Iberian and Moroccan Meseta topographies, channel way West of Gibraltar).

1.4 THE ALBORAN BLOCK

Introduction

From 1971, the concept of the "Alboran microplate" (Andrieux et al.,1971) or "Alboran block" was used in most of the tectonic reconstructions of the Western Mediterranean. The reasons are the following :

- The displacement from West towards East of the European and African plates, no more than their N-S convergence could explain alone the curvature of Gibraltar. On the other hand if we admit the existence of a small intermediate plate shifted eastwards, but less than the main adjacent plates and facing them at their boundary, we could explain the curvature.

- The Betic-Rifian internal zones are bounded northward and southward by large strike-slip faults, which suggest strongly a mainly autonomous shift of these Internal zones with respect to the Betic External zones (belonging to Iberia) and to the Rifian External zones (belonging to Africa).

Boundary faults of the Alboran block (fig.3)

A critical review of the main faults of the Betic-Rifian area of the Chain shows (Leblanc and Olivier, 1984) that four major strike-slip faults played an important role during the evolution of the Alboran block. Two of these until the Middle-Late Burdigalian, and the two others until the beginning of the Late Miocene (early Tortonian).

The Burdigalian strike-slip faults

- The Jebha fault which strikes from N60E to N80E marks out southward the main part of the Rifian Internal zones. It is a left-lateral strike-slip fault with at least 50 km of shifting. It continues with the Alboran ridge in the Mediterranean sea ; towards the WSW, it reaches the Atlantic coast through the Exter-

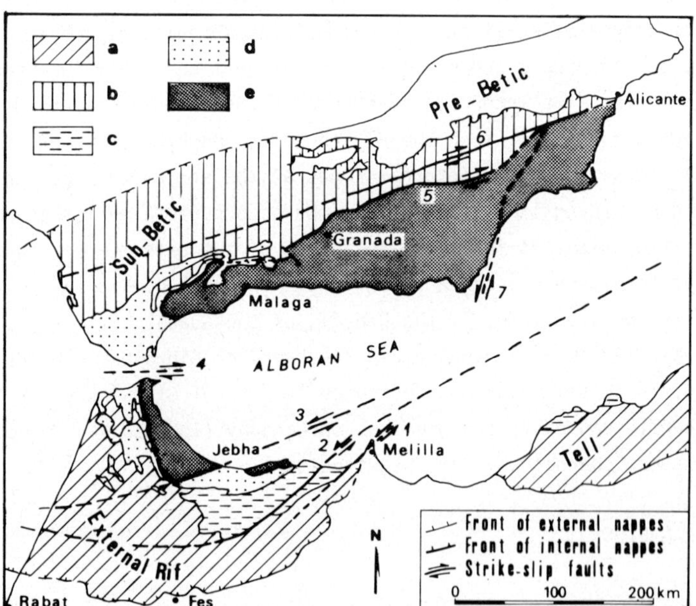

Fig. 3.- The main strike-slip faults in the Betic-Rifian Orogeny. -(a) Maghrebide External non-schist zones ; (b) Betic Ext. non schist zones ; (c) Maghrebides Ext. schist zones ; (d) Flyschs and Pre-Dorsalian area ; (e) Internal zones. - Strike-Slip faults : (1) Melilla fault ; (2) Nekor f. ; (3) Jebha f. ; (4) Straits of Gibraltar f. ; (5) North Betic f. ; (6) Crevillente f. ; (7) Carboneras f.

nal Rifian Zones (fig. 3). We think that the last major movement of the Jebha fault occurred during the Middle/Late Burdigalian, but previous movements (during Eocene or Oligocene) have probably occurred.

- The North-Betic fault separates the Betic Internal and External Zones. It is oriented N 60 E to N80E. Its continuation both westward and eastward is unknown. Most of the authors agree to consider the North-Betic fault as a right-lateral strike-slip fault with several hundred km of horizontal shift. The age of the last major strike-slip movement along this fault seems to be in the Middle to Late Burdigalian. But taking into account the amount of displacement along this fault we have obviously to consider older movements.

Early Tortonian strike-slip faults

- The Nekor fault, which roughly trends N 45 E, cuts the Rifian External zones. It corresponds to a left-lateral strike-slip fault of at least 50 km. It continues probably in the Mediterranean sea following a N 60 direction. Westward this fault becomes bent and follows an E-W direction. The last major strike-slip movement was during the earliest Tortonian.
- The Crevillente fault cuts the Betic external zones along a roughly parallel direction to their structure. It is a right-lateral strike-slip fault with at least 20 km of horizontal offset. This fault is the most evident in the Eastern part of the Betic Cordillera. However it seems to continue westward in the Cadiz area. Its Eastern continuation in the Mediterranean sea is not known. The age of the last major strike-slip movement along the Crevillente fault seems to be located at the Middle Miocene - Upper Miocene boundary.

Paleogeographic origin of the Alboran block

The lack of sufficient paleomagnetic data hampers the accurate definition of the initial location, before the Alpine displacements, of the paleogeographic domains which will constitute the Alboran block. Thus we must use indirect methods. First we could notice that several convergent data lead to reject the Alboran block in a more Eastern position : - the Middle Upper Triassic carbonated rocks of the internal zones contrast strongly with the "Germano-Andalusian" gypsiferous facies of the Betic external zones ; - at the end of the Lias, before the opening of the Middle Atlantic, the Southern part of the future Iberian plate was located in front of Algeria (Olivet et al.,1982) ; - the directions of the strike-slip faults seen in the previous paragraph imply a whole displacement of the Alboran block from East towards West. The stretching lineations, WSW-ENE trending, in the deepest internal Betic nappes (Campos et al.,1984) would be related to this motion.

Moreover the Betic-Rifian internal zones have a practically common history with the Kabylian, Sicilian and Calabrian internal zones. Thus we may locate the subsequent Betic-Rifian internal zones eastwards with respect to the present Betic Cordillera, probably in front of Tunisia at the end of the Lias.

If it is possible to give some precisions about the relative paleolongitude of the Alboran domain and consequently of the microplate built by the Alboran-Kabylian-Peloritan-Calabrian Internal zones (that we will now call ALKAPECA), it is much more difficult to estimate its relative paleolatitude : we wonder whether the Alboran block was located northward (Europe s.l.) or southward (Africa s.l.) with respect to the left lateral strike-slip boundary which was necessarily present between Europe and Africa after the beginning, during the Dogger, of the Middle Atlantic opening. We may however consider the following data. The left lateral strike-slip fault between Europe and Africa acted mainly during the Dogger, the Malm and the Neocomian. Indeed, we know (Olivet et al.,1982) that the Northern edge of Morocco and the Southern edge of Iberia were likely in front of each other, near their present position at the end of the lower Cretaceous. This major strike-slip fault, comparable to a transform fault, must be located in a thinned, perhaps oceanized crust zone. The only possible oceanic zones known between Western Europe and Africa, at the Jurassic time, correspond to the basement of the Flyschs which was located southward of Alkapeca. The width of this oceanization remains debated. However we could assume it was rather narrow as related to the scarcity of ophiolites in the Chain.

All that is tantamount to admitting that the ALKAPECA domain was bound to Europe but in a completely Southern location, that is to say very close to Africa (fig. 4). About the same opinion is followed by Dercourt et al. (1985) in their paleogeographic reconstruction of the Tethyan domain.

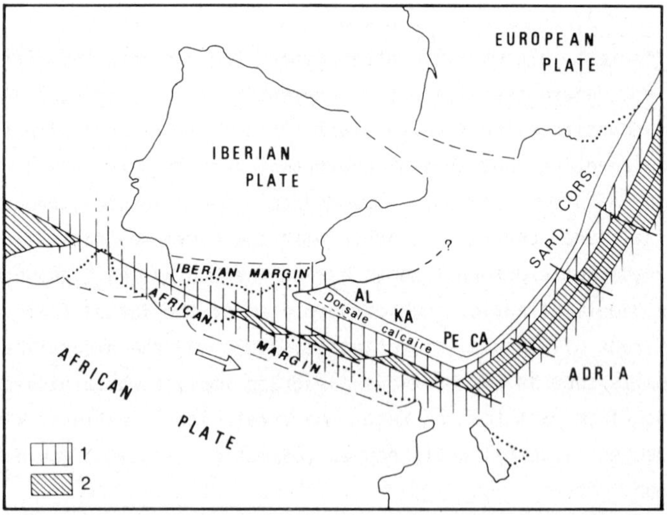

Fig. 4.- A possible location of the Alkapeca domain during the Malm. (1) Thin continental crust ; (2) Oceanic crust ; Al, Alboran (= Betic-Rifian) domain ; Ka, Kabylia ; Pe, Peloritan domain ; Ca, Calabria.

The nearness of Africa explains that, especially for Cretaceous time, when the transform fault was nearly sealed, we can find all the paleogeographic transitions from the Southern margin of ALKAPECA until the African margin. The remaining problem is that of the Northern boundary of the ALKAPECA microplate. In the reconstruction of Dercourt et al., this microcontinent continues in fact towards the NE with Sardinia and Corsica (before their rotation) until the Briançonnais. This scheme is plausible but remains difficult to be proved at the present time. One has to account for the fact that the Northern boundary of the Alboran block corresponds with the North Betic right lateral strike-slip fault and that an important break appears when crossing this fault from South (Alboran) to North (Subbetic). This is equivalent to saying that, even if we admit the original connection of Alboran with Europe, this block became independent from this large plate at a time not accurately known, perhaps the end of Cretaceous. The North Betic fault is probably the new boundary between Europe and Africa since the end of Mesozoic because the strike-slip zone following the Mograbin flyschs basin is then inactive.

This Mesozoic history, still poorly known, obviously led to the specification of a peculiar domain which subsequent Tertiary evolution gave rise to the Arc of Gibraltar.

The formation of the Arc of Gibraltar

- The roughly N-S convergence of the European and African plates has occurred from the Upper Cretaceous. It was a "scissor move" around a pole of rotation located somewhat westward of the area of Gibraltar. The related shortening of the intermediate space induced nappes emplacements in the internal zones and several phases of deformation during Eocene and Oligocene. Then, as the convergence of the two main plates continued, the deformation inside the internal zones could not accomodate alone the stress. Thus there would be a phase of blockage of the system, which would be cancelled by the breaking up of the ALKAPECA microplate and the formation of the Alboran block (Betic and Rifian internal zones) bounded northward by the right lateral North Betic fault and southward by the left lateral Jebha fault (fig. 5-2).

Such an explanation implies that the Alboran block became independent only at the latest Oligocene, after the end of the internal zones structuration. This opinion remains questionable as among others we do not know the time when the boundary fault began to act. In the case of the North Betic fault, at least, an older history seems acceptable. Thus one could in fact also admit that the independence of the Alboran block began before the end of the internal zones structuration.

Independently from the selected scheme, it remains that after its individualization the Alboran block moved towards WSW with respect to the African and European major plates. Backward of this block, that is to say in the most internal

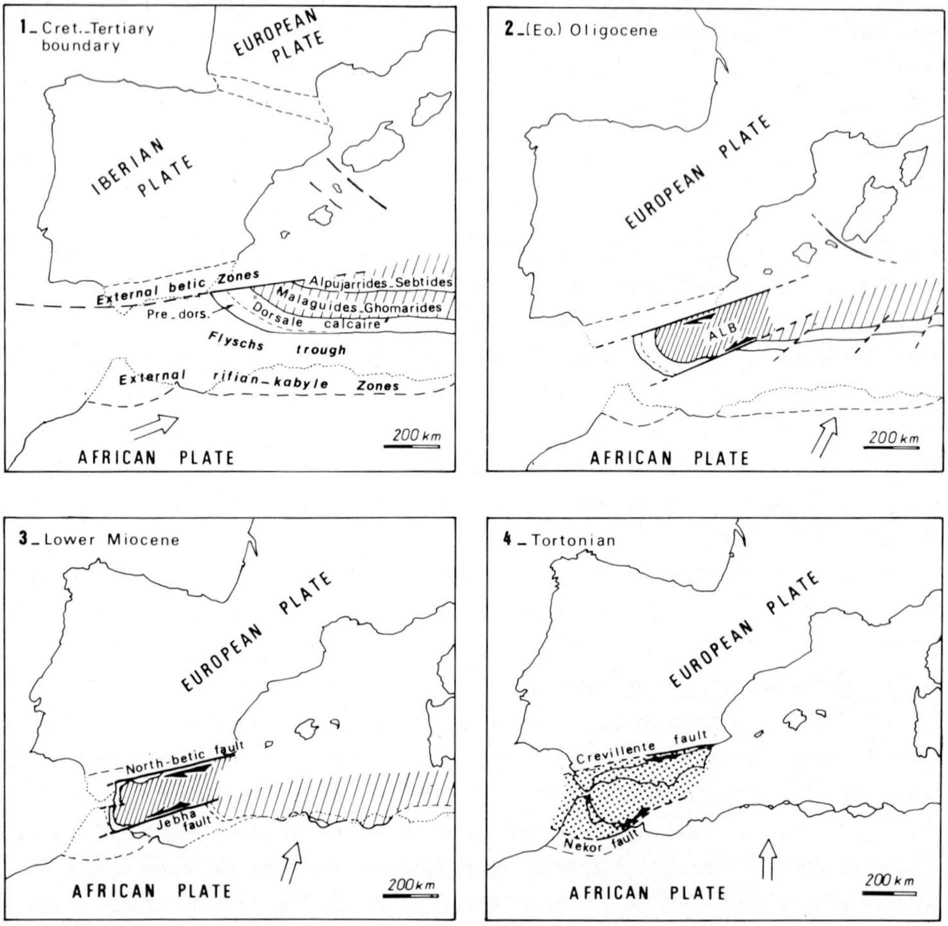

Fig. 5 - Sketch maps showing the evolution of the Western Mediterranean sea from the Upper Cretaceous to the Upper Miocene.

domains, collapsings occurred since the latest Oligocene and mainly during Aquitanian and Lower Burdigalian. These collapsings allowed the arrival of a general transgression, fore-shadowing the present Western Mediterranean which takes, by this way, the value of a back-arc basin with a thinned crust (Rehault et al., 1984). It is acceptable to suggest a reduced motion of the lithospheric block during this part of the Lower Miocene because we do not know any compressive event of such an age.

It is in fact during the Middle to Upper Burdigalian that the Alboran block started again. Then, about rigid, this block collided and overthrust more and more external zones : Predorsalian domain, Mauretanian then Massylian flyschs (fig. 5-3).

The idea of expulsion towards the WSW of the Alboran block is taken, with modification, from the hypothesis of the "ejected triangular wedge" of P. Tapponnier (1977). It allows us not only to propose a model for the stress redistribution in the space between the two converging plates but also it explains the building of the Arc of Gibraltar. At its Western end this ejected block faces the Southern part of Iberia and the Northern part of Africa. The front of this "buffer" comprises the part of the Betic-Rifian internal zones, now oriented NW-SE. In this area, the thrustings were generally of a westward vergence. Nearer the faults which bound the Alboran block, the various units underwent other stress : on one hand the N-S. compression which induced thrustings mainly towards the North along the North Betic fault and towards the South along the Jebha fault, and on the other hand the stretching which explains the scattered and reduced character of the units along these two boundaries of the Alboran block. In this way the Alboran block came in connection with more and more external zones which plastic covers were ejected on its Northern and Southern borders. So, the present bend is gradually acquired.

By the end of the Burdigalian time, the North Betic and the Jebha fault were definitively locked and the part of the Arc comprising the internal zones and the Flyschs has acquired its definitive shape.

The Middle Miocene (Langhian, Serravallian) corresponds to a period of comparative tectonic rest.

At the end of the Middle Miocene, the renewal of the N-S convergence of Europe and Africa induced the development of nappes in the most external zones, with northward vergence in the Betic Cordillera (Subbetic) and southward vergence in the Rif (Mesorif to Prerif). Once again this formation of nappes would be locked. A renewed movement would be allowed by the ejection, once again towards the WSW, of an Alboran block with a size and a shape different from the Lower Miocene ones. The new boundaries of the block would be the right lateral Crevillente fault to the North and the left lateral Nekor fault to the South (fig. 5-4). The collision of the most external zones by this new Alboran block could induce the formation of the curved part of the Chain comprising the Guadalquivir basin, the Horse Shoe plain and the Prerif.

The Arc of Gibraltar, curved structure with a relatively high level of symmetry, thus possesses a mainly tectonic origin. It arose from a whole N-S stress applied to paleogeographic domains with an oblique orientation (perhaps NW-SE). Such a device may have induced first the inner deformations of these domains, then the segmentation and the individualization of a moving block, lateraly ejected towards the WSW. The discontinuous collision with more and more external zones by this block would have formed the curvature of Gibraltar, before the Middle Miocene with respect to the Internal domain and at the beginning of the Upper Miocene with respect to the most external domains.

2. THE TYRRHENIAN ARC

2.1. GENERAL STATEMENT

In the Southeastern part of the Western Mediterranean, the Maghrebides chain, E-W trending, and the Apenninic chain, NNW-SSE trending, are linked by the Tyrrhenian Arc (fig. 1). In that manner this arc holds a symmetrical position with respect to the Arc of Gibraltar. The geological features of the Tyrrhenian Arc have been described in various syntheses (Caire, 1973; Ogniben, 1973; Vezzani, 1973; Amodio-Morelli et al., 1976; Grandjacquet and Mascle, 1978; Scandone, 1979; Wezel, 1982; Moussat, 1983).

The internal zones comprise, from top to base : a) large basement nappes consisting of old material; b) ophiolitic units. The whole overthrusts the external zones. Units comprising Cretaceous and Tertiary flyschs overlie both the external and internal zones. Since the Upper Oligocene, various series have played locally and temporarily the role of post-nappe formations. Lastly, the Arc shows a marked volcanism.

2.2 GEOLOGICAL SETTING

Internal zones

Large basement nappes characterize these zones. The basement consists of granulite, gneisses, micaschists, weakly metamorphozed deposits of Cambro-Ordovician to Carboniferous age and Hercynian granites. In Calabria, we believe, after R. Dubois (1976), that the basement-bearing units marked under the names of "Dioritico-kinzigitic nappe", "Polia-Copanello unit","Longobucco unit", "Monte Gariglione unit", "Stilo unit" belong in fact to only one main Upper basement unit. It is the Sila nappe (fig. 6 and 7b), with a cover formed by the Triassic and Jurassic sequences of Longobucco and Stilo. This nappe, with a normal position, overthrusts the Bagni unit, overturned as a whole, which comprises a basement including in our opinion the Castagna unit and a metamorphozed Mesozoic cover. We admit that the Serre and Aspromonte basements belong to the Sila nappe. In Sicily (fig. 7c), the basement of the Peloritan Mounts would belong mainly to the Sila nappe, while a part of the Peloritan phyllites and the Ali unit terranes

Fig. 6. Sketch-map of the Tyrrhenian Arc. 1. Upper Miocene, Pliocene and Quaternary deposits; 2. Pliocene and Quaternary volcanism; 3. Stilo-Capo d'Orlando formation; 4. Silaides : 4a Mesozoic and Paleogene outcrops of the Peloritan Dorsale calcaire and of the Stilo, Longobucco, Caloveto and Paludi sequences; 4b basement. 5. Bagnides : basement and metamorphozed sedimentary cover; 6. ophiolites and associated sedimentary deposits (Frido flysch); 7. Monte Soro, Reitano and Troina flyschs (Sicily); Cilento flyschs (Northern Calabria); 8. Argille Varicolori and related deposits (Numidian flysch of Lucania, etc.); 9. Sicilian Numidian flysch; 10. Platform external units : Panormid (Sicily), Calabro-Campania platform and Monte Alpi (Lucania); 11. basin external units : Sclafani, Vicari (Sicily) and Lagonegro (Lucania); 12. Foreland.

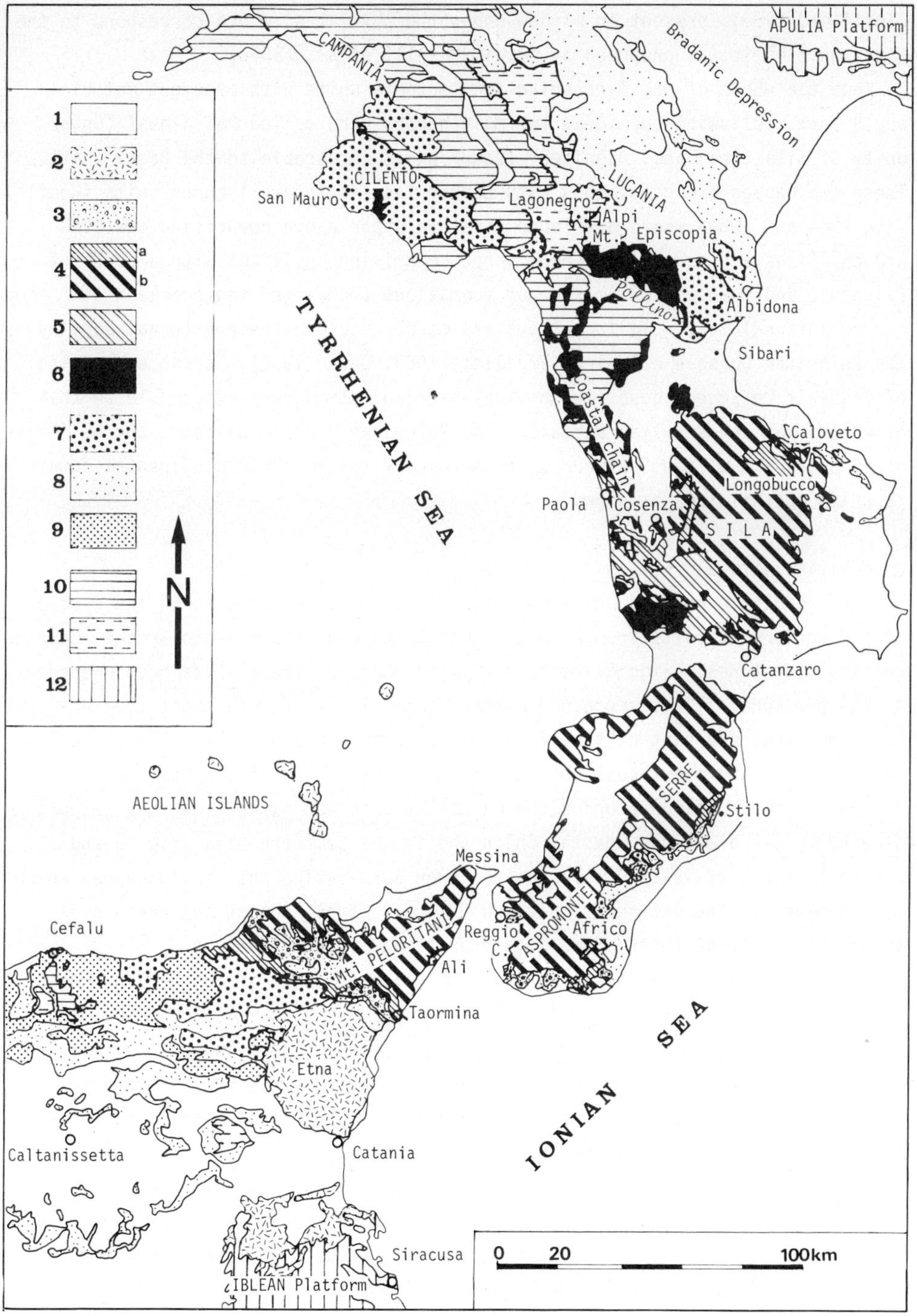
1
2
3
4
a
b
5
6
7
8
9
10
11
12
N
APULIA Platform
Bradanic Depression
CAMPANIA
CILENTO
San Mauro
Lagonegro
LUCANIA
Alpi
Mt
Episcopia
Pollino
Albidona
Sibari
TYRRHENIAN SEA
Coastal chain
Caloveto
Longobucco
Paola
Cosenza
S I L A
Catanzaro
SERRE
Stilo
AEOLIAN ISLANDS
Messina
Reggio
C.
ASPROMONTE
Africo
Cefalu
Mti PELORITANI
Ali
Taormina
Etna
IONIAN SEA
Caltanissetta
Catania
Siracusa
IBLEAN Platform
0
20
100km

(Truillet, 1968), present in window under the phyllites, would correspond to the Bagni unit as it was proposed by Amodio-Morelli et al. (1976).

Thus the whole of the Tyrrhenian Arc tectonic units with basement material would form, following our idea, two main nappes here called "Silaides" (Upper units of Sila type) and "Bagnides" (Lower units comparable to the Bagni unit). These two nappes are homologous with Lesser Kabylia internal zones units (Bouillin, 1984 and 1986) where are found first an Upper nappe comprising gneisses and phyllites and secondly a Lower nappe comprising phyllites with a cover of Paleozoic and Mesozoic metamorphozed formations (Achaiches sequence).

The internal zones old formations are bordered along their external edge by the Peloritan Dorsale calcaire (Truillet, 1968; Duée, 1969). It concerns units of Triassic to Eocene deposits showing an organization very comparable to that of the Rifian and Kabylian Dorsales. The Peloritan Dorsale calcaire can be considered as a part of the Southern paleo-margin of the ALKAPECA continental domain to which the Silaides belonged.

Ophiolitic units and Flyschs

The Ophiolitic-bearing units comprise mafic and ultramafic rocks and a cover of sedimentary deposits from Malm to Lower Cretaceous (Malm radiolarites, Calpionellids limestones, Frido quartzitic flysch). Some of these units have undergone a high pressure - low temperature metamorphism. It is admitted that they originate from a Southern part of the Ligurian-Piedmontese basin (also called Ligurian Tethys). They are overthrust onto the Southern Calcareous Apennines. More Southerly, they disappear under the Calabrian basement nappes but they reappear in windows all along the Coastal Chain and in the Southern Sila (fig. 6 and 7).

The flyschs, of lower Cretaceous to Lower Burdigalian age, build nappes which are involved in the Internal zones and External zones contact and which overthrust the External zones. Some flysch units have been reactivated, often as olistostromes, by late gravity slide over the Internal zones since the Uppermost Oligocene.

The Sicilian flyschs are identical to those cropping out in North Africa (Monte Soro Mauretanian flysch; "Argille Scagliose Varicolori", very comparable with some Mauretanian distal sequences; Massylian-like Troina and Capizzi sequences; Reitano sandy-micaceous flysch; Numidian flysch). Towards the Tyrrhenian Arc Northeastern edge, adding to the Argille Varicolori and the Numidian flysch, outcrops the Cilento flysch (fig. 7a) with facies differing from the Maghrebides flysch ones. Lastly the pelitic, quartzitic sequences of the Frido flysch are associated with ophiolite-bearing formations; they show similarities with the Lower Cretaceous North-African flyschs.

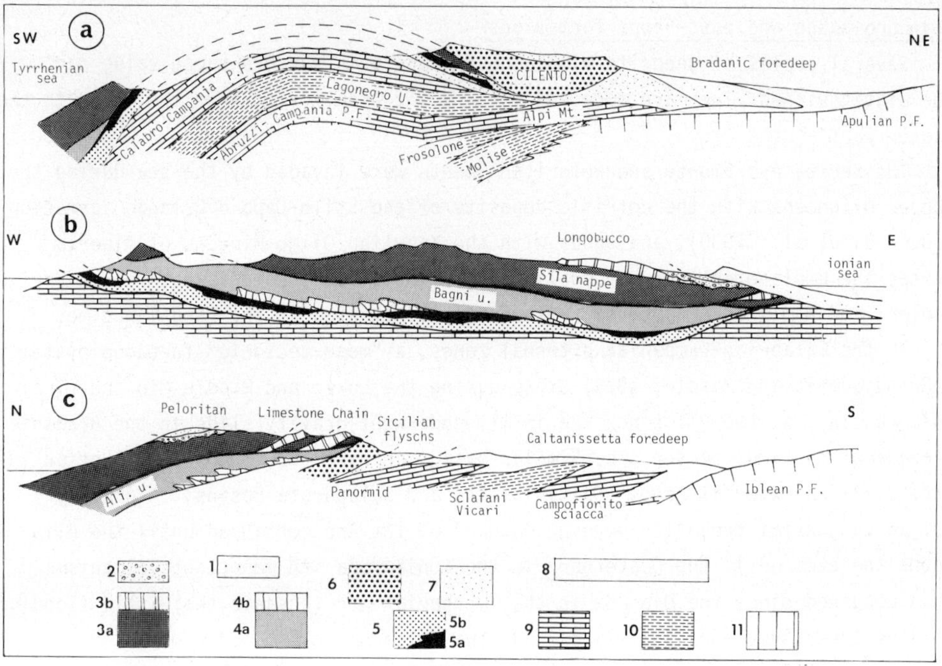

Fig. 7. Schematic and interpretative cross-sections of the Tyrrhenian Arc. a : Cross section of the Northern branch, after Elter and Scandone, 1980, simplified; b : Section across the Central Calabria; c : section across the Southern (Sicilian) branch. - 1.Miocene-Pliocene deposits of the Internal zones; 2. Stilo-Capo d'Orlando sequence; 3. Silaides : 3a basement, 3b Mesozoic-Paleogene cover; 4.Bagnides : 4a basement, 4b Mesozoic cover; 5.Ophiolite-bearing units : 5a ophiolites, 5b sedimentary cover of the ophiolites and Frido flysch; 6.Flyschs linked to the Silaides : Monte Soro, Reitano (Sicily) and Cilento flyschs; 7. Argille Varicolori and related formations; 8.Mio-Pliocene foredeep; 9.Platform external units; 10.Basin external units; 11.Foreland.

External zones

The external zones comprise formations from Permian to Lower Miocene, deposited before the Apenninic tangential tectonics, and synchro-nappe terranes. They outcrop in Sicily, Southward of the Peloritan Mounts, and in Northern Calabria, in Lucania and in Campania; they reappear in windows under the Mid Calabrian internal zones thrusting. The external zones formed nappes overthrusting the foreland : Iblean platform (Sicily) and Apulian platform.

The external zone sequences show, especially during the Upper Cretaceous, a paleogeographic segmentation with calcareous platforms (Campanian-Lucanian platform and Panormide sequence of Sicily; Campanian-Abbruzian platform) and basins with a marly-cretaceous sedimentation (Frosolone unit of Apennines; Vicari, Campofiorito and Sciacca sequences of Sicily), or a siliceous one (Lagonegro sequences in the Apennines and Sclafani sequence in Sicily).

Synchro-nappe and post-nappe formations

Several cycles of deposits with a post-nappe or a synchro-nappe value can be distinguished. They are more and more recent when moving from the inner to the outer part of the Arc.

The Serre, Aspromonte and Peloritan Mounts were invaded by the sea during the Upper Oligocene with the detritic deposits of the Stilo-Capo d'Orlando formation (Bonardi et al., 1980), analogous with the Kabylian Oligo-Miocene of Algeria. After the emplacement, at Burdigalian, of olistostromes comprising Argille Varicolori and Numidian sandstones, a second cycle started at the Langhian time.

In the Calabrian-Campanian external zones, a "meso-tectonic" foredeep system (Grandjacquet and Mascle, 1978) arose during the Lower and Middle Miocene (Irpinian basin); at the Pliocene, the infill underwent gravity slide in the Bradanic foredeep; outwards of the Arc. Similar basins appeared at Tortonian-Helvetian in the Sicilian external zones (Caltanissetta and Salaparuta basins). A gravity slide tangential tectonics verging outward of the Arc continued until the Pliocene and even until the Quaternary in the Ionian Sea. An important volcanism has occurred since the Miocene in the Tyrrhenian Arc (oceanic tholeiitic flooding of the inner Arc, Aeolian Islands and Etna volcano).

2.3 PALEOGEOGRAPHIC INTERPRETATIONS

Many authors agree with the idea that the Calabrian basement nappes originated from the African/Insubrian margin of the Tethys. They would have been emplaced, during Alpine verging movements (directed towards Europe) on terranes originating from the Ligurian basin. The pile of nappes would have been subsequently transported, in the opposite direction, over the Southern Apennines (Apenninic - Maghrebides verging movements). Taking in account the sequences and structural homologies between the Calabrian structural domain and the Lesser Kabylian one, we think, on the contrary, that the Silaides and Bagnides basement originated from a continental block with a European origin. The Silaides would belong to the most internal domain, with emersions and a continental shelf type sedimentation from Dogger to Oligocene. The more external Bagnides would correspond to a more external crust area, thinned since the Dogger, and marked by a deep sedimentation; they would be intercalated between the Silaides margin (with the Peloritan Dorsale) and the ophiolitic domain. In this frame, the European plate widely overthrusts the African plate and the Adria. Thus the structuration of Calabria, Sicily and Kabylia would originate from an hypercollision phenomena comparable with the Mid and Eastern Alps one, but oppositely verging.

2.4 HOMOLOGIES AND SYMMETRIES IN THE TYRRHENIAN ARC

The external zones

There is a lack of observation in the Tyrrhenian Arc external zones near Mid and Southern Calabria. Nevertheless the units and sequences seem to correspond well to each other between the Sicilian and the South-Apenninic external zones. Despite this fact, some authors (Dercourt et al., 1985) insert a break, from the Upper Jurassic, between the Sicilian external zones and the Apulian block (Adria). We rather agree with the interpretation of Adria as an African promontory (D'Argenio et al., 1980) and with the idea of a continuity between the Sicilian and Apenninic internal zones. Thus the external domain could show a marked bending around the Tyrrhenian Arc.

The Flyschs

The flyschs show two types of position.

- The Argille scagliose varicolori and the Numidian flysch are known in Sicily, Calabria and Lucania, with various stages of resedimentation in the Miocene sequences. These deep deposits, very comparable to some Algerian sequences (Ziane flysch from Constantine area) indicate a linkage and a symmetry between the two branches of the Tyrrhenian Arc.

- The other flyschs are settled in an asymmetrical way. The Mauretanian and the Massylian flyschs, widespread in Sicily, are no longer found in the Northern part of the Tyrrhenian Arc. There, outcrop the Frido flysch, the Cilento flysch and, restricted to the Sila Northern boundary, the Paludi Eocene formation. The Frido flysch and associated Calpionellids-bearing calcareous flysch form the stratigraphic cover of the ophiolites and radiolarites with Ligurian (s.l.) origin. The Cilento flyschs might have been the stratigraphic complement of a part of the Frido flysch; they differ markedly from the Sicilian and Maghrebide flyschs. Their age is still debated : frequently attributed to the interval from Albian to Late Eocene, they might be in fact of Eocene to Lower Miocene age.

The Eocene Paludi formation and the Cilento flysch highest formation (San Mauro-Albidona flysch) are fed by the internal zones (Silaides). Furthermore the Mauretanian flyschs and the sandy-micaceous flysch (Reitano) are linked likewise to the internal zones for they have reworked the Dorsale calcaire deposits and the Kabylian-Peloritan basement, at least since the Middle Cretaceous.

Thus the subsequent Internal domain might have been the source area on the Maghrebide side (Sicily) and on the Apennine side (Northeastern Calabria) of two very different sedimentary sequences (mainly since Middle Cretaceous). The occurrence of the Argille scagliose varicolori all around the Tyrrhenian curvature might be induced by their link with the external zones.

The internal zones

In Sicily, the Peloritan Dorsale shows a structure with tectonic slices of heteropic terranes which is very comparable to the structure of the Kabylian Dorsale calcaire. In the Northern Sila, the Liassic thick, detritic Longobucco formation is transgressive over Triassic sandstones and conglomerates and over a Paleozoic sequence. The whole is segmented in flat slices overthrusting the Caloveto sequence, characterized during the Lias by Ammonitico rosso facies and during the Malm by radiolarites. The structure and the facies are clearly different from the Peloritan Dorsale ones. This asymmetry is weakened however by the fact that the Trias and Lower Lias of the Longi unit (Peloritan chain) and of the Longobucco sequence show the same facies. In a more internal position than the Dorsale, the Clypeina-bearing Jurassic limestones of the Novara sequence, in the Peloritan Mounts, are identical to those which constitute the cover of the Paleozoic from Aspromonte, Serre and Southern Sila. These deposits are shallower than the Dorsale ones and correspond with a continental shelf. From this point of view they are comparable with the Sardinian sequences. Moreover the Dogger sandstones of Tiriolo, in the Southern Sila, are similar to the Laconi sandstones, in Central Sardinia.

Thus, the outcropping Jurassic facies (fig. 8) are probably homogenous in the most inner zones while they are asymmetrically differentiated on the internal block margins (Dorsale calcaire on one hand, Longobucco and Caloveto sequences on the other). The facies asymmetry of the flyschs originally the nearest from the internal zones (in the selected paleogeographic restoration) is likely related with the asymmetry of their margins which were the source area of the clastic material. Lastly, the lack of outcrops from the Stilo-Capo d'Orlando formation North of the Catanzaro graben is noticeable, the Upper Miocene levels being directly transgressive on the Silaides.

In conclusion, there is, on one hand, an apparent continuity and symmetry of the external zones of the Tyrrhenian Arc and, on the other hand, some asymmetry between both sides of the Arc with respect to the sedimentary facies at the boundary between the internal and external zones.

This asymmetry is more conspicuous if we take into account the extensions of the two branchs of the Arc. In fact the two basement nappes and their content, the sequences and the structures of the Peloritan Dorsale calcaire, the Sicilian flyschs, the Stilo-Capo d'Orlando formation and the overlying olistostromes have their strict equivalents in Lesser Kabylia, at 600 km from Sicily. Thus the Tyrrhenian Arc Western branch is linked to the whole of the Maghrebides, the Southwestern boundary of the Tyrrhenian Arc near the internal zones (Taormina fault) being subsequent to the main structures. On the contrary, Northwards, the internal zone materials disappear and no equivalent could be found in the Northern Apennines (the ophiolites and the Alpine metamorphism excepted). For us, the Northeastern boundary of the Calabrian internal zones is nothing more than the Eastern boundary of the Maghrebides. It corresponds to a thrusting (as indicated by the Episcopia basement klippes lying on the Lucanian flyschs) but probably also to a major transcurrent fault.

2.5 PALEOGEOGRAPHIC ORIGIN AND STRUCTURATION OF THE TYRRHENIAN ARC

From the point of view of the structural evolution we are led to consider, on one hand, an internal composite block located in the Tyrrhenian Arc central part and, on the other hand, the external zones and the foreland, overthrust and collided by the internal block.

The internal block is globally built with the basement nappes (Silaides and Bagnides) and with the Peloritan Dorsale calcaire. During the tectonic evolution of the Tyrrhenian Arc, this internal domain gradually incorporated the ophiolite-bearing units and a part of the Flyschs by overthrusting them and then by carrying them onto the external zones. After that, a part of the external zones probably moved towards the SE with the internal block, through the action of transcurrent faults located at the Arc boundaries. We propose the name "Aeolian block" for the domain comprising the Calabrian-Sicilian internal zones and the units which, throughout time, have accreted to these. According to the diverse epochs the composition of the block and the location of its limits could have been somewhat different.

Perhaps the Aeolian block was fore-shaded as early as the Triassic time, as the so-called "Verrucano" of its cover indicates a high area with respect to some Sicilian and South-Apenninic external zones which show a pelagic sedimentation and a volcanic activity during the Upper Trias. But this block was mainly individualized at the Dogger-Malm. We have seen that the Northeastern boundary of the Calabrian internal zones is correlated with a major geological break, of tectonic and paleogeographic order. One of us has stressed the fact that this limit probably continues towards the SE the Paul-Fallot fault along which has occurred, since the Upper Oligocene, the counterclockwise rotation of the Corsican-Sardinian block (Durand Delga, 1980). Another scheme is to consider the Aeolian block Northeastern boundary as the inheritance of a "Cevenol" or "Atlantic" trend, along which the Ligurian Tethys would have opened parallelly with the Mid Atlantic ridge as proposed by many authors (Bernoulli and Lemoine, 1980). This trend could have made near Calabria an angle of about 120° with the Maghrebides E-W trending. The Peloritan Dorsale sequences would correspond to the Aeolian block margin with a Maghrebide trending while the Longobucco, Caloveto and Paludi sequences would correspond to its Ligurian margin, with possible extension in Eastern Sardinia (fig. 8).

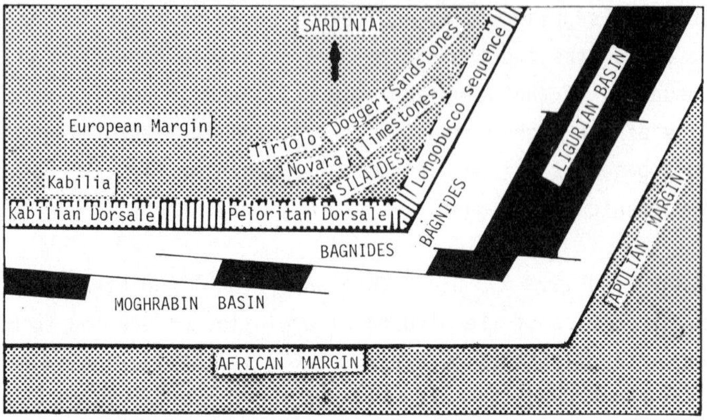

Fig. 8. Reconstruction of the Calabrian-Peloritan internal zones during the Malm.

Such an interpretation means that the Tyrrhenian Arc might have been prepared since the Jurassic and could have originated from Jurassic distensive structures before being formed by a mainly compressive tectonic development.

The structuration of the Calabrian-Sicilian internal zones corresponds with the thrusting of the Silaides over the Bagnides deeper domain, comprising a thinned crust. This last unit itself overthrusts an oceanic domain, probably rather narrow. The Cretaceous to Eocene age of the metamorphism and of the structuratic of the Bagni and ophiolitic units is generally accepted. However, despite the fact that radiometric data from the basement units point out Cretaceous events, we still need direct evidence for the occurrence of important thrustings of the same age and some data lead to the re-examination of this question : it is the occurrence of Eocene flysch slices involved in a contact between two ophiolitic units in the Northern Sila (Bouillin, 1984) and, also, the similarities between the Bagnides and the Kabylian Achaiches unit, where metamorphic Eocene has been reported.

If, as we think, the Silaides comprise two sides with an angle facing Africa, strike-slip movements along both sides might have occurred. One has mainly to consider a major left-lateral transcurrent movement located in the area between the Eastern extremity of the Silaides domain and the Adria, with a trending nearly parallel to the Ligurian basin direction. The Tertiary flyschs (Eocene-Oligocene or Miocene) of San Mauro - Albidona may have recorded such a shift. These flyschs have been fed with blocks originating from a Liassic sequence comparable to the Longobucco one and with basement fragments of Sila type. Their sedimentary characters differ from the Mograbin and Ligurian flysch ones and indicate that they were deposited in a deep, narrow trench, with active margins. These characters agree with the scheme of a trench located at the boundary of a major transcurrent zone.

Most of the structuration of the Internal zones occurred before the beginning of the deposit of the Stilo - Capo d'Orlando formation (Uppermost Oligocene).

The thrusting, on the external zones, of the Aeolian block, comprising now the ophiolite-bearing units, predates the Upper Tortonian; it occurs after the Aquitanian-Burdigalian deposits, which are involved in the tectonic and metamorphism of some external zone units (Verbicaro and San Donato units, in Middle and Northern Calabria).

Since this time, the Aeolian block would have incorporated the overthrust part of the external zones and its advancement towards the SE, in relationship with the opening of a back-arc basin in the Tyrrhenian area, would have occurred mainly by the frontal subduction of the Ionian basin crust under the Tyrrhenian Arc and by major lateral transcurrent movements. Towards SW the Taormina fault may have played such a role. Towards NE, we can estimate that a transcurrent zone, with a NW-SE trending, follows the NE edge of the Sila Mounts and extends through a net of faults corresponding to the boundary of the Pollino Mounts, at the Calabrian-Lucanian border. These faults might have shifted the Sila basement towards the SE with respect to the bench-mark represented by the Episcopia klippe. Lastly the Aeolian block would have incorporated a larger part of the near external zones by new thrustings, until the Middle Pliocene; its lateral boundaries are probably now the Bradanic basin and the Caltanissetta-Salaparuta basin.

At the present time, the Aeolian block is completely separated from Sardinia, to which it was probably linked before the Tyrrhenian basin opening. Furthermore similarities of materials and of structures suggest that, until the Uppermost Oligocene, the Aeolian block and the Lesser Kabylia internal zones were closer than now. One can believe that the Aeolian block, then, came apart from the other ALKAPECA fragments and that it has moved Eastward with respect to the Kabylia, by slipping along the external zones, in relationship with the Tyrrhenian basin opening.

CONCLUSION

Both the Arc of Gibraltar and the Tyrrhenian Arc have a tectonic origin. Both have been caused by the motion of Internal blocks, constituted by basement nappes. These blocks, the Alboran block and the Aeolian block, were originally a part of a larger domain (ALKAPECA and probably Sardinia) which formed locally the Southern border of the European plate. This domain constituted a high paleogeographic zone from the Lower Mesozoic to the Upper Oligocene. The history of the Internal blocks motion is strongly related to the European and African plates tectonics. Distensional phenomena and strike-slip movements between Europe and Africa during the Jurassic time gave rise to the Mograbin and to the Ligurian-Piedmontese basins, which were the initial boundaries of the ALKAPECA domain. Furtherly the Europe-Africa convergence led, during the Tertiary, to the individualization of the Internal blocks which collided, overthrust and dragged with them the External zones belonging to the South Iberian, African, and Apulian margins. The collisions led to the bending of some internal and external domains and to form the two arcs.

In the Arc of Gibraltar, there is no paleogeographic nor tectonic continuity from the Betic external zones (Iberian margin) to the Rifian external zones (African margin) which are both weakly curved. On the other hand, the Internal zones and the Flyschs have suffered the strongest curvature of the Arc (180°). The latter was formed during the Oligocene and especially the Miocene, in a discontinuous manner, by the relative motion towards the WSW of the Alboran block, bounded by major strike-slip faults, in a space reducing as the European and African plates were converging. The Arc of Gibraltar is located in a region in which paleogeographic evolution was greatly marked by the successive stages of the opening of the Atlantic Ocean.

In the Tyrrhenian Arc, on the contrary, the External zones seems to have been initially continuous; they would have been curved by the motion (thrusting, acting of strike-slip faults) of the Aeolian block. In the latter the bending deformation would have been less important than in the Alboran block. Indeed the Aeolian block possesses two faces with dissimilar Mesozoic and Paleogene series. So it is possible that the initial position of this block was in the angle of about 120° made, from the Lias, by the Ligurian-Piedmontese basin and the Mograbin one. The location of the Tyrrhenian Arc would have been determinated by the Jurassic distensional tectonics.

Thus the Arcs which form the two ends of the Maghrebide chain are each one issued from a combination of paleogeographic inheritance (contemporaneous with a rather distensional Mesozoic phase) and of compressional deformations, associated with transcurrent movements, Tertiary in age.

ACKNOWLEDGMENTS

The authors wish to thank H. Feinberg and R.D. Hester for translating into English their paper and C. Cavaré for the drawings.

REFERENCES

Amodio-Morelli, L., Bonardi, G., Colonna, V., Dietrich, D., Giunta, G., Ippolito, F., Liguori, V., Lorenzoni, S., Paglionico, A., Perrone, V., Piccarreta, G., Russo, M., Scandone, P., Zanettin-Lorenzoni, E. and Zuppetta, A., 1976. L'arco Calabro-Peloritano nell'orogene appenninico-maghrebide. Mem. Soc. Geol. It., 17 : 1-60.

Andrieux,J., Fontbote, J.-M., and Mattauer, M., 1971. Sur un modèle explicatif de l'Arc de Gibraltar. Earth Planet. Sci. Lett., 12 (2) : 191-198.

d'Argenio, B., Horvath, F. and Channell, J.E.T., 1980. Paleotectonic evolution of Adria, the African promontory. 26e Congr. géol. Intern., Paris, Coll. C5 "Géologie des chaînes alpines issues de la Téthys", p. 331-351.

Bernoulli, D. and Lemoine, M., 1980. Birth and early evolution of the Tethys : the overall situation. 26e Congr. géol. Intern., Paris, Coll. C5 "Géologie des chaînes alpines issues de la Téthys", p. 168-179.

Bonardi, G., Giunta, G., Perrone, V., Russo, M., Zuppetta, A. and Ciampo, G. 1980. Osservazioni sull'evoluzione dell'arco Calabro-Peloritano nel Miocene inferiore : la formazione di Stilo - Capo d'Orlando. Boll. Soc. Geol. It., 99 : 365-393.

Bouillin, J.-P., 1984. Nouvelle interpretation de la liaison Apennin-Maghrebides en Calabre : conséquences sur la paléogéographie téthysienne entre Gibraltar et les Alpes. Rev. Géol. dyn. Géogr. phys., 25 : 321-338.

Bouillin, J.-P., 1986. Le "bassin maghrébin" : une ancienne limite entre l'Europe et l'Afrique à l'Ouest des Alpes. Bull. Soc. géol. Fr. (in press).

Bourgois, J., 1978. La transversale de Ronda (Cordillères bétiques, Espagne). Données géologiques pour un modèle d'évolution de l'arc de Gibraltar. Ann. Sci. Univ. Besançon, 30, 445 p.

Caire A., 1973. The Calabro-Sicilian Arc. in "Gravity and Tectonics", De Jong, K.A. and Scholten, R. ed., p. 157-173; Wiley & Sons.

Campos, J., Garcia-Dueñas, V., Gonzalez-Lodeiro, F. and Aldaya, F., 1984. Direcciones de traslación y apilamiento de unidades en los mantos Alpujarrides centrales y orientales. In "El Borde Mediterraneo español : evolución del Orógeno Bético y geodinámica de las depresiones neogenas"; ed. Lopez Ruiz, p. 15-17. Granada.

Dercourt, J., Zonenshain, L.P., Ricou, L.E., Kazmin, V.G., Le Pichon, X., Knipper, A.L., Grandjacquet, C., Sborshchikov, I.M., Boulin, J., Sorokhtin, O., Geyssant, J., Lepvrier, C., Biju-Duval, B., Sibuet, J.-C., Savostin, L.A., Westphal, M. and Lauer, J.P., 1985. Présentation de 9 cartes paléogéographiques au 1/20 000 000 s'étendant de l'Atlantique au Pamir pour la période du Lias à l'Actuel. Bull. Soc. géol. Fr., 1 (5) : 637-652.

Didon, J., Durand Delga, M. and Kornprobst, J., 1973. Homologies géologiques entre les deux rives du détroit de Gibraltar. Bull. Soc. géol. Fr., 15 (2) : 77-105.

Dubois, R., 1976. La suture calabro-apenninique crétacé-éocène et l'ouverture tyrrhénienne néogène; étude pétrographique et structurale de la Calabre centrale. Thèse Sc. Université Paris-VI, 567 p.

Duée, G., 1969. Etude géologique des Monts Nebrodi (Sicile). Thèse Sc. Fac. Sc. Paris, 421 p.

Durand Delga, M., 1980. La Méditerranée occidentale : étapes de sa genèse et problèmes structuraux liés à celle-ci. Soc. géol. Fr., Mém. h. sér., 10 : 203-224.

Elter, P. and Scandone, P., 1980. Les Apennins. 26e Congr. géol. intern., Paris, Coll. C5 "Géologie des chaînes alpines issues de la Téthys", p. 99-102.

Grandjacquet, C. and Mascle, G., 1978. The structure of the Ionian sea, Sicily and Calabria-Lucania. In "The Ocean Basins and Margins", Nairn, A.E.M., Kanes, W.H. and Stehli, F.G. ed. Plenum publ. corpor., 4b : 257-329.

Lajat, D., Biju-Duval, B., Gonnard, R., Letouzey, J. and Winnock, E., 1975. Prolongement dans l'Atlantique de la partie externe de l'Arc bético-rifain. Bull. Soc. géol. Fr., 17 (4) : 481-485.

Leblanc, D. and Olivier, Ph., 1984. Role of strike-slip faults in the Betic-Rifian orogeny. Tectonophysics, 101 : 345-355.
Michard, A., Chalouan, A., Montigny, R. and Ouazzani-Touhami, M., 1983. Les nappes cristallophylliennes du Rif (Sebtides, Maroc), témoins d'un édifice alpin de type pennique incluant le manteau supérieur. C.R. Acad. Sc. Paris, sér. 2, 296 : 1337-1340.
Moussat, E., 1983. Evolution de la mer Tyrrhénienne centrale et orientale et de ses marges septentrionales en relation avec la néotectonique dans l'arc calabrais. Thèse 3e cycle, Université Paris-VI, 122 p.
Ogniben, L., 1973. Schema geologico della Calabria in base ai dati odierni. Geol. Romana, 12 : 243-585.
Olivet, J.L., Bonnin, J., Beuzart, P. and Auzende, J.M.., 1982. Cinématique des plaques et paléogéographie : une revue. Bull. Soc. géol. Fr., 24 (5-6) : 875-892.
Rehault, J.-P., Boillot, G. and Mauffret, A., 1984. The western Mediterranean basin geological evolution. Marine Geology, 55 : 447-477.
Scandone, P., 1979. Origin of the Tyrrhenian sea and Calabria Arc. Boll. Soc. Geol. It., (98) : 27-34.
Tapponnier, P., 1977. Evolution tectonique du système alpin en Méditerranée : poinçonnement et écrasement rigide-plastique. Bull. Soc. géol. Fr., 19 (3) : 437-460.
Vezzani, L., 1973. L'Appennino siculo-calabro-lucano. Att. Acc. Naz. Lincei, (183) : 15-37.
Wezel, F.C., 1982. The Tyrrhenian sea : a rifted krikogenic-swell basin. Mem. Soc. Geol. It., (24) : 531-568.
Wildi, W., 1983. La chaîne tello-rifaine (Algérie, Maroc, Tunisie) : structure, stratigraphie et évolution du Trias au Miocène. Rev. Géol. dyn. Géogr. phys., (24) : 201-297.

NEOTECTONICS OF THE CALABRIAN ARC AND APENNINES (ITALY) : AN EXAMPLE OF PLIO-QUATERNARY EVOLUTION FROM ISLAND ARCS TO COLLISIONAL STAGES.

J.C. BOUSQUET and H. PHILIP
Laboratoire de Géologie Structurale and E.R.A. n° 266 Université des Sciences et Techniques du Languedoc 34060 MONTPELLIER-Cédex (France)

ABSTRACT.

We define different types of compressive and distensive deformations within the framework of an evolution from oceanic subduction to continental subduction and collision (s.l.). The variations of these deformations are illustrated by the geodynamic evolution of the Apennines and Calabrian arc deduced from neotectonic and microtectonic observations.

INTRODUCTION

The general dynamic characteristics of the Plio-Quaternary and present day evolution of most of the Mediterranean basin are now relatively well known (Philip, 1983, 1986). However, both the neotectonics and seismotectonics of the Apennines and Calabrian arc present some problems. Several papers have already summarized the evolution of various sectors of the Appennines or the Calabrian arc. But they only offer regional interpretation which can hardly be used in other sectors. Some authors choose to reject any interpretation based on the general concepts of plate tectonics. Others introduce "des variations si peu orthodoxes que l'Italie devient un cas unique tout à fait particulier" (Praturlon, 1980). In fact, Italy illustrates the passage from subduction to collision, thus a complex and progressive phenomenon for which no simple and unique scheme can be found. New concepts must be superimposed on those of earlier plate tectonics such as, for instance the ideas of "ensialic subduction", "A subduction", "intracontinental subduction", "crustal accretion prism", and "continental lithosphere subduction" (Bird et al., 1975 ; Amstuz, 1976 ; Bird, 1978 ; Molnar and Gray, 1979 ; Davis et al., 1983 ; Mattauer, 1983, 1986). These concepts allow the reconciliation of most observations and are the framework on which is based our description and evolutionary model of the Plio-Quaternary deformations affecting the Italian peninsula.

NEOTECTONIC MEDITERRANEAN FRAMEWORK AND THE GENERAL GEODYNAMIC PROCESSES

In the Mediterranean region, neotectonics are ruled by the convergence of the African and Arabian plates with the Eurasian plate (along a NW-SE direction around Gibraltar, and a N-S to NE-SW direction in the East of the Mediterranean with a higher convergence rate ; Mc Kenzie, 1972). The convergence is absorbed through different mechanisms (fig. 1), due to the presence of large sectors of oceanic or thinned continental crust in between regions with true continental crust.

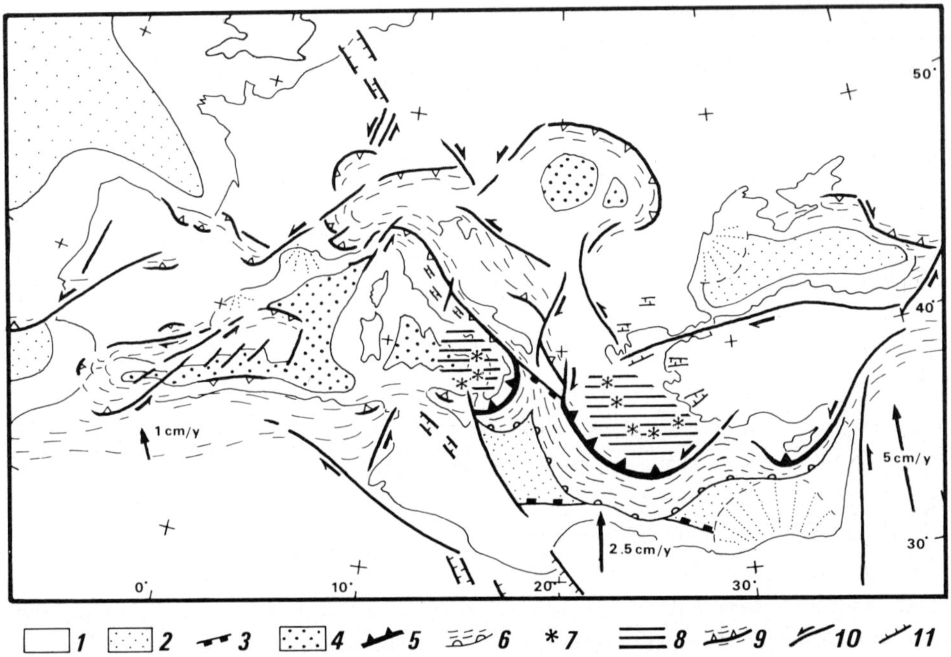

Fig. 1 - Present-day geodynamic framework of the Mediterranean basin.
1 - Continental crust ; 2 - Oceanic or thinned crust of Mesozoic age ; 3 - Passive margin of Mesozoic age ; 4 - Oceanic or thinned crust of Cenozoic age ; 5 - Oceanic subduction ; 6 -Accretional prism of the Mediterranean and Calabrian ridge ; 7 - Calc-alkaline volcanism ; 8 - Back-arc basin with present-day extension ; 9 - Main continental thrusts and compressive structures ; 10 - Main strike-slip faults ; 11 - Main normal faults.

Collision (s.l.)

The first and dominant mechanism since the Pliocene falls in the general category of "collision tectonics". It particulary affects sectors characterized by rather old plate collision (for instance the Alps) and without recent (Plio--Quaternary) processes of "continental subduction". But it also affects orogenic zones related to large lithospheric faults (Pyrenees, High Atlas) and large regions which do not belong to any peri-mediterranean alpine chain and thus correspond to intraplate deformation. In any case this deformation is either compressive or distensive and it may be accompanied by alkaline volcanism.

Oceanic subduction

This second mechanism corresponds to an oceanic subduction active in limited arcs. However, the present day subductions underneath the Aegean and Calabrian arcs present some problems having to do mostly with the subduction mechanism itself (Tapponnier, 1977 ; Angelier, 1977 ; Mc Kenzie 1978 ; Dewey and Sengör 1979 ; Angelier et Le Pichon 1980) and with the nature of the subducted crust (oceanic or intermediate crust). Indeed, the crust subducted under these arcs belongs to the Ionian basin which is known (Finetti, 1980) to exhibit the characteristics of an oceanic or thinned continental crust overlain by a thick Mesozoic series.

The Apennines and continental subduction

Other mechanisms help keep the convergence active. This is the case of the Apennines where the passage from oceanic subduction to continental subduction is evidenced by the following observations :

- seismicity, heat flow, calc-alkaline volcanism of the Eolian islands (Barberi et al. 1974 ; Beccaluva et al. 1982), gravity, crustal thickness etc ..., all imply a limited subduction at the Calabrian arc (e.g. review in Bousquet 1973, Bocaletti et Manetti 1978).
- a continuity (Auroux, 1983) between the Plio-Quaternary compressive structures of the foreland of the Calabrian arc or Calabrian ridge, (Belderson et al. 1974) and the Mio-Pliocene or even Quaternary compressive structures of the eastern side of the bradanic trough (Richetti et al. 1980) to the Po plain (cf. fig. 1).
- recent deep seismic profils across Calabria and the Apeninnes (Colombi et al., 1973 ; Morelli et al., 1975 ; Schutte, 1978 ; Panza, 1979 ; Cassinis et al., 1979) shows a doubling of continental crust, with a sinking from East to West.

Thus are heeded the concepts introduced above in order to allow the maintenance of plate convergence once the oceanic subduction has ceased. Some authors (Castellarin et al., 1978 ; Scandone, 1979 ; Reuter et al. 1980 ; Boccaletti et al., 1981) already rely on various of these new mechanisms (especially on ensialic subduction) to describe the structure and evolution of the Italian peninsula. Thus it seems more appropriate to consider the Apennines as a typical example of passage from oceanic subduction to collision (s.l.) rather than as a péculiar oddity escaping all known model. Note that they may have acted sequentially, the "flaking" of the continental crust which led more or less to the formation of a "crustal accretionary prism' being accompanied, at first, by a limited sinking of the continental crust into the mantle (ensialic subduction) and then by its decoupling from the mantle lithosphere (subduction of continental lithosphere). Our schematic figures (fig. 3, 4 and 5) with different types of superficial deformations (0 - 15 km) are not intended to support ~~such or~~ such deep mechanism. We use the term "continental subduction" with the sole purpose of differentiating this subduction from the classical oceanic subduction.

support such or such deep mechanism. We use the term "continental subduction" with the sole purpose of differentiating this subduction from the classical oceanic subduction.

The problem of the alkali-potassic volcanism in the Apennines

Another fundamental concept of plate tectonics, i.e. the close relationship between oceanic subduction and calc-alkaline volcanism or "andesitic" volcanism, must also be revised. The problem arose from the Oligocene calc-alkaline volcanism of the Aegean domain (Bellon et al. 1979) and of the Alps (Dal Piaz et al. 1979) which are not related to oceanic subduction but are part of the context of large intracontinental thrusting. The same problem occurs with the Miocene calc-alkaline volcanism recorded anywhere from Tunisia to Southern Spain. As for Italy, its Neogene and Quaternary alkali-potassic volcanism is also rather difficult to integrate into the different classical mechanisms of plate tectonics. The discovery of Quaternary calc-alkaline andesites on the Tyrrhenian side of the Apennines (Di Girolamo et al., 1976) as well as many geochemical data indicate that it might stem from a calc-alkaline magma of mantle origin which would have absorbed some crustal material (e.g. review in Pecerillo, 1984). Clearly, the definition of processes (acting in continental subduction) which best explain both the crustal anatexy and the production of calc-alkaline magma is by far preferrable to local explanations (relationship between strike-slip faulting and calc-alkaline volcanism, Bellon et al. 1983, or existence of a hot-spot in the Northern Apennines, Civetta et al., 1978).

THE PLIOQUATERNARY DEFORMATIONS AND GENERAL GEODYNAMIC PROCESSES

Types of deformations according to microtectonic analysis

The study of microstructures (microfaults, tension gashes, stylolithes) and of their respective arrangements clarifies and completes the suggestions brought about by general structural analyses. Today, several methods can be used to define the mean stress tensor(s) from a given striation population. With the latter methods (Etchecopar et al., 1981 ; Etchecopar 1984) can be defined the direction of σ_1, σ_2, σ_3 and their relative amplitudes may be represented by the ratio $R = (\sigma_2-\sigma_3)/(\sigma_1-\sigma_3)$, where R varies from 1 ($\sigma_2=\sigma_1$) to $0 (\sigma_2=\sigma_3)$. To the three classical cases defined by Anderson (1942 ; $\sigma_1 > \sigma_2 > \sigma_3$ with either σ_3 vertical (case B, Fig. 2), σ_2 vertical (case D) or σ_1 vertical (case F)) must be added four other limiting stress ellipsoïds (cases A, C, E and G). Then, one may define various types of deformation corresponding to either the model cases (R = 0, R = 0,5, R = 1) or to any intermediate state (e.g. 0 < R < 0,5). Note that within a same region several neighbouring but different stress tensors can sometimes be defined (e.g. cases B, C and D corresponding to the passage from pure reverse to pure strike-slip faulting).

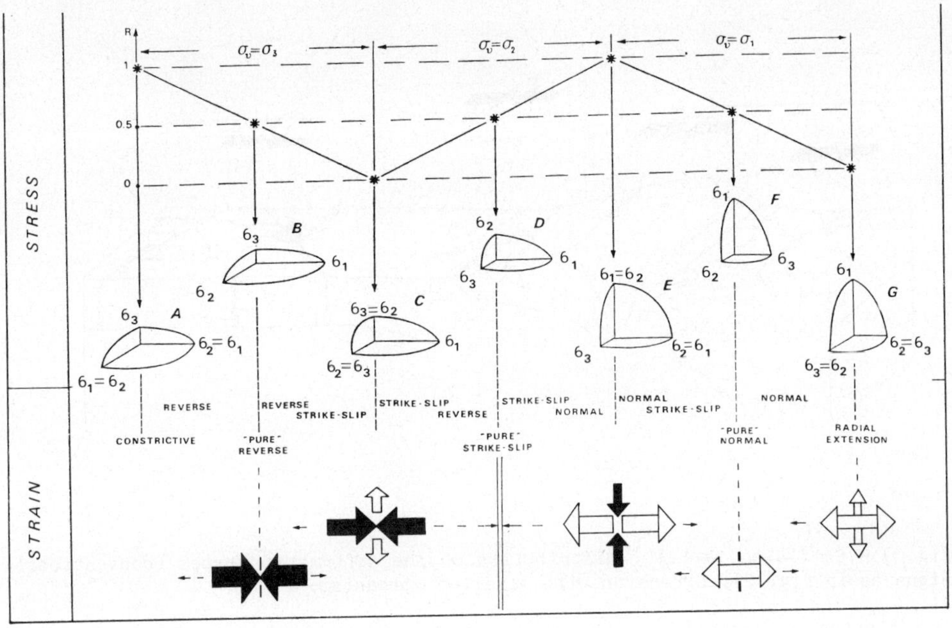

Fig. 2 - Relations between stress and strain in brittle tectonic with conventional signs used for deformation types as in fig. 3 to 8 according to Philip, 1986.

In our description of deformations we use the term "compression" for all deformations of type A to D and the term 'extension" for all deformations ranging from D to G. Compression, itself is divided in two sub-classes depending on whether the major structures correspond to reverse or strike-slip faulting. By extension, it is also possible to differentiate a sub-classe with strike-slip faults associated with normal faulting from another with only a majority of normal faults. The extreme case of radial extension (case G) has been retained because of its common occurence (e.g. extension of back-arc basins) while the compressive deformation known as constrictive deformation (case A) is very rare.

Relation between the distribution of deformation types and various geodynamic processes according to Mediterranean examples

(i) Deformations related to collision s.l. (Fig. 3) The general characteristics of these tectonics are well known (Bousquet et Philip, 1980) and we only recall here the two resultant types of deformation :

- compressive tectonics, with fold and reverse faults more or less associated with strike slip faulting (cases B to D). They result in a NW-SE to N-S shortening in the western Mediterranean (except for a few sectors, e.g. the Alps).

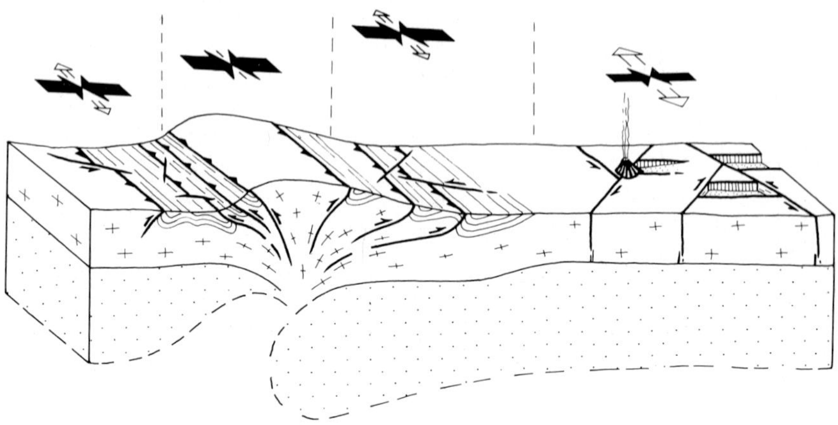

Fig. 3 - Collision (s.l.) : distribution of the deformation types (conventional signs as in fig. 2). Extension with alkaline volcanism.

- distensive tectonics, with normal faults more or less associated with strike slip faulting (cases F to D) and corresponding to a SWW-NEE to W-E extension in the western Mediterranean, thus in a direction perpendicular to the shortening direction. The major normal faults usually point NW-SE to N-S.

The distribution of these deformations related to collision (s.l.) shows that the maximum horizontal stress (σ_H maximum) which corresponds either to σ_1 or to σ_2, usually retains the same direction which is the plates convergence direction.

(ii) <u>Deformation related to oceanic subduction (based on the Aegean arc example)</u>(Fig. 4). Tectonics related to oceanic subduction have been studied mostly from earthquake focal mechanisms. Different conditions can exist above the subduction (where one can find either compression or extension perpendicular or parallel to the direction of subduction, Nakamura and Uyeda, 1980). These differences might be related to the characteristics of the subduction, e.g., speed, dip, etc. (Mégard et Philip, 1975 ;Uyeda et Kanamori, 1979).

In the Mediterranean, the Aegean arc reveals (Mercier, 1976 ; Mercier et al., 1976 ; Angelier, 1977) :

- compressive tectonics (with folds, reverse faulting and a few strike-slip faults) are active at the front of the arc. Folds and reverse faults follow a direction parallel to the arc curvature. The shortening direction displays a radial distribution and may thus differ markedly from the general direction of plate convergence.

- in the external arc and back-arc regions, normal faults characterise quite exclusively the distensive tectonics. The extension directions may be

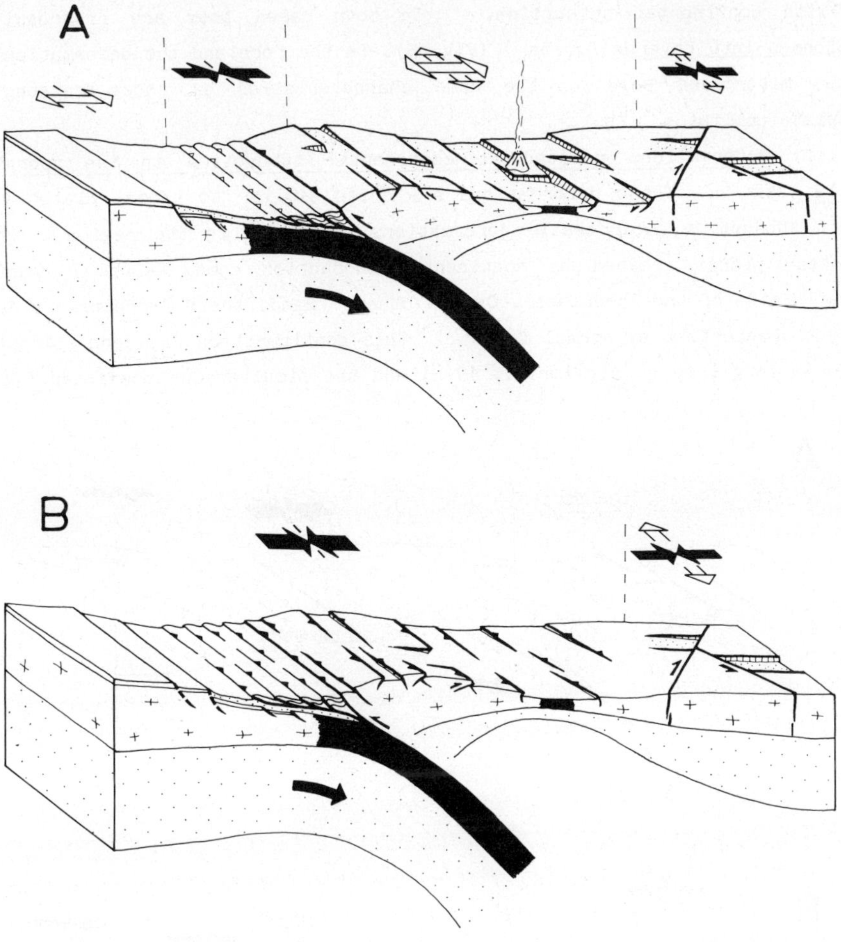

Fig. 4 - Oceanic subduction : distribution of the deformation types according to the Aegean example. A - Late Pliocene to Lower Quaternary stage. Calc-alkaline volcanism. B - Lower Quaternary compression stage (conventional signs as in fig. 2).

quite variable : therefore, the opening of the Aegean sea (a small marginal basin) may occur along different directions. The microtectonic measurements used with numerical estimate often reveal that the horizontal stresses (σ_2 and σ_3) are very close and may easily be interchanged (e.g. radial extension, case G, Fig. 2).

To these two major types of deformation, we may add some distensive deformations of limited importance located at the flexing level of the subducted plate. There, normal faults appear parallel to the direction of the flexure. This type of extension has been observed directly in the Hellenic trench

(Huchon et al., 1982). Note that these faults may also be formed in association with continental subduction. In both cases they are progressively transformed into reverse faults (Fig. 5). In the foreland the deformations are usually minor and have the same characteristics as those related to intraplate collision (s.l.).

(iii) Deformations related to continental subduction in the Apennines example (Fig. 5). These deformations are quite similar to those related to oceanic subduction. Compressive deformations are found in the region where the shear planes created by continental subduction come to the surface (Adriatic side of the Apennines). On the other hand, their Tyrrhenian side is mostly characterized by normal faults. This configuration has been acknowledged for a long time (Migliorini, 1942) and the simultaneous coexistence of

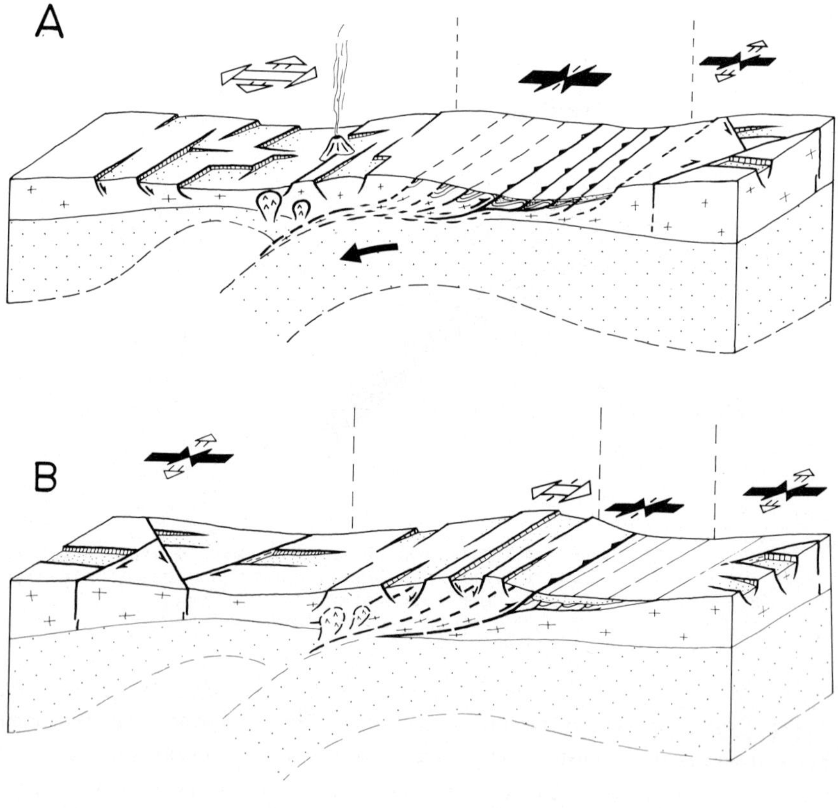

Fig. 5 - Continental subduction : distribution of the deformation types according to the Apennine example. A - Opening stage of a barck-arc basin associated with calc-alkaline volcanism and anatexy. B - Blocking stage (Conventional signs as in fig. 2).

compression and extension (Migliorini, 1942 ; Elter et al., 1980) was confirmed for the present day by earthquake focal mechanisms (Cagnetti et al., 1978 ; Gasparini et al., 1980). This situation of Italy has led to numerous interpretations (see Colacicchi, 1980), the extension on the Tyrrhenian sea being presented as a consequence of subduction (Scandone, 1979) or the subduction as passive (Boccaletti and Guazzone, 1974) and a consequence of the opening of the Tyrrhenian basin. A closer look at the distribution and characteristics of the deformations brings the following comments :

- compressive deformations are recognized by reverse faults related to folds at the compressive front of the Apennines. The deformations are more or less associated with strike-slip faulting which may become dominant (cases C to D) on either side of this compressive fault. These deformations usually give a NE-SW shortening direction for the Apennines.
- distensive deformations in the foreland of the Apennines, may show characteristics similar to those listed for the oceanic subduction and thus related to the curvature of the dipping "plate". Other distensive deformations have characteristics similar to those of collision (s.l.) deformations. They correspond to normal faulting more or less associated with strike slip faults and causing a NW-SE extension perpendicular to the direction of compression.
- distensive deformations in the Apennines and the Tyrrrhenian basin : behind the compressive front one may rapidly pass to distensive deformations. Most of the faults run parallel to the Apennines (the so called "Apennines faults") and correspond thus to an extension in the same direction as the shortening of the neighbouring compression. At a greater distance, they may be replaced by normal faults more or less associated with strike slip faulting giving a extension direction almost perpendicular to the direction of the closer ones and corresponding to the classical intraplate deformations associated with collision (s.l.). Note that this type of distension has also been observed right in the back of a compressive front. This is the case of the North-Western extremity of the Aegean arc when its appearance during the Quaternary goes together with a reduction of the Aegean oceanic subduction (Lyberis et Lallemant, 1985).

Thus the problem arises from extension parallel rather than perpendicular to compression. We believe with Scandone (1979), that continental subduction explains this problem, and can be linked to the problem of calc-alkaline volcanism and anatexy in the back-side of the compressive front. However in sectors of hightly marked topographic features located right behind the compressive front, other mechanisms could be called on to explain the presence of grabens parallel to the compressive front in the Apennines (for instance grabens of Norcia and Cascia, actually active in the Umbria-Marche Apennines, Deiana et al., 1980 ; Calamita et al., 1982). For instance, the deformation mechanism would be quite comparable to that explaining the distensive deformation observed, at another scale, behind the active reverse fault after the El Asnam earthquake of 10.10.1980 (Philip and Meghraoui, 1983). Some modelling effort is required in order to test the hypothesis at the Apennines scale.

The problem of the tectonic phases and of the diachronism of the geodynamic processes in the Apennines.

A close look at compressive and distensive tectonics shows that for each general geodynamic context, they are not random, either in their directions or with respect to one another. In particular, collision (s.l.) is characterized by shortening (compression) and distensive (extension) directions perpendicular to each other. But must these considerations on extension and compression make us reject the notion of tectonic phase ? The answer will be found in the Plio-Quaternary evolution of the Mediterranean since compressive tectonics invaded twice domains previously in extension.

The first "phase" dated to the end of Lower-Pliocene and the beginning of Middle Pliocene, has often been described in peninsular Italy and Sicily. It may have been slightly precedeed by the commencement of subduction in the Aegean arc (Lower Pliocene, Sorel, 1976).

A second "phase" of compression invasion in sectors previously subjected to extension seems to have affected most of the Mediterranean domain in the Lower Quaternary (between 1 Ma and 700.000 years ?). It is known as the "Quaternary compressive phase" (Bousquet, 1977 ; Bousquet et Philip, 1980 ; Mercier, 1981), an event which affected sectors in active oceanic subduction as well as sectors in collision s.l. In both sectors, compression may have temporarily replaced extension. In most cases this compressive deformation being very small is only revealed by microtectonic observations. In the Aegean arc (Fig. 4B) the compression invades the back-arc basin (Aegean sea) with identical shortening directions (i.e. perpendicular to the arc curvature) close to the front of the marginal basin, but with quite different directions in the remainder of the basin and its borders. This brief intrusion of compressive phenomena in different geodynamic contexts indicates that these were of Mediterranean scale and shows that the general conditions governing a given stress field may vary rather quickly (t < 500.000 years).

The distribution of distensive and compressive deformations may thus be briefly modified by "tectonic phases". Were these due to a temporary acceleration in convergence rate between Europa and Africa or on the contrary to its blocking in some part of the Mediterranean leading to a general "counter reaction" ?

EXAMPLES OF EVOLUTION OF THE PLIO-QUATERNARY DEFORMATIONS IN THE APENNINES AND CALABRIAN ARC

The successive Plio-Quaternary distensive phase along the Tyrrhenian side of the Apennines

(i) Neotectonic and microtectonic results in the Campano-Lucanian Apennines. A good example of Quaternary evolution in a sector located on the Tyrrhenian side of the Southern Apennines is given by the Eboli region (Gars 1983, Gars et Lipman 1984), which shows three consecutive deformations :

- the first deformation dated to the Lower Quaternary corresponds to a N20 extension due to faults of N80 to N135 direction. - the second deformation is compressive and is only evidenced by microtectonic observations (strike-slip striations on previous normal faults and stylolithes on pebbles). It occurs towards the end of the Lower Quaternary and corresponds to a N20 shortening.
- the third deformation corresponds again to extension with a N140 direction.

This sequence of Quaternary deformations leads to the following conclusions : an extension of the type "opening of a marginal basin" is still affecting this part of the Apennines during the Lower Quaternary whilst compressive deformations are acting more to the East in the bradanic trough. The extension is interrupted by a brief compressive episod. The shortening direction (N20) is identical to that of extension, a fact which forbids any interpretation in terms of slight modification in the stress fields (σ^1-σ^2 interchanged). The extension starts again afterwards but along N140 direction i.e. 60° off the previous extension, which shows that it cannot have the same origin as the previous extension. In the context of the compressive event that just preceded it, it can be explained by a σ_1-σ_2 interchange, accompanied by a change of direction of 30° at most.

Thus this example shows the replacement of an extension of the type "marginal basin" by a collision (s.l.) extension, the two events being separated by a brief compressive event also observed in other sectors of the Mediterranean (Quaternary compressive "phase").

For the present day the collision (s.l.) extension does not affect yet the whole of the Southern Apennines from West to East. For instance it does not seem compatible with the extension related to the NW-SE fault of the Irpinia earthquake (23.11.80 ; Deschamp et King 1983).

(ii) Interpretation of the Plio-Quaternary evolution on the Tyrrhenian side of the Apennines. The Tyrrhenian side of the Apennines shows the presence of two sets of roughly orthogonal faults in the Central Apennines. These sets are commonly known as the "Apennine" and "anti-Apennine" systems, the latter probably being the most recent. The same thing is observed in Tuscany, where Plio-Quaternary basins with Apennine direction are cut by perpendicular anti-Apennine faults (Bartoloni et al. 1983).

Finally on the South-Eastern margin of the Tyrrhenian basin, the collision (s.l.) extension is particurlaly well illustrated by NE-SW faults (fig. 6 in Bartoli et al. 1984). This implies that neither the formation of the Tyrrhenian basin nor its distensive features can be related to a single phenomenon (oceanic subduction or opening of "a distensive megafracture" due to collision (Boccaletti et al. 1984).

The Plio-Quaternary evolution of the compressive front and the foreland of the Apennines.

Deep sismic profils at the Po plain (Rocco, 1952 and 1955 ; Pieri and Groppi, 1983) as well as numerous geological studies indicate that the orogenic

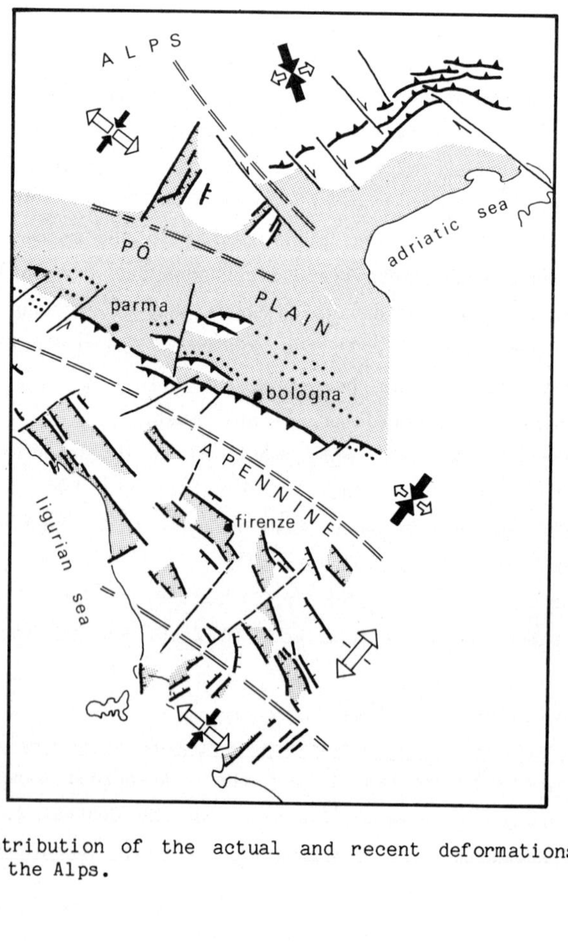

Fig. 6 - Distribution of the actual and recent deformations in the Northern Apennines and the Alps.

migration towards the East and the North-East continues during the Pliocene and the Quaternary. The following facts may be considered :

- in the Northern Apennines (Fig. 6), the compressive deformations are quite visible at the foot of the Apennines but continue within buried structures of the Po plain (Marchetti et al., 1978 ; Bartoloni et al., 1983). The little (or un-)deformed sector caught between the Apennines and the Alps progressively reduces in size. Between the two compressive fronts the N.E normal faults extension (collision s.l.) were active on a larger sector in the Pliocene to the Lower Quaternary, according to maps of Zanferrari et al., 1982.

- in the central Apennines (fig. 7) the migration of the compressive deformation can also be followed towards the foreland of the chain (Boccaletti et al., 1983). The major tangential phases are quite recent here (Lower to Middle-Pliocene), but only affect the Adriatic half of the chain whilst the Tyrrhenian side is under extension with faults mostly in the Apennine direction. During the Upper Pliocene and the Quaternary, this extension migrates quite markedley towards the East, where it overlies compressive structures dated to the Lower and Middle Pliocene, whilst the compression remains active in the most external zones between Pescara and Ancona.

- in the Southern Apennines, the compressive events following the major Miocene tangential movements (and thus older than in the Central Apennines) are not very much marked. A general NE.SW extension affects the Apennines and their foreland during the Lower Quaternary.

On the Tyrrhenian side, this extension is related to the opening of the basin, but in the foreland (Puglia) it may correspond to an extension related to the plunge of the Adriatic lithosphere under the Apennines (Richetti and Mongelli, 1980). Nevertheless the compression "phase" of the Lower Quaternary may be responsible for the activation of numerous reverse faults associated folds posterior to the Lower Quaternary in the bradanic trough. Surface observations lead Gars (1983) to reinterprate some deep structures (Carissimo et al., 1963) commonly considered as normal faults in terms of compression.

Minor strike-slip faults in the foreland (Puglia, Gargano) at the same epoch give a NE-SW shortening direction identical to that of the compressive front. Today, this front seems to play a minor role compared to that of the northern Apennines, normal faulting earthquakes such as that of Irpinia (13/11/80) being located on the Eastern limit of the Apennines.

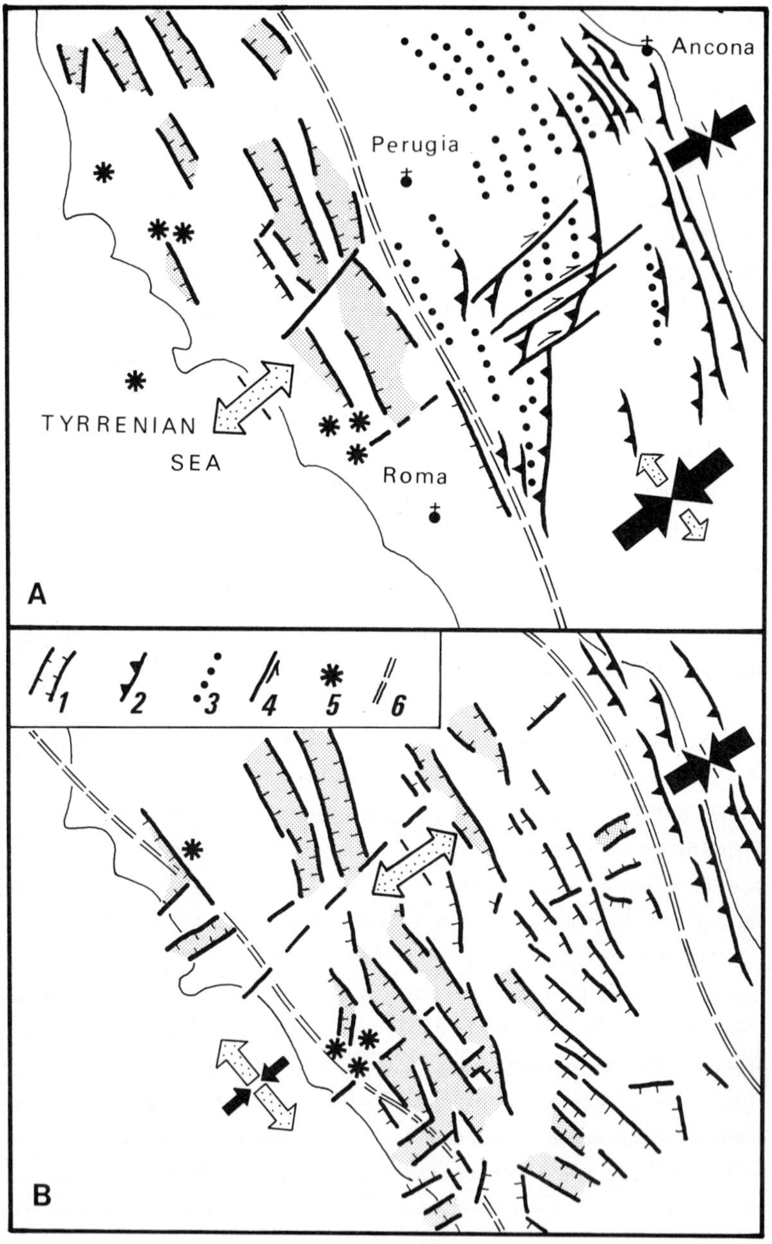

Fig. 7 - Distribution of the deformation types in the Central Apennines : A - Lower and mid Pliocene. B - Late Pliocene and Lower Quaternary
1 - Normal faults. 2 - Reverse faults. 3 - Folds. 4 - Strike-slip faults - 5 Alkalino-potassic volcanism. 6 - Boundaries between different domains of deformation (conventional signs as in fig. 2).

The Plio-Quaternary evolution of the Calabrian arc (Fig. 8)

Calabria shows features quite similar of the Apennines (rapid changes in deformation) regimes but which correspond to a situation closer to what is known in the Aegean arc during the Plio-Quaternary. Recent neotectonic studies (Philip and Tortorici, 1980 ; Tortorici 1980, Lanzafame and Tortorici, 1981 ; Ghisetti and Vezzani, 1981 ; Moussat, 1983) summarized by Ciaranfi et al. 1983 ; Philip (1983), Chabellard (1984), Auroux (1984), Moussat et al. (1985), imply that :

- the Middle Pliocene compressive phase as a shortening direction orthogonal to the general curvature of the arc. It is expressed by major structures (folds, reverse faults) on the Ionian side of the arc but only displays minor structures within the arc itself.
- the Upper Pliocene-Lower Quaternary extension deeply affects the whole arc which is almost completely immerged. This extension is characterized by variable radial directions with respect to the arc.
- at the end of the Lower Quaternary, a compression briefly interrupts the extension in the Calabrian arc. It is expressed by folds at the extremities of the arc (Sicily, Bradanic trough) but can only be recognized by microtectonics within its central part. Contrary to that of the Middle Pliocene compressions, its NE-SW shortening direction is relatively constant.
- the extension which resumed activity later on in Calabria and which is accompanied by a strong uplifting (up to 1 000 m), is generaly NW.W - SE.E to E.W.

Thus the first two events seem to belong to the same scheme as that observed in the Aegean arc and related to oceanic subduction. They are organised with respect to the arc curvature ; the extension helps the opening of the Tyrrhenian basin which at this time is increasing in size towards the East. The Quaternary compression on the other hand corresponds to a general blocking with shortening directions which may be independant of the arc curvature.

The extension which resumes activity afterwards is more difficult to interprete. If one assumes that subduction resumes activity after a brief interruption, it must correspond to an extension parallel to the shortening direction of the compressive front and related to subduction in a manner similar to that observed in the Apennines or in the Aegean arc (Crete). On the other hand, if one consider that this extension is perpendicular to the previous compression, it could correspond to the continuation of a blocking of the Calabrian subduction. This hypothesis agrees quite well with the northward migration of the alkaline volcanism in north-east Sicily which has been located for about 200 000 years at Mount Etna, on the compressive front of the

Apennine-Maghreb chain. In fact, it is the very existence of such a volcanism as well as the small dimension of the arc and its orientation with respect to Europa-Africa convergence vector which indicates that the Calabrian subduction has stopped or is on the verge of stopping (Tapponnier, 1977 ; Bousquet, 1977).

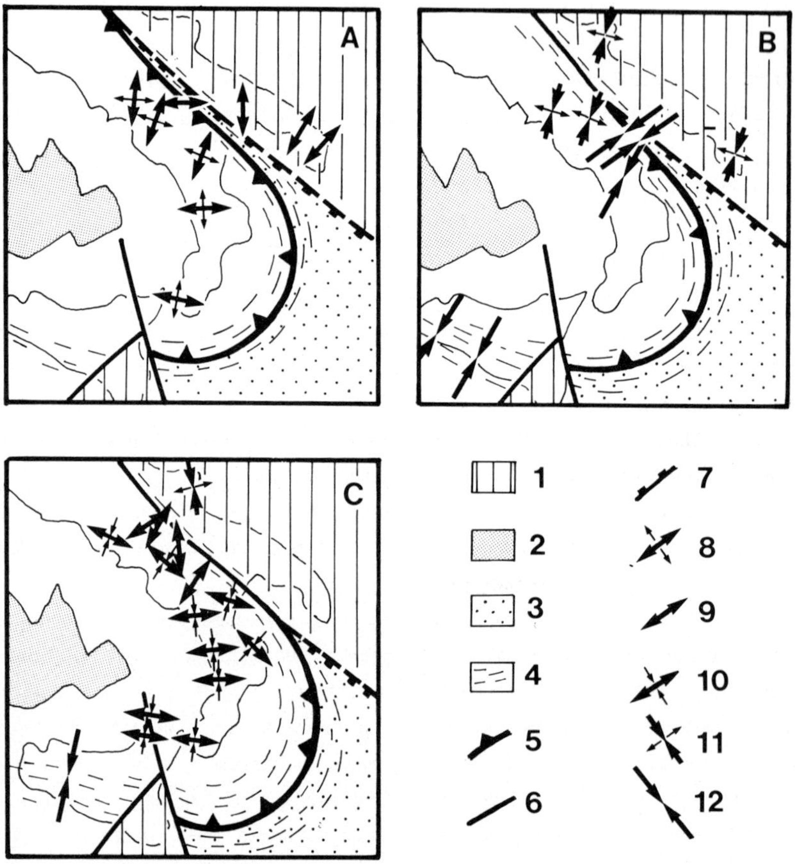

Fig. 8 - Evolution of the deformation in the Southern Apennines and Calabrian arc from late Pliocene to present day. A - Late Pliocene to Lower Quaternary. B - End of Lower Quaternary. C - Recent Quaternary to Actual.
1 - Mesozoic carbonate plateforms of the foreland. 2 - Oceanic or thinned crust of the Tyrrhenian sea. 3 - Oceanic or thinned crust of the Ionian sea. 4 - Compressive structures. 5 - Subduction. 6 - Main faults. 7 - Apulian passive margin. 8 - Radial extensive deformation. 9 - "Pure" extension deformation. 10 - Normal-strike-slip faulting deformation. 11 - Strike-slip - Normal faulting deformation. 12 - "Pure" compressive deformation (from 8 to 12 see also Fig. 2).

CONCLUSIONS

These examples of Plio-Quaternary deformations in the Apennines and on the Calabrian arc give a good idea of the difficulty in any attempt to synthetise or fit the deformation within general processes. A first step is to define schemes recorded from field data in each individual sector, that can be used in other sectors. However, a major problem, in this respect comes from the fact that the deformations are polyphased and diachronous with respect to other sectors. The evolution, oceanic subduction-continental subduction, collision (s.l.), provides the features necessary for numerous situations with, every time, the existence of synchronous extension and compression of non-random distribution. The sectors under compression are of variable extents and compressive phases may invade regions previously under extension. When this type of situation is recorded in various contexts the phases must correspond to major phenomena at the Mediterranean scale such as, for instance, an improvement in the Europe-Africa convergence.

But the processes that we gathered under the words "continental subduction" may, themselves, cause various sequences of "phases". In some sectors, the shortening of the continental crust correponds to a progressive flaking with an outwards migration of the compressive front. Major phases may be well established at some epochs : crustal shearing is then followed by magmatic phenomena (crustal anatexy and calc-alkaline volcanism). This is the case of the northern Apennines during the Miocene and of the whole Alpine chain of northern Africa and southern Spain. "Phases" like these, which are accompanied by a shortening of usually more than several tens of kilometers, must be due to some continental subduction mechanism different from that leading to simple progressive flaking. It must allow the sinking of part of the continental crust as well as its anatexy. This ensialic subduction is accompanied by a calc-alkaline volcanism and by the opening of a basin with thinned continental crust leading eventually to the formation of oceanic crust. In regions such as Southern Spain, the continental subduction did not continue and was followed from the end of the Miocene, by collision tectonics (s.l.) : the E.W extension with alkaline volcanism of the end of the Pliocene has been progressively replaced by an important N.S to N.W-S.E compression (kilometric scale folds). Strike-slip faults play an important role in the region (Groupe de recherche néotectonique de l'Arc de Gibraltar 1977, Bousquet 1979) and helps expel the continental crust towards the Algerian-Provence basin and the Atlantic ocean. In the Apennines, on the other hand, the continental subduction continued after the main Miocene phases (Northern and Southern Apennines) and the compressive deformations progressively moved eastwards.

Finally, at a larger scale (Philip 1983), the direction of maximum horizontal stress (σ^1 or σ^2) remains remarkably constant in the western mediterranean basin where collision (s.l.) is dominant ; it follows the NW-SE direction of the plate convergence vector. In the Apennines, however, the directions of σH maximum seem to be orthogonal to the compressive fronts of the continental subduction. This distribution is reminiscent of the Aegean subduction where σH maximum points southwards in Crete but is E.W at the North-West arc extremity (Zante and Cephalonie, Libeyris and Lallemant, 1985).

The Plio-Quaternary evolution of the Apennines thus fits within a general framework which is beginning to be well known. To us, it does not constitute any "particular case" even if some problems remain that neotectonics alone cannot explain.

ACKNOWLEDGEMENTS. We are very grateful to M. Mattauer and R. Gaulon for critically reading this paper. We are very much indebt to G. Lanzafame, G. Gars and J.G. Chabellard for their help during the field work in the Southern Apennines and the Calabrian arc. We also thank all those who helped us by their comments and valuable discussions, in particular M. Boccaletti, A. Etchecopar, G. Papani, P.L. Rossi, C.F. Wezel. Finally we thank Mmes Faure et Roch who did the typing and M. Mareschal who translated the manuscript.

REFERENCES

Amstuz, J., 1976. Subductions cisaillantes et création de nappes sans racine dès l'origine. C.R. Acad. Sci., Paris, 283, 1277-1280.

Angelier, J., 1977. Sur l'évolution tectonique depuis le Miocène d'un arc insulaire méditerranéen : l'arc égéen. Rev. Geogr. Phys. et Geol. Dyn., 19, 271-292.

Angelier, J. and Le Pichon, X., 1980. La subduction hellenique et l'expansion égéenne : reconstitution cinématique et interprétation dynamique. C.R. Som. Soc. Geol. Fr., 5, 158-161.

Auroux, C., 1984 - Evolution néotectonique de la dorsale apulienne et de ses bordures. Thèse de 3ème cycle, Univ. de Nice, 138 p.

Bally, A.W., 1981. Thoughts on the tectonics of folded belts. Geol. Soc. London. Thrust and nappe tectonics, 13-32.

Barbieri, F., Innocenti, F., Ferrara, G., Keller, J. and Villari, L., 1974. Evolution of Eolian arc volcanism (Southern Tyrrhenian sea). Earth and Plan. Sci. Lett. 21, 269-276.

Bartole, R., Savelli, D., Tramontana, M. and Wezel, F.C., 1984. Structural and sedimentary features in the Tyrrhenian margin off Campania, Southern Italy. Marine Geology, 55, 163-180.

Bartoloni, C., Bernini, M., Carloni, G.C., Costantini, A., Federici, P.R., Gasperi, G., Lazzarotto, A., Marchetti, G., Mazzanti, R., Papani, G., Pranzini, G., Rau, A., Sandrelli, F., Verlesi, P.L., Castaldini, D. and Francavilla, F., 1983. Carta neottectonica dell'Appennino settentrionale. Nota illustrative. Boll. Soc. Geol. It., 101, 523-549.

Beccaluva, L., Rossi, P.L. and Serri, G., 1982. Neogene to recent volcanism of the Southern Tyrrhenian-Sicilian area : implications for the geodynamic evolution of the Calabrian arc. Earth. Evol. Sciences, 3, 222-238.

Belderson, R.H., Kenyon, N.H. and Stride, A.H., 1974. Calabrian ridge, a newly discovered branch of the Mediterranean ridge. Nature, 247, 453-455.

Bellon, H, Bordet, P. and Montenat Ch., 1983. Chronologie du magmatisme néogène de Cordillères bétiques (Espagne méridionale). Bull. Soc. Géol. Fr., XXV, 205-217.

Bellon, H., Jarrige, J.J., and Sorel D., 1979. Les activités magmatiques égéennes de l'Oligocène à nos jours et leurs cadres géodynamiques. Rev. Geol. Dyn. et Geogr. Phys., XXI, 41-45.

Bird, P., Toksöz, M.N. and Sleep, N.H., 1975. Thermal and mechanical models of continent-continent convergent zone. J. Geophys. Res., 80, 4405-4416.

Bird, P., 1978. Finite element modeling of lithosphere deformation : the Zagros collision orogeny. Tectonophysics, 50, 307-306.

Boccaletti, M., Calamita, F., Centamore, E., Deiana, G., and Dramis, F., 1983. The Umbria-Marche Apennine : an example of thrusts and wrenching tectonics in a model of ensialic neogenic-quaternary deformation. Boll. Soc. Geol. It. 102, 1-12.

Boccaletti, M., Coli, M., De Candia, F.A., Giannini, E., and Lazzarotto, A., 1981. Evoluzione dell'Apennino settentrionale secondo un nuovo modello strutturale. Mem. Soc. Geol. It., 21, 359-373.

Boccaletti, M., and Manetti P., 1978. The Tyrrhenian sea and adjoining regions. In Oceans Basins and Margins, vol. 4b, Nairn A.E. and Stehli Ed., 149-200.

Boccaletti, M., Nicolich, R., and Tortorici L., 1984. The Calabrian arc and the Ionian sea in the dynamic evolution of the central Mediterranean. Marine Geology, 55, 219-245.

Bousquet, J.C., 1973. La tectonique récente de l'Apennin Calabro-Lucanien dans son cadre géologique et géophysique. Geologica Romana, XII, 1-104.

Bousquet, J.C., 1977. Contribution à l'étude de la tectonique récente en Méditerranée occidentale : les données de la néotectonique dans l'Arc de Gibraltar et dans l'Arc tyrrhénien. In Biju-Duval, B. and Montadert, L. Edit. : Structural history of the Mediterranean basins, Ed. Technip Paris, 199-214.

Bousquet, J.C., 1979. Quaternary strike-slip faults in Southeastern Spain. Tectonophysics, 52, 277-286.

Bousquet, J.C., and Philip H., 1980. Les caractéristiques néotectoniques en Méditerranée occidentale. In Sed. basins of Med. Margins, Wegel F.C. Edit., Tecnoprint Bologna, 389-405.

Cagnetti, V., Pascuale, V., Polinari, S., 1976. Focal mechanisms of earthquakes in Italy and adjacent regions. Com. Nat. Energia nucleare. RT/AMB, 14.

Calamita, F., Cottorni, M., Deina, G., Dramis, F., and Pambianchi G., 1982. Neotectonic evolution and geomorphology of the Cascia and Norcia depressions (Umbria-Marche Apennine). Geofis. Dinam. Quat., 5, 263-276.

Carissimo, L., D'Agostino, D., Loddo, C., Pieri, M., 1963. Petroleum exploration by Agip Mineralia and new geological information in Central and Southern Italy from the Abruzzi to the Taranto Gulf. VIth International Petroleum Congress, Frankfurt/Main, Sect. 1, 267-292.

Cassinis, R., Franciosi, R., Scarascia, S., 1979. The structure of the earth's crust in Italy on preliminary typology based on seismic data. Boll. Geofis. Teor. Appl., 21, (82), 105-126.

Castellarin, A., Colacicchi, R. and Praturlon A., 1978. Fase distensive, trascorrenze e sovrascovimenti lungo la linea Ancona-Anzio, dal Lias medio al Pliocene. Geologica Romana, 12, 161-189.

Cello, G., Guerra, I., Tortorici, L. and Turco, E., 1982. Geometry of the neotectonic stress field in Southern Italy : geological and seismological evidence. J. of Struct. Geol, 4, 385-393.

Chabellard, J.G., 1984. Evolution néotectonique du secteur NE du détroit de Messine dans le cadre de l'Arc tyrrhénien. Thèse de 3ème cycle, Montpellier, U.S.T.L.

Ciaranfi, N., Ghizetti, F., Guida, H., Jaccarino, G., Lambiase, S., Pieri, P., Rapisardi, L., Ricchetti, G., Torre, H., Tortorici, L. and Vezzani, L., 1983. Carta neotettonica dell'Italia méridionale. CNR Publ. n° 5 del P.F. Geodinamica. Bari, 62 p.

Civetta, L., Orsi, G., Scandone, P. and Pece, R., 1978. Eastwards migration of the Tuscan anatectic magmatism due to anticlockwise rotation of the Apennines. Nature, 26, 604-606.

Colombi, B., Giese, P., Luongo, G., Morelli, C., Riuscetti, M., Scarascia, S., Schutte, K.G., Sirowald, J. and De Visintini, G., 1973. Preliminary report on the profile : Gargano-Salerno-Palermo-Pantelleria (1971) Boll. Geof. teor. ed appl., vol. XV, 59, 225-254.

Colacicchi, R., 1980. Evoluzione geodinamica neogenica del margine appenninico periadriatico. Societa Geologica Italiana. Cento anni di geologia it., 297-301.

Dal Piaz, G.V., Venturelli, G., Scolari, A., 1979. Calc-alkaline to ultrapotassic post-collisional volcanic activity in the internal northwestern Alps. Mem. Sc. Geologiche, 32, 4-16.

Davis, D., Suppe, J. and Dalhen, F.A., 1983. Mechanic of fold and thrust belts and accretionary wedges. J. Geophys. Res., 88, 1153-1172.

Deschamps, A. and King, G.C.P., 1983. The Campanian-Lucanian (Southern Italy) earthquake of November 1980. Earth and Plan. Sc. Letters, 62, 296-304.

Deiana, G., Dramis, F., Lavecchia, G. and Pialli, G., 1980. Schema geologico dell'area Nursina ed eventi sismici, in Intervento a seguito del terremoto di Norcia del 1979. C.N.R. Italy, 40-48.

Dewey, J.F. and Celâl Sengör, A.M., 1979. Aegean and surrounding regions : complex multiplate and continuum tectonics in a convergent zone. Geol. Soc. of America Bull., 90, 84-92.

Di Girolamo, P., Nardi, G., Rolandi, G. and Stanzione, D., 1976. Occurence of Calc-alkaline two pyroxene andesites from deep boreholes in the Phlegrean fields. Rend. Accad. Sc. Fis. e Nat., Napoli, 43, 45-73.

Elter, P., Giglia, G., Tongiorgi, M. and Trevisan, L., 1975. Tensional and compressional areas in the recent (Tortonian to present evolution of the Northern Appennins). Boll. Geof. Teor. ed. appl., 17, 3-18.

Etchecopar, A., Vasseur, G. and Daignières, M., 1981. An inverse problem in microtectonics for the determination of stress tensors from fault striation analysis. J. Struct. Geol., 3, 51-55.

Etchecopar, A., 1984. Etude des états de contrainte en Tectonique cassante et simulations de déformations plastiques (approche mathématique). Thèse doctorat d'Etat. U.S.T.L. Montpellier, 269 p.

Finetti, I., 1980. Geophysical study on the evolution of the Ionian Sea. Sed. basins of the Medit. Sea. F.C. Wezel Ed., 465-484.Gars, G., 1983. Etudes sismotectoniques en Méditerranée centrale et orientale. I. La Néotectonique de l'Apennin méridional et le séisme de l'Irpinia (Italie). II. Les failles activées par les séismes de Corinthe (Grèce). Thèse 3ème cycle. Université de Paris-Sud, 226 p.

Gars, G. and Lipmann, M., 1984. Nouvelles données néotectoniques dans l'Apennin Campanian (Italie du Sud). C.R.Acad.Sc.Paris, 298, 495-500.

Gasparini, C., Iannaccone, G. and Scarpa, R., 1980. On the focal mechanism of Italian earthquakes. Rock Mechanics, suppl. 9, 85-91.

Ghizetti, F. and Vezzani L., 1981. Contribution of structural analysis to understanding the geodynamic evolution of the Calabrian arc (Southern Italy). J. Struct. Geol., 3, 371-381.

Groupe de recherche néotectonique de l'Arc de Gibraltar, 1977. L'histoire tectonique récente (Tortonien à Quaternaire) de l'Arc de Gibraltar et des bordures de la mer d'Alboran. Bull. Soc. Géol. France, 25, 575-614.

Huchon Ph., Angelier, J., Le Pichon, X. Lybéris, N. and Ricon, L.E., 1980. Les structures tectoniques observées en plongées dans les fosses helléniques. C. Rend. Som. Soc. Géol. France, 5, 162-166.

Lanzafame, G. and Tortorici, L., 1981. La tettonica recente della valle del F. Crati (Calabria). Geogr. Fis. e Din. Quat., 4, 11-21.

Libeyris, N. and Lallemant, S., 1985. La transition subduction-collision le long de l'arg égéen externe. C.R.Acad.Sc. Paris, 300, 885-890.

Marchetti, G., Papani, G. and Sgavetti, M., 1978. Evidence of Neotectonics in the North-West Apennines-Po side. In Alps, Apennines, Hellinides. H. Closs et al. Edit. Stuttgart, 283-288.

Mattauer, M., 1983. Subduction de lithosphère continentale, décollement croûte manteau et chevauchements d'échelle crustale dans la chaîne collision himalayenne. C.R.Acad.Sc.Paris, 296, 481-486.

Mattauer, M., 1986. Les subductions intracontinentales des chaînes tertiaires d'Asie, leurs relations avec les décrochements. Bull. Soc. Géol. France, 2, 143-157.

Mc Kenzie, D.P., 1972. Active tectonics of the Mediterranean region. Geophys. J.R. Astron. Soc., London, 30, 109-185.

Mc Kenzie, D.P., 1978. Active tectonics of the Alpine-Himalayan belt : the Aegean sea and surrounding regions. Geophys. J. R. aström. Soc. London, 55, 217-154.

Megard, F. and Philip., H., 1976. Plio-quaternary tectono-magmatic zonation and plate tectonics in the central Andes. Earth Plan. Sc. Lett., 33, 231-238.

Mercier, J.L., 1976. L'arc égéen, une bordure déformée de la plaque eurasiatique ; réflexion sur un exemple d'étude néotectonique. Bull. Soc. Géol. Fr., 19, 663-672.

Mercier, J.L., Carey, E., Philip, H. and Sorel, D., 1976. La Néotectonique plio-quaternaire de l'arc égéen externe et de la Mer égée et ses relations avec la séismicité. Bull. Soc. Géol. Fr., 355-372.

Mercier, J.L., Delibassis, N., Gauthier, A., Jarrige, J.J., Lemeille, F., Philip. H., Sebrier, H. and Sorel, D., 1979. La néotectonique de l'Arc égéen. Rev. Géol. Dyn. Géogr. Phys., 21, 67-92.

Mercier, J.L., 1981. Extensional-compressional tectonics associated with the Aegean arc : comparison with the Andean Cordillera of South-Peru-North Bolivia. Philos. Trans. Roy. Soc. London, 300, 337-355.

Migliorini, C., 1948. I cunei compsti nell'orogenesi. Boll. Soc. Géol. It., 67, 29-142.

Molnar, P. and Gray, D., 1979. Subduction of continental lithosphere some constraints and uncertainties. Geology, 7, 58-62.

Morelli, C., Giese, P., Cassinis, R., Colombi, B., Guena, I., Luongo, G., Scarascia, S. and Schutte, J.G., 1975. Crustal structure of Southern Italy. A seismic refraction profile between Puglia-Calabria-Sicily. Boll. Geof. Teor. Appl., 67.

Moussat, E., 1983. Evolution de la mer tyrrhénienne centrale et orientale et de ses marges septentrionales en relation avec la néotectonique dans l'arc calabrais. Thèse 3ème cycle. Univ. de Paris VI, 332 p.

Moussat, E., Mascle, G. and Angelier, J., 1985. Regimes de paléocontraintes et déformation de l'arc tyrrhénien. C.R.Acad.Sc. Paris, 300, 417-422.

Nakamura, K. and Uyeda, S., 1980. Stress gradient in arc-back arc regions and plate subduction. J. Geophys. Res., 85, 6419-6428.

Panza, G.F., 1979. The crust and upper mantle in Southern Italy from geophysical data. Riv. italiania di Geof. e Sc. aff., 5, 17-22.

Peccerillo, A., 1985. Roman comagmatic province (central Italy) : evidence for subduction-related magma genesis. Geology, 13, 103-106.

Philip, H., 1983. La Tectonique actuelle et récente dans le domaine méditerranéen et ses bordures, ses relations avec la seismicité. Thèse doctorat d'Etat, U.S.T.L. Montpellier, 240 p.

Philip, H., 1986. Plio-quaternary evolution of the stress field in Mediterranean zones of subduction and collision. (submited to Annales Geophysicae).

Philip, H. and Tortorici, L., 1980. Tectonique superposée dans les sédiments Miocène supérieur à Pleistocène de la Calabre centrale et septentrionale (Italie méridionale). C.R.Som.Soc.Géol.Fr., 5, 191-194.

Philip, H. and Meghraoui, M., 1983. Structural analysis and interpretation of surface deformations of the El Asnam earthquake of October 10, 1980. Tectonics, 2, 17-49.

Pieri, M. and Groppi, G., 1983. Subsurface geological structure of the Po Plain, Italy. Publ. 414. Progetto Fin. Geodinamica Italy.

Praturlon, A., 1980. Cadre géologique général de l'Italie. Sa place dans le cadre européen et méditerranéen. In Geologie des pays européens. Dunod Ed. France, 177-190.

Reuter, K.J., Giese, P. and Closs, H., 1980. Lithospheric split in descending plate : observations from the northern Apennines. Tectonophysics, 64, T1-T9.

Richetti, G. and Mongelli, F., 1980. Flessione e campo gravimetrico della microplastra apula. Boll. Soc. Geol. It., 99, 431-436.

Rocco, T., 1952. Progressi della sismica a reflessione nella coplorazione petrolifera del AGIP nella Pianura padana. Riv. Ital. Petr. Roma.

Rocco, T., 1955. Comparative geological study and geophysical study of the Pô bassin. IV. Congr. Mond. du Pétrole. Sect. IE Roma, 14 p.

Scandone, P., 1979. Origin of the Tyrrhenian sea and Calabrian arc. Boll. Soc. Geol. It., 98, 27-34.

Schutte, K.G., 1978. Crustal structure of Southern Italy. In. Alps, Apennines, Hellenides, H. Closs et al. Edit., Stuttgart, 315-321.

Sorel, D., 1976. Etude néotectonique de l'arc égéen externe occidental. Thèse 3ème cycle. Univ. de Paris XI.

Tapponnier, P., 1977. Evolution tectonique du système alpin en Méditerranée, poinçonnement et écrasement rigide plastique. Bull. Soc. Géol., Fr., 19, 437-460.

Tortorici, L., 1981. Analisi delle deformazioni fragili dei sedimenti postorogeni della Calabria settentrionale. Boll. Soc. Geol. It., 100, 291-308.

Zanferrari, A., Bollettinari, G., Carobene, L., Carton, A., Carulli, G.B., Castaldini, D., Cavallin, A., Panizza, A., Pellegrini, G.B., Pianetti, F. and Sauro, U., 1982. Evoluzione neotettonica dell'Italia nord-orientale. Mem. Sc. Geol. Ist. Geol. Mineral. Padova, 35, 355-376.

EVENT STRATIGRAPHY, BASIN EVOLUTION AND TECTONICS OF THE HELLENIC AND CALABRO-SICILIAN ARCS

J.E. Meulenkamp and F.J. Hilgen
Dept. Stratigraphy and Micropaleontology
Institute of Earth Sciences, University of Utrecht
P.O. Box 80.021, 3508 TA Utrecht, The Netherlands

ABSTRACT

The sedimentary records of the Hellenic and Calabro-Sicilian arcs reflect the impact of a sequence of contemporaneous tectonic events which controlled the origin and evolution of Late Cenozoic basins. These events were the response to changes in regional stress fields resulting in time-equivalent onlap and offlap trends along the margins of the basins. The changes are considered to be a regional overprint on the development of the two arcs.

The evolution of the Hellenic arc is defined by a SW - NE compressional regime. Fragmentation of the Southern Aegean landmass and the collapse of remnants of the Apulian platform initiated about 11 Ma ago the development of the arc proper. The arc got its final shape through the combined effects of the southward migration of the Cretan block and Late Miocene - Pleistocene vertical movements.

Changes in the direction of thrusting along shear-zones intersecting the Calabro-Sicilian arc mirror the interference of two stress fields, associated with approximately NW - SE and ENE - WSW compression. This interference generated opposite directions of relative horizontal displacements along faults connected with NW - SE shear-zones in Calabria, which may account for the bending of the arc. The four sedimentary-tectonic units constituting the arc were sutured by NE - SW and NW - SE shortening, resulting from intra-Pliocene and Early Pleistocene compressional phases.

INTRODUCTION

Recent models for the kinematic and dynamic evolution of the Hellenic and Calabro-Sicilian arcs are primarily based on the evaluation of structural and geophysical data. For the Hellenic arc such models were established by Le Pichon and Angelier (1981) and Angelier et al. (1982). Diverging opinions on the development of the Calabro-Sicilian arc can be found in the papers by for instance Scandone(1979, 1982), Ghisetti and Vezzani (1981, 1982), Moussat (1983), Boccaletti et al. (1984) and Wezel (1984). None of these models, however, is based on comprehensive stratigraphical studies of the Late Cenozoic basins.

High-resolution stratigraphic scales allow to reconstruct in considerable detail the timing and the impact of tectonic events which controlled the Late Cenozoic basin development in the two arcs. In previous papers (Meulenkamp, 1982 a, b) we argued for a pulsating evolution of the arcs. This was based on the episodic recurrence of tectonically-defined, contemporaneous changes in paleogeography, basin configurations and sedimentation patterns. The nature of such changes and of their effects were shown to vary strongly. There is, however, a remarkable correspondence in time between large-scale paleogeographic reorganizations and fundamental basin rearrangements in the two arcs. This correspondence seems not to be confined to the Hellenic and Calabro- Sicilian arcs. Paleogeographic reconstructions (Rögl and Steininger, 1983) and composite stratigraphic columns (Steininger et al., 1985) indicate that most, if not all, tectonically-defined major changes had time-equivalent counterparts elsewhere in the Mediterranean.

In this paper we will elaborate the sequence of tectonic events reconstructed from the analysis of Late Cenozoic basins of the Hellenic and Calabro-Sicilian arcs, which form part of the Dinarid - Hellenic - Taurid and Apenninic - Maghrebian chains, respectively (fig. 1). Subsequently, problems pertinent to the origin of the arcs proper will be discussed. The results are considered to be of importance for a better understanding of the geodynamics of Mediterranean young orogenic arcs and they will be calibrated with new models (Cloetingh et al., 1985; Cloetingh, 1986) for regional, tectonically-defined relative changes of sealevel.

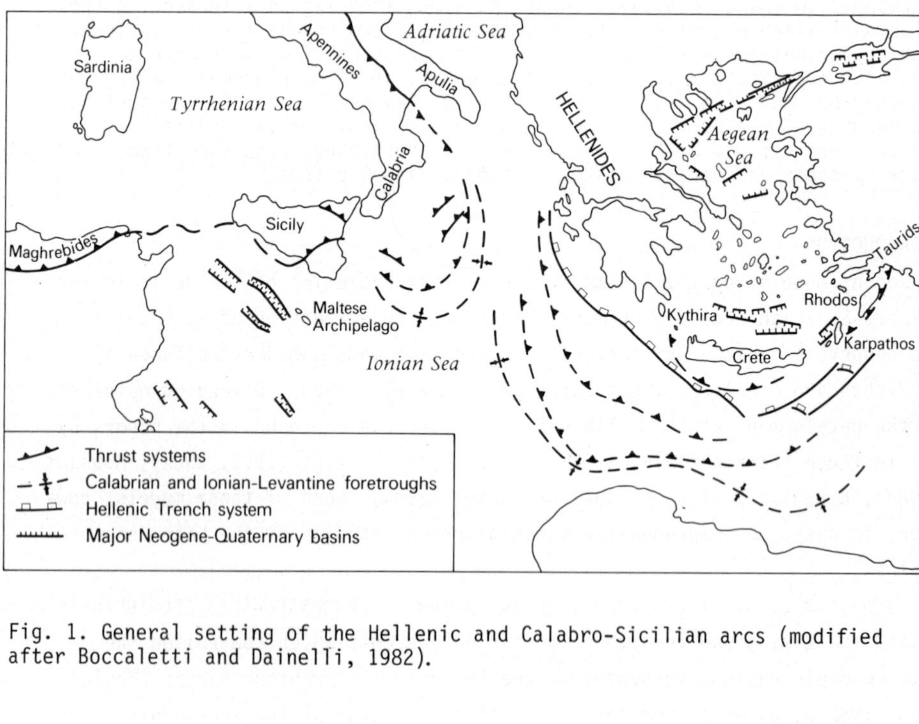

Fig. 1. General setting of the Hellenic and Calabro-Sicilian arcs (modified after Boccaletti and Dainelli, 1982).

THE SEQUENCE OF TECTONIC EVENTS

In figure 2 the timing of tectonic events in Sicily, Calabria and the Hellenides is depicted relative to planktonic zones, Mediterranean stages and the numerical time scale. During the Neogene, fundamental changes in basin configurations and paleogeography occurred in the Late Burdigalian, the Serravallian, the Serravallian - Tortonian boundary interval, the Late Miocene (intra - Tortonian, Tortonian - Messinian transition and intra-Messinian) and in the late Early to early Late Pliocene. Paleogeographic reconstructions by Rögl and Steininger (1983) suggest that some of these changes were related to plate reorganizations.

During the Late Paleogene, major changes occurred at about the transition from the Eocene to the Oligocene, in the Early-Late Oligocene boundary interval and during the Late Oligocene. Another event straddles the Oligocene - Miocene boundary. Correlations of Late Paleogene tectonic events, however, are hampered by a much lower time-resolution than achieved for the Neogene. This is mainly due to the often poor biostratigraphic control of Late Paleogene clastic successions.

In the following sections we will summarize the effects of the most important tectonic events for six time-slices. These include the latest Eocene - Early Miocene, the late Early Miocene - early Middle Miocene, the Middle - Late Miocene, the latest Miocene, the Early Pliocene - early Late Pliocene, and the Late Pliocene - Early Pleistocene.

Latest Eocene - Early Miocene (± 37 - 20 Ma)

Detailed reconstructions can be made for the external Ionian and Preapulian (Paxos) zones of the Hellenides (figure 3). The Ionian zone may be considered a forearc basin, the Preapulian zone representing the slope of the Apulian platform (African foreland). The Late Paleogene - Neogene record mirrors a pulsating subsidence of carbonate platforms, accompanied by the synchronous progradation of turbidite basins from the external Ionian to the Preapulian zone, i.e. in the direction of the external platform (I.F.P., 1966; de Mulder, 1975; Richter et al., 1978; Meulenkamp, 1982 a, b, see also figure 3). Tectonic events during the latest Eocene - Early Miocene time-span triggered the episodic supply of coarse debris from nearby located carbonate platforms. This supply was superimposed upon turbidite accumulation by longitudinal infill.

It is more difficult to unravel the relation between tectonics and time in the Calabro-Sicilian and Southern Aegean arcs, partly because Oligocene - Lower Miocene successions are incorporated in allochthonous units. This is particularly the case in Sicily, Calabria and Crete. Wezel (1974, 1975) argued for a pulsating evolution of Oligocene - Early Miocene flysch basins in the Calabro-Sicilian chain. Meulenkamp et al. (in press) reported a new phase of basin

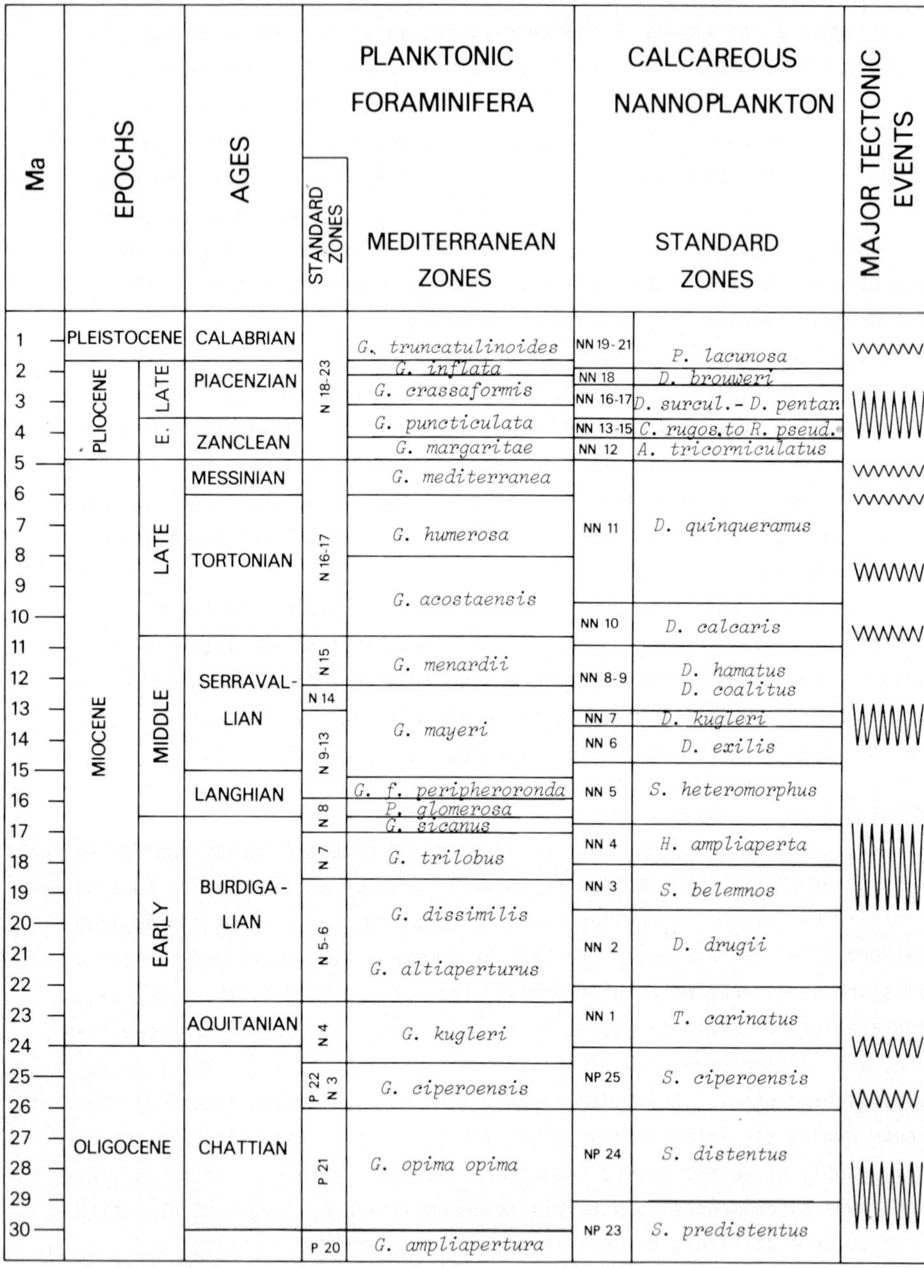

Fig. 2. Position of major tectonic events relative to planktonic zones, Mediterranean stages and the numerical time-scale (modified after Meulenkamp, 1982 a; new data on the Tortonian - Messinian and Messinian-Zanclean boundaries after Langereis et al., 1984 and Zijderveld et al., in press)

development during the transition from the Early to the Late Oligocene (about 30 Ma ago) and a pronounced tectonic rejuvenation of reliefs around the Oligocene - Miocene boundary, 24 Ma ago.

On the external platform ,the counterparts of the Late Paleogene - Early Miocene events were recurrent differential, vertical movements of blocks, which caused opposite trends in depth of depositional environments. For instance, the uppermost Oligocene - lowermost Miocene of the Maltese archipelago shows a facies-transition from platform carbonates to calcareous muds, deposited in deeper environments (Felix, 1973; Bennett, 1979; Drooger and Roelofsen, 1982; Rose, 1985). In contrast, successions along the margin of the Iblean platform, Southeastern Sicily, indicate a shallowing.

Late Early Miocene - early Middle Miocene (± 20 - 15 Ma)

Intra - Burdigalian compressional tectonics, clustering around 18 Ma ago, caused the emplacement of nappes in most of the Alpine chains of the Mediterranean and the Paratethys. These movements resulted in large-scale, fundamental changes in the paleogeography. According to Adams et al. (1983) the marine connections between the Mediterranean proper and the Indo-Pacific across the Middle East platform were definitely interrupted at the time.

In the Hellenic and Calabro-Sicilian arcs the compression culminated in the Late Burdigalian. Allochthonous units were emplaced in Sicily, Calabria and probably also in Crete (Meulenkamp et al., in press, see also figure 4). Thrusting and overthrusting of the Pindos relative to the Ionian zone in Western Greece caused folding and intensive deformation of the Ionian sequences (cf. I.F.P., 1966; de Mulder, 1975; Dercourt et al., 1977; Bizon and Bizon, 1985), whereas the sediments of the more external Preapulian zone were much less affected.

The data from Western Greece and the Iblean platform reflect the effects compressional tectonics had along a transect from forearc basins to the external platform. Precise datings indicate that folding, uplift and partial erosion of the pre - latest Burdigalian sediments in the Ionian zone corresponded to isoclinal folding along the eastern margin of the Preapulian zone. Intraformational folds observed in a monoclinal sequence represent the only visible effect of the compressional event in a more external setting (figure 3). Simultaneously, low rates of deposition or uplift (resulting in local hiatuses) characterized the Iblean platform (Meulenkamp, 1982 b).

Regional transgressions and fundamental basin rearrangements marked the shift from compressional to tensional regimes, which took place in latest Burdigalian to Early Langhian time (between ± 17 and 15.5 Ma ago). This reorganization can be recognized all over the Mediterranean and the Paratethys (Rögl and Steininger, 1983; Steininger et al., 1985).

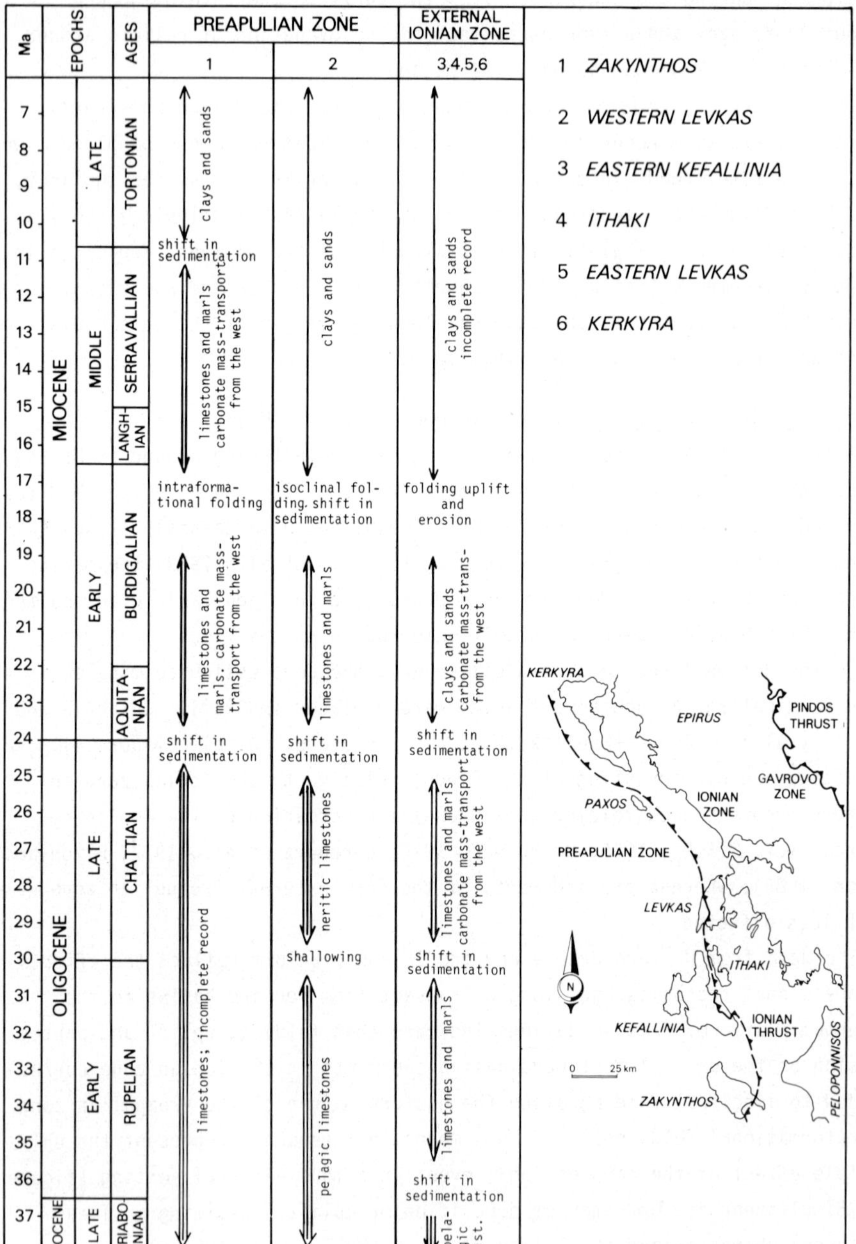

Fig. 3. Generalized scheme of the impact of the most important tectonic events on sedimentation patterns in the external Ionian and Preapulian zones of the Western Hellenides. Note pulsating shift in limestone, limestone - marl and clay - sand sedimentation. For further explanation, see text.

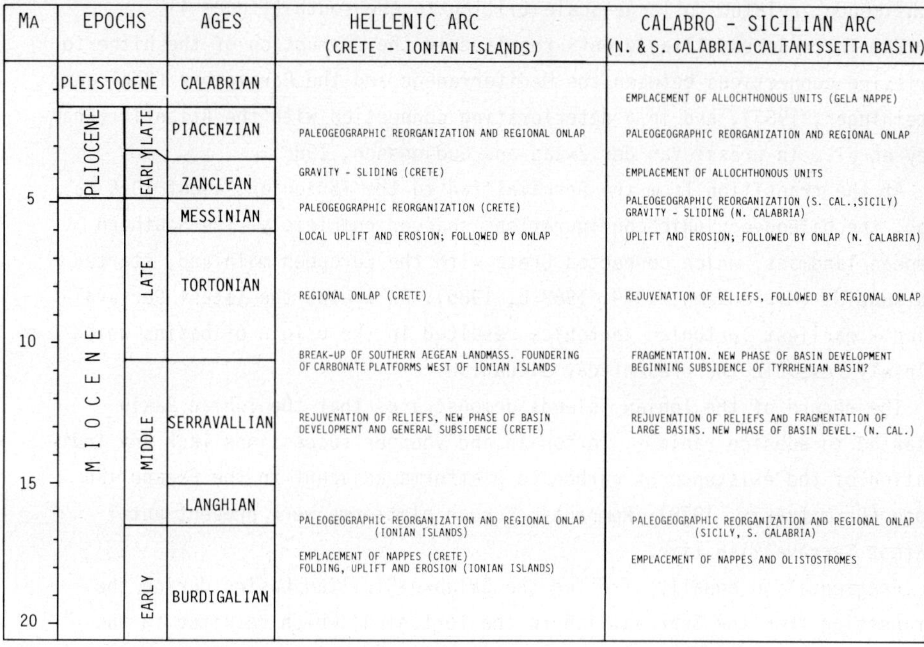

Fig. 4. Compilation of the effects of some of the major, contemporaneous tectonic events on the development of Neogene - Early Pleistocene basins of parts of the Hellenic and Calabro-Sicilian arcs.

In Calabria and Sicily uppermost Burdigalian to Lower Langhian successions unconformably overlie the argille scagliose, the highest allochthonous unit. In Western Greece sediments of the same age were deposited upon the older, folded sequences in the Ionian zone, while the shift from compression to tension along the margin of the Preapulian zone was connected with a sudden change from limestone - marl to clay sedimentation (figure 3).

The relative rise of the sealevel, evidenced by onlapping successions in Calabria, Northern Sicily and Western Greece, was of the magnitude of some tens of metres. On the other hand, benthic foraminiferal associations from clays overlying the Lower Miocene carbonates on the Iblean platform suggest a relative sealevel rise of several hundred metres in the course of the latest Burdigalian - Langhian time-span (Van der Zwaan, pers. comm., 1986).

Middle Miocene - Late Miocene (± 15 - 6 Ma)

Serravallian tectonic events, around 14 and 13 Ma ago (figure 2) caused a conspicuous rejuvenation of reliefs. In Calabria and Sicily they initiated the fragmentation of large basins which originated in the latest Burdigalian - Early Langhian. At this time, Crete became incorporated in a regime of general subsidence, related to large-scale tilting to the south (figure 4).

The intra-Serravallian events resulted in the disruption of the hitherto existing connections between the Mediterranean and the Paratethys (Rögl and Steininger, 1983), and in a deteriorating connection with the Atlantic (Chamley et al., in press; Van der Zwaan and Gudjonsson, 1986).

At the transition from the Serravallian to the Tortonian, about 10,6 Ma ago, the paleogeographic configurations changed completely. The Southern Aegean landmass, which connected Crete with the European mainland, started to subside (Meulenkamp, 1979, 1982 b, 1985). On Crete, the latest Serravallian - earliest Tortonian tectonics resulted in the origin of basins which closely resemble the present-day situation.

The record of the Ionian islands demonstrates that the Ionian Basin started to subside rapidly. Tortonian and younger successions lack any indication of the existence of carbonate platforms adjacent to the Preapulian zone (Dermitzakis, 1978). Remnants of such platforms were present until latest Serravallian time.

Fragmentation equally affected the Calabro-Sicilian basins during the transition from the Serravallian to the Tortonian, which resulted in the beginning of differentiation between various parts of the Caltanissetta Basin on Sicily and in a new phase of basin development in Northern Calabria. Probably (parts of) the Tyrrhenian Basin began to subside, although the general subsidence of the basin apparently started later, in the Late Tortonian - Early Messinian. Our data indicate that the events in the Serravallian - Tortonian boundary interval gave rise to paleogeographic configurations which can be considered the embryonic stage of the present Hellenic and Calabro-Sicilian arcs.

Intra - Tortonian tectonics (between 8 and 9 Ma ago) accentuated the earlier fragmentation. This fragmentation was folllowed by regional transgressions. In Southern Calabria, for instance, Upper Tortonian successions unconformably overlie the older Neogene or the crystalline basement (Meulenkamp et al., in press; see also figure 4). On Crete time-equivalent successions locally onlap on various units of the Alpine nappes. In general, the Late Tortonian episode was characterized by a flattening of reliefs. This conclusion is supported by clay mineral data (Chamley et al., in press).

Latest Miocene (± 6 - 5 Ma)

Block faulting, locally accompanied by tilting, uplift and erosion caused a further differentiation between (sub)basins in the Hellenides and the Calabro-Sicilian arcs at about the transition from the Tortonian to the Messinian (5,6 Ma after Langereis et al., 1984). These differential tectonics were followed by a regional submergence. Shallow marine (generally calcareous) successions of the lowermost Messinian unconformably overlie the older Neogene or the pre-Neogene basement (Angelier, 1975; Meulenkamp, 1979, 1985; Meulenkamp et al., 1981).

A major tectonic reorganization occurred in Messinian time (see also Roda, 1964; Decima and Wezel, 1972). The distribution of Upper Messinian sediments on Crete and in Calabria (Meulenkamp et al., 1979; Meulenkamp et al., in press) indicate new phases of basin development and pronounced changes in paleogeography. This intra-Messinian event most probably defined the setting of the Early Pliocene basins; it may have been related to compressional tectonics (see also Di Nocera et al., 1975; Amodio-Morelli et al., 1976; Ortolani et al., 1979; Moussat, 1983). Presumed glacio-eustatically controlled rejuvenation of reliefs during the Messinian salinity crisis (Hsü et al., 1973) is superimposed on these intra-Messinian, regional tectonics.

Early Pliocene - early Late Pliocene (± 5 - 2,5 Ma)

Intra-Pliocene compressional tectonics culminated between 4 and 2,5 Ma ago (figure 2). A sequence of events, which started shortly after the beginning of the Pliocene induced the uplift of the Hellenic and Calabro-Sicilian arcs and the pronounced increase of subsidence rates of the Aegean and Tyrrhenian "back-arc" basins. The compression was followed by regional transgressions and new phases of basin development in the Late Pliocene.

In Sicily (Caltanissetta Basin) and in Calabria, the impact of compression caused a conspicuous steepening of basin slope gradients (Meulenkamp et al., 1981; Meulenkamp et al., in press). This triggered the downslope transport of Early and pre-Pliocene packets of strata, which locally resulted in the emplacement of nappe-like units. The ensuing general submergence and basin rearrangements are mirrored by early Late Pliocene shallow marine successions which are onlapping over tilted, older Neogene beds, allochthonous units or the Paleozoic, crystalline basement. In the Tyrrhenian Basin the transition from the 'Lower' to the 'Upper' Pliocene is marked by a clearly detectable, regional unconformity (Selli and Fabbri, 1971; Barone et al., 1982).

Thrusting of the Ionian relative to the Preapulian zone (figure 3) reflects the effect of the intra-Pliocene compressional phase on the external Hellenides (Sorel, 1976; Dercourt et al., 1977; Mercier et al., 1979). The compressional phase was about time-equivalent to the beginning of a second phase of the clockwise rotation of the Ionian islands (20 - 25° during the Pliocene - Quaternary; Laj et al., 1982; Kissel et al., 1985). This rotation contributed to the extension of the basins between the Ionian islands and the Greek mainland.

The shift from a compressional to a tensional regime in the early part of the Late Pliocene interval initiated a new phase of basin development in Western Greece, as evidenced by the Upper Pliocene - Pleistocene of the Western

Peloponnisos (Hageman, 1977, 1979; Meulenkamp, 1985).

Differential vertical movements on Crete triggered large-scale gravity sliding of lowermost Pliocene and older Neogene sediments during the Early Pliocene ("marl-breccias" of Fortuin, 1977). Similar features were recognized on the island of Karpathos (Meulenkamp, 1985). These processes formed the prelude to the emergence of the Southern Aegean island arc by overall tilting to the north and to the increased subsidence rate of the Aegean Basin. On Crete the uplift was interrupted by a short, transgressive episode in the early Late Pliocene, which, however, only effected parts of the island. Precise dating shows that local hiatuses observed between Early or early Late and Late Pliocene successions correspond in time to those inferred from the Calabro - Sicilian record.

Late Pliocene - Early Pleistocene (± 2,5 - 1 Ma)

Late Pliocene - Early Pleistocene tectonics shaped the present-day contours of the Hellenic and Calabro-Sicilian arcs. To date, we have no reliable data on the precise ages of basin rearrangements and local inversions of reliefs which, for instance, occurred in the Western Peloponnisos and on Rhodos (Meulenkamp et al., 1972; Hageman, 1979). Data from the Ionian islands demonstrate repeated eastward tilting during the Early Pleistocene (Dermitzakis et al., 1979; Tsapralis, 1981). These movements, possibly related to the continuing clockwise rotation of the Ionian islands, caused a conspicuous deepening of the hitherto shallow seas between the island of Zakynthos and the Greek mainland. A short compressional episode affected the Hellenic arc during the Early Pleistocene (Sorel, 1976; Angelier, 1979).

The impact of Late Pliocene - Early Pleistocene movements in Calabria and Sicily was much greater. The staggering uplift of the axis of the arc corresponded to increased rates of subsidence of the forearc basins (Caltanissetta through, Sicily; Crotone - Spartivento Basin, Calabria). A new episode of large-scale gravity sliding, culminating in the Early Pleistocene, resulted in the emplacement of allochthonous units (Gela nappe, Ogniben, 1973), which can be traced all along the external side of the Calabro-Sicilian arc (Calabrian Ridge, see also Amodio-Morelli et al., 1976; Finetti, 1982). This process was related to compression, which caused, for instance, overthrusting of the successions of the Caltanissetta Basin on the Iblean platform (Patacca et al., 1979).

BASIN EVOLUTION, FAULT PATTERNS AND TECTONICS

The tectonic events discussed above probably reflect the effects of contemporaneous changes in regional stress fields on the evolution of the Hellenic and Calabro-Sicilian arcs. This conclusion is corraborated by literature data (Steininger et al., 1985), which indicate that the most important events inferred from the Late Cenozoic record of the arcs had time-equivalent counterparts elsewhere in the Mediterranean and the Paratethys.

Differences in the development of the Hellenic and Calabro-Sicilian arcs may be attributed to their different geodynamical setting. Some aspects concerning the sedimentary-tectonic histories of the arcs, in particular the orientation of and movements along synsedimentary-active faults will be discussed below.

The Hellenic Arc

The evolution of the Hellenic arc was controlled by a SW - NE compressional regime, related to the motions of the Aegean relative to the African plate (McKenzie, 1972; Le Pichon and Angelier, 1979, 1981; Angelier and Le Pichon, 1980; Angelier et al., 1982). In a forthcoming paper we will treat in detail the impact of this process on the orientation of and movements along major faults intersecting the arc. Here, we confine ourselves to a discussion of some general aspects.

Most, if not all major faults recognized were active throughout the Late Cenozoic. In the Western Hellenides movements along NNW - SSE faults delimiting or running parallel to the isopic zones (figure 3) defined basin configurations and sedimentation patterns. The Late Burdigalian, intra-Pliocene and Early Pleistocene reversed faulting or overthrusting took place along these faults. WSW - ENE and SW - NE faults, oriented roughly perpendicular to the general strike, defined the differentiation between subbasins along strike during the Late Paleogene and Early Miocene. Displacements along such faults continued in the younger Neogene; they contributed, for instance, to the origin and facies distribution of Late Pliocene basins in the Western Peloponnisos (Hageman, 1979).

SW- NE trends defined the evolution of Late Cenozoic basins in the Southeastern Hellenic arc. Although the record is very incomplete, we know that faults perpendicular to these trends had a pronounced impact. This is evidenced by the WNW - ESE fault zone across the island of Rhodos (Mutti et al., 1970; Meulenkamp et al., 1972). Fundamental differences in the sedimentary-tectonic histories of Rhodos and the island of Karpathos (Meulenkamp, 1985) suggest the presence of a major fault in between both islands, which intersects the general SW - NE trend.

Neogene basin analysis on Crete (located in the bending of the arc) permits the detailed reconstruction of synsedimentary-active faults (Drooger and Meulenkamp, 1973; Fortuin, 1977; Fortuin and Peters, 1984; Meulenkamp, 1979, 1982 b, 1985). These faults are depicted in figure 5. From the origin (or reactivation) of and displacements along these faults we can infer three major phases in the evolution of Crete and adjacent parts of the Southern Aegean

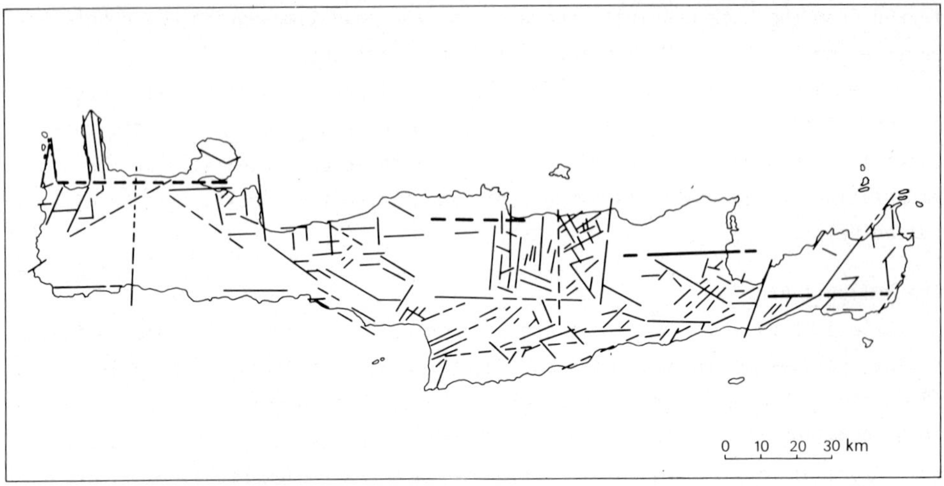

Fig. 5. Schematic representation of fault trends which defined Late Neogene basin configurations and sedimentation on Crete. Heavy lines indicate Northern margins of Middle Miocene basins.

(see also figure 6). The first phase (Late Burdigalian to latest Serravallian, i.e. from about 18 to 11 Ma ago) was characterized by general tilting to the south or southwest along faults bordering the Southern Aegean landmass.

There was no preferential direction of tilting during the second phase, which lasted from the Early Tortonian to about the Early - Late Pliocene boundary interval (± 3,5 Ma). This phase was marked by the fragmentation of Crete into a mosaic of horsts and grabens, delimited by the faults indicated in figure 5. The fragmentation was initiated by gravity-sliding in latest Serravallian time, which resulted in the emplacement of allochthonous slabs of Preneogene rocks (figure 6). The intra-Pliocene event initiated general tilting to the north or northeast. These data point to a large-scale inversion of reliefs; Crete was located along the hinge line. This inversion was probably related to the extension of the Southern Aegean Basin and the southward migration of Crete.

Fig. 6. Outlines of Middle Miocene to Early Pliocene basin configurations on Crete, based on data in Freudenthal, 1969; Meulenkamp, 1969; Gradstein, 1973; Fortuin, 1977; Meulenkamp et al., 1979; Meulenkamp, 1985, and unpublished reports. Vertical shading: emerged; horizontal shading: partly submerged or affected by onlap in the course of time-slices indicated. A: Late Middle Miocene (Serravallian). B: latest Middle Miocene (latest Serravallian). Triangles indicate distribution of allochthonous slabs. C: Tortonian. Areas with horizontal shading started to submerge in the Late Tortonian. D: Early Messinian. Note onlap of Lower Messinian successions over the margins of Tortonian subbasins. Horizontal shading: partly submerged platforms. E: Late Early Pliocene.

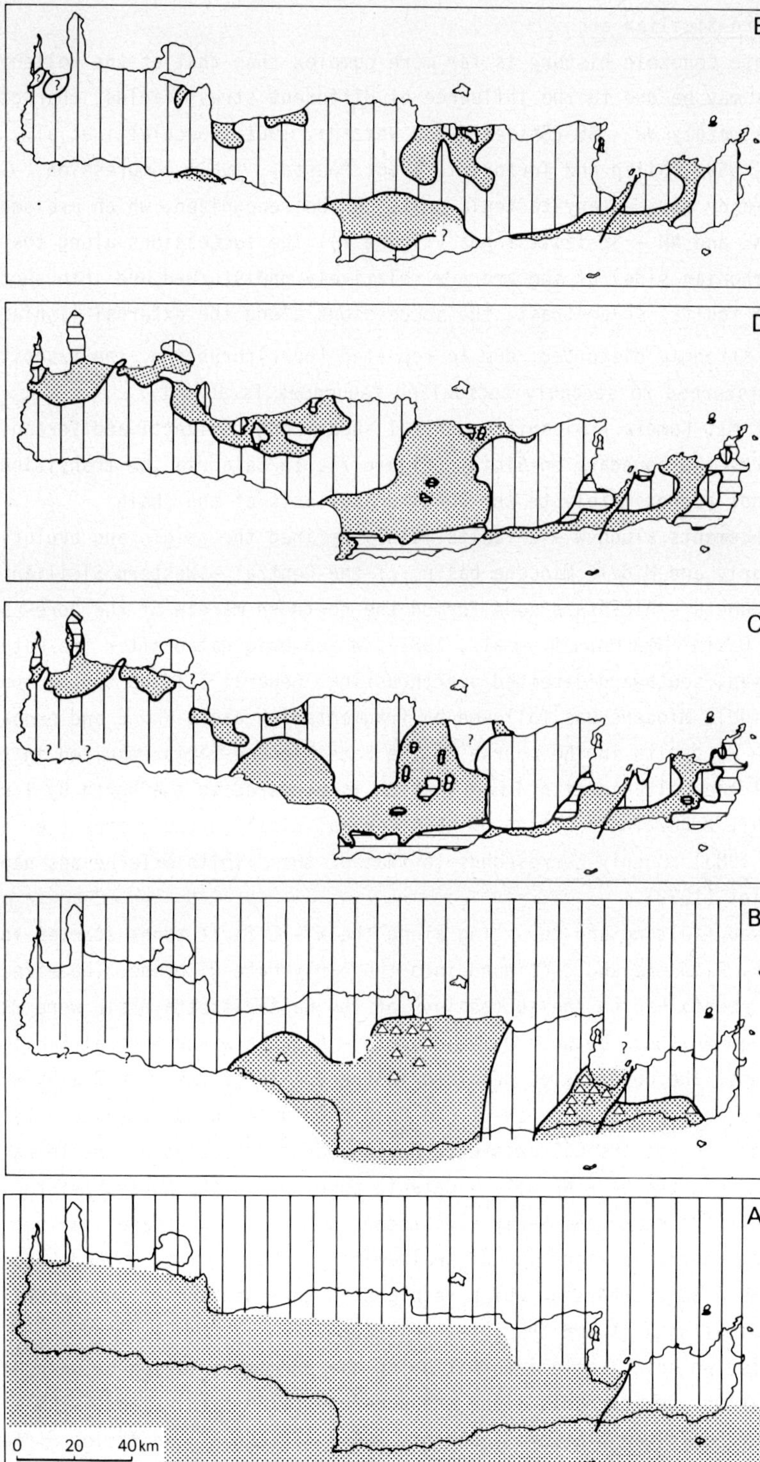
E
D
C
B
A
0 20 40 km

The Calabro-Sicilian arc

The Late Cenozoic history is far more complex than that of the Hellenic arc. This may be due to the influence of different stress fields, characterized by approximately NW - SE (Ghizetti and Vezzani, 1981; Boccaletti et al., 1982) and ENE - WSW (Philip and Tortorici, 1980; Moussat, 1983) compression.

Four major sedimentary-tectonic units can be recognized, which are separated by SE - NE and NW - SE fault zones (figure 7). The successions along the internal (Tyrrhenian side) of the arc are relatively undisturbed and intersected by normal faults. In contrast, the successions along the external (Ionian) side are strongly disturbed, due to repeated (over)thrusting. The transition from undisturbed to strongly tectonized sequences is abrupt.
The W - E Mt. Kumeta - Alcantara dextral shear-zone (Ghizetti and Vezzani, 1981) marks the boundary on Sicily (figure 7). In Calabria the transition corresponds approximately to the longitudinal axis of the chain.

Displacements along W - E faults mainly defined the origin and evolution of the Early and Middle Miocene basins of the Central - Western Sicilian unit. The Mt. Kumeta - Alcantara zone formed the northern margin of the "pre-Caltanissetta Basin" (Meulenkamp et al., 1981), which originated after the Late Burdigalian, southward-directed overthrusting. General tilting to the south in the Middle Miocene was followed by fragmentation along W - E and probably also SW - NE faults in the course of the Late Miocene. This resulted in the origin of the Caltanissetta Basin, which is bordered to the north by a complex, W - E fault zone. The position of this Central Sicilian fault zone (see also Moussat, 1983) roughly corresponds to that of the "limite externe des nappes" of Burollet (1979).

Reversed faulting and thrusting along the W - E fault zones started in the late Early Pliocene and continued into the early Late Pliocene. The effects these movements had on the successions of the Caltanissetta Basin were much larger than on those between the Mt. Kumeta - Alcantara and Central Sicilian fault zones. The compression resulted in the uplift of Northern Sicily and a corresponding rapid subsidence of the Gela - Catania foredeep (figure 7).
The successions of the Caltanissetta Basin were overthrusted on the Iblean platform along the SW - NE Gela - Catania zone during the Early Pleistocene.

The intra-Pliocene and Early Pleistocene compressional phases defined the pulsating, southward migration of shallowing sequences. They caused a considerable NW - SE shortening, which resulted in the suturing of the Central - Western Sicilian unit and the Iblean platform.

To date, we have little information on the precise relation between Neogene sedimentation and movements along the numerous (S)SW - (N)NE and (W)NW - (E)SE faults on the Iblean platform (Lentini, 1984). Thickness and facies differences

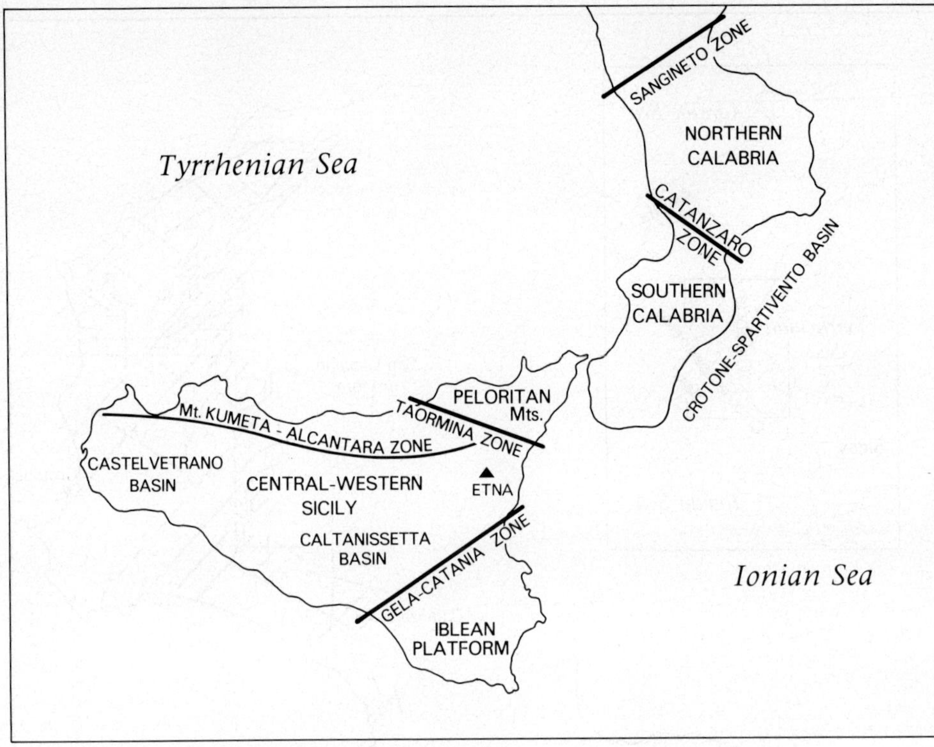

Fig. 7. Map showing the position of the four major sedimentary-tectonic units constituting the Calabro-Sicilian arc.

of the Middle Miocene - Lower Pliocene (Di Grande, 1985) seem to be related to synsedimentary, normal faulting. Much of the once comprehensive sediment cover was eroded during the Late Pliocene - Pleistocene uplift. Remnants of Middle to Late Miocene, relatively deep-water sediments (Theodorides, 1984) were preserved in downfaulted blocks.

The faults defining basin configurations and sedimentation along the external, Ionian side of the Northern and Southern Calabrian units are schematically represented in figure 8. Most of these faults were active since at least the Early - Late Oligocene boundary interval, i.e. since about 30 Ma ago. Meulenkamp et al. (in press) argued that these faults are related to a conjugate system of SW - NE (N 50° - 60°E) and NW - SE (N 120° - 140°E) basement wrench faults. The most important fault zones (shear-zones) are indicated in figure 9. The NW - SE zones form the transition between segments with different types of sediment succession, superimposed upon the large-scale differences between the Northern and Southern Calabrian megasegments (Meulenkamp et al., in press).

Relative horizontal displacements along the Northern Calabrian NW - SE

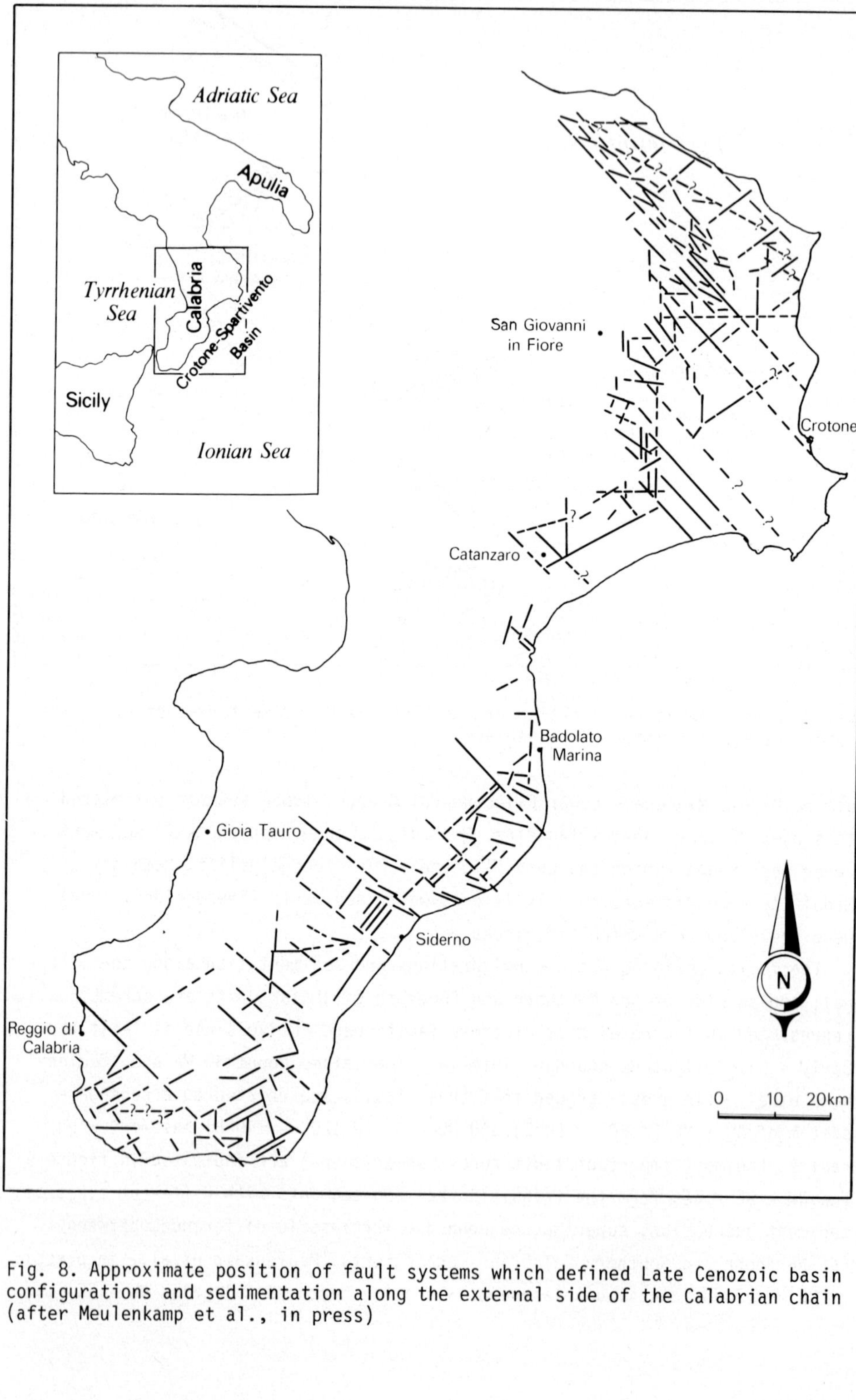

Fig. 8. Approximate position of fault systems which defined Late Cenozoic basin configurations and sedimentation along the external side of the Calabrian chain (after Meulenkamp et al., in press)

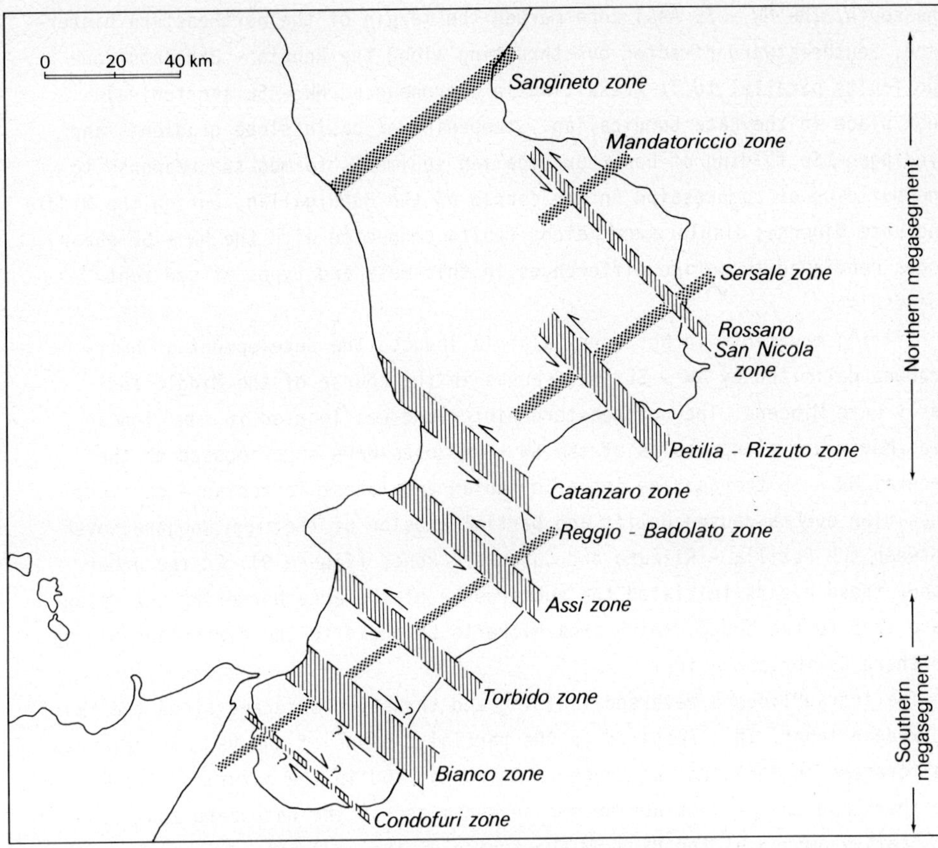

Fig. 9. Position of the most important wrench zones intersecting the Calabrian chain. These zones represent the transition between segments with different sedimentary-tectonic histories, superimposed on the differentiation of the chain into the Northern and Southern Calabrian megasegments along the Catanzaro zone (modified after Meulenkamp et al., in press).

shear-zones and the Assi and Torbido zones of Southern Calabria probably were left-lateral (cf. Moussat, 1983). Those along the Bianco and Condofuri zones may have been right-lateral, which concurs with the relative direction of motions inferred from the WNW - ESE and NW - SE faults bordering and intersecting the Peloritan block, Northeastern Sicily (Ghisetti and Vezzani, 1981; Boccaletti and Dainelli, 1982). Displacements along the SW - NE zones probably were right-lateral.

The SW - NE Reggio - Badolato zone (figure 9) corresponds to the transition from undisturbed Tyrrhenian to strongly tectonized Ionian successions in Southern Calabria. The zone formed the western margin of the Oligocene - Early Miocene basins, which were arranged in a set of halfgrabens, separated by NW - SE and WNW - ESE faults. The net result of tilting movements was to the south. The NW - SE Assi zone marked the margin of the northeastern hinterland. Southeastward-directed overthrusting along the Reggio - Badolato zone and faults parallel to it (resulting in a pronounced NW - SE shortening) took place in the Late Burdigalian. Steepening of basin slope gradients and syndiagenetic folding of Lower Burdigalian sediments formed the response to the build-up of compression in the course of the Burdigalian. During the Middle and Late Miocene, displacements along faults connected with the NW - SE shear-zones continued to define differences in thickness and types of sediment succession.

Tilting movements in Northern Calabria induced the development of half-grabens delimited by NW - SE shear-zones in the course of the Middle and early Late Miocene. The northeastern hinterland was located in the Ionian Sea. Movements along faults of the SW - NE zones were superimposed on the general NW - SE trends. The intra-Tortonian and latest Tortonian - earliest Messinian events caused uplift and partial erosion of the older Neogene cover between the Petilia - Rizzuto and Catanzaro zones (figure 9). On the other hand, these events initiated the submergence of the area bordering the Catanzaro zone to the south, which area hitherto belonged to the hinterland of the Southern Calabrian basins.

The intra-Pliocene reversed faulting and thrusting occurred along the NW - SE wrench zones. This resulted in the partial closure of the NW - SE oriented halfgrabens in Northern and Southern Calabria, and in the suturing of the Northern and Southern Calabrian megasegments across the Catanzaro zone. The cumulative amount of the NE - SW shortening of the sediment cover may have been as much as 100 km.

We postulate that the shortening began already during the Messinian. Relative uplift of the hinterland and steepening of basin-slope gradients by tilting to the southwest caused large-scale gravity sliding and the emplacement of allochthonous units in Northern Calabria (Cariatides of Roda, 1967; Crotonide Complex of Ogniben, 1973) in the latest Messinian - early Late Pliocene time-span. These processes formed the sedimentary expression of the build-up of compression, which culminated in the early Late Pliocene. The Southern Calabrian counterpart is reflected by the onlap of Upper Messinian conglomerates over the Assi and Reggio - Badolato zones. These clastics were probably deposited in synclines originating from low-amplitude basement folding (van Wamel, pers. comm., 1984).

The Early Pleistocene reversed faulting and overthrusting were mainly directed towards the southeast, which corresponds to the direction of Late Burdigalian thrusting. Early Pleistocene compressional tectonics seem to have had a larger impact in Southern than in Northern Calabria; the opposite was true for the intra-Pliocene compressional phase.

We conclude that the Calabro-Sicilian arc originated from horizontal and vertical displacements along W - E, SW - NE and NW - SE shear zones. The four major sedimentary-tectonic units were sutured in response to the intra-Pliocene and Early Pleistocene compressional events. The approach of the Northern relative to the Southern megasegments of the Calabrian chain was reflected by the break-up of their respective northeastern hinterlands in the course of the Late Miocene - Early Pliocene. The Southern Calabrian hinterland has to be looked for northeast of the Assi zone. Part of it probably became incorporated in the rapidly subsiding, southern margin of the Catanzaro depression. The hinterland of the Northern Calabrian unit was located in the Ionian Sea and subsided in the course of the Pliocene.

DISCUSSION AND CONCLUSIONS

The data and interpretations summarized in the preceding chapters reveal similarities as well as fundamental differences in the origin and evolution of the Hellenic and Calabro-Sicilian arcs. The development of the Hellenic arc was controlled by a SW - NE compressional regime, generated by the motions of the Aegean area relative to the African plate (McKenzie, 1972; Le Pichon and Angelier, 1979, 1981; Angelier and Le Pichon, 1980; Angelier et al., 1982). The extension of the Aegean ultimately shaped the Southern Aegean island arc through the southward migration of Crete. This process probably started in the Serravallian - Tortonian boundary interval, when the Southern Aegean landmass began to subside and (remnants of) carbonate platforms west of the Ionian islands collapsed.

Extension of the Tyrrhenian Basin had a large impact on the development of the Calabro-Sicilian arc, but the features observed by us cannot merely be explained by this process. In fact, the Late Cenozoic record mirrors the influence of two compressional regimes, characterized by approximately NW - SE (Ghisetti and Vezzani, 1981; Boccaletti et al., 1982) and ENE - WSW (Philip and Tortorici, 1980; Moussat, 1983) directed compressional forces. We postulate that the interference of the two regimes caused the changes in the direction of thrusting in Calabria during the Late Cenozoic. Shortening along strike of the chain played a pronounced part and resulted in the suturing of different sedimentary-tectonic units. Unfortunately, there are no high-resolution seismic profiles along NE - SW transects adjacent to Calabria to test our hypothesis on the large impact the NE - SW shortening had on the origin of the arc. Such profiles have been published only for transects perpendicular to the arc (Finetti, 1982).

The interference of the two compressional regimes, evidenced for the Pliocene (see also data in Philip and Tortorici, 1980; Ghisetti and Vezzani, 1981;

Moussat, 1983) may as well account for the shift in the direction of relative horizontal displacements along faults connected with the NW - SE shear-zones of Southern Calabria, from left-lateral to right-lateral. This shift has also been reported by Boccaletti and Dainelli (1982) and Boccaletti et al., (1982). We postulate that the bending of the Calabro-Sicilian arc is related to this shift in the direction of relative horizontal motion in Southern Calabria. It is remarkable that the bisector of the Calabrian Ridge (Ionian Sea) intersects the Calabrian chain approximately via the Bianco zone of Southern Calabria. In our opinion, the strikingly consistent directions of the Calabrian shear-zones preclude major block rotations as a mechanism for the bending of the arc, as suggested by Ghisetti and Vezzani (1981).

We ascribe the similarities in the sedimentary-tectonic histories of the Hellenic and Calabro-Sicilian arcs to the effects of contemporaneous changes in regional stress fields. The similarities concerning the episodic recurrence of regional onlaps and offlaps in the two arcs can be explained by the tectonic mechanism for regional sealevel fluctuations proposed by Cloetingh et al., (1985) and Cloetingh (1986), which is based on the impact of changes in intra-plate stresses on basin evolution.

In this model relaxation of compressional forces or a shift from compression to tension would result in a relative sealevel rise. Sealevel falls are connected with a change from tension to compression. The magnitude of the induced sealevel fluctuations is generally some tens of metres, but can be higher if changes in intra-plate stress fields are related to major plate reorganizations or to plate fragmentation (see for example Wortel and Cloetingh, 1981). The model was established for passive margins, but can as well be applied to active margins, such as the Hellenic and Calabro-Sicilian arcs (Cloetingh, pers. comm., 1986). It also accounts for the correspondence in time between the majority of tectonically-defined onlaps and offlaps we reconstructed and events of Vail et al.'s (1977) chart of global, eustatic sealevel changes (Cloetingh et al., in prep.).

Regressive trends, local uplift and erosion are found to be connected with the Late Burdigalian compression. The ensuing general submergence mirrors the abrupt shift from compression to tension in latest Burdigalian - Early Langhian time. Shallowing and offlap in the course of the Pliocene were the response to the change from tension to compression. The return to tensional regimes caused the onlap of Late Pliocene sequences over older Neogene deposits or the basement.

The Late Burdigalian - Early Langhian and intra-Pliocene changes in stress fields were related to major plate reorganizations resulting in fundamental basin rearrangements. This explains the regional, Mediterranean-wide impact of the latest Burdigalian - Early Langhian and Late Pliocene transgressions. Fragmentation in the Serravallian - Tortonian boundary interval, expressed by the break-up of the Southern Aegean landmass, the foundering of the carbonate platform west of the Ionian islands, and the fragmentation of the Calabro-Sicilian basins, accounts for the regional transgression in earliest Tortonian time. Current investigations concentrate on a more precise correlation of the tectonic events with regional changes in intraplate stresses.

ACKNOWLEDGEMENTS

We thank S. Cloetingh, J.P. van Dijk, P. van Wamel, R. Wortel, W.J. Zachariasse and G.J. van der Zwaan for stimulating discussions and for valuable suggestions to improve the manuscript. The illustrations were skilfully made by T. van Hinte and W. den Hartog; A. Pouw provided indispensable assistance in the preparation of the manuscript.

REFERENCES

Adams, C.G., Gentry, A.W. and Whybrow, P.J., 1983. Dating the terminal Tethyan event. In: J.E. Meulenkamp (Editor), Reconstruction of marine paleoenvironments. Utrecht Micropal. Bull., 30: 273-298.

Amodio-Morelli, L., Bonardi, G., Colonna, V., Dietrich, D., Giunta, G., Liguori, V., Lorenzoni, S., Paglionico, A., Perrone, V., Piccarreta, G., Russo, M., Scandone, P., Zanettin-Lorenzoni, E. and Zuppetta, A., 1976. L'Arco Calabro-Peloritano nell'orogene Appenninico-Maghrebide. Mem. Soc. Geol. It., 17: 1-60.

Angelier, J., 1975. Sur l'analyse des phases superposées de la tectonique cassante: la néotectonique et les failles du massif de l'Ida (Crête, Grèce). Ann. Soc. Géol. Nord, 95: 183-200.

Angelier, J., 1979. Néotectonique de l'arc égéen. Soc. Géol. Nord, 3: 418 pp.

Angelier, J. and Le Pichon, X., 1980. La subduction hellénique et l'expansion égéenne: reconstruction cinématique et interpretation dynamique. C.R. Somm. Soc. Geol. France, 5: 158-161.

Angelier, J., Lyberis, N., Le Pichon, X., Barrier, E. and Huchon, P., 1982. The tectonic development of the Hellenic arc and the Sea of Crete: a synthesis. In: X. Le Pichon, S.S. Augustithis and J. Mascle (Editors), Geodynamics of the Hellenic Arc and Trench. Tectonophysics, 86: 159-196.

Barone, A., Fabbri, A., Rossi, S. and Sartori, R., 1982. Geological structure and evolution of the marine areas adjacent to the Calabrian Arc. Earth. Evol. Sci., 3: 207-221.

Bennett, S.M., 1979. A transgressive carbonate sequence spanning the Paleogene - Neogene boundary on the Maltese islands. Ann. Géol. Pays Hellén., 1: 71-86.

Bizon, G. and Bizon, J.J., 1985. Area no 33b-d: Ionian zone, Area 37a: Corfu, Area 37b: Lefkas, Area 37c: Kefallinia, Area 37d: Zakinthos. In: F.F. Steininger et al., (Editors), Neogene of the Mediterranean Tethys and Paratethys. Inst. Paleont. Vienna, I: 29-31; II: 94-100.

Boccaletti, M. and Dainelli, P., 1982. Il sistema regmatico Neogenico - Quaternario nell'area Mediterranea: Esempio di deformazione plastico - rigida post-collisionale. Mem. Soc. Geol. It., 24: 465-482.

Boccaletti, M., Conedera, C., Dainelli, P. and Gocev, P., 1982. The recent (Miocene - Quaternary) regmatic system of the Western Mediterranean region. J. Petrol. Geol.., 5: 31-49.

Boccaletti, M., Nicolich, R. and Tortorici, L., 1984. The Calabrian Arc and

the Ionian Sea in the geodynamic evolution of the Central Mediterranean. Mar. Geol., 55: 219-245.
Burollet, P.F., 1979. A propos de la mer Tyrrhénienne. Rapp. Proc. Verb. Réun. Comm. Intern. Explor. Scient. Mer Méd., 25/26: 63-64.
Chamley, H., Meulenkamp, J.E., Zachariasse, W.J. and Van der Zwaan, G.J. (in press). Middle to Late Miocene marine ecostratigraphy: clay minerals, planktonic foraminifera and stable isotopes from the Caltanissetta Basin, Sicily.
Cloetingh, S., 1986. Intraplate stresses: A new tectonic mechanism for fluctuations of relative sea level. Geology, 14 (in press).
Cloetingh, S., McQueen, H. and Lambeck, K., 1985. On a tectonic mechanism for relative sealevel variations. Earth. Planet. Sci. Lett., 75: 157-166.
Cloetingh, S., Meulenkamp, J.E. and Wortel, R. (in prep.). On Neogene tectonics and relative sea-level fluctuations in the Mediterranean.
Decima, A. and Wezel, F.C., 1972. Osservazioni sulle evaporiti messiniane della Sicilia centro-meridionale. Riv. Min. Sic., 22: 172-187.
De Mulder, E.F.J., 1975. Microfauna and sedimentary-tectonic history of the Oligo-Miocene of the Ionian Islands and western Epirus (Greece). Utrecht Micropal. Bull., 13: 140 pp.
Dercourt, J., Aubouin, J., Savoyat, E. et al., 1977. Réunion extraordinaire de la Société géologique de France en Grèce. Bull. Soc. Géol. France, ser. 7, 19: 5-70.
Dermitzakis, M.D., 1978. Stratigraphy and sedimentary history of the Miocene of Zakinthos (Ionian islands, Greece). Ann. Géol. Pays Hellén., 29: 47-186.
Dermitzakis, M.D., Papanikolaou, D. and Karotsieris, Z., 1979. The marine Quaternary formations of SE Zakinthos and their paleogeographic implications. Proc. 6th Coll. Geol. Aeg. Reg., Athene, 1977, I: 407-415.
Di Grande, I., 1985. Area no 25, 26, 27: Sicily. In: F.F. Steininger et al. (Editors), Neogene of the Mediterranean Tethys and Paratethys. Inst. Paleontol., Vienna, I: 28, 29; II: 81-82.
Di Nocera, S., Ortolani, F. and Torre, M., 1975. Fase tettonica messiniana nell' Appennino meridionale. Boll. Soc. Natur. Napoli, 84: 499-515.
Drooger, C.W. and Meulenkamp, J.E., 1973. Stratigraphic contributions to geodynamics in the Mediterranean area: Crete as a case history. Bull. Geol. Soc. Greece, 10: 193-200.
Drooger, C.W. and Roelofsen, J.W., 1982. Cycloclypeus from Ghar Hassan, Malta. Proc. Kon. Ned. Ak. Wetensch., ser. B, 85: 203-218.
Felix, R., 1973. Oligo-Miocene stratigraphy of Malta and Gozo. Meded. Landbouwhogeschool Wageningen, 73-20: 104 pp.
Finetti, I., 1982. Structure, stratigraphy and evolution of Central Mediterranean. Boll. Geof. Teor. Applic., 24: 247-312.
Fortuin, A.R., 1977. Stratigraphy and sedimentary history of the Neogene deposits in the Ierapetra region, eastern Crete. GUA Pap. Geol., ser. 1, 8: 164 pp.
Fortuin, A.R. and Peters, J.M., 1984. The Prina Complex in eastern Crete and its relationship to possible Miocene strikeslip tectonics. J. Struct. Geol., 6: 459-476.
Freudenthal, T., 1969. Stratigraphy of Neogene deposits in the Khania province, Crete, with special reference to foraminifera of the family Planorbulinidae and the genus Heterostegina. Utrecht Micropal. Bull., 1: 208 pp.
Ghisetti, F. and Vezzani, L., 1981. Contribution of structural analysis to understanding the geodynamic evolution of the Calabrian Arc (Southern Italy). J. Struct. Geol., 3: 371-381.
Ghisetti, F. and Vezzani, L., 1982. Different styles of deformation in the Calabrian Arc (Southern Italy): implications for a seismotectonic zoning. Tectonophysics, 85: 149-165.
Gradstein, F.M., 1973. The Neogene and Quaternary deposits in the Sitia district of eastern Crete. Ann. Géol. Pays Hellén., 24: 527-572.
Hageman, J., 1977. Stratigraphy and sedimentary history of the Upper Cenozoic of the Pyrgos-area (western Peloponnesus), Greece. Ann. Géol. Pays Hellén., 28: 299-333.

Hageman, J., 1979. Benthic foraminiferal assemblages from Plio-Pleistocene open bay to lagoonal sediments of the western Peloponnesus (Greece). Utrecht Micropal. Bull., 20: 171 pp.

Hsü., K.J., Cita, M.B. and Ryan, W.B.F., 1973. The origin of the Mediterranean evaporites. In: W.B.F. Ryan, K.J. Hsü et al. (Editors), Init. Repts. DSDP, 13: 1203-1231.

I.F.P., 1966. Etude géologique de l'Epire (Grèce nord-occidental). Publ. Inst. Français du Pétrole, Ed. Techniq. Paris, 306 pp.

Kissel, C., Laj, C. and Muller, C., 1985. Tertiary geodynamical evolution of northwestern Greece; paleomagnetic results. Earth Planet. Sci. Lett., 72: 190-204.

Laj, C., Jamet, M., Sorel, D. and Valente, J.P., 1982. First paleomagnetic results from the Mio-Pliocene series of the Hellenic sedimentary arc. In: X. Le Pichon, S.S. Augustithis and J. Mascle (Editors), Geodynamics of the Hellenic Arc and Trench. Tectonophysics, 86: 45-67.

Langereis, C.G., Zachariasse, W.J. and Zijderveld, J.D.A., 1984. Late Miocene magnetostratigraphy of Crete. Mar. Micropal., 8: 261-281.

Lentini, F., 1984. Carta Geologica della Sicilia sud-orientale, 1: 100.000. Univ. di Catania.

Le Pichon, X. and Angelier, J., 1979. The Hellenic Arc and Trench system: a key to the neotectonic evolution of the eastern Mediterranean area. Tectonophysics, 60: 1-42.

Le Pichon, X. and Angelier, J., 1981. The Aegean Sea. Philos. Trans. R. Soc. London, ser. A, 300: 357-372.

McKenzie, D., 1972. Active tectonics of the Mediterranean region. Geoph. J.R. Astron. Soc., 30: 109-185.

Mercier, J.L., Delibassis, M., Gauthier, A., Jarrige, J.J., Lemeille, F., Philip, H., Sebrier, M. and Sorel, D., 1979. La néotectonique de l'Arc Egéen. Rev. Géol. Dyn. Géogr. Phys., 21: 61-72.

Meulenkamp, J.E., 1969. Stratigraphy of Neogene deposits in the Rethymnon Province, Crete, with special reference to the phylogeny of uniserial Uvigerina from the Mediterranean region. Utrecht Micropal. Bull., 2: 172 pp.

Meulenkamp, J.E., 1979. Field guide to the Neogene of Crete, Publ. Dept. Geol. Pal. Univ. Athens, ser. A; 32: 1-32.

Meulenkamp, J.E., 1982a. On the pulsating evolution of the Mediterranean. Episodes, 5: 13-16.

Meulenkamp, J.E., 1982b. The Hellenic and Calabro-Sicilian Arcs - a comparative study. Proc. Intern. Symp. Hell. Arc Trench, Athens, 1981, 2: 40-56.

Meulenkamp, J.E., 1985. Aspects of the Late Cenozoic evolution of the Aegean region. In: D.J. Stanley and F.C. Wezel (Editors), Geological evolution of the Mediterranean Basin, Springer-Verlag, New York, 307-321.

Meulenkamp, J.E., de Mulder, E.F.J. and van de Weerd, A., 1972. Sedimentary history and paleogeography of the Late Cenozoic of the island of Rhodos. Z. Dt. Geol. Ges., 123: 541-553.

Meulenkamp, J.E., Hilgen F.J. and Voogt, E. (in press). Late Cenozoic sedimentary-tectonic history of the Calabrian Arc. In: M. Boccaletti and F. Ricchi Lucchi (Editors), Paleogeography and geodynamics of the peri-Tyrrhenian area, Boll. Soc. Geol. It., Bologna.

Meulenkamp, J.E., Jonkers, H.A., Balder, H., et al., 1981. Middle Miocene-Pleistocene sedimentary-tectonic history of Sicily. Rapp. Proc. Verb. Réun. Comm. Intern. Explor. Scient. Mer Méd., 27: 149-150.

Meulenkamp, J.E., Jonkers, H.A. and Spaak, P., 1979. Late Miocene to Early Pliocene development of Crete. Proc. 6th Coll. Geol. Aeg. Reg. Athens, 1977, 1: 137-149.

Moussat, E., 1983. Evolution de la Mer Tyrrhénienne centrale et orientale et de ses marges septentrionales en rélation avec la néotectonique dans l'Arc Calabrais. Thèse 3e Cycle, Univ. P. et M. Curie, Paris, 122 pp.

Mutti, E., Orombelli, G. and Pozzi, R., 1970. Geological studies on the Dodecanese islands (Aegean Sea). IX. Geological map of the island of Rhodos (Greece). Explanatory notes. Ann. Géol. Pays Hellén., 22: 77-226.

Ogniben, L., 1973. Schema geologico della Calabria in base ai dato odierni. Geol. Romana, 12: 243-587.
Ortolani, F., Torre, M., and Di Nocera, S., 1979. I depositi altomiocenici del Bacino di Amantea (catana Costiera Calabria). Boll. Soc. Geol. It., 98: 559-587.
Patacca, E., Scandone, P., Giunta, G. and Liquori, V., 1979. Mesozoic paleotectonic evolution of the Ragusa zone (Southeastern Sicily). Geol. Romana, 18: 331-369.
Philip, H. and Tortorici, L., 1980. Tectonique superposée dans les sédiments Miocène supérieur à Pleistocène de la Calabre centrale et septentrionale (Italie méridionale). C.R. somm. Soc. géol. Fr., 5: 191-194.
Richter, D., Mariolakos, I. and Risch, H., 1978. The main flysch stages of the Hellenides. In: H. Closs et al.(Editors), Alps, Apennines, Hellenides. Schweizerb. Verlagsbuchh., Stuttgart, 434-438.
Roda, C., 1967. Distribuzione e facies dei sedimenti neogenici nel Bacino Crotonese. Geol. Romana, 3: 319-366.
Roda, C., 1967. I sedimenti neogenici et allochtoni della zona di Ciro-Cariati (Catanzora et Cosenza). Mem. Soc. Geol. It., 6: 137-149.
Rögl, F. and Steininger, F.F., 1983. Vom Zerfall der Tethys zu Mediterran und Paratethys. Die neogene Paläogeographie und Palinspastik des zirkum-mediterranen Raumes. Ann. Naturh. Mus. Wien, 85/A: 135-163.
Rose, E.P.F., 1985. Area no. 136: Maltese islands. In: F.F. Steininger et al. (Editors), Neogene of the Mediterranean Tethys and Paratethys. Inst. Paleontol., Vienna, I: 50, 51; II: 315.
Scandone, P., 1979. Origin of the Tyrrhenian Sea and Calabrian Arc. Boll. Soc. Geol. It., 98: 27-34.
Scandone, P., 1982. Structure and evolution of the Calabrian Arc. Earth Evol. Sci., 3: 172-179.
Selli, R. and Fabbri, A., 1971. Tyrrhenian: a Pliocene deep sea. Acc. Naz. Lincei. Rend. Cl. Sci. Mat. Fis. Nat., ser. 8, 50: 104-116.
Sorel, D., 1976. Etude néotectonique dans l'Arc Egéen externe occidental. Les îles Ioniennes de Kephallinia et Zakinthos et l'Elide occidentale. Thèse, Univ. de Paris, 196 pp.
Steininger, F.F., Senes, J., Kleemann, K. and Rögl, F. (Editors), 1985. Neogene of the Mediterranean Tethys and Paratethys. Stratigraphic correlation tables and sediment distribution maps. Inst. Paleontol., Vienna, I: 189 pp.; II: 536 pp.
Theodorides, S., 1984. Calcareous nannofossil biozonation of the Miocene and revision of the Helicoliths and Discoasters. Utrecht Micropal. Bull., 32: 271 pp.
Tsapralis, V., 1981. Contribution to the study of Pleistocene of Zakynthos island, W. Greece. Thesis, Univ. Patras, 134 pp.
Van der Zwaan, G.J. and Gudjonsson, L., 1986. Middle Miocene - Pliocene stable isotope stratigraphy and paleoceanography of the Mediterranean. Mar. Micropal., 10: 71-90.
Vail, P.R., Mitchum, R.M. and Thompson, III, S, 1977. Global cycles of relative changes of sealevel. Am. Assoc. Petrol. Geol., Mem., 26: 83-97.
Wezel, F.C., 1974. Flysch successions and the tectonic evolution of Sicily during the Oligocene and Early Miocene. In: C.H. Squyres (Editor), Geology of Italy, Earth Sci. Soc. Libyan Arab Rep., Tripoli, 2: 105-127.
Wezel, F.C., 1975. Diachronism of depositional and diastrophic events. Nature, 253: 255-257.
Wezel, F.C., 1982. The structure of the Calabro-Sicilian Arc: result of a post-orogenetic intra-plate deformation. In: J.K. Legget (Editor), Trench-forearc geology: sedimentation and tectonics on modern and ancient active plate margins. Geol. Soc. London, Spec. Publ., 10: 345-354.
Wortel, R. and Cloetingh, S., 1981. On the origin of the Cocos-Nazca spreading centre. Geology, 9: 425-430.
Zijderveld, J.D.A., Zachariasse, W.J., Verhallen, P.J.J.M. and Hilgen, F.J. (in press). The age of the Miocene - Pliocene boundary. Newsl. Stratigr.

TYRRHENIAN VOLCANIC ARCS: VOLCANO-TECTONICS, PETROGENESIS AND ECONOMIC ASPECTS

ENZO LOCARDI

ENEA, C.R.E. Casaccia, 00060 Roma (Italy)

ABSTRACT

The tectonics of the Mediterranean basin is governed by a series of mantle uplifts. On top of them collapse and oceanization occur coupled with rotational movements of crustal blocks. As a consequence of rotations, sphenocasms like the Algero-Provençal and the Tyrrhenian basins form. This is the tectonic environment hosting the highly potassic, "Mediterranean" magmatism.

During Pleistocene times two volcanic arcs formed in the Tyrrhenian area and in the Italian peninsula. One volcanic arc, developing about 450 km long on the Apenninic coastal range, delivered potassic magmatism. The other arc, on the southeastern slope of the Tyrrhenian basin, delivered calcalkaline volcanism followed by progressively more potassic magmatism of "Mediterranean" character. The calkalcaline volcanism is very short-lived and locally erupts from the same volcanic apparata that formerly were fed by tholeiitic, intra-plate, volcanism. Tectonics by subcrustal transfer may account for this peculiar volcanic pattern.

Mantle updoming has been enhanced by the metasomatic activity of mantle-derived fluids which are very rich in incompatible elements. Fluid supply does not only occur in the upper mantle, but in the crust also. These separate contributions have been identified and evaluated in the mantle partial melts and in melts stored in shallow magmatic reservoirs. The components of the mantle-derived fluids are the same as the ones forming a huge mineralogenetic province on top of a mantle dome in the Tuscan-Latial region.

INTRODUCTION

The main geodynamic feature during Neogene and Quaternary times in the Tyrrhenian area is the upwelling of an asthenolithic mantle-dome. This mantle upwelling is part of a more extended mantle uplift area that gave rise to the Mediterranean basin on top of it (Makris, 1985; Locardi, 1985; Wezel, 1985). Fig. 1 shows the

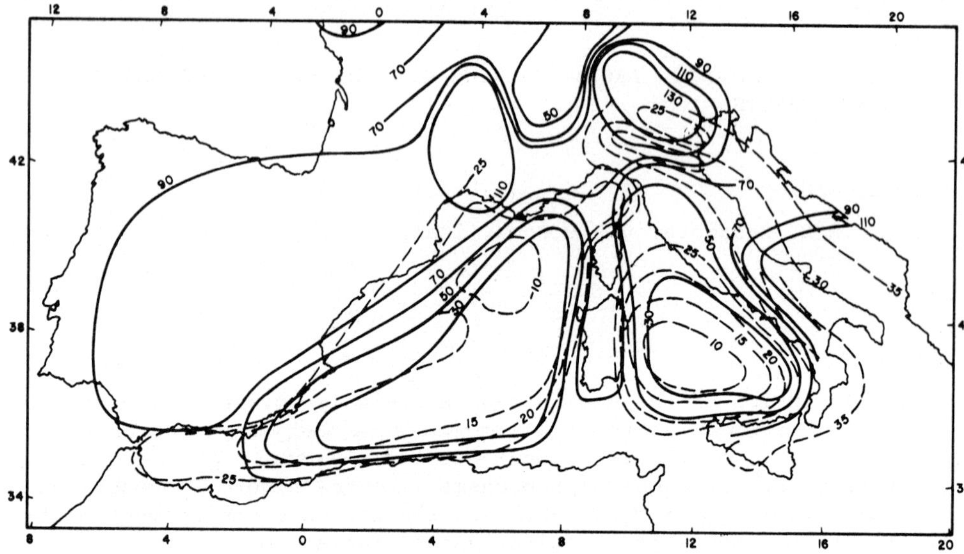

Fig. 1. Lithospheric (solid line) and crustal (dashed line) thickness in Western-Central Mediterranean area (modified from Panza et al., 1980).

distribution of those single mantle uplifts divided by crustal units showing "normal" lithospheric character.

The mantle uplifts show decreasing ages from the western Mediterranean basin to the east, from lower Miocene in the Alboran and Algero-Provençal basins, to upper Miocene in the Tyrrhenian area, to the Present in the Aegean basin. In correspondence of the single mantle domes, during their ascent, crustal blocks like the "Sardo-Corso" and the Italian peninsula rotated counterclockwise giving rise to sphenocasms like the Algero-Provençal basin and the Tyrrhenian one. Even the tectonic movements in the Hellenic areas are more connected with mantle uplifts than with subduction processes (Makris, 1985).

From the geophysical viewpoint the mantle uplifts show an

asthenospheric character, but particularly "soft" and "hot". In the Tyrrhenian area, at present, mantle-crust transitions accur at values of 7.5-7.8 km/s and the geothermal gradient is exceptionally high.

To this geophysical anomaly a geochemical anomaly corresponds in the composition of the melts extracted from the mantle. The magmatic products are so peculiar as to be grouped in an independent suite, the "Mediterranean" one, being completely different from the more represented "Atlantic" and "Pacific" suites.

Several hundreds of petrological papers have been devoted to this potassic, "Mediterranean" magmatism, and several tens of different genetic hypotheses have been suggested. It is noteworthy that very few papers deal with the structural geology and the tectonic evolution related to this kind of mantle uplifting and related magmatism.

Presently, geochemical and isotopic data support mantle metasomatism at the origin of mantle upwelling and magma generation.

Aqueous supercritical fluids, very CO2-, F-, Cl-, S-rich, bringing in solution incompatible elements, rise from deep mantle regions and react with the upper mantle. The original mineralogic constituents of the mantle, olivine and orto-pyroxene, become thus metasomatically transformed in hydrated minerals (phlogopite, k-richterite). These reactions bring to a volume increase that explains the mantle upwelling, and the supply of hot fluids from the depths enhances partial melting in the metasomatized mantle. This accounts for the peculiar potassic "Mediterranean" magmatism, with its characteristic high incompatible elements content and isotopic disequilibria (Capaldi et al., 1982; Cortini, 1981; Cortini and Don Hermes, 1981).

A similar mantle-fluid supply has been proposed by the British school after a series of studies on the genesis of alkaline magmas in cratonic "rift" areas (Bailey, 1970, 1972, 1982). The fluids represent the volatile fraction of the mantle (C-O-H compounds) that can exolve and migrate upwards driven by a "lesion" in the lithosphere. The chemical character of the supercritical watery fluids is such as to leach easily from the soaked rocks the incompatible elements. Therefore the magmas, derived from partial melting of this metasomatized mantle, are so rich in alkalies and other incompatibles.

But the Mediterranean basin was not formed on a craton. It developed along the axis of an orogenetic belt during its tectonic evolution, and the oceanization processes are there coupled with compressional pulses. These are quite peculiar situations if we consider the general mechanism controlling the basin formation.

Another possibility for driving the mantle metasomatism is the activity of fluids liberated by dehydration of subducted oceanic crust and sediments.

Most of the geochemical papers support the mechanism of fluid supply from subducted crust, even if "rift" geochemical characters are present and the entity of enrichment of some incompatible elements is absolutely unique.

Some connection with orogenic situations should exist if the only analogue of the Mediterranean magmatism has been recognized in a leucitic volcano of the Sunda arc in Indonesia (Foden and Varne, 1980). But for the same reason this connection should not be generalized. The thousands of orogenic volcanoes with shoshonitic trend do not show the characters of the Mediterranean volcanism.

The last results from a geochemical and isotopic work on alkaline potassic volcanism in Italy (Rogers et al., 1985) show that, to a "normal" MOR-type mantle, two single and distinct fluids have been added, each one with its particular content of incompatible elements and particular isotopic character. From the mantle regions affected by these additions, the magmatic series of the well known Low K and High K chemistry have been extracted.

Actually, a minimum of four independent magmatic suites have been identified in the frame of the Mediterranean magmatism (Locardi, 1981) pointing out very nonhomogeneous conditions of mantle metasomatism.

VOLCANO-TECTONIC EVOLUTION OF THE TYRRHENIAN AREA DURING THE NEOGENE AND QUATERNARY TIMES

The Tyrrhenian basin is peculiar because it formed by thinning without extension (Smith and Woodcock, 1982). Thinning without extension can occur by sideways transport of continental crust by an uprising mantle diapir that sweeps crust laterally away from under the basin site. The subsidence effects in the thinned zone will be similar to those of a passive margin, but on its edges mountain belts will form. This is exactly what occurs in the Tyrrhenian basin and in the Apenninic belt from upper Miocene on.

This mechanism of subcrustal transfer has been proposed by Van Bemmelen (1972) and has been accepted by Selli (1974, 1981) because most of the data of marine geology and of Apenninic tectonics fit with it. The Tuscan geologists (Elter et al., 1975) came to a similar conclusion outlining the fact that in Italy, along the Apennine chain, there is a sharp change in the tectonic style, which is extensional to the west turning to compressional to the east. Only a mechanism of east oriented convection in the asthenospheric mantle can account for it, according to the Authors.

For the southern Tyrrhenian a mechanism of plate subduction has been proposed by Barberi et al. (1973, 1974), mostly because of the presence of andesites. The Tyrrhenian basin just represents a marginal, back-arc basin. The timing of magmatic events in respect to the tectonic evolution of the basin, the lack of an oceanic trough, the seismic data and the magmatic composition themselves weaken very much the efficiency of this proposal.

The last suggestion comes from Scandone, 1979, who recognizes a tectonic regime of rifting in the Tyrrhenian basin as the result of a counterclockwise rotation of the Peninsula, as formerly suggested by Locardi et al., 1976.

According to the writer's analysis (Locardi, 1982, 1985) it is very improbable that the Italian peninsula rotated as a single block. The structural differences between the northern Tyrrhenian and Apennines and the southern Tyrrhenian and Apennines are such as to envisage some differential mechanism in the geodynamic evolution of the area.

The driving mechanism of the tectonic in these two sectors of the Tyrrhenian-Apennine system has been recognized in the evolution of two single mantle domes, one in the North, one in the South (Fig. 2). There is evidence that the two mantle domes migrated from West to East, leading to the opening of the basins to their back and to compressional tectonics in the Apennines on their front.

This conception is basically the same as that of Van Bemmelen, Selli, Elter et al., Wezel, bringing some more detail in the mechanism of tectonic evolution of the area and, principally, the causes of the driving forces of this kind of tectonics of subcrustal transfer. As a matter of fact, one of the perplexities which arose in accepting this tectonic hypothesis was the unexplained origin of the high geothermal anomaly in the area, enhancing this kind of movement. With the help of the ascending deep mantle fluids we are now able to explain the high heat flow, the reasons why the asthenospheric mantle is so upwelling and "soft", the reasons why the most fluidized upper portions of the mantle reach a critical point where uprising is no longer possible, and therefore they collapse laterally to the mantle dome. The thinned lithosphere overlaying this "soft" asthenospheric mantle may passively follow its flowage.

These mantle processes have a pulsatory character, because they depend on the recharge time of the fluids, before collapsing. Pulsations in volcanic activity have been recognized in the Basin during Messinian, Lower Pliocene, Upper-Middle Pliocene, Lower Pleistocene times coupled with very rapid phases of gravitational tectonics in the Apennine belt.

The magmatism is of crustal anatectic origin in the northern part of the Tyrrhenian, where the basin forms on crustal continental basement rocks (Fig. 3). This anatectic magmatism migrated stepwise towards East along bands which correspond to the axis of the mantle dome.

In the southern Tyrrhenian Basin, oceanization and eruption along fissural volcanic seamounts evolve radially. In the center

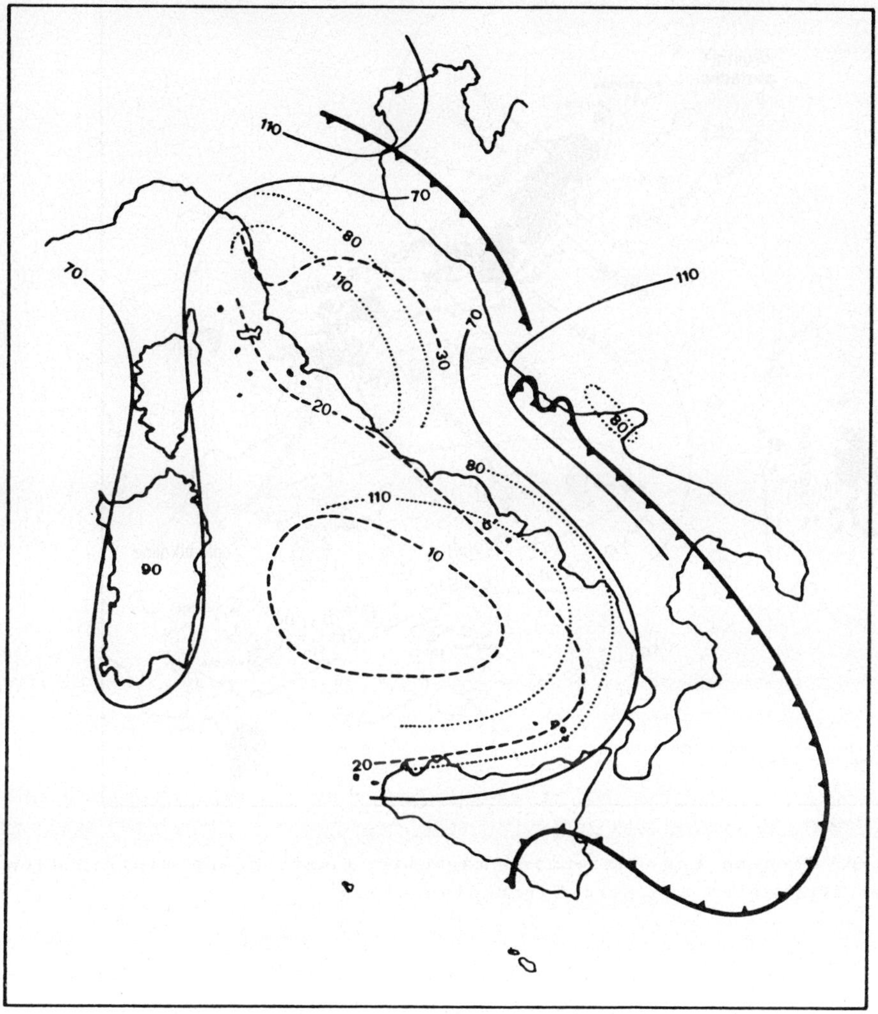

Fig. 2. Map showing two distinct mantle-domes in the Tyrrhenian area and their relationship with the compressional fronts. Solid lines indicate lithospheric thickness, dashed lines crustal thickness, and dotted lines heat flow values expressed in mWm^{-2} (data from Panza et al., 1980 and from Nicolich, 1981). The thick line marks the external compressional front.

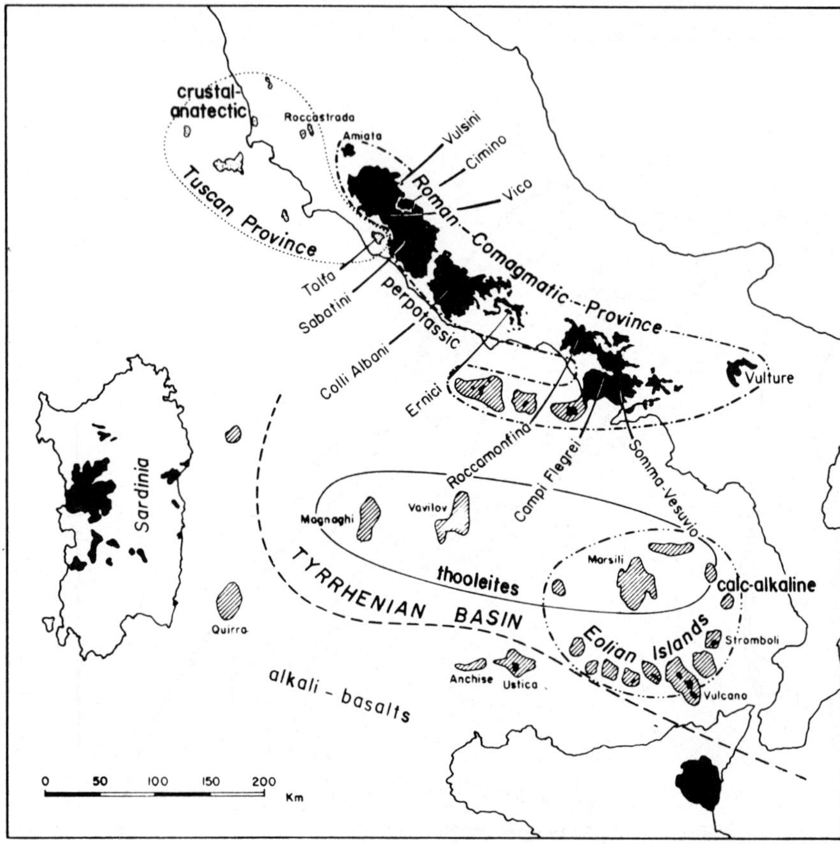

Fig. 3. Neogene and Quaternary magmatic zonation and distribution in the Tyrrhenian and peri-Tyrrhenian areas.

tholeiitic volcanism prevails bordered by alkaline olivine-basalts that outcrop even in Sardinia and Sicily.

This happened from the Tortonian to the Pliocene times, in association with the opening of the basin. In Pleistocene the tectonic style changed abruptly, and a new kind of volcanism developed along arcs.

VOLCANIC ARCS

The collapse of the Tyrrhenian Basin and the emersion of the Apennines correspond to the Pliocene-Pleistocene transition. Most of the volcanic activity on land occurred during the Pleistocene, and the characteristic potassic Mediterranean melts erupted during this period, as well as the Aeolian volcanism in the southeast Tyrrhenian. Both form arcs whose genesis is quite different even if related to the same major tectonic event, the foundering of the Tyrrhenian Basin.

The potassic volcanoes extend from southern Tuscany to Basilicata (Vulture volcano), forming a volcanic arc that is convex towards the Tyrrhenian (Fig. 4). This arc, about 450 km long, bends in an opposite direction with respect to the Apennine orogenic front. It belongs to an extensional range whose features cut the earlier extensional faults that developed during the Neogene between the Tyrrhenian Basin and the advancing orogenic front (Locardi et al., 1976). The volcanism cannot be considered, therefore, as post-orogenic because in this case the volcanics should be emplaced in the extensional features that form behind the orogenic arc and parallel to it. On the contrary, this newly formed graben system developed parallel to the present continental slope formed as a consequence of the Tyrrhenian collapse, and the associated volcanism is directly connected with this tectonic event. The central volcanoes are differently spaced according to the local conditions of crustal and lithospheric thickness. Very sudden tensional forces were able to displace the crust along new tectonic directions and to spill mantle-derived melts of different origin which became intimately mixed. The time span of volcanic activity is about 1 Ma, demonstrating continuous feeding of deep-derived melts to the upper magmatic chambers.

Along this arc the most alkalic and incompatible elements--enriched magmas have been erupted, and therefore the mantle-source

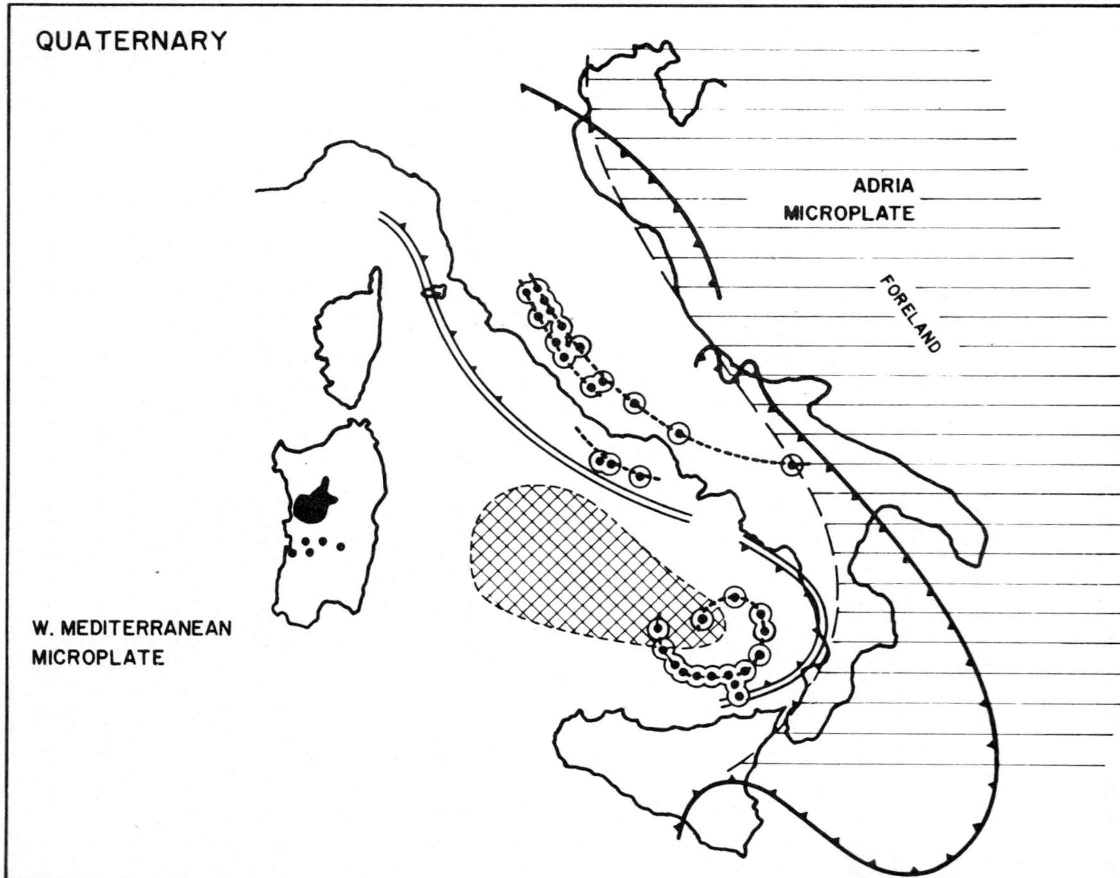

Fig. 4. Volcanic arcs and main tectonic features during the Quaternary. The oceanized part of the Basin is represented by cross-hatching. The double lines represent the contacts between different crustal domains.

regions of the melts should correspond to the same area. In the present interpretation (Locardi, 1982, 1985) based on subcrustal transfer, this fluid-enriched mantle region corresponds to the one that formerly accumulated on top of the Tyrrhenian mantle dome, and afterwards flowed laterally to the dome. In the denudated part, the pressure drop, due to the fissuring and to crustal transport, favoured tholeiitic volcanism, because the metasomatized part was flowing away. During the flow the stresses were of compressional type, and the eruption on the front of the transported crustal block was inhibited. Only during the Pleistocenic extensional tectonic regime the newly formed fissures allowed the eruptions to occur.

The structural meaning of this arc is therefore a superimposed tectonic feature due to a main phase of the basin collapse. Through these new fissure systems alkalic magmas have been spilled from different reservoirs corresponding to different situations in the mantle.

On the opposite side of the Tyrrhenian trough, on its southeast flank, the Aeolian Island volcanic belt forms behind and parallel to the orogenic wave that rapidly propagated to the southeast during the Pleistocene (Ghisetti and Vezzani, 1982). The Calabrian Arc represents the uplifted part of this orogen. The associated volcanism is indicative of an unstable compressional regime, suggested by the calcalkaline, orogenic affinity of the magma.

Looking nearer to the structure of this volcanic arc, we see that it forms a kind of ring corresponding to the top of a mantle uplift centered on a former tholeiitic sea-mount. Suddenly the tholeiitic volcano (Marsili) delivering magmas of classic intra-plate character, changes its composition in a calcalkaline, orogenic magmatism. The ring of calcalkaline volcanoes formed around it, but this activity lasted less than 0.5 Ma (Beccaluva et al, 1985) giving rapid transitions to more potassic and incompatible elements-rich magmatism whose final products are leucite tephrites very similar to the ones erupting along the Apenninic arc, of typical Mediterranean character. During this transition, the inner part of this volcanic ring collapsed into cauldron-like structures (Wezel, 1982). The age of the single volcanoes composing the ring becomes younger in counterclockwise direction, as if it were the response of a rotation of something plastic on top of the mantle uplift, followed by collapse.

This feature is generally interpreted as an andesitic volcanic arc formed in front of a subduction zone.

An andesitic volcanic arc forms on a section of thickened crust, but in the case under examination we find a volcanic ring on top of a rounded mantle dome covered by thinned crust. The duration of an andesitic volcanism is in the order of tens of million years, and in the Aeolian case it is in the order of a hundred of thousand years. From the same volcanic apparatus (Marsili) delivering typical intraplate, MORB-type volcanism, suddenly andesitic orogenetic volcanism erupts. It is hard to believe that a transition from extensional intra-plate volcanism and orogenic active margin volcanism can occur in the same point and from the same volcanic vent.

It is more probable that the change responsible for the genesis of the different kind of magmas did affect the mantle, instead of the crust. The present interpretation is that a fluid-enriched asthenospheric mantle sector flowed below the Aeolian Island area in Pleistocene time. The flow of this mass from the mantle dome towards the southeast caused the Quaternary orogenic wave of the Calabrian Arc. The reduced lithospheric sector below the Aeolian Islands has thus been underthrust by hot, plastic asthenospheric material. Partial melting of the upper mantle wedge region occurred delivering a magmatism of calcalkalic affinity. The torsional movements of this vortexlike moving asthenospheric material is imprinted in the Aeolian volcanic pattern. The orogenic well moves rapidly towards the southeast, the Calabrian block is shifted away and the magma belonging to the asthenospheric mass with its typical "Mediterranean" character can erupt through the new fissures (Locardi, 1985).

A geophysical survey (Panza et al., 1980) recognized a mass of

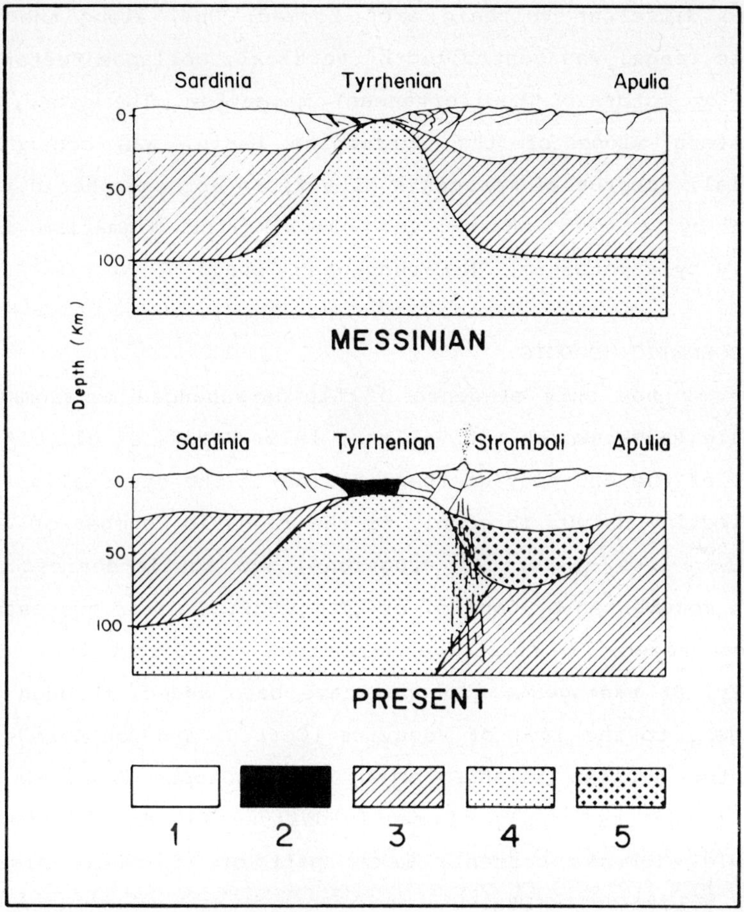

Fig. 5. The position of the "soft" asthenospheric mass as inferred by Panza et al., 1980. In the present interpretation this mass corresponds to the mantle-fluid activated upper portion of the dome that slid laterally. 1, crust; 2, oceanized crust; 3, lithospheric mantle; 4, asthenosphere; 5, "soft" asthenosphere.

"soft" asthenospheric material lying laterally to normal asthenospheric dome but on a lithospheric mantle. They depicted (Fig. 5) this situation but could not find the reason for the inversion in the position of asthenosphere in respect to lithosphere. On the contrary, this demonstrates very nicely the modalities of the subcrustal transfer tectonics and the features of the Aeolian and Calabrian arcs.

In conclusion, in the same structural domain and at the same time two different volcanic arcs formed. One, along the coastal Apenninic range, was controlled by vertical, collapse tectonics and was fed by potassic "Mediterranean" magmatism. The other, on the southeastern slope of the Tyrrhenian Basin, was controlled by tangential, subcrustal movements of a fluid asthenospheric flow and was fed by a very short-lived calcalkaline magmatism followed gradually by a potassic, "Mediterranean" one.

MINERALOGENETIC ASPECTS

There is not only evidence of fluids-enhanced metasomatism in the mantle (Hawkesworth and Vollmer, 1979; Rogers et al., 1985) but evidence of the activity of these fluids in the crust also.

A selective Ra, U, Th supply to the magmatic chamber of Vesuvius just before its historical eruptions, via mantle-derived fluids, has been documented by Capaldi et al., 1982. Similar processes have been recognized in zero age eruptions of Stromboli and Etna. Similarly, Sr radiogenic isotopes have been added, through different pulses, to the lava of Vesuvius (Cortini and Don Hermes, 1981) and of the Aeolian Islands, with lead isotopes (Cortini, 1981).

This pulsatory supply of mantle-derived fluids with their incompatible elements content, makes part of the whole history of alkaline volcanism in Italy.

The crustal anatectic volcanoes (Cimino and Amiata) which have been erupted on the front of the ascending alkalic magmas, show the same petrographic composition of the former "Tuscan" anatectic magmas, but the same incompatible elements and isotopic character of the mafic alkalic magmatism. The crustal melting occurred in a bath of the same fluids which imprinted the isotopic and incompatible elements content on the mantle-derived magmas. From a comparison with the incompatible elements content of the former "Tuscan" magmatism, it is easy to show that, in this case, it is the mantle material that delivers incompatible elements to the crust, and not viceversa as generally is the case elsewhere on the Earth (Locardi and Rotach, in prep.).

In the alkalic suite itself the distribution of the incompatible elements seems peculiar. In the most primitive magmas, closer to the composition of the source region in the mantle, the content of incompatible elements is very high but variable as a function of the degree of metasomatism which occurred in the source region before extractions of the melts. Their amount then increases according to the element partition laws in crystal liquid fractionation processes. Locally, a further increase which does not fit into any law of a magmatic differentiation is observed. A very strong increase in incompatible elements takes place without a parallel increase of the major elements (such as alkalies and silica) which are normally enriched during fractional crystallization.

Fig. 6 exemplifies the linear correlations between Th, Zr, Sb, and silica. They follow the rule of crystal fractionation process, with increasing of both incompatible elements and silica, but suddenly they show a very rapid increase whilst the chemical composition of the rock remains the same. This might be due to gas enrichment on top of the magmatic chamber (Locardi, 1967), but it is noteworthy that lavas, which are degassed magma, behave the same as the ignimbrites, which are magmas with a good part of their volatile fraction in them.

Moreover, the incompatible elements with a very low solid-liquid partition coefficient are progressively concentrated during the crystallization in the liquid fraction , mantaining a constant ratio. Such a law applies to our case, but together with a constant ratio trend very strong increases may be observed of some of the incompatible elements in respect to others, though their partition coefficients are all similar (f.i. U, Sb vs Ta, Fig. 6). This contrasts with the geochemical laws governing element partition in magmatic melts, and shows that the selective supply of some incompatible elements comes from outside the shallow magmatic reservoir.

Such increase is most probably related to a mantle-derived volatile phase which introduces its solute into the melt without modifying the major element content.

A question arises: if these fluids manage to reach the

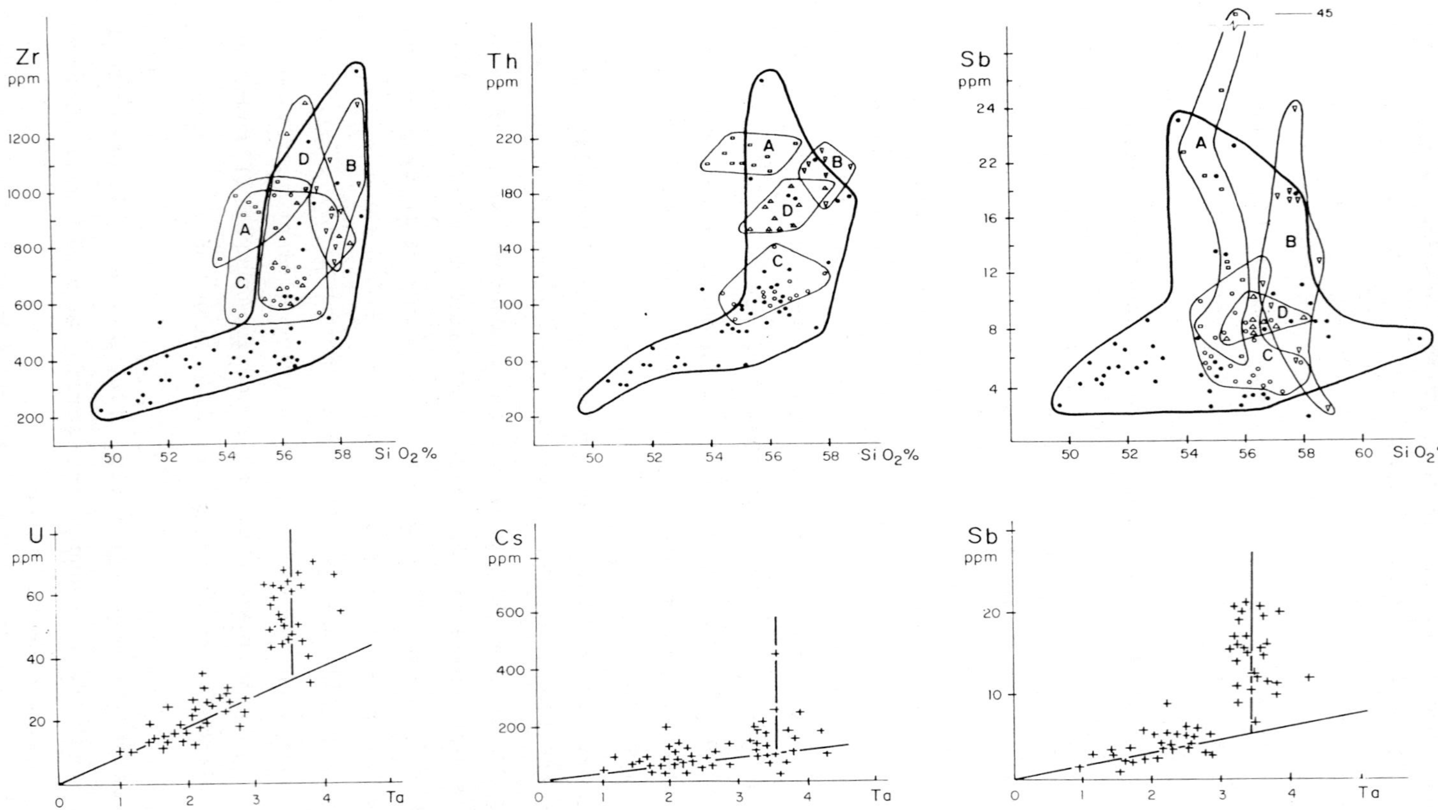

Fig. 6. Zr, Th, Sb versus silica, and U, Cs, Sb versus Ta diagrams of the Vico potassic volcano. Letters A to D correspond to single ignimbritic units, by order of eruption; full circles to lava samples. The rules of element partition during fractional crystallization are only partly respected. A selective inflow of some incompatible elements is evidenced. (Data from Palacin, 1985).

superficial magma chambers, may they also spread away from the magmatic system, interacting with the hydrosphere and the uppermost geological environment and deposit mineralizations?

If the answer is positive, the genesis of the mineralizations enhanced by supply of mantle-derived fluids may be evidenced by the mineral association of the metalliferous province, i.e. by the type of element which was added by this mechanism both to the source region of the magmatism (metasomatized mantle) and to the magmatic chambers already located in the superficial crust.

The metasomatic supply to the mantle can be evaluated by comparing the composition of the most primitive magmas extracted at the beginning of the metasomatic process (such as the 7 Ma old Tyrrhenian basalts, DSDP drilling 373A) and the composition of the most primitive magmas extracted from the mantle when the metasomatic process was very active (f.i. the less than 1 Ma old olivine--leucitites of the Vulsini volcanic group).

The comparison between columns 1 and columns 2 in Table 1, shows the enrichment in some incompatible elements due to mantle metasomatism. This does not correspond to the maximum enrichment because the 7 Ma old Tyrrhenian basalts were already enriched in LREE, Ba, Rb, Sr, Th, U in respect to the mean oceanic rocks. Furthermore, the available basalt samples are less primitive than the most primitive olivine leucitites selected in column 2. If we consider the olivine leucitites which show the same evolutionary stage of the Tyrrhenian basalts (same values of Mg, Ni, Co, Cr, Sc) we come nearer to the real effect of mantle metasomatism. From a comparison between column 1 and column 3 we get enrichment factors of more than 300 for elements like Rb, Cs, of about 100 for Th and 30 for U, 20 for K and high values for LREE, Ba and Sr.

Let us now see what happens in the crust. The data on rocks related to fractional crystallization due to cooling of the magma into the shallow reservoir are listed in column 4, and show a further increase of some incompatible elements up to 10 compared to the most primitive magmas. Other enrichments which cannot be referred to fractional crystallization processes, but to direct

TABLE 1

	Tyrrhenian basalts	Olivine-leucitites	E.F.	Olivine-leucitites	E.F.	Trachytes	E.F.	Trachytes	E.F.
	1	2		3		4		5	
SiO_2	47 %	47,5		48,5		60		56	
MgO	8	11		7,5					
Cr	250 ppm	600		240					
Co	35	40		33					
Ni	90	180		90					
Sc	28	33		28					
K_2O	0,25%	3,5%	(14)	5,2	(20)	8,5	(2,4)	7,7	--
U	0,2 ppm	5	(25)	5,7	(28)	23	(4,6)	61	(2,7)
Th	0,4	22	(55)	38	(95)	90	(4)	258	(2,8)
Zr	70	140	(2)	250	(3,5)	700	(5)	1023	(1,5)
Hf	1,6	3,8	(2,4)	6	(3,6)	15	(4)	21	(1,4)
Ta	0,24	0,4	(1,6)	0,7	(3)	2,3	(9,5)	4	(1,7)
Cs	0,1	20	(200)	30	(300)	60	(3)	210	(3,5)
Rb	1	250	(250)	380	(380)	550	(2)	630	(1,1)
Sb	0,06	0,4	(6,6)	0,5	(8)	2	(5)	14	(7)
La	5	50	(10)	60	(12)	145	(3)	250	(1,8)
Tb	0,5	0,8	(1,6)	1	(2)	1,4	(1,8)	0,6	--
Ba	65	500	(7,6)	1000	(15)	360		590	(1,4)
Sr	220	1000	(4,5)	1300	(6)	500		690	(1,4)

Table 1. Indicative values of incompatible element enrichment due to mantle metasomatism, to fractional crystallization, and to direct supply of fluids to the magmatic chamber before eruption. Explanations in the text. (Data from Rotach, 1983; Palacin, 1985; Dietrich et al., 1977; Hamelin et al., 1979).

supply of fluids from depth, are listed in column 5. These are the elements which have the highest probability of forming economic concentrations at the surface. From column 5 it appears that the maximum concentrations are those of Sb (Hg, As), Cs (Li), U. These elements represent the solute of the CO_2-F-Cl-S-rich aqueous fluids.

If the elements characterizing the Tuscan-Latial metalliferous province are considered, strong similarities may be noted with those found in the metasomatizing fluid phase.

CO_2 and H_2S quantities mark a clear geochemical anomaly in the Tuscan-Latial province (Fig. 7). Each kind of mineralization, notwithstanding the type of element concentrated, is associated with exhalations of these gases. They partecipate in enormous quantities in the composition of the geothermal fluids. The same province contains a huge quantity of travertines which are connected with excess CO_2. Thermodynamic studies on the CO_2 (Atkinson et al., 1978) show that only part of it could be derived by metamorphism of the basament carbonate rocks. Travertines show, furthermore, peculiar chemical characters with their very high contents of Sr, As, Hg, Be.

H_2S is responsible for the huge alteration haloes which, like the travertines, are associated with all the different lithologies and magmatic suites of the area. It gives rise to the enormous concentrations of iron sulphide in the same area.

Fluor, the other element making part of the mantle-fluids, is highly concentrated in the ground water and forms the biggest fluorine deposit in the world, associated with baritine and apatite.

Considering the incompatible elements, uranium forms big concentrations in volcano-sedimentary basins. Tens of thousands of U have been ascertained in low grade stratiform concentrations, but the total quantity contained in the basins is of some order of magnitude higher.

Mercury forms deposits which are the biggest in Europe, even if not the most economic. Antimony is associated with the mercury in many areas, or it forms independent deposits.

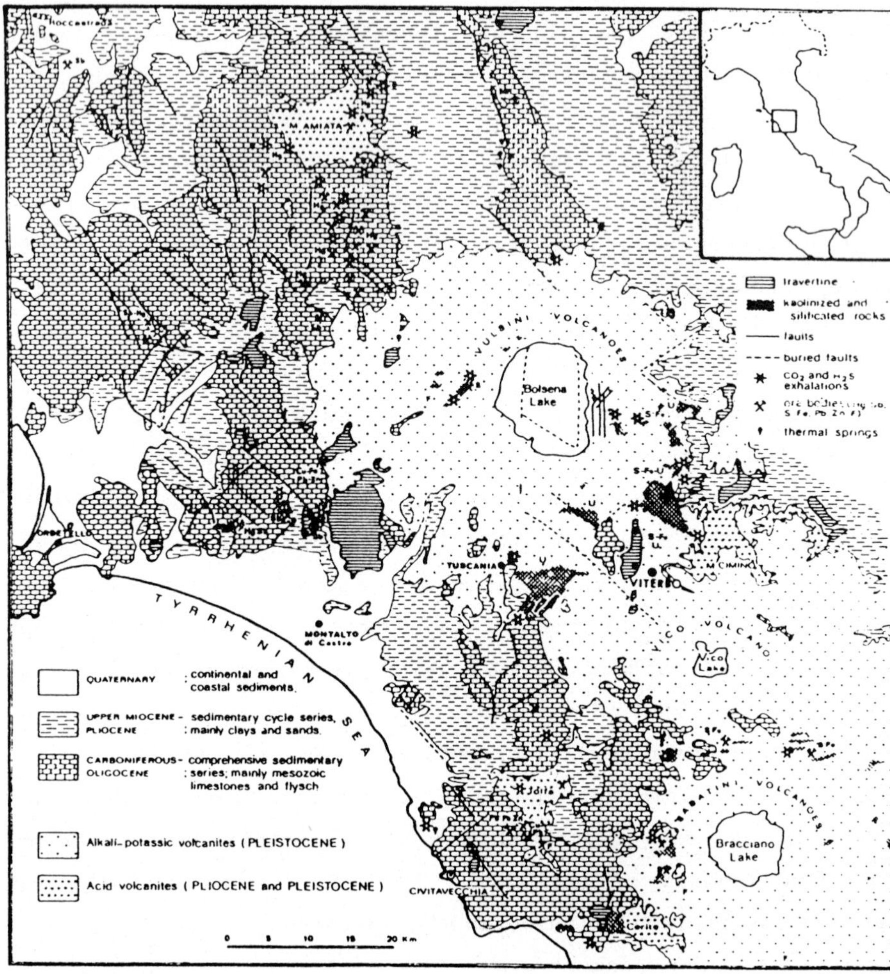

Fig. 7. Structural map of the Tuscan-Latial province showing the distribution of ore deposits, alteration zones, major travertine deposits, thermal springs and gas exhalations.

Barium is not only associated with fluor, but forms huge mineralizations with strontium.

Dealing with the economic aspects, the geothermal fields should be mentioned, forming one of the most important geothermal areas in the world.

These fields contain high quantities of 3He of mantle origin (Hooker et al., 1985), and the Authors observe that its inflow has a pulsatory character, associated with radon, CH4 and H2.

The conclusion is that the geothermal fields derive from the interaction between mantle derived fluids and shallow aquifers. In the brines associated with the geothermal field of Cesano (Rome) the geochemical anomalies characterizing the mantle derived fluids is imprinted: high contents of CO2, F, Cl, S, As, B, Rb, Li. The evaluated temperature of 500°C or more testify the presence of hypercritic fluids.

REFERENCES

Atkinson, P., Celati, R., Corsi, R., Kucuk, P. and Ramey, H.I.Jr., 1978. Thermodynamic behaviour of the Bagnore Geothermal Field. Geothermics, 7, 185-208.

Bailey, D.K., 1970. Volatile flux, heat focussing and the generation of magma. Geol. J., Spec. Iss. 2: 177-186.

Bailey, D.K., 1972. Uplift, rifting and magmatism in continental plates. J. Earth Sci. (Leeds), 8: 225-239.

Bailey, D.K., 1982. Mantle metasomatism - continuing chemical change within the Earth. Nature, 296: 525-530.

Barberi, F., Gasparini, P., Innocenti, F. and Villari, L., 1973. Volcanism of the Southern Tyrrhenian Sea and its geodynamic implication. J. Geophys. Res., 78: 5221-5232.

Barberi, F., Ferrara, G., Innocenti, F., Keller, J., and Villari, L., 1974. Evolution of the Aeolian arc volcanism (Southern Tyrrhenian Sea). Earth Planet. Sci. Lett., 21: 269-276.

Beccaluva, L., Gabbianell, G., Lucchini, F., Rossi, P.L. and Savelli, C., 1985. Petrology and K/Ar ages of volcanics dredged from the Eolian seamounts: implications for geodynamic evolution of the Southern Tyrrhenian basin. Earth Plan. Sci. Lett., 74, 187-208.

Capaldi, G., Cortini, M. and Pece, R., 1982. The isotopes at Vesuvius: evidence for open-system behaviour of partial melting process. J. Volc. Geotherm. Res., 14: 247-260.

Cortini, M., 1981. Eolian Island arc (S. Tyrrhenian Sea) magma heterogeneities in historical lavas: Sr and Pb isotopic evidence. Bull. Volc., 44: 711-722.

Cortini, M. and Don Hermes, O., 1981. Sr isotopic evidence for a multisource origin of the potassic magmas in the Neapolitan area (S. Italy). Contrib. Mineral. Petrol., 77: 47-55.

Dietrich, V., Emmermann, R., Keller, J. and Puchelt, H., 1977. Tholeiitic basalts from the Tyrrhenian Sea floor. Earth Plan. Sci. Lett., 36, 285-296.

Elter, P., Giglia, G., Tongiorgi, M. and Trevisan, L., 1975. Tensional and compressional areas in the recent (Tortonian to present) evolution of the northern Apennines. Boll. Geof. Teor. Appl., 56: 3-18.

Foden, J.D. and Varne, R., 1980. The petrology and tectonic setting of Quaternary-Recent volcanic centres of Lombok and Sumbawa, Sunda arc. Chem. Geol., 30, 201-226.

Ghisetti, F. and Vezzani, L., 1982. Different styles of deformation in the Calabrian Arc (Southern Italy): implications for a seismotectonic zoning. Tectonophysics, 85: 149-165.

Hamelin, B., Lambert, B., Joron, J.L., Treuil, M. and Allegre, C.J. 1979. Geochemistry of basalts from the Tyrrhenian Sea. Nature, 278, 832-834.

Hawkesworth, C.J. and Vollmer, R., 1979. Crustal contamination versus enriched mantle: 143Nd/144Nd and 87Sr/86Sr evidence from the Italian volcanics. Contrib. Mineral. Petrol., 69: 151-165.

Hooker, P.J., Bertrami, R., Lombardi, S., O'Nions, R.K. and Oxburgh E.R., 1985. Helium-3 anomalies and crust-mantle interaction in Italy. Geoch. Cosmoch. Acta, 49, 2505-2513.

Locardi, E., 1967. Uranium and Thorium in Volcanic Processes. Bull. Volc., 31, 235-260.

Locardi, E., 1981. La provincia comagmatica romana e l'evoluzione geodinamica dell'area tirrenica. CNEN-RT/AMB Roma, (81) 12: 1--73.

Locardi, E., 1982. Tectonic control of volcanism in an uranium province. Proc. Symp. "Uranium Exploration Methods". NEA, Paris, pp. 249-262.

Locardi, E., 1985. Neogene and Quaternary Mediterranean Volcanism: The Tyrrhenian Example. In: "Geological Evolution of the Mediterranean Basin", D.J. Stanley and F.C. Wezel editors, Springer Verlag, 273-291.

Locardi, E., Funiciello, R., Parotto, M. and Lombardi, G., 1976. The main volcanic groups of Latium (Italy): relation between structural evolution and petrogenesis. Geol. Rom., 15: 279-300.

Nicolich, R., 1981. Crustal structures in the Italian Peninsula and surrounding seas: a review of DSS data. In: F.C. Wezel (Editor), Sedimentary Basins of Mediterranean Margins. C.N.R. Italian Project of Oceanography, Tecnoprint, Bologna, pp. 3-17.

Makris, J., 1985. Geophysics and Geodynamic Implications for the Evolution of the Hellenides. In: "Geological Evolution of the Mediterranean Basin", D.J. Stanley and F.C. Wezel editors, Springer Verlag, 231-248.

Palacin, P., 1985. Etude geologique et geochimique du volcanisme du Latium (Italie): Exemples du Vico et du Vulsini. Thése de 3è Cycle, Univ. Paris VI, I-242.

Panza, G.F., Calcagnile, G., Scandone, P. and Mueller, S., 1980. La struttura profonda dell'area mediterranea. Le Scienze, 24, 141: 60-69.

Rogers, N.W., Hawkesworth, C.J., Parker, R.J. and Marsh, J.S., 1985. The geochemistry of potassic lavas from Vulsini, Central Italy and implications for mantle enrichment processes beneath the Roman Region. Contrib. Mineral. Petrol., 90, 244-257.

Rotach, N., 1983. Approche geochimique de la Pétrogénèse des séries magmatiques latiales et toscanes: répartition de l'Uranium et des éléments en traces associés. Thése de 3è Cycle, Univ. Paris VI, 1-268.

Scandone, P., 1979. Origin of the Tyrrhenian Sea and Calabrian Arc. Boll. Soc. Geol. It., 98: 27-34.

Selli, R., 1974. Appunti sulla geologia del Mar Tirreno. Rend. Sem. Fac. Sci. Univ. Cagliari, (suppl.), 43: 327-351.

Selli, R., 1981. Thoughts on the geology of the Mediterranean region. In: F.C. Wezel (Editor), Sedimentary Basins of Mediterranean Margins. C.N.R. Italian Project of Oceanography, Tecnoprint, Bologna, pp. 489-501.

Smith, A.G. and Woodcock, N.H., 1982. Tectonic syntheses of the Alpine-Mediterranean Region: a review. In: H. Berckhemer and K. Hsü (Editors), Alpine Mediterranean Geodynamics. Geodynamics Series, Am. Geophys. Un., 7: 15-38.

Van Bemmelen, R.W., 1972. Driving forces of Mediterranean orogeny (Tyrrhenian test-case). Geol. Mijn., 51: 548-573.

Wezel, F.C., 1982. The structure of the Calabro-Sicilian Arc: result of a post-orogenic intra-plate deformation. In: J.K. Leggett (Editor), Trench-Forearc Geology: Sedimentation and Tectonics on Modern and Ancient Active Plate Margins. Geol. Soc. London, Spec. Publ. 10: 345-354.

Wezel, F.C., 1985. Structural Features and Basin Tectonics of the Tyrrhenian Sea. In: "Geological Evolution of the Mediterranean Basin", D.J. Stanley and F.C. Wezel editors, Springer Verlag, 153-194.

IV. THE CARIBBEAN

GROWTH OF ACCRETIONARY PRISMS: TECTONIC PROCESSES FROM CARIBBEAN EXAMPLES

A. MASCLE[1], B. BIJU-DUVAL[2], P. de CLARENS[3], H. MUNSCH[4]
[1]IFP, Boite Postale 311, 92506 Rueil Malmaison CEDEX, France
[2]IFREMER, 66 avenue d'Iéna, 75763 Paris CEDEX, France
[3]SNEA(P), Tour Elf, 92078 Paris la Défense CEDEX 45, France
[4]Total CFP, 92069 Paris la Défense CEDEX 47, France

ABSTRACT

Accretionary prisms will develop in front of active margins when some amount of sediments is present on the subducting plate. Accretion will first consist of sediment offscraping at the deformation front. Then will follow incorporation of sediments at the top and at the back of the prism, as well as below it (underplating). Secondary processes include gravity sliding and mud diapirism. All these features have led to the growth of large accretionary prisms such as the Barbados Ridge in front of the Lesser Antilles Island Arc. Extensive seismic profiling, a few Seabeam maps, direct control on the field on Barbados Island and deep-sea drilling during the DSDP leg 78A have been used to document all these tectonic processes.

INTRODUCTION

Fifty percent of the present continental margin can be classified in the active margin category. Three main types can be distinguished:

. Those associated with the consumption of an oceanic crust by subduction under another plate;

. Those marked by collision processes between two plates with a thick or light crust (continental crust/island arc/thick oceanic crust);

. Those linked with dominantly horizontal strike-slip movement between two plates.

We will consider only the first type here. Subduction of oceanic material is mainly defined by the seismically active Benioff zone and by a calco-alkaline volcanic arc. Tectonic deformation of superficial structures is not so spectacular. Nevertheless, stress constraints due to plate convergence are such that understanding superficial deformation is a key point in evaluating young and old active margins.

The variety of structures and associated sedimentary basins is then very great. Very fast progress in our understanding of them has been made possible by:

. The acquisition of numerous and good quality multichannel seismic profiles along transects.

. Very detailed site surveys of the sea-bottom topography by using a high-resolution multibeam sounder (Seabeam). These bathymetric maps are similar in quality to continental topographic maps. They give an accurate description of the most recent sedimentary and tectonic processes.

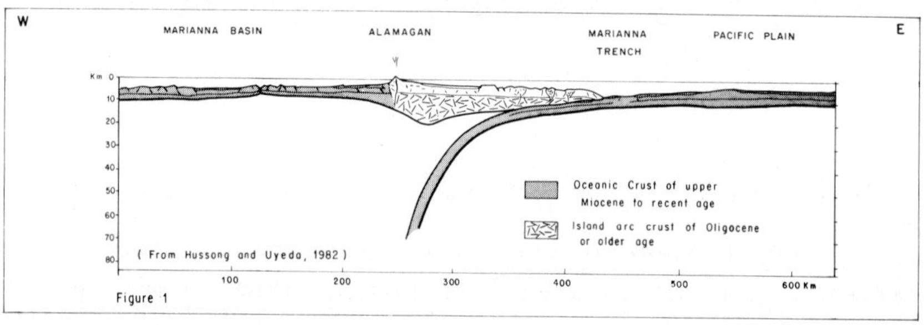

Figure 1

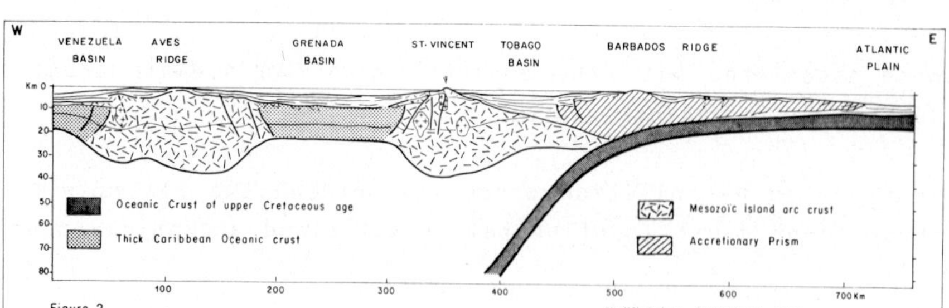

Figure 2

Fig. 1 - Schematic section across the Marianna Margin.

Fig. 2 - Schematic section across the Lesser Antilles Margin.

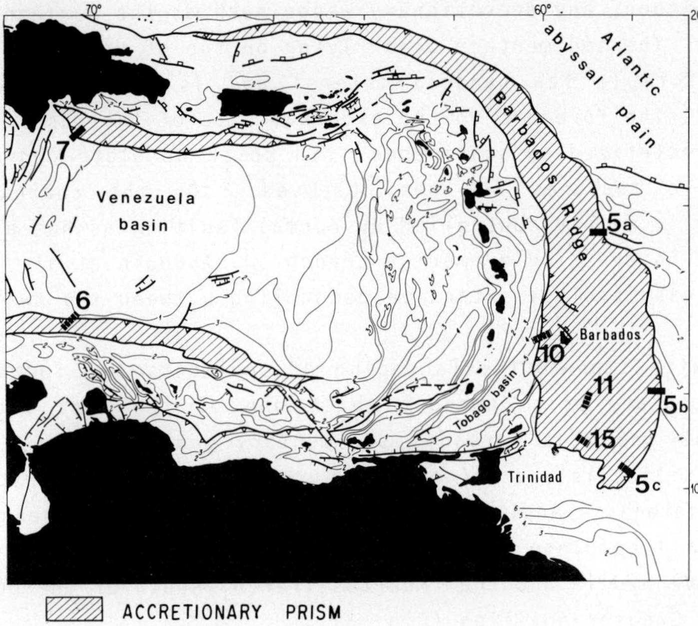

Fig. 3 - Eastern Caribbean margins. Small numbers and countours are isopachs of undeformed sediments (from Mascle et al., 1985); large numbers refer to figures in the text.

. Deep drilling along different transects during the DSDP-IPOD history: Marianna, Japan, Middle America Trench, Barbados (respectively Legs 59, 56-57-87, 66-67-84 and 78A).
. More recently, by direct in situ observation at depth from submersibles (see particularly the spectacular results obtained during the Franco-Japanese Kaïko project, Boulegue et al., 1985).

It can be seen that the developpment of new tools based on state-of-the-art technology has played a great role in obtaining new insights into the tectonic processes that occur on the active margins. We must also keep in mind that direct approaches in the field also provide a lot of very useful information.

In the fore-arc segment the tectonic processes progressively control the infilling and structure of sedimentary basins. Two main types of fore-arc region may be distinguished:

- Those without any accretionary wedge such as the Marianna trench (Fig. 1). The sedimentary cover lying on top of the oceanic crust is very thin (a few hundred meters) and is subducted with its basement. The fore-arc on the external site of the insular arc is then characterized by subsidence. In spite of plate convergence, compressive features are not observed. On the contrary, the structure is mainly controlled by normal faulting as has also been described in the Middle America trench (J. Aubouin et al., 1983). This process involves extensive decoupling between the two plates.

- Those with important offscraping of sediments and formation of an accretionary prism amongst which one of the best understood examples is the wide Barbados sedimentary wedge in front of the Lesser Antilles island arc (Fig. 2). This will be described here in some detail. Additional examples on other Caribbean active margins will complete the description. They are the Curaçao prism north of Venezuela and the "Muertos trench" south of the Dominican Republic, Greater Antilles (Fig. 3).

The description will be based on MCS profiles shot by the French oil industry (see synthesis in A. Mascle et al., 1985) in collaboration with IFREMER (ex CNEXO) for Seabeam acquisition (Biju-Duval et al., 1982-c; Valery et al., 1985) and based on IPOD data (Sites 541, 542 and 543, Leg 78A) at the toe of the prism (Biju-Duval, Moore et al., 1984) Field studies have been carried out on Barbados island (Biju-Duval et al., 1985) as well as in the Dominican Republic (Biju-Duval et al., 1983-a).

LESSER ANTILLES ISLAND ARC
GENERAL FRAMEWORK

The Lesser Antilles Island Arc is the result of the subduction of the Mesozoic Atlantic oceanic crust beneath the Caribbean plate (Fig. 3). The location and age of volcanism indicate that the subduction process is as old as Middle Eocene. Therefore, older volcanic rock locally outcropping on Desirade

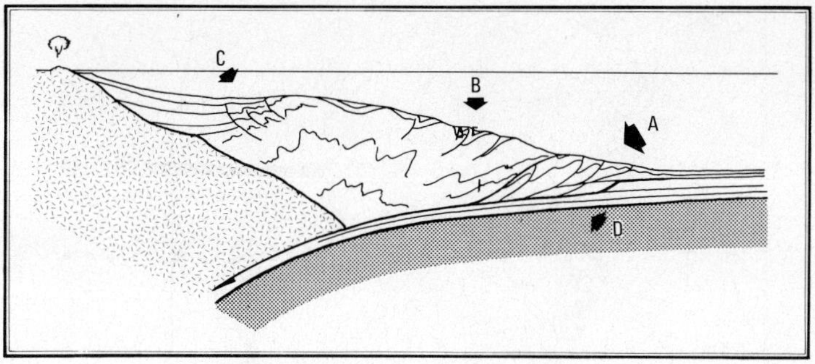

Fig. 4 - Schematic section across an accretionary prism. Accretion can occur simultaneously at the toe of the prism (A), on its top (B), its back (C) or under it (D).

island or dredged up on the Aves ridge (remnant arc) and plate kinematic constraints lead us to presume that subduction has been active since the late Jurassic, at least temporarily (Bouysse, 1984).

The fore-arc region is characterized by a very large accretionary prism, the Barbados Ridge. Its width is 300 km in the south and only about 50 km in the north (Westbrook, 1982). This decrease is due to the northward diminution of sediment thickness on the oceanic crust because of the clastic influx source from the South American continent. Similarly the fore-arc basin infilling (Tobago and Lesser Antilles Basins) and the sedimentary draping on the prism are also characterized by a rapid northward decrease in thickness, i.e. more than 5 km south of 16°N and practically a few dozen meters to the north where the external wedge of the volcanic arc forms very steep scarps several km high (see synthesis in Westbrook et al., 1984).

The main but not the only factor leading to sediment deformation and to prism formation is tectonic shortening linked to subduction and convergence at the Atlantic-Caribbean plate boundary. We will now examine the various modes of this tectonic deformation, and then we will look at secondary processes such as gravity sliding and mud diapirism.

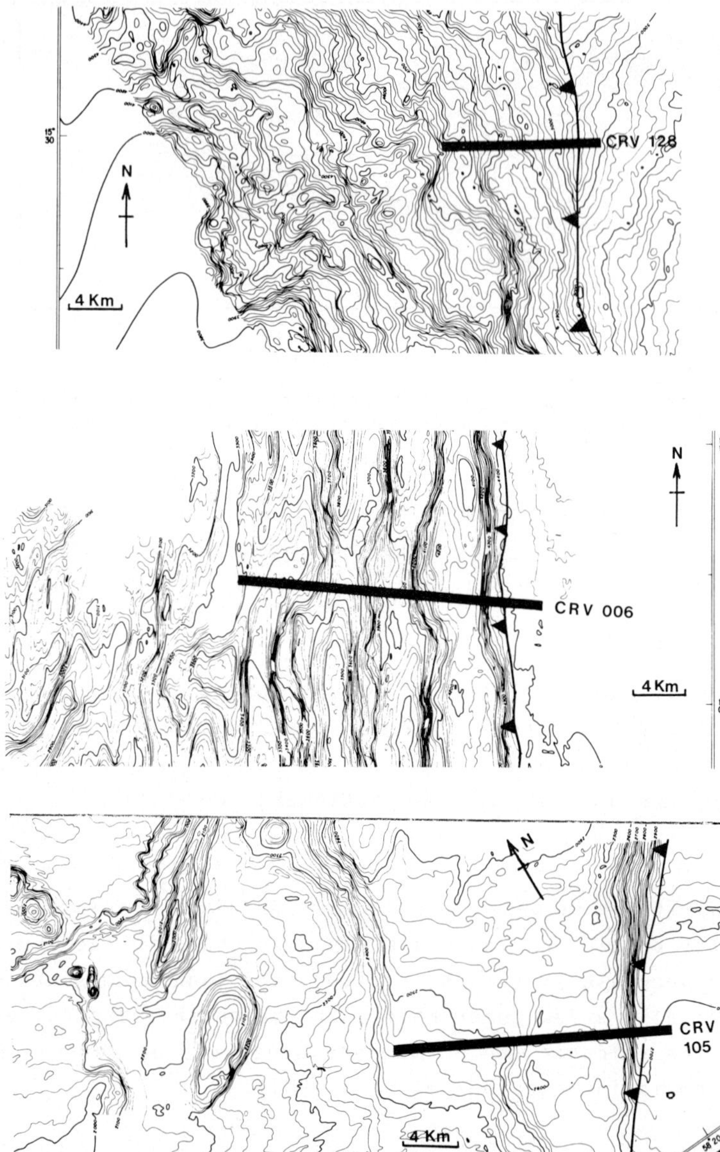

Fig. 5 - **Right**: Seismic expression of structural styles at the front of deformation (CEPM multichannel and migrated seismic profiles)
Left: Corresponding morphology of the sea bottom (CEPM-IFREMER Sea beam maps).
Location in Figure 3.

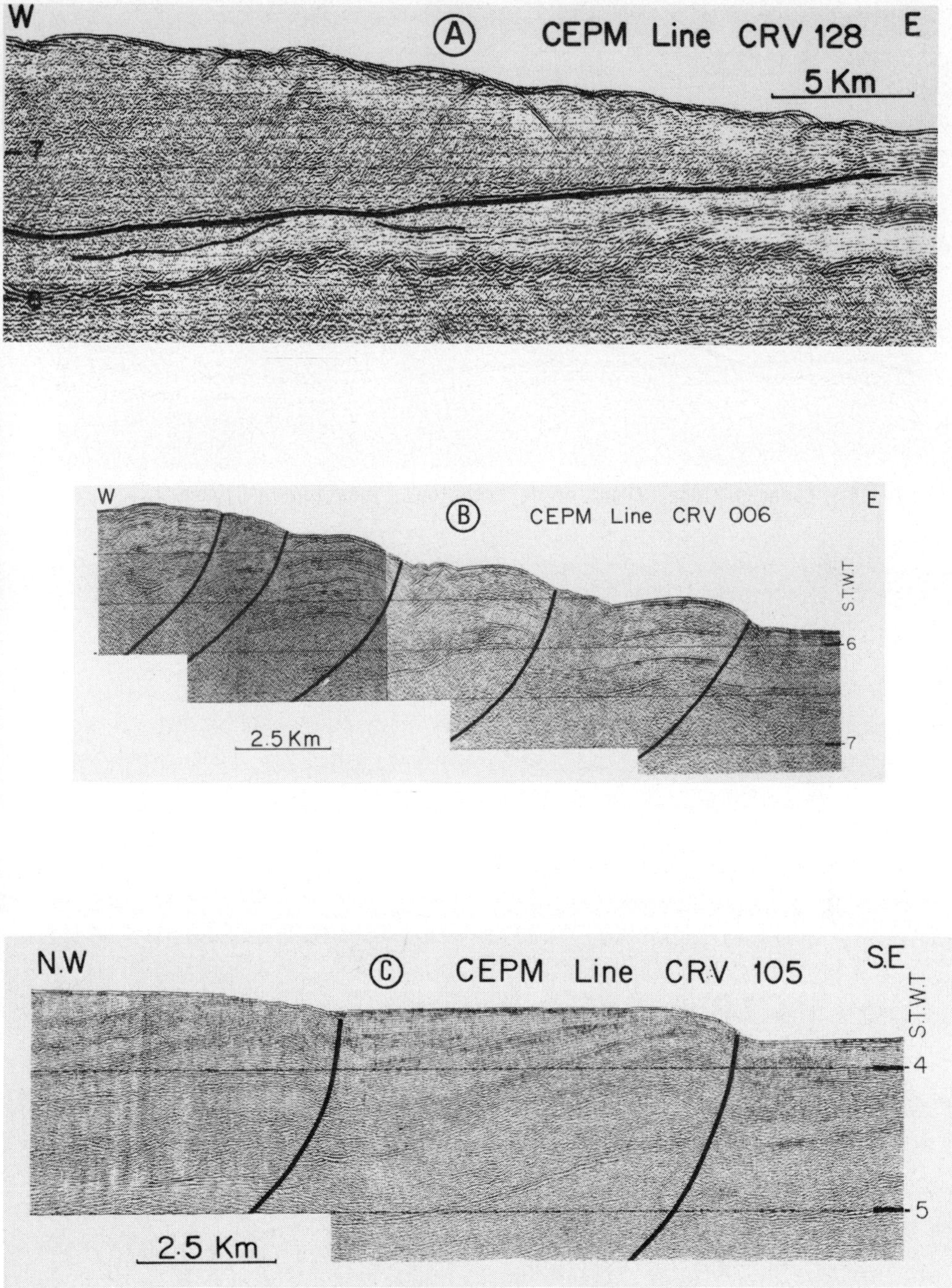
W
A
CEPM Line CRV 128
E
5 Km
7
8
W
B
CEPM Line CRV 006
E
S.T.W.T
6
7
2.5 Km
N.W
C
CEPM Line CRV 105
S.E
S.T.W.T
4
5
2.5 Km

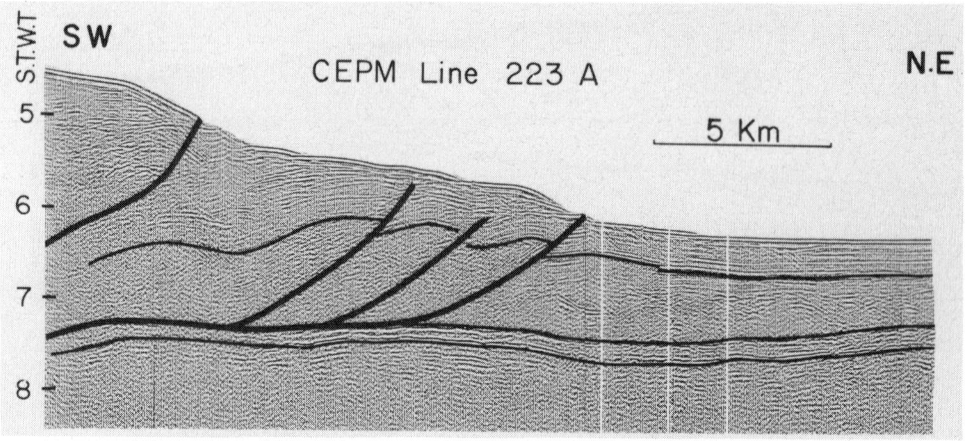

Fig. 6 - Curaçao Ridge, front of deformation. Location in Figure 3.

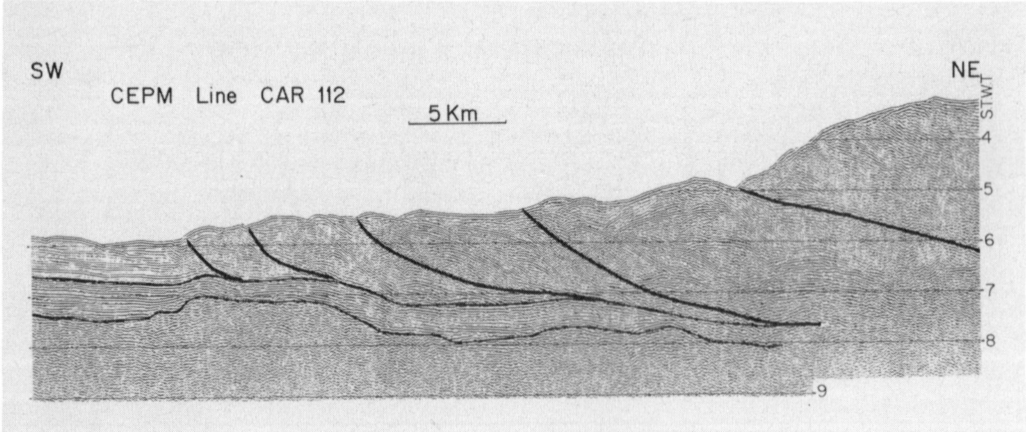

Fig. 7 - South Dominican margin, front of deformation. Location in Fig. 3.

TECTONIC SHORTENING AND SEDIMENT DEFORMATION

An accretionary wedge is a volume bounded by a décollement at the bottom with a relatively flat surface, a deformation front at the toe seaward of which the abyssal plain is in a "foreland" position, the rough surface of the sea bottom at the top, and in the rear a complex zone of transition to the fore-arc basin. Tectonic stresses in a compressive regime occur in these different boundaries, and so new material can be incorporated in the prism on practically every surface (Fig. 4).

Frontal Accretion

Part of the sediments deposited in the oceanic domain is incorporated in the prism. Three different tectonic styles are illustrated in Figure 5. Small-scale Seabeam documents from published maps at 1:200,000 scale (Valery et al., 1985) and corresponding transects illustrated by high resolution multichannel seismic-reflection profiles give some idea of the variety of structures in three areas where the sedimentary volumes are quite different.

In the central part of the deformation front, from 11°N to 13°N, about 5000 meters of sediments cover and move with the Atlantic Oceanic crust towards the subduction zone. A large volume of them is incorporated in the prism. Several dissymmetric anticlines, bounded in the east by listric reverse faults, are followed from south to north over distances up to 10 and 30 km. They have a width of about 5 km (CRV 006 profile, Fig. 5B). The sea-bottom morphology clearly expresses the tectonic style. The level of décollement is not observed on the profiles but is assumed to occur at depth. In the northern part, north of the Tiburon rise, less than one thousand meters of sediments lies on the oceanic crust, here of Senonian age (see profile CRV 128, Fig. 5A). A very well defined and relatively flat level of décollement corresponds to a stratigraphic interval of the uppermost Lower Miocene. Senonian to early Miocene sediments are underthrust with the oceanic crust and are disharmonically separated from deformed sediments of mid-Miocene to recent ages (Moore and Biju-Duval,

1984). The tectonic style is not obvious on the seismic profiles. This can be explained since the amplitude of the folded structures is less than 1000 meters, as indicated by the Seabeam map which shows small-scale irregularities (Fontas et al., 1984), also observed on the Gloria data (Belderson et al., 1984). Sites 541 and 542 of DSDP Leg 78A enable us to propose a progressive offscraping of several layers of sediments. They were probably separated by reverse faults forming successive slices on top of the décollement. In situ measurements of high pore fluids pressure have been found to be equivalant to lithostatic loading above the décollement. Rapid changes in the physical properties of the sediments and positive anomalies in the temperature distribution (14° to 16°C between 49 and 150 meters depth) indicate the prominent role of the high permeability and high pressure zone which probably facilitated the décollement and thrustings (Moore and Biju-Duval, 1984).

Moreover, the deformation above the décollement is not limited to the simple slicing off of successive layers of offscrapped sediments. It can include secondary imbrications

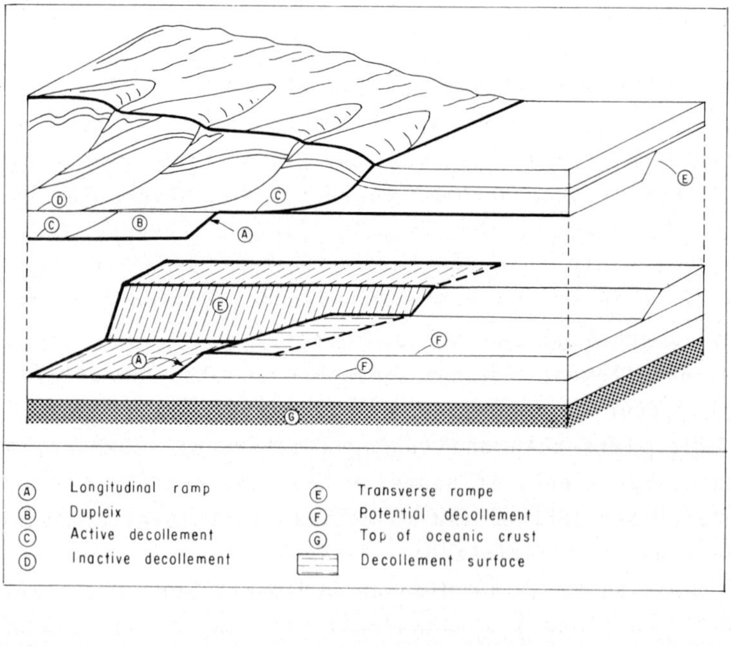

Fig. 8 - Schematic diagram of flat and ramp geometry below the accretionary wedge.

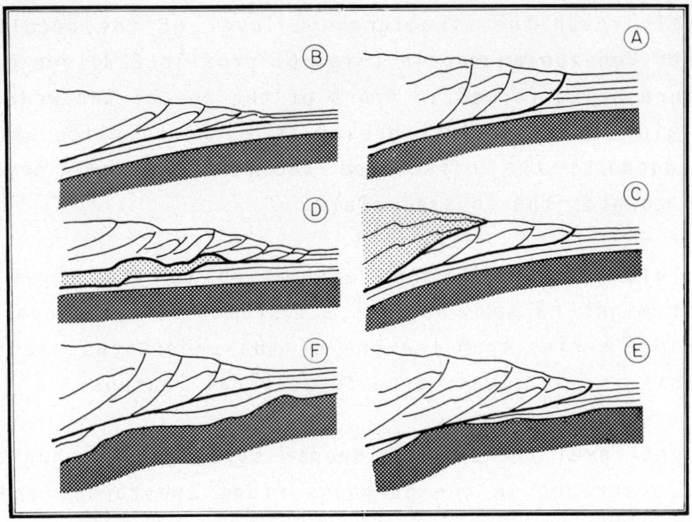

Fig. 9 - Summary of a few tectonic styles at the deformation front of an accretionary prism.

along shear zones which can be considered as internal décollements (see figure in Moore and Biju-Duval, 1984).

In the southward (profile CRV 105, Fig. 5C), the oceanic basement is buried by at least 7000 meters of sediments, a large portion of which corresponds to the Orinoco Neogene deepsea fan. The amplitude of the folds is around 8 km. Because of the very large clastic influx from the South American margin, tectonic structures are not well marked. Moreover, deep currents cause erosion on top of the anticlines. Seismic profiles indicate that progressive folding induces rapid variations and fan structures in the synclinal deeps.

In conclusion, these data on the Barbados deformation front do not clearly show any relationships between reverse faults in the offscraped series and the décollement level. Such relations are obviously illustrated across the deformation front of the North Venezuelan margin (Fig. 6) and of the South Dominican margin (Fig. 7) where wide accretionary wedges have also been described (Biju-Duval et al., 1983a and b).

In these two examples, reverse listric faults progressively reach the stratigraphic level of the décollement. On the North Venezuelan margin (Fig. 6, profile 223) one can see a new structure being formed in front of the toe of the wedge in the abyssal plain. This is interpreted as a future slice which will soon be added to the offscraped sediments as the décollement progresses towards the abyssal plain.

This control of the deformation by the level of the décollement might normally suggest a systematically younger age of the offscraped series than the one of the underthrust sediments. However, this is not always true for several reasons:

. Décollement level transfer to deeper stratigraphic horizons via ramps as described in the Barbados ridge (Westbrook and Smith, 1983, see below),

. Internal development of thrusting into the wedge as suspected in the South Dominican accretionary prism. Here internal thrusting observed on the multichannel seismic reflection profile (Fig. 7, profile 112) can be compared with the flat tectonic contact observed nearby on land in the Sierra de Neiba (Biju-Duval et al., 1983-a) when tectonized pelagic and clastic rocks from Paleocene to Miocene age directly overthrust the moderately deformed late Neogene Enriquillo trough infilling.

In conclusion, sediment thickness and sedimentary facies of the subducted plate have a major role in the tectonic style of folds and frontal thrusts. The location of the décollement level and the thickness of sediments above this level will determine the size, the amplitude of induced anticlines and the dips of the reserve faults which separate the different slices. The location of the level of décollement in the stratigraphic column is partly constrained by the physical properties of the sediments. Therefore, in hole 543 (DSDP, Leg 78A) the uppermost part of the Lower Miocene, where the décollement will propagate, corresponds to a transition between radiolarian muds at the bottom and smectitic muds with associated increase of porosity, decrease of density and shear strength (Moore, Biju-Duval at al., 1984). Fluid circulation is certainly essential in the dynamics of the

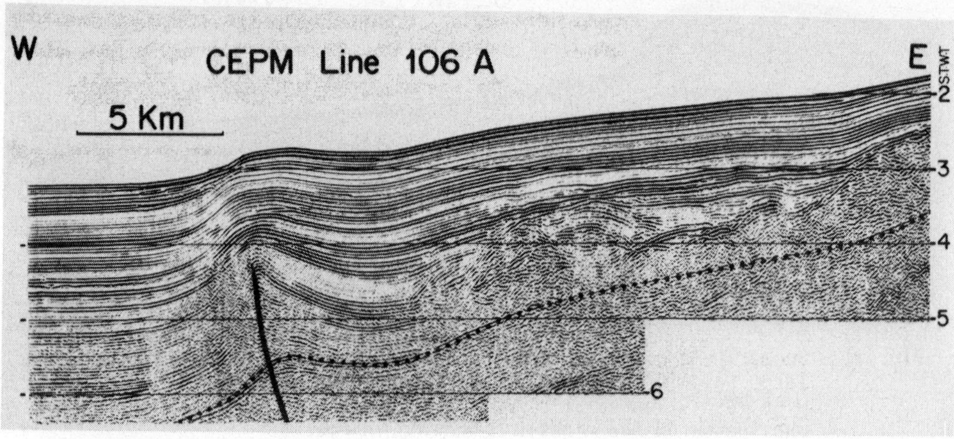

Fig. 10 - Seismic expression of deformation at the back of the accretionary prism. Location in Figure 3.

deformation. The origin of the fluids could be due to the occurrence of poorly consolidated pelagic sediments with low initial permeability. By development of horizontal constraints, the water is expelled and the phenomenon will be accentuated at depth with the increase in the lithostatic pressure. A preferential pathway for water migration could be large or small thrust faults. The presence of water in such a shear zone could in turn develop changes in the physical properties of the parent rocks (acoustics impendance for instance) which could explain some dipping reflectors commonly observed in accretionary prism (Cloos, 1984). Part of the water is not necessarily entirely expelled but could be deeply buried and so participate in crustal metamorphism (Fryer et al., 1985). It could also be involved in the magma generation leading to the calco-alkaline volcanism together with subducted sediments (Dupre et al., 1985).

The above-described shortening represents only part of the total deformation due to plate convergence. Seismic data show thickening above the décollement, which implies the large-scale addition of material on the same vertical. This thickening increases in the prism by internal secondary stacking and addition of older material at the bottom.

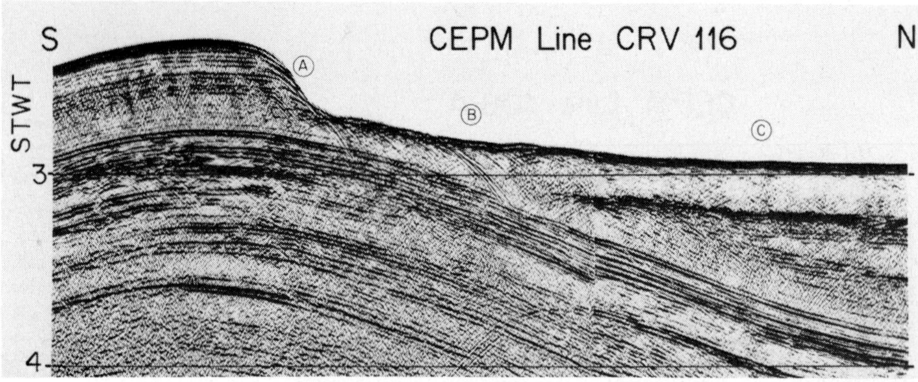

Fig. 11 - Submarine erosion, sliding and resedimentation along the flank of a large fold in the central part of the Barbados Ridge.
High resolution seismic profile (above) and corresponding seabeam map (below on the right). Location in Fig. 3.

Accretion below the prism: underplating

In the Barbados accretionary prism, depth variations in the décollement level from east to west but also from north to south lead to incorporation below the prism of part of the initially underthrust sediments. These depth variations of the décollement occur along ramps as depicted in Figure 8. Longitudinal ramps (A) parallel to the deformation front are observed on deep seismic profiles across the prism (Westbrook and Smith, 1983). Transverse ramps (E) are not so easy to observe, mainly due to the lack of seismic profils parallel to the deformation front. Nevertheless, analysis of Seabeam map and seismic profiles around 12°50N (Biju-Duval et al., 1982) enable us to propose such transverse E-W ramps. Very rapid variations in the bottom topography which, as seen above, expresses the tectonic structure, also correspond to a radical jump in the level of décollement which rises from south to north.

Frontal accretion and underplating processes are summarized in Figure 9. The simplest are types A and B. They differ according to the outcrop (A) or not (B) as a reverse fault of the décollement level at the deformation front. Landward dipping is the most common, but sometimes seaward dipping faults have been described, i.e. Oregon and North Columbian margin (Bally, 1983). In the second type (B) the décollement can progress seaward and lead to the formation of mud diapirism

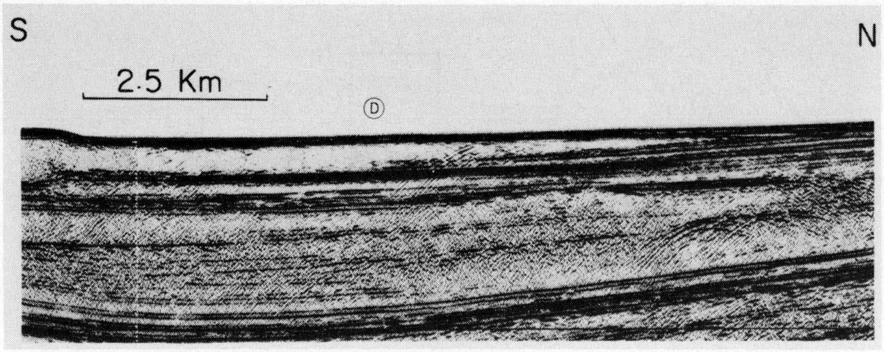

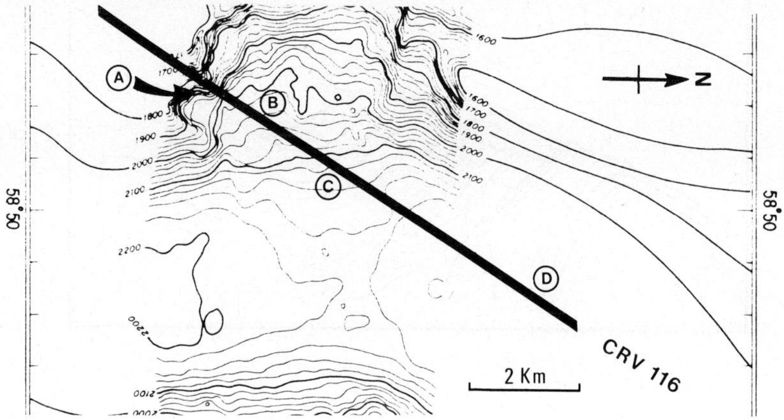

outside of the prism as described east of the Barbados Ridge where a mud volcano has been observed 10 km east of the deformation front (Westbrook and Smith, 1983). The other types are more complicated, i.e. internal thrusting (C) or deep duplex (D).

Another type of underplating, not yet discussed, corresponds to when the disharmonic level between underthrust and sliced sediments does not stand parallel to the bedding but cuts across it on its positive rough ondulations. We can thus expect truncation of some part of the underthrust sediments and their addition to the prism (E). The shear level will later act as a décollement level. One could also propose, although without any evidence up to now, that some highs of the oceanic crust could be cut across and incorporated with the sediments in the prism.

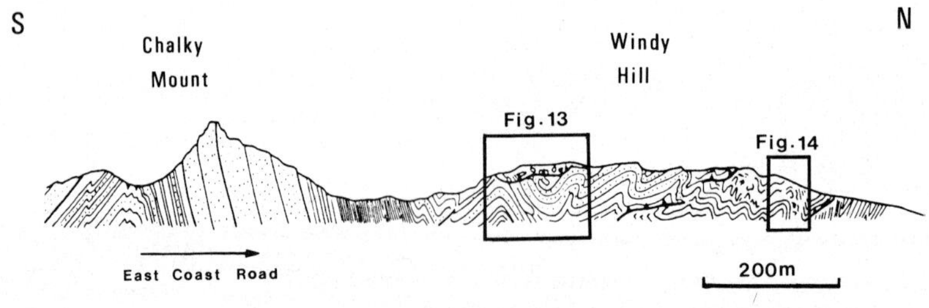

Fig. 12 - General section of the Chalky Mount - Windy Hill area, Scotland district, Barbados. No vertical exaggeration.

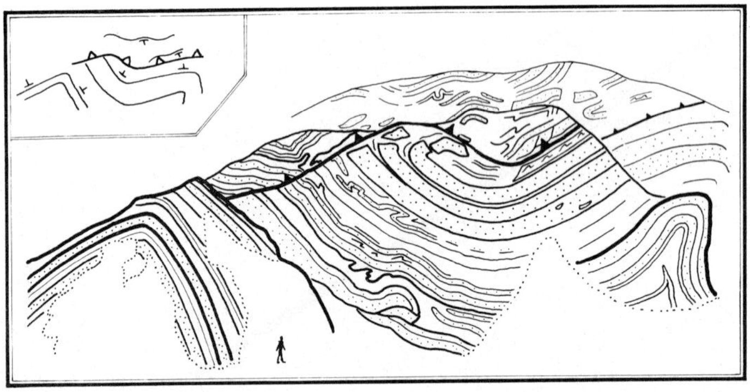

Fig. 13 - Detail of Figure 12.

Accretion above the prism

Stress induced by convergence and subduction is also transmitted to the internal part of the prism, following a predictible distribution depending on sediment facies and thickness (N'Gokwey, 1984). Consequently, the bottom topography is strongly progressively deformed and syntectonic basins can rapidly develop (Biju-Duval et al. 1982). With time this state of constraints enables slope basins and hemipelagic draping sediments to be incorporated in the underlying deformed series. New stacking along secondary layers of décollement can then occur as noticed previously on the rear of the deformation front.

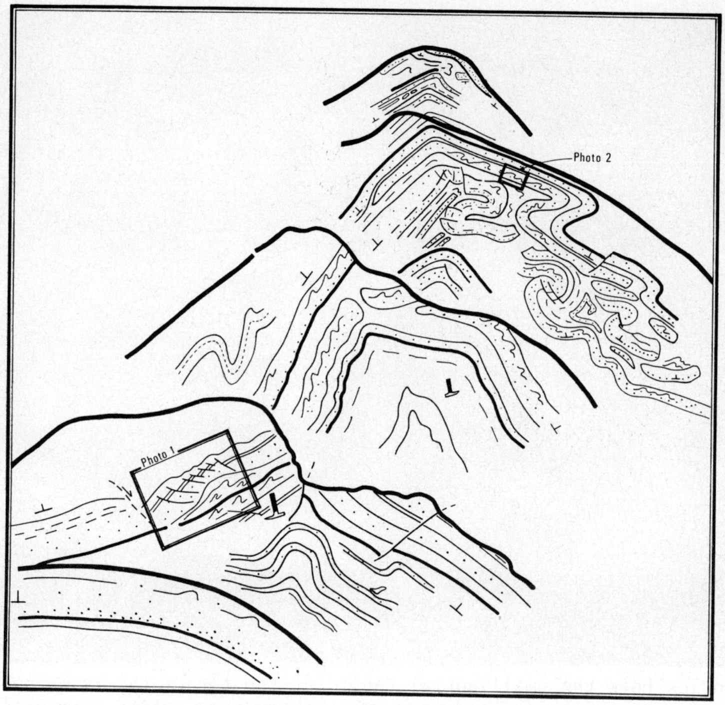

Fig. 14 - Detail of Figure 12 with small-scale illustrations of sliding (Photo 1 and 2).

Accretion at the back of the prism

Finally, the last area where tectonic deformations are recorded across an active margin corresponds to the boundary between the subsident fore-arc basin and the more internal part of the prism. Even if this is not always the case in all accretionary systems, this has been observed all along the Tobago fore-arc basin, as illustrated in Figure 10. As seen on the profile, deformation is very recent and cannot be due to older different processes. Similar structures have also been described on the North Venezuelan and South Dominican margin (Biju-Duval et al., 1983 a and b). Folds and reverse faults may be spectacular, but the amplitude of shortening (difficult to quantify) is probably small with regard to the plate convergence rate. Nevertheless, because of the very high rate of sedimentation and subsidence in those fore-arc basins, one can expect incorporation of a certain amount of sediments in the prism. As will be discussed

Photograph 1 - Note the small normal faults at the top of the sandstone. They root in contorted clays which have been acting as a small décollement level. Clays and sandstones have been later folded.

later on, these sediments will have to be differentiated from the ones accreted at the front.

DEFORMATIONS DUE TO GRAVITY SLIDING

As a result of these various forms of tectonic deformations, the genesis of the Barbados accretionary wedge induced a considerable uplift as well as local subsidence. The consequences are (1) the emergence of Barbados island where good outcrops allow direct observation of past tectonic events, and (2)at any place in the prism the existence of steep slopes on which unconsolidated sediments can easily move for different reasons (earthquakes, deformation, additional loading, etc). Gravity sliding occurs on different scales and with varying intensity and amplitude.

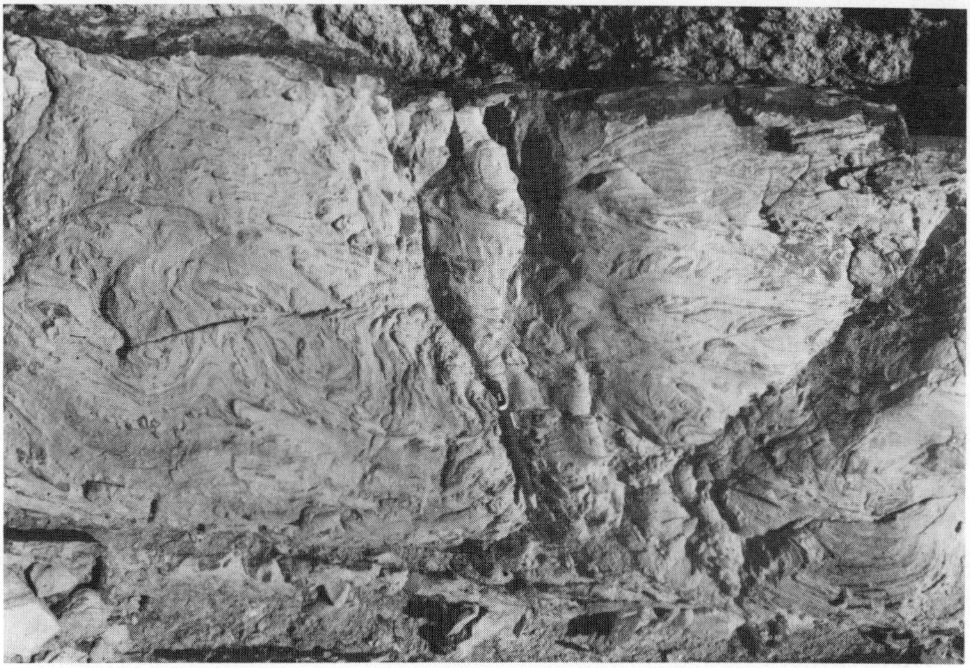
Photograph 2 - Small-scale slumping in a homogeneous sandstone layer

On the scale of the accretionary prism itself, the uplift of the central part is marked by Barbados island with an elevation of 2.5 kilometers above the subsiding Tobago basin. The present morphology results from an equilibrium between the horizontal constraints due to the shortening, the vertical constraint due to loading by sediments, and finally the sediment cohesion (Westbrook et al., 1984; N'Gokwey, 1984). For any decrease in the horizontal constraint, the uplifted part of the prism should subside. This will induce new unbalance and also new deformation and folding by excess of mass and pressure at the boundaries of the wedge. This hypothesis would need to be quantified.

Fine-scale studies carried out on the biostratigraphic material collected in the cores of Leg 78A holes have enabled us to propose the formation of gravity flows at the Pliocene-Pleistocene boundary and in the Pliocene (Hemleben and Auras, 1984). Observations of the microstructures of the sediments (Cowan et al., 1984) have not shown any chaotic structures or mélange zones which would have reinforced the hypothesis of massive gravity sliding, one of the initial explanations for the chaotic seismic aspect on the reflection profiles (Biju-Duval et al., 1978).

Nevertheless, as observed on intermediate scales, gravity sliding on steep slopes is quite common:

Sonar imagery recorded by the Gloria survey south of Tiburon Rise shows an area of debris-flows about 100 km^2 wide in front of a steep bathymetric slope (Belderson et al., 1984).

Seismic imagery recorded by the high-resolution MCS system also shows spectacular deformation due to gravity sliding (Figure 11). A wide collapse structure of about 20 km^2 glided along the slope to the bottom of the basin. It is now internally layered on top of the Plio-Quaternary sequence as a "chaotic" seismic unit which pinches out towards the north. The resulting scarp is about 300 meters high (A in Figure 11).

Smaller slidings, difficult to see by conventional seismic surveys are likely to occur on the whole prism. From the bottom topography and the type of sediments, they can form at any stage of growth of the prism.

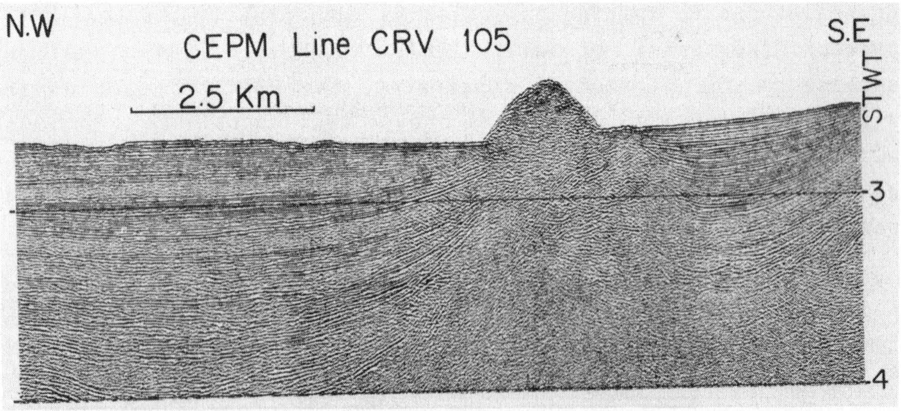

Fig. 15 - Seismic expression of mud volcanism in the southern Barbados Ridge. Location in Fig. 3.

Field studies on Barbados island provide additional information from older series at the core of the prism (Figure 3). A terrigenous suite as old as the Paleocene (Saunders, 1979; Reading and Pudsey, 1981) is highly folded along E-W trends (Speed, 1981; Biju-Duval et al., 1985). A good section is exposed along the east-coast road, Scotland district, between Chalky Mount and Windy Hill. The general structure is a large and dissymmetric anticline followed by a complex synclinorium (Fig. 12). Debris flows and large slumps have been observed in Chalky Mount; in the Windy Hill area some small features are interpreted as gravity sliding during the early phase of folding. The overturned series observed on Figure 13, show internal dislocations and thrustings. They are interpreted as olistolites at the foot of the Chalky Mount anticline. Wide internal slidings observed immediately beneath reinforce the hypothesis. On a smaller scale (few meters) the dislocated and contorted sandstone beds observed on the northern edge of the Windy Hill structures (Figure 14) also indicate such gravity phenomena. Also included within these folds, some small-scale features show a typical gravity print, i.e. bed-to-bed motion (photo 1) or microslumping into beds of sand (photo 2).

Gravity tectonics then played an active role in the deformation of the clastic series of Barbados island. One could even propose, as an extreme hypothesis, that the whole Windy Hill structure is an olistostrome with a possible décollement over the massive sandy Chalky Mount series at the level of plastic clays which separate the two stratigraphic units (Figure 12).

These different examples show the difficulty in finding evidence for such deformation related to gravity but also emphasize the numerous reasons for such a development in unconsolidated and water-filled sediments. The role of undercompacted clays is probably very important as it is in the next type of deformation observed in accretionary prisms.

MUD DIAPIRISM

Mud diapirism is obvious on the southern Barbados Ridge in the form of numerous mud volcanoes. They are clearly depicted on Seabeam maps (Valery et al., 1985) and on Gloria imagery (Stride et al., 1982). These volcanoes could form prominent sea-mounts (more than 200 meters high in Fig. 15) when they are isolated in a sedimentary basin. However, they are usually formed on top of anticlinal structures as they are found on Trinidad island for instance (Higgins and Saunders., 1974). These volcanoes are related to undercompacted clays in the prism and the mud extrusion itself is probably a result of the propagation of tectonic constraints. This led to large-scale reworking of old sediments in more recent ones. For instance in Trinidad mud volcanoes, Upper Cretaceous to Pliocene boulders have been collected in a matrix of Miocene to Pliocene age. Two origins have been proposed for these exotic pebbles. Rugged blocks of Neogene age have probably been scraped off from the upper sedimentary section by the ascending mud. Older blocks, some of them well rounded, would be secondary reworked pebbles from any olistoliths or previous mud extrusions of Miocene to Pliocene age (Higgins and Saunders, 1974). The presence of undercompacted clays at depth is probably related to the rapid deposition and covering of sediments with initial low permeability. Such conditions are thought to occur preferentially along margins with high sedimentary influxes.

Recent and present examples are well documented because of their spectacular occurrence. However, their recognizion in past orogenic belts is much more difficult because basically they are temporary structures that are easily destroyed by subsequent sedimentary or tectonic events. It is also quite difficult to distinguish them from other chaotic series of tectonic or gravitational origin. This is the case for the "Joe's River" formation on Barbados island which has been successively interpreted as a shear zone breccia, a submarine land slide or a mud volcano (see discussion and references in Saunders, 1979). The last hypothesis is the one we presently favor (Biju-Duval et al., 1985).

DISCUSSION

The tectonic building of the Barbados accretionary prism is characterized by numerous types of deformation along the different part of the wedge transversely as well as longitudinally. A prominent parameter in all the tectonic processes is the occurrence of fluids and overpressurized sediments. Pore fluid pressure will determine the location of the main décollements, will facilitate gravity sliding and will be a leading factor in the diapirism. The pore pressure distribution is directly related to the sediment distribution in the stratigraphic pile of the oceanic series. So the geological history of the oceanic sedimentary basin, i.e. environments of deposition, rates of accumulation, lateral change of facies, will finally explain the distribution of physical properties which in turn will control the tectonic development of the prism itself.

In addition to the variety of tectonic styles, a wide range of sedimentary facies can also be observed in the prism. Two large classes of sediments can be distinguished: (1) those whose origin is outside of the prism,i.e. offscraped oceanic sediments, direct pelagic blanketing and terrigenous influxes from the island arc or the continent; (2) those formed by internal reworking (gravity sliding, shale diapirism) or those which have been deposited along a margin before the development of active subduction.

This kind of juxtaposition is interesting to note for the interpretion of past accretionary complexes. Particularly, all deformed sediments are not necessarily derived from offscraped oceanic sequences.

From this rapid review of tectonic processes it appears that the location, style and intensity of deformation are not uniform along the prism in spite of a single geodynamic setting. For instance wide and large undisturbed basins can simultaneously develop close to highly disturbed series. When ancient orogenic belts are discussed, such closeness of different style and intensity in the deformation history is not necessarily evidence of any subsequent tectonic transport and juxtaposition. It could also just reflect the original depositional setting. The notion of tectonic phase should also be used carefully, keeping in mind this type of progressive tectonic development and syntectonic basin formation.

REFERENCES

Aubouin, J., Von Huene, R. et al., 1982 - Initial reports of DSDP, 67, Washington (U.S. Gov. Printing office).

Bally, A.W. ed., 1983 - Seismic expression of structural styles. A picture and work atlas AAPG studies in geology, series 15, V. 3.

Belderson, R.H., Kenyon, N.H., Stride, A.H., 1984 - Morphology and structural trends of the Barbados Ridge complex in the vicinity of DSDP sites 541, 542 and 543 as revealed by Gloria long-Range Side-Sean Sonar. In: Biju-Duval, B., Moore, J.C. et al., Initial report of the DSDP, 78 A.

Biju-Duval, B., Mascle, A., Montadert, L., Wannesson, J., 1978 - Seismic investigation in the Columbia, Venezuela and Grenada Basins, and on the Barbados Ridge for future IPOD drillings. Geol. Mijnbouw, v. 57.

Biju-Duval, B., Bizon, G., Mascle, A., Muller, C., 1982-A - Active margin processes: field observations in southern Hispaniola. AAPG memoir 34.

Biju-Duval, B., Mascle, A., Rosales, H., Young, G., 1982-B - Episutural oligo-miocene basins along the North Venezuelan Margin. AAPG, memoir 34.

Biju-Duval, B., Le Quellec, P., Mascle, A., Renard, V., Valery, P., 1982-C - Multibeam bathymetric survey and high resolution seismic investigations on the Barbados Ridge complex (eastern Caribbean): a key to the knowledge and interpretation of an accretionary wedge. Tectonophysics, v. 80.

Biju-Duval, B., Moore, J.C., 1984 - Initial Reports or DSDP, 78A: Washington (U.S. Gov. Printing office).

Biju-Duval, B., Caulet, J.P., Dufaure, Ph., Mascle, A., Muller, C., Richert, J.P., Valery, P., 1985 - The terrigenous and pelagic series of Barbados Island. In: Symposium Geodynamique des Caraïbes, Mascle, A. (ed.), Editions Technip.

Boulegue, J., le Pichon, X., Iiyama, J.T., 1985 - Prevision des tremblements de terre dans la région de Tokaï (Japon). C.R. Acad. Sc. Paris, v. 301.

Bouysse, Ph., 1984 - The Lesser Antilles Island Arc, structure and geodynamic evolution. In: Biju-Duval, B., Moore, J.C. et al., Initial Reports of the DSDP, 78A.

Cloos, M., 1984 - Landward dipping reflectors in accretionary wedges: Active dewatering conduits? Geology, V. 12/9.

Cowan, D.S., Moore, J.C., Roeske, S.M., Lundberg, N., Lucas, S.E., 1984 - Structural features at the deformation front of the Barbados Ridge

Complex, DSDP-78A. In: Biju-Duval, B., Moore, J.C. et al., Initial Reports of the DSDP, 78A.

Dupré, B., White, W.M., Vidal, Ph., Maury, R.C., 1985 - Utilisation des traceurs couplés (Pb-Sr-Nd) pour déterminer le role des sédiments dans la genèse des basaltes de l'arc des Antilles. In: Symposium Geodynamique des Caraïbes, Mascle, A., (ed.), Editions Technip.

Fontas, P., Valery, P., Le Quellec, P., Mascle, A., Renard, V., Tardy, M., Biju-Duval, B., 1984 - Multibeam bathymetric survey of the Leg 78-A drilling area and comparison with the southern part of the Barbados Ridge deformation front. In: Biju-Duval, B., Moore, J.C. et al., Initial Report of the DSDP, 78A.

Fryer, P., Ambos, E.L., Hussong, D.M., 1985 - Origin and emplacement of Mariana forearc seamounts. Geology, v. 13/11.

Hemleben, C., Auras, A., 1984 - Variation in the calcite dissolution pattern on the Barbados Ridge complex at Sites 541 and 543, Deep Sea Drilling Project Leg 78-A. In: Biju-Duval, B., Moore, J.C. et al., Initial Reports of the DSDP, 78-A.

Higgins, G.E., Saunders, J.B., 1974 - Mud volcanoes, their nature and their origin. In: Contributions to the geology and paleobiology of the Caribbean and adjacent areas. Verhanlunger der Naturforschenden Geselfchaft in Basel, v. 84.

Hussong, D., Uyeda, S. et al., 1982 - Initial Reports or the DSDP 60, Washington (U.S. Gov. Printing office).

Mascle, A., Cazes, M., Le Quellec, P., 1985 - Structure des marges et bassins caraïbes, une revue. In: Symposium Geodynamique des Caraïbes, Mascle, A., (ed.), Editions Technip.

Moore, J.C., Biju-Duval, B., 1984 - Tectonic synthesis, DSDP, Leg 78-A. Structural evolution of offscraped and underthrust sediment, Northern Barbados Ridge Complex. In: Biju-Duval, B., Moore, J.C. et al., Initial Report of the DSDP, 78-A.

N'Gokwey, K., 1984 - Numerical modeling of sediment deformation linked to subduction: mechanical model and comparison with Barbados Ridge Complex. In: Biju-Duval, B., Moore, J.C. et al., Initial Reports of the DSDP, 78-A.

Pudsey, C.J., Reading, H.C., 1982 - Sedimentology and structure of the Scotland Group, Barbados. In: Trench fore-arc Geology, Leggeth ed., Geological Society of London special publication n° 10.

Saunders, J.P., 1979 - Field guide to Trinidad, Tobago and Barbados, 4th Latin American Geological Congress, Port of Spain, Trinidad and Tobago.

Speed, R.S., 1981 - Geology of Barbados: implication for an accretionary origin. International Geological Congress, Paris, Colloquium G 3.6, Oceanologica Acta, v. 4.

Stride, A., Belderson, R., Kenyon, N., 1982 - Structural grain, mud volcanoes and other features on the Barbados Ridge complex revealed by Gloria long-range side-sean sonar. Marine Geology, v. 49.

Valery, P., Nely, G., Mascle, A., Biju-Duval, B., Le Quellec, P., Berthon, J.L., 1985 - Structure et croissance d'un prisme d'accretion tectonique proche d'un continent : la ride de la Barbade au sud de l'arc antillais. In: Symposium Geodynamique des Caraïbes, Mascle, A. (ed.), Editions Technip.

Westbrook, G.K., 1982 - The Barbados Ridge Complex: tectonic of a mature-arc system. In: Trench and fore-arc geology, Leggett, J.K. (ed.), Geological Society of London special publication, v. 10.

Westbrook, G.K. and Smith, M.J., 1983 - Long décollements and muds volcanoes: evidence from the Barbados Ridge Complex for the role of high pore fluid pressure in the development of an accretionary complex. Geology, v. 11.

Westbrook, G.K., Mascle, A., Biju-Duval, B., 1984 - Geophysics and the structure of Lesser Antilles fore-arc. In: Biju-Duval B., Moore, J.C. et al., Initial Reports of the DSDP, 78-A.

NORTHERN AND SOUTHERN CARIBBEAN FESTOONS (PANAMA, COLOMBIA-VENEZUELA AND HISPANIOLA - PUERTO RICO), INTERPRETED AS PSEUDOSUBDUCTIONS INDUCED BY THE EAST-WEST SHORTENING OF THE PERICARIBBEAN CONTINENTAL FRAME.

J.F. STEPHAN[1], R. BLANCHET[1] and B. MERCIER DE LEPINAY[2]
(1 : *GIS "Océanologie et Géodynamique", Université de Bretagne Occidentale, Brest, France.* 2 : *UA 215, U.P.M.C., Paris, France).*

ABSTRACT

The Caribbean plate is moving eastward, relatively to and between the two Americas, at about 2 cm/year. This motion is expressed : - laterally by large strike-slip movements, sinistral on the North side,dextral on the right one ; - at the leading edge, to the East, by the Lesser Antilles subduction zone and the Barbados accretionary prism.

The Caribbean plate is made of a central domain with oceanic crust, more or less thickened, surrounded by a continental frame. This frame is deformed and draws arched structures ("festoons") with a Caribbean convexity : the Panama and the Colombia - Venezuela festoons to the South ; the Hispaniola -Puerto-Rico (= Los Muertos) festoon, to the North. These structures orientated at 90° to the Lesser Antilles Arc, geometrically mime subduction zones, with their frontal accretionary prism similar to the Barbados or Nankai one. Nevertheless, they do not have the seismicity nor the volcanism tied to subductions : we propose to call them "pseudosubductions".

These arched structures are classically interpreted by the authors as the result of the light convergence (<1 cm/an) between the North and South American plates. Structural and seismotectonics data allow us to propose a different mechanism. The Caribbean festoons could be the result of the bending of the Caribbean continental frame related to an approximatly E.W. shortening. Due to the geometry of the continental borders (Panama and Hispaniola Puerto-Rico) and/or to the displacement along intra-continental strike slip faults (Colombia - Venezuela, with the northern extrusion of the Maracaibo block between the Bocono and the Santa Marta - Bucaramanga faults), this bending creates the pseudosubductions.

INTRODUCTION

The Caribbean plate is moving eastward with a rate of about 2 cm/year (fig. 1) relatively to the North and South American plates (Mac Donald and Holcombe, 1978, Minster and Jordan, 1978 and Plate tectonic map of the circum-Pacific region, NE quadrant, 1981 indicating all rates used in this paper). This movement is

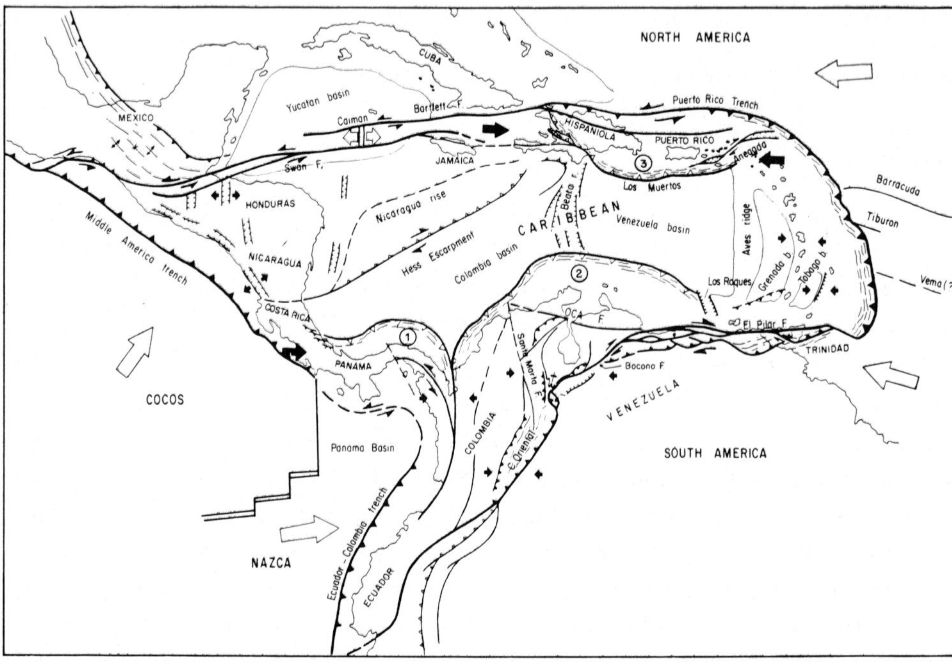

Figure 1 : *The Caribbean plate with Northern and Southern Caribbean festoons : 1 - Panama ; 2 - Colombia - Venezuela ; 3 - Hispaniola - Puerto-Rico.*

displayed along the east front by the Lesser Antilles subduction zone where Atlantic oceanic crust is absorbed. The lateral boundaries are characterized by strike slip faulting with a more complex setting in the South.

Along the Pacific side the three plates (Caribbean, North and South American plates) are limited by a subduction zone almost continous from the South of the Gulf of California all along Central and South America. This subduction zone comprises :

1. to the North, the Central American trench in front of the North American and Caribbean plates where the oceanic crust of the Cocos plate is subducted with a rate of 7 to 9 cm/year to the NE.

2. to the South, the Colombia-Equator trench facing South America where the oceanic crust of the Nazca plate is subducted with a rate of 8 cm/year towards ENE.

The northern and southern boundaries of the Caribbean plate act as transform faults between the W vergent subduction of the Lesser Antilles and the E-NE vergent subduction of the Pacific front.

The northern boundary (fig. 1) of about 3000 km length is essentially a broad zone of left lateral strike slip faulting including from W to E :

- the Polochic and Motagua strike slip faults in Guatemala and Honduras ;
- the submarine Swan and Bartlett/Oriente strike slip faults which enclose the narrow Cayman oceanic basin (120 km from N to S) ;
- and the Puerto Rico trench E of Cuba (maximum depth 8.269 m) characterized by very oblique subduction between the pure strike slip Bartlett fault and the frontal subduction of the Lesser Antilles.

The southern boundary is entirely intracontinental (fig. 1) including the northern and northwest side of South America. It comprises two segments (Stephan, 1982) :

1. the eastern segment (= Caribbean) is dominated by the right lateral El Pilar and Bocono strike slip faults (about 1400 km long) ;
2. the western segment (= Andean), where tectonic movements are marked by folds and overthrusts along the front of the Andes in Colombia and Ecuador. The transition between these two types of deformation is located at the contact between the East Cordillera of Colombia and the Merida Andes of Venezuela.

The Andean segment represents a triple junction between the Caribbean, the South American and the Nazca plates.

I .- THE MAIN GEOLOGICAL AND GEODYNAMIC FEATURES OF THE CARIBBEAN PLATE (fig. 2)

The Caribbean plate is of very particular constitution. It consists of oceanic crust in the central zone, and of a continental frame along which the present boundaries are located.

A. The center of the plate :

The central part of the plate corresponds to the Colombia and Venezuela basins, the Beata ridge and the southern part of the Nicaragua ridge. The substratum of these different domains is of thickened oceanic crust of late Cretaceous age or older (Saunders et al., 1973, Christofferson, 1976, Lu and Mc Millen, 1983). This region was only slightly deformed with exception of 1. the Beata ridge where post-Eocene (post A") uplift of at least 1000 m of the basement took place (Roemer et al., 1976) ; 2. the Hess escarpment (Case and Holcombe, 1980) where the recent sediments of the Colombia basin are folded (profiles of the University of Texas, Austin, unpublished) ; 3. the Pedro escarpment (Case, 1981).

B. The continental frame :

The continental frame consists of Central America, the southern Nicaragua ridge, the north Caribbean orogen (Jamaica, Hispaniola, Puerto Rico, Virgin islands), the Lesser Antilles arc *sensu lato* (from the Aves ridge to the Tobago basin except the Barbados ridge), the south Caribbean orogen and the north Andes Cordillera (Trinidad, Venezuela, Colombia).

The north Caribbean orogen, the south Caribbean range and its prolongation in Colombia, as well as the western Cordillera comprise piles of nappes made of ophiolites and/or Jurassic - Cretaceous volcanic arcs (Stephan et al., 1980 ; Bourgois et al., 1982a). Their formation (overthrust, metamorphism) took place mainly between Upper Cretaceous (about 80 Ma) and Upper Eocene (about 35 Ma). These orogens mark the north and the south paleo-boundaries of the Caribbean plate.

The Andes *sensu lato* (Central and Eastern Cordillera, Santander and Santa Marta ranges in Colombia, the Merida Andes in Venezuela) are a "liminaire" chain (as defined by Aubouin, 1972) created along the Pacific front of the South American continent. Volcanic-plutonic arc during Triassic - Jurassic time, and marine shelf during the Cretaceous, it has been folded and uplifted only since Middle Miocene (Faucher and Savoyat, 1973 ; Julivert, 1973 ; Anonyme, 1972 ; Stephan, 1982, 1985).

The Lesser Antilles arc s.s. has been active at least since Late Eocene - Oligocene time (Westercamp and Tazieff, 1980). It was established upon a Jurassic - Upper Cretaceous arc which is exposed at La Desirade and on Tobago (Girard, 1981 ; Girard and Maury, 1983). The Aves ridge was also an arc before (Upper Cretaceous) and after (Oligocene) the opening of the Grenada basin (Nagle, 1972 ; Richard, 1983 ; Pinet et al., 1985).

The southern Nicaragua ridge and continental Central America (Guatemala, Honduras, Nicaragua) correspond to a continental microblock belonging to North America (Tardy, 1980 ; Wadge and Burke, 1983). It is the substratum of the Tertiary volcanic arc of Central America partially superposed upon a tectonized Cretaceous arc (Tardy and Azema, 1985).

The Central American Isthmus (Costa Rica, Panama, NW Colombia) seems to have a predominantly ophiolitic substratum obducted during late Cretaceous time (about 80 Ma., Bourgois et al., 1982b). This suture can be followed north of Costa Rica in the inner wall of the Central American trench (Aubouin, Von Huene et al., 1982). It marks the western paleo-boundary of the Caribbean plate.

In conclusion, the present boundaries of the Caribbean plate follow in general the old ones. The orogens, created by successive interactions (obduction, collision, strike slip) between the North and South American plates since 80 Ma., have been incorporated in the peri-Caribbean continental frame. Only Cuba has been "welded" on to North America by the opening of the Yucatan basin at the beginning of the Tertiary (Hall and Young, 1980). The present boundaries represented by the Bartlett and Swan strike slip faults existed since the end of the Eocene. They cut the NW part of the Caribbean paleo-plate.

C. Recent deformation of the peri-Caribbean frame :

Deformations since Oligocene time are reflected in the continental frame of the Caribbean plate. Their importance varies depending from the region. Deformations are minor in continental Central America, southern Nicaragua ridge and Lesser Antilles arc *sensu lato* (with exception of the Barbados ridge which does not belong to the continental frame). In the first area they are restricted to N-S or NW-SE trending grabens locally associated with alkaline volcanism (Wadge and Wooden, 1982). Apparently this extentional regime can be followed to the inner wall of the Central American trench (Aubouin, Bourgois et al., 1982 ; Bourgois et al., 1984). Deformation of the Lesser Antilles arc is limited to an important fracturation of the volcanic arc. It seems to be mainly the result of an almost E-W compression (Bonneton, 1982). In general, the continental frame is deformed by bending towards the Caribbean basins which are overthrust. These festoons are the Hispaniola - Puerto-Rico festoon in the North and the Panama and Colombia - Venezuela festoon in the South (fig. 1).

From a geometrical point of view, these festoons imitate subduction zones. Indeed some of them display a deformation front or an active accretionary prism. However, the characteristics of active subduction zones are missing. They have neither seismicity which would allow us to define the depth of the Wadati-Benioff plane nor an active arc volcanism. It has to be also noted that the direction of these festoons is 90° to the Lesser Antilles arc, and that their convexity is opposite.

In the following the main structural and seismo-tectonic characteristics of each festoon are described.

II .- THE PANAMA FESTOON.

It is the smallest and with the most pronounced convexity. It extends from W to E from the Panama - Costa Rica boundary to the Uraba Gulf and Atrato - San Juan depression in Colombia.

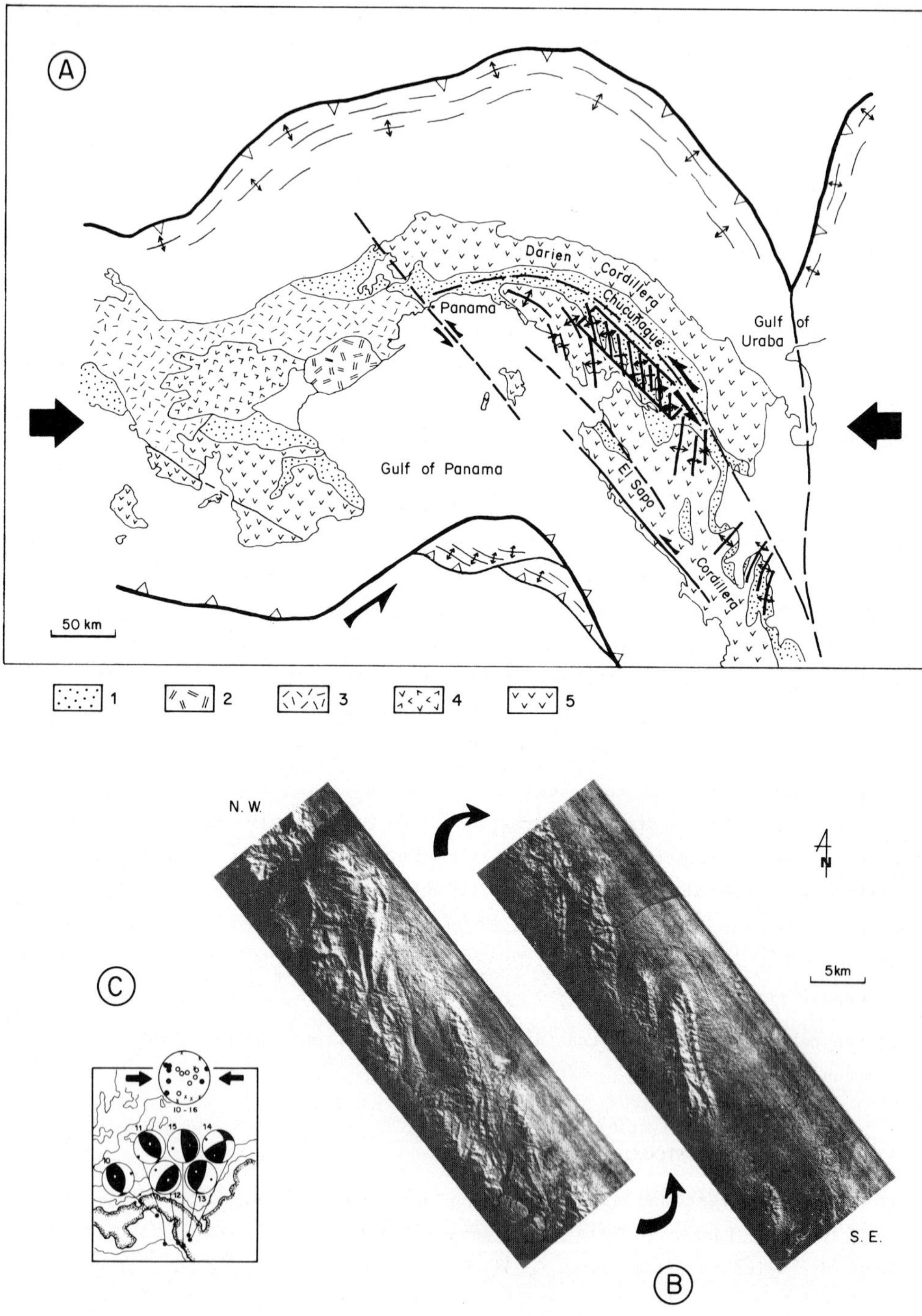
A
Darien
Cordillera
Chucunaque
Panama
Gulf of Uraba
Gulf of Panama
El Sapo
Cordillera
50 km
1
2
3
4
5
N. W.
S. E.
N
5km
C
B
10-16
11
15
14
10
12
13

A. The North Panama accretionary prism.

The prism is about 300 km wide and 700 to 800 km long (curved). It is an active structure as shown by multichannel seismic profiles from the University of Texas (Lu and Mc Millen, 1983) and the sea-beam survey by the Jean Charcot (unpublished). The sediments of the Colombia basin are incorporated in the accretionary prism. The decollement lies above the reflector B" which corresponds to the Upper Cretaceous oceanic basement of the basin (Lu and Mc Millen, 1983).

To the East, in the area of the Gulf of Urata, the North Panama accretionary prism faces the Colombia - Venezuela prism (Case and Holcombe, 1980). However, the detailed structures of this area are unknown. The deformation front disappears progressively westwards. It terminates onshore with the recent overthrusts of the outermost parts of Panama and Costa Rica (Case and Holcombe, 1980).

B. The continental substratum of Panama.

It is strongly curved and consists of two parts (Case and Holcombe, 1980) :
1. a broad central zone, west of the Panama strait, which is composed of Tertiary volcanics, and a Pacific peninsular zone where the Mesozoic substratum outcrops. This picture is similar to Costa Rica located outside of the festoon ;
2. a region east of the strait consisting of the curved Neogene Chucunaque depression which is surrounded by the Darian Cordillera in the North, and the El Sapo Cordillera in the South with a Cretaceous - Paleogene substratum. The latter turns into the coastal Baudo Cordillera in Colombia. The Chucunaque depression prolongates in the depression of Atrato. These features change from an E-W direction near Panama to a N-S direction near Baudo.

The contact between the Chucunaque depression and the El Sapo Cordillera (fig. 2) is characterized by a series of N-S to NNW-SSE en échelons folds (Case and Holcombe, 1980). These folds are very recent as confirmed by the young morphology on radar images (Wilcox et al., 1973). The obliquity of these folds in respect to the orientation of the El Sapo Cordillera and the Chucunaque depression is indicative of left lateral strike slip movements (NW-SE) in the substratum (Wilcox et al., 1973).

Figure 2 : *Panama festoon : A - Schematic geologic - tectonic map (from Case and Holcombe, 1980, simplified ; marine data south of Panama Gulf after T. Hilde, pers. comm. 1984) : 1. Post-Eocene sediments ; 2. Quaternary and Tertiary volcanics ; 3. Tertiary volcanics ; 4. Miocene and Oligocene volcanics ; 5. Upper Cretaceous and Lower Tertiary volcanics and intrusives.*

B - En échelon folds along the Chucunaque depression, radar imagery (Wilcox et al., 1973) : see location on fig. 2A.

C - Focal mechanisms in the Panama area showing East - West compression (from Pennington, 1981).

Further NW-SE left lateral strike slip faults have been recognized in Panama. One of them could control the Panama strait (Lowrie et al., 1982).

The N-S oriented folds of this festoon which belongs to Colombia have the same direction as the Atrato depression and the Baudo Cordillera (fig. 2).

The El Sapo Cordillera is characterized by an intense seismo-tectonic activity. The focal mechanisms determined by Pennington (1981) indicate an almost E-W compression. These geophysical data are in good agreement with the described structural observations.

It seems that the bending of the Panama festoon (at least E of the strait) is predominantly the result of an E-W compression of the continental frame (Stephan, 1982). A slightly different hypothesis is also given by Wadge and Burke (1983).

C. The Caribbean / Nazca boundary along the southern rim of the festoon.

The subduction is interrupted south of Panama between the Central America and the Colombia - Venezuela trench (fig. 1). The contact between the continental frame of Panama and the Miocene oceanic crust of Nazca is marked by a series of E-W orientated ridges and troughs (Lonsdale and Klitgord, 1978). The turbiditic infilling is only slightly deformed (Lowrie, 1978). The seismicity is not remarkable compared with the neighbouring zones (Pennington, 1981). However, this part of the Caribbean / Nazca boundary is characterized by left lateral strike slip movements with a rate of 5-6 cm/year (Plate tectonic map of the circum-Pacific region, NE quadrant 1981 ; remark : the South Panama strike slip fault shown on this map can be followed eastward according to the authors, where it cuts the eastern part of the Panama festoon ; the structure of this region given in the map (Case and Holcombe, 1980) however, is opposite to this interpretation). It is certain that neither a N-S convergence nor a N-S shortening occurs within this region.

III .- THE COLOMBIA - VENEZUELA FESTOON.

It is the widest and most complex. It extends from the Gulf of Uraba and the Atrato - San Juan depression in the West till the Los Roques Canyon in the East (fig. 1). It models itself on the Andes / South Caribbean range contact.

A. The Colombian - Venezuelian accretionary prism.

This prism is 1800 km long and 50 to 130 km wide. Its western end is exposed in the area of Sinu in Colombia, east of the Uraba Gulf (Duque-Caro, 1979), where it faces its Panama equivalent. To the East it ends at the Los Roques Canyon, a NW-SE trending graben of unknown significance.

The Colombian - Venezuelian prism is also an active feature as displayed by numerous seismic reflection profiles, and by some available sea-beam surveys (Jean Charcot, unpublished). The prism is made of the sedimentary cover "supra B" " or "supra A" " of the Colombia and Venezuela basins (Talwani et al., 1977 ; Ladd and Watkins, 1979 ; Biju-Duval et al., 1982a, b).
The accretionary wedge (Curaçao ridge) north of Venezuela is flanked by a "fore-arc" basin (Los Roques basin) where the negative gravity anomaly is localized (Bowin, 1976).

B. The continental substratum.

The continental part of the festoon corresponds to the link between the Andes and the South Caribbean range. The contact between these two ranges is marked by the pre-Oligocene Barquisimeto Transverse (Aubouin, 1975 ; Stephan, 1982 ; 1985). According to the type of deformation it is possible to distinguish 4 zones within the continental festoon (Stephan, 1982) :

1. *The Maracaibo block* (= Santa Marta - Maracaibo block of Kellog and Bonini, 1982, fig. 3) of triangular shape is located in the center of the festoon (fig. 3). It cuts the Caribbean allochton and its Andean fore-land. It is bounded by the Oca fault in the North (Rod, 1956 ; Feo-Codecido, 1972), the Santa-Marta - Bucaramanga fault (S.M.B.) in the West (Campell, 1968 ; Irving, 1971), and by the active Bocono strike slip fault in the East (Rod, 1956 ; Giegengack et al., 1976 ; Schubert, 1982).

The seismicity at the Oca and S.M.B. fault (N100/110° respectively N165°) is almost non existent compared with the activity along the Bocono fault. The predominantly vertical Plio-Quaternary movements along these two faults on both sides of the Santa-Marta range explain the important uplift to the range (more than 6 km above the Magdalena delta in the West and the Rio Hacha plain in the North). The Oca and S.M.B. faults have acted as strike slip faults especially from Middle to Late Miocene time : right lateral strike slip movements for the Oca fault provoking the "en échelon" folding of the Oligocene - Miocene Falcon pull-apart basin (Vasquez and Dickey, 1972 ; Stephan, 1982) ; left lateral strike slip movements of the S.M.B. fault associated with folding of the Colombia eastern Cordillera.

The Bocono fault (N50°) represents along its almost entire length (450 km) the backbone of the Venezuelian Merida Andes. It is an active right lateral strike slip fault as confirmed by geological observations and seismo-tectonic data (Giegengack et al., 1976 ; Schubert, 1982). The Bocono fault turns northeastward into the system of the also right lateral El Pilar strike slip fault.

The Merida Andes modeled itself on the Bocono fault,corresponds to a large doming of the South American basement surrounded by reverse faults. The mountain chain

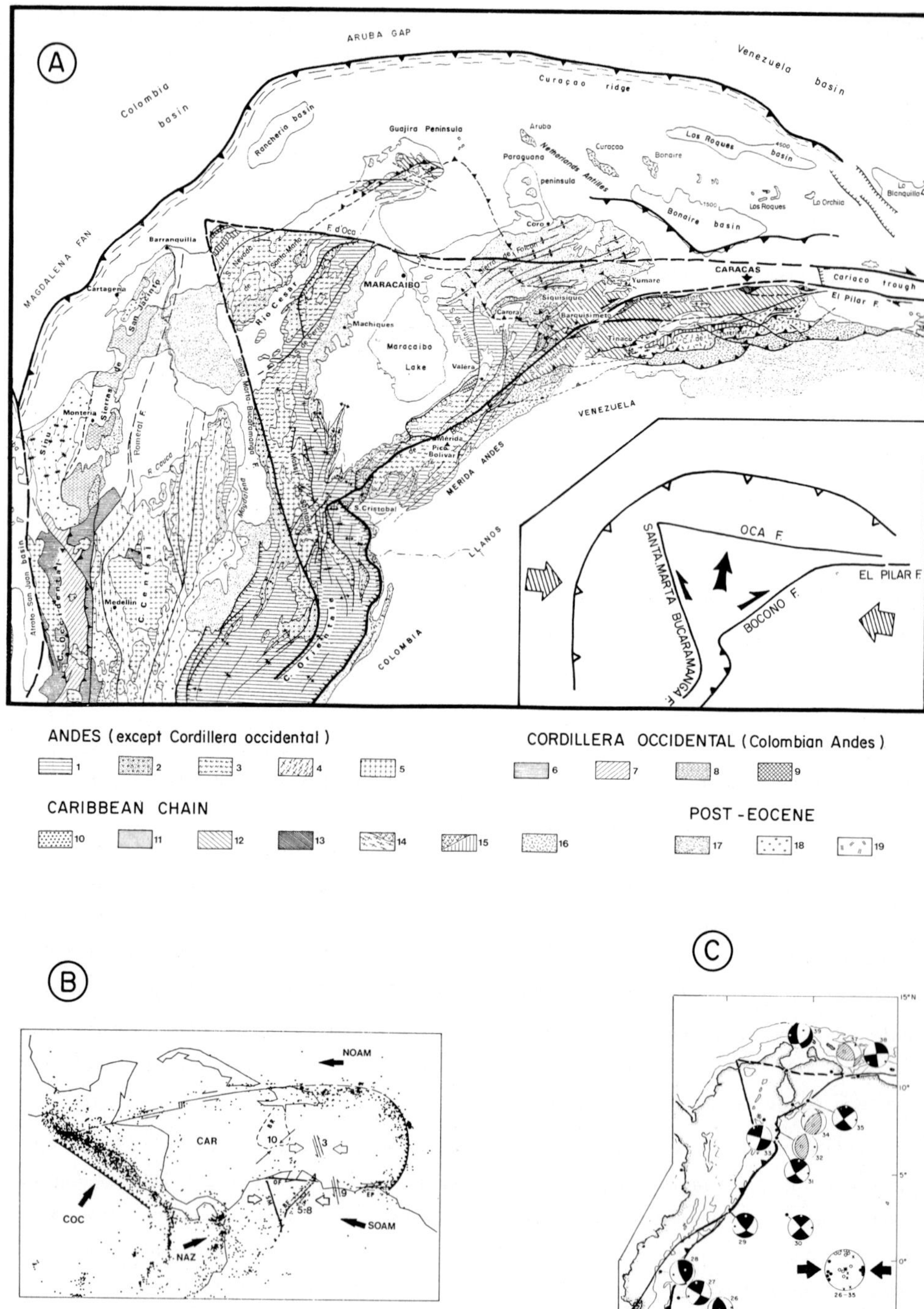

Figure 3 : *Caption next page.*

reaching its highest point in the Pic Bolivar (5007 m), is of fan or mushroom shape in cross sections. The reverse faults in the superstructure turn into overthrusts of more than 5 km width along the NW flank.

The shortening of the Andes and the right lateral movements along the Bocono fault are the two components of an almost E-W compression. This is inferred from seismo-tectonic data (Kafka and Weidner, 1981).

Analysis of the Plio-Quaternary structures near the Bocono fault confirm this statement. Also the disappearance of the strike slip fault in the zone of Tachira (Stephan, 1982), where the Merida Andes, the eastern Cordillera, and the mountain range of Santander come together, coincides with the development of N-S structures (folds, reverse faults, overthrusts : Macellari, 1982 ; Gonzalez de Juana et al., 1980, Boinet et al., 1982, in press, geological map of Colombia and Venezuela).

The Andean autochton / Caribbean allochton block is affected by N-S to N20° orientated folds and NE- SW right lateral strike slip faults in the regions of Lara and Portuguesa at the NE branch of the Merida Andes. They prove an almost E-W shortening direction (N105°, Stephan, 1982). These deformations nevertheless older, are post-Eocene to pre-Upper Miocene.

The post-Eocene displacement created by the Bocono fault has been estimated from the offset of the front of the Caribbean nappes. The NE-SW horizontal offset is about 80 km, anyway less than 100 km (Stephan, 1982).

2. *The Colombian Andes compartment.*

The Colombian Andes, South of the Maracaibo block, are located between the Atrato - San Juan depression in the West and the Llanos fore-deep in the East. The N-S to N20° oriented western central and eastern Cordillera as well as the Canca and Magdalena depressions are Plio-Quaternary mega-anticlines and -synclines. The eastern Cordillera has a fan like structure and overthrusts the Llanos

---

Figure 3 : *Colombia - Venezuela festoon : A - Tectonic map (from Stephan, 1985) with interpretative cartoon ; Autochton : 1, 2 - Mesozoic and Lower Tertiary sediments ; 3, 4 - Precambrian and Paleozoic basement ; 5 - Mesozoic and Lower Tertiary intrusives ; Allochton : 6, 7 - Ophiolitic nappes ; 8, 9 - Upper Cretaceous and Lower Tertiary sediments ; 12, 13, 14 - Epimetamorphic and metamorphic Mesozoic ; 15 - Upper Jurassic - Lower Cretaceous ophiolites ; 16 - Cretaceous arc ; Post-Eocene : 17, 18 - Oligo-mio-pliocene sediments ; 19 - Plio-quaternary volcanism.*

B - Structural interpretation of seismic data (from Kafka and Weidner, 1981).

C - Focal mechanisms in the Northern Andes (Equator, Colombia, Venezuela)(from Pennington, 1981) showing East-West compression.

fore-deep towards the East. Uplift and overthrust of the Andes are the reponse to an almost E-W shortening as inferred from the seismicity in the foothills of the eastern Cordillera and below the Llanos fore-deep (Pennington, 1981). Probably, Middle to Upper Miocene folding of the eastern Cordillera was followed by recent deformations. The eastward subduction of the Nazca plate is reflected in the Colombian Andes by the Upper Miocene to recent volcanism in the Cauca valley and in the Central Cordillera (Radelli, 1967 ; Irving, 1971) as well as by shallow to deep seismicity (Pennington, 1981).

3. The Bonaire compartment (partially equivalent to the "Bonaire plate" of Silver et al., 1975).
Located north of the Maracaibo block and the El Pilar strike slip fault, it is cut in the south Caribbean allochton. The post-Eocene deformations in this area are generally weak except for the Falcon domain which is folded along the Oca fault. The only significant deformation is observed north of the Falcon at the contact with the Paraguana shelf. Vertical or overturned Plio-Pleistocene series indicate Recent N-S compression (Wozniak and Wozniak, 1979 ; Biju-Duval et al., 1982b).

4. The central Caribbean range compartment.
It is situated between the El Pilar and Bocono strike slip faults, in the North and the West respectively, and the front of the Caribbean nappes, in the South. The Plio-Quaternary tectonic in this region is controlled by the El Pilar and Victoria strike slip faults. The latter separates the northern coastal Cordillera with its highest point of 2500 m near Caracas from the southern Interior range.

Along the Victoria fault several grabens occur (= pull-apart or rhomb-graben) at different evolutionary stages : presently subsiding is the Valencia Quaternary depression (Schubert and Laredo,1979), uplifted and in compression is the Upper Miocene - Pliocene Tuy graben (Beck, 1979).

The N-S shortening in this region since late Miocene time is weak regarding the slight deformation of the Tuy graben (Beck, 1979) and the front of the mountain range (Beck, 1977, 1978). The E-W horizontal displacement is therefore dominant during this time interval. This type of deformation was following a Middle Miocene episode during which the Oligocene - Miocene basin in front of the Caribbean range was folded and overthrust by the latter (Beck, 1977, 1978).

The eastern Interior chain, located east of this compartment was created during the same episode (Lamb and Sulek, 1968, Rossi et al., 1985) of horizontal displacement and compression (= transpression) between the Caribbean and the South American plates (Lamb and Sulek, 1968 ; Rossi et al., 1985). The N-S component of the movement was greater than at present time. This "transpressional" tectonic is illustrated by the "en échelon" Miocene folds (N50° / N60°, whereas the mountain range has a general E-W trend) and by secondary strike slip faults (=riedel R) like the Guarico, Urica, San Francisco and Los Bajos faults (Stephan, 1982).

Conclusion : The bending of the Colombian - Venezuelian festoon.

The dominant Plio-Quaternary deformation in the center of the Colombian - Venezuelian festoon are E-W shortenings or E-W strike slip faulting. These deformations provoked or increased the bending of the festoon. Because of the geometry of the boundary, there occur at the same time an E-W shortening and a N-S stretching in front of the Maracaibo block due to right lateral movements along the Bocono fault.

Compression was stronger and more of WNW-ESE (N100/110°) direction during early middle to late Miocene time resulting in folding and overthrusting in numerous areas (eastern Cordillera, Merida Andes, Falcon, central and east Caribbean range). The Colombia - Venezuela festoon was most probably created during this time interval.

IV .- THE HISPANIOLA - PUERTO RICO FESTOON.

This festoon, the less convex, extends about 1000 km between the Canal du Vent (between Cuba and Hispaniola), in the West, and the Sombrero graben, in the East, between the southern end of the Lesser Antilles arc and the Virgin islands plateau.

A.The trench and the Muertos accretionary prism south of the festoon, represent an E-W elongated system, 600 km long and 90 km wide. In the East it disappears progressively south of the Ste-Croix island, whereas in the West it seems to extend onshore on Hispaniola farther than the vertical faults which prolong the eastern edge of the Beata ridge.

Seismic reflection profiles (Garrison et al., 1972 ; Matthews and Holcombe, 1976 ; Ladd et al., 1979, 1981) show clearly the reflectors A" and B" in the Venezuela basin (corresponding respectively to the Middle Eocene and Coniacian) extending about 15 km from the trench axis below the deformed sediments of the prism. These profiles show also small tilted tensional basins upon the prism itself. These are classical features suggesting subduction of the Venezuela basin below the southern margin of the Greater Antilles ridge (Ladd et al., 1979, 1981). Basins developed offshore on top of the prism are interpreted as "forearc" basins (Ladd et al., 1979, 1981 ; Biju-Duval et al., 1982b) like the San Pedro basin south of Hispaniola. However, characteristic seismicity is not associated with this system. Moreover there is no gravity anomaly as should be expected from subduction of the Venezuela basin crust.

B. The westward onshore prolongation of the Muertos system (fig. 4).

The Muertos trench extends on Hispaniola island in the Enriquillo graben - Cul du Sac plain, an elongated WNW-ESE depression. It lies 40 m below sea-level, and is characterized by important subsidence, more than 3000 m of non-turbiditic sediments overlying the Upper Miocene. On the contrary, seismic profiles show offshore only 400-500 m of infilling above horizon A" (Middle Eocene). The deformation front of the inner wall of the trench continues in the Sierra Martin Gracia and Sierra de Neiba where south vergent imbrications of late Miocene age have been described (Bourgois et al., 1979).

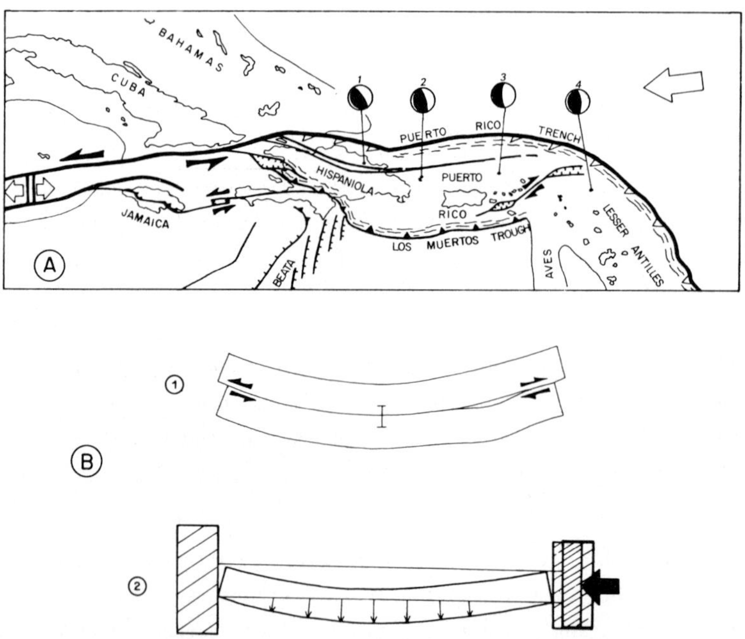

Figure 4 : *Hispaniola - Puerto Rico festoon. A - **Tectonic map** (from Houlgatte et al., 1984 and Stephan et al., 1985) ; focal mechanisms : 1. from Molnar and Sykes, 1969 ; 2, 3, 4, from Kafka, 1980.*

*B - **Schematic cartoons showing** : **in** B1, bending and concomitant right and left lateral strike-slips,and releasing bend ; **in B2**, bending and concomitant displacement (small arrows) perpendicular to the compression axis (black arrow). These cartoons can explain the structure and genesis of Panama and Hispaniola - Puerto Rico festoons.*

This emerged system is very different from the Muertos system, because in the South there is the "continental" block of the Sierra De Baoruco representing the eastern termination of the South Haiti peninsula. The latter unit was attached to the main block of Hispaniola probably prior to the Lower Miocene by left lateral strike slip movements (Calmus, 1984). Its presence modified the orientation of the deformations along the fractures in this region. A part of the E-W stress has been "absorbed" by reactivation of pre-existing faults.

The direction of the fractures on marine Plio-Quaternary terrasses in the southern part of the Baoruco peninsula are also in agreement with a regional SSW-NNE stress (Calmus, 1984).

C. The connection with the Anegada system.

On the east side, the Muertos trench and prism disappear in the SE of Ste Croix island. North of this island toward the East the graben system of Anegada, Virgin islands and Sombrero are linked to right lateral strike slip faulting (Houlgatte et al.,1984). The two latter grabens are considered as "pull-apart" basins. Taking in account its SW-NE orientation the Anegada strait can be explained by strike slip faulting only in the context of a regional almost E-W compression, but not N-S.

D. The continental basement.

It consists of a ridge comprising Hispaniola island (except the Sierra De Baoruco - South Haiti Peninsula), Puerto Rico and the Virgin islands plateau. This complex has been successively tectonized, incorporating parts of volcanic arcs, oceanic crust and even continental crust. This ridge was cut by important faults linked to the Cayman - Bartlett system. The main part of this ridge is represented by the Hispaniola island crossed by a series of NW-SE trending mountain ridges and valleys. The northern Cordillera is limited to the North by N110° faults (Cibao fault) or E-W faults (south of Samana) with left lateral displacement (Bowin, 1975).

The Dominican central Cordillera, the north chain of Haiti as well as the eastern Cordillera have been subjected to an intense N-S shortening during the Upper Eocene. Since this time the movements have been controlled by the left lateral NW-SE strike slip faults which produced a series of "en échelon" folds. In the NW of Haiti, the left lateral strike slip faults cut the Pliocene deposits. (Mercier de Lepinay et al., 1985). Some N-S folds (Anticline of Gros Morne) have been interpreted as "en échelon" folds linked with these strike slip faults. The Yabon fault in the eastern Dominican Cordillera, an active left lateral N130° strike slip fault with N100° relays indicates a regional E-W shortening

(Bourdon et al., 1985).

In the southern part of the Dominican Republic, NNW-SSE folds were observed as well as small ESE-WNW vergent overthrusts in the San Cristobal basin (Biju-Duval et al., 1982b). They affect at least the Middle Miocene.

Inactive E-W to ENE-WNW faults are known on Puerto Rico island. Present movements seem to be only vertical and are most probably related to the Puerto Rico trench in the North (Horsfield, 1974).

E. The northern boundary of the festoon :

The Oriente fault (Bartlett fault), SE of Cuba, linked up the vertical faults of the Old Bahama Channel. These are left lateral strike slip faults which turn eastward into a very oblique SW dipping mini-subduction (Bracey and Vogt, 1970 ; Schell and Torr, 1978), active since the Miocene (Nesteroff, 1982). This subduction is associated with some volcanic activities south of the central Cordillera (San Juan valley, northern boundary of the Cul du Sac plain, Mc Donald and Melson, 1969 ; Wedge and Wooden, 1984). Seismo-tectonic data confirm an E-W compression in this region (Molnar and Sykes, 1969 ; Kafka, 1980 ; Sykes et al., 1982). The inner wall of the trench is fractured by vertical faults with large throws shown by the existence of Pleistocene reefs now 3650 m deep (Nesteroff, 1982 ; Heezen et al., 1985).

The Puerto Rico trench s.s. develops more to the East. It is associated with left lateral strike slip faulting with very active seismicity. It turns progressively into a very oblique subduction north of the Virgin islands plateau as inferred from the distribution of epicenters (Frankel et al., 1980) and then to the frontal subduction of the Lesser Antilles.

All structures observed in the festoon and created since late Miocene time (left lateral NW-SE and right lateral NE-SW strike slip faults, folds ...) can be explained by an E-W compression. The weak "transpressive" component which exists at the North Caribbean boundary is not sufficient to provoke a real subduction below the southern margin of the Greater Antilles. But it is sufficient to give the impulse towards the Venezuela basin allowing the North Caribbean bending. However, the Muertos system could be an old feature (strike slip fault ? transform ? subduction ?) which could have concentrated the deformations during the bending of the continental substratum.

INTERPRETATION

The North and South Caribbean festoons are considered by most of the authors as slow subduction zones, due to the weak convergence between North and South America (Ladd and Watkins, 1979). The low subduction rate would explain the absence of typical seismicity and arc volcanism.

What are the relations between North and South America ? In the kinematic model adopted by the authors of the "Circum-Pacific maps" (NE quadrant, 1981), the North and South American plates are converging each other with a rate of about 0.6 cm/year) following the NW-SE direction. Therefore, a weak N-S convergence (0.4 cm/year) between the two plates exists. The North/South America boundary is located east of the Lesser Antilles arc somewhere between the Barracuda ridge and the hypothetical prolongation of the Vema fracture zone (cf. fig. 1 in Mc Cann and Sykes, 1984). It is aseismic for the greatest part. The only structures which affect the Atlantic crust in this region are the Barracuda and the Tiburon ridges. No major tectonic disturbance seems to be associated with those ridges, although the thickness of the sedimentary cover (more than 500 m cf. map of Case and Holcombe, 1980) is sufficient to record the slightest compression of the oceanic substratum. It seems therefore difficult to relate the genesis of the northern and southern Caribbean festoons only to the small convergence between North and South America.

Concerning the deformation of the continental substratum inside the festoons, geological and geophysical data show that the bending is more likely due to a global E-W to ESE-WNW shortening. Finally the deformations recorded by the Lesser Antilles arc s.l. although weak, correspond essentially to an almost E-W shortening ; this is the case for the volcanic arc itself (Bonneton, 1982) and for the fore-arc zone (reverse faulting at the contact Barbados ridge / Tobago basin, Biju-Duval et al., 1978).

Therefore, it is suggested that the North and South Caribbean festoons resulted from a bending of the continental frame due to an almost E-W shortening of the whole system. This is also the case for the Panama and Hispaniola - Puerto Rico festoons which are continental slices surrounded by oceanic terrain.

Their original trend was almost E-W parallel to the suggested shortening. In consequence bending towards the adjacent oceanic basin is the more logical effect.

This bending is accompanied by bed to bed slidings inside of the continental slice. This could explain the left lateral movement of the East Panama strike slip fault or the right lateral movements of the Anegada strike slip fault.

Two factors could make easier the genesis of the Colombia - Venezuela festoon moulded on the NW edge of the South American continent :

1. NNE-SSW and E-W trends inferred from the Pacific and Caribbean sides of South America.

2. the displacement of the Maracaibo block along the Bocono fault. Thus, the

North and South Caribbean festoons can be considered as "pseudo-subductions" caused by the bending of the peri-Caribbean continental frame resulting from a global E-W shortening.

The accretionary prisms characterizing each of these festoons were created by folding of the sedimentary cover in the Caribbean basins around the continental frame during the bending. The folds of the prisms are on each point sub-perpendicular to the local relative movement. There is no shearing, that should be the case on the east and west flanks of the Panama and Colombia - Venezuela festoons, if a N-S compression had caused their formation.

In conclusion, a relatively modest E-W shortening of the Caribbean frame seems sufficient to create the described festoons. Indeed if the festoons are approximately considered as segments of sinusoîdes, the length being constant during deformation, a very small shortening at the beginning can produce a noticable bending. Therefore, one can suggest that deformations of the Caribbean oceanic basin were related to this small E-W shortening.

This interpretation is proposed to stimulate new research and discussions in this part of the world.

Acknoledgments :

The ideas mentioned in this paper have been discussed since 1982 with J. Aubouin, X. Le Pichon, M. Mattauer, B. Biju-Duval, A. Mascle, J.M. Vila, M. Tardy, J.C. Carfantan. We thank them very much for help, and stimulating suggestions. We thank also C. Muller for translations, G. Quentel and E. Ruellan for the final version.

REFERENCES :

ANONYME, 1972. "Guide-book to the Geology of the Eastern Cordillera between Aguazul-Sogamoso-Villa de Leiva". Colomb.Soc. Petrol. Geol. Geophys., XIIIth Annual field Conf., inédit.

AUBOUIN, J., 1972. Chaînes liminaires (andines) et chaînes géosynclinales (alpines). Cong. Géol. Intern. Montréal, section 3, p. 438-461.

AUBOUIN, J., 1975. Réflexion sur les bordures pacifiques, l'exemple des cordillères sud-américaines. C.R. Acad. Sc., Paris, p. 280, série D, p. 2633-2636.

AUBOUIN, J., VON HUENE, R. et al., 1982. Leg 84 of the D.S.D.P. Project, subduction without accretion : Middle America Trench off Guatemala. Nature, 297, p. 458-460.

BECK, C.M., 1977. Geologia de la Faja Piemontina y del Frente de Montana en el Noreste del Estado Guarico, Venezuela septentrional. Mem. V, Congres. Geol. Venez. Caracas, t. II, p. 759-787.

BECK, C.M., 1978. Polyphasic tertiary tectonics of the Interior Range in the Central part of the Western Caribbean Chain, Guarico State, Northern Venezuela, in H.J. Mac Gillavry and D.J. Beets (eds) : The VIIIth Caribb. Geol. Conf., Willemstad, 1977 ; Geol. Mijnbouw, 57, n° 2, p. 99-104.

BECK, C.M., 1979. New data about recent tectonics in the central part of the Caribbean Chain : the Santa Lucia Ocumare del Tuy graben, Miranda State, Venezuela. IVth Latina-mer. Geol. Congr. Trinidad (1979). Mém. à paraître.

BIJU-DUVAL, B., MASCLE, A., MONTADERT, L. and WANNESSON, J., 1978. Seismic investigations in the Colombia, Venezuela and Grenada Basins, and on the Barbados Ridge for future IPOD drilling. In : Mac Gillavry H.J. et Beets D.J., édit. Géol. en Mijnbouw, vol. 57 (2), p. 105-116.

BIJU-DUVAL, B., BIZON, G., MASCLE, A. and MULLER, C., 1982a. Active Margin Processes : Field Observations in South Hispaniola. Amer. Assoc. Petrol. Geol. Mem., n° 34, p. 325-344.

BIJU-DUVAL, B., MASCLE, A., ROSALES, H. and YOUNG, G., 1982b. Episutural oligo-miocene Basins along the North venezuelan margin. Amer. Assoc. Petrol. Geol. Mém. 34, p. 347-358.

BOINET, T., BOURGOIS, J., MENDOZA, H.F., 1982. Tectonica de sobrecorrimiento y sus implicaciones estructurales en el area Pamplona-Labateca (Cordillera Oriental de Colombia). Ingeominas, Bucaramanga, rapport inédit.

BONNETON, J.R., 1982. Fracturation séismicité et tectonique des Plaques dans l'Arc insulaire des Petites Antilles. IXème Réun. Ann. des Sc. de la Terre, Paris, 1982. Soc. Géol. de France édit., p. 71.

BOURGOIS, J., GLACON, G., TAVARES, I. et VILA, J.M., 1979 .- Découverte d'une tectonique tangentielle récente à vergence Sud dans la Sierra de Neiba (République Dominicaine, Grandes Antilles). C.R. Sci. Paris, t. 289, p. 257-260.

BOURGOIS, J., CALLE, B., TOURNON, J., TOUSSAINT, J.F., 1982a. The Andean ophiolitic megastructures on the Buga-Buenaventura transverse (Western Cordillera-Valle Colombia). Tectonophysics, 82, p. 207-229.

BOURGOIS, J., AZEMA, J., TOURNON, J. et al., 1982b. Ages et structures des complexes basiques et ultrabasiques de la façade pacifique entre 3°N et 12°N (Colombie, Panama et Costa Rica). Bull. Soc. Géol. France (7), t. XXIV, n° 3, p. 545-554.

BOURGOIS, J., AZEMA, J., BAUMGARTNER, P.O., TOURNON, J., DESMET, A., AUBOUIN, J., 1984. The geologic history of the Caribbean - Cocos plate boundary with special reference to the Nicoya Complexe (Costa Rica) and D.S.D.P. results (legs 67 and 84 off Guatemala) : synthesis. Tectonophysics, 108, p. 1-32.

BOWIN, C.O., 1975 .- The Geology of Hispaniola. In : The Ocean Basins and Margins, vol. 3, A.E.M. Nairn et F.G. Stehli ed., Plenum Press, N.Y., p. 501-552.

BOWIN, C., 1976. Caribbean gravity field and plate tectonics. G.S.A., Special papers, p. 169.

BRACEY, D.R. and VOGT, P.R., 1970. Plate tectonics in the Hispaniola area. Geol. Soc. of Amer. Bull., vol. 81, n° 9, p. 2855-2859.

CALMUS, T., 1984. Décrochement senestre sud-haïtien : analyses et conséquences paléogéographiques dans la région de Camp - Perrin (Massif de Macaya, presqu'île du Sud d'Haïti). Ann. Soc. Géol. du Nord, t. CIII, séance spéc. Lille 10/06/83, p. 309-316.

CAMPBELL, C.J., 1968 .- The Santa Marta Wrench Fault of Colombia, its Regional Setting. IV Carib. Geol. Conf. Port of Spain, Trinidad - Tobago, 1965, trans, p. 247-261.

CASE, J.E. et HOLCOMBE, T.L., 1980. Geologic - tectonic map of the Caribbean region. U.S. Geol. Survey, Map., 1-1100.

CHRISTOFFERSON, E., 1976 .- Colombia Basin magnetism and Caribbean plate tectonics. Geol. Soc. Amer. Bull., vol. 87, p. 1255-1258.

DUQUE-CARO, H., 1979. Major structural elements and evolution of northwestern Colombia. Am. Assoc. Petrol. Geol. Mem. 29, p. 329-352.

FAUCHER, B. et SAVOYAT, E., 1973. Esquisse géologique des Andes de l'équateur. Revue de Géographie physique et de Géologie dynamique, Paris, 1973, (2), vol. XV, fasc. 1-2, p. 115-142.

FEO-CODECIDO, G., 1972. Breves ideas sobre la estructura de la falla de Oca, Venezuela. VI. Conf. Geol. Caraïbes, Porlamar (1971), Mem., p. 184-190.

FRANKEL, A., McCANN, W.R. and MURPHY, A.J., 1980. Observations from a seismic network in the Virgin Islands region : Tectonic structures and earthquake swarms. Jour. Geophys. Res., 85, p. 2669-2678.

GARRISON, L.E., 1972. Acoustic-reflection profiles, eastern Greater Antilles. U.S. Geological Survey Pub., n° USGS-GD-72-004.

GIEGENGACK, R., GRAUCH, R. and SHAGAM, R., 1976. Geometry of late cenozoic displacement along the Bocono fault, Venezuelan Andes. Mem. II, Cong. Latinoamericano Geol., Caracas, t. 2, p. 1201-1225.

GIRARD, D., 1981. Pétrologie de quelques séries spilitiques mésozoïques du domaine caraïbe et des ensembles magmatiques de l'île de Tobago : implication géodynamique. Thèse 3ème cycle, Brest.

GONZALEZ de JUANA, C., ITURRALDE de AREZONA, J., PICARD CADILLAT, X., 1980. Geologia de Venezuela y sus cuencas petroliferas (2 tomes). Ediciones Foninves, Caracas.

HALL, S.A., YEUNG, T., 1980. A study of magnetic anomalies in the Yucatan basin. 9ème Conf. Geol. Caraïbes, Santo Domingo (1980), t. 2, p. 519-526.

HEEZEN, B.C., NESTEROFF, W.D., RAWSON, M., FREEMAN-LYNDE, R.P., 1985. Visual evidence for subduction in the western Puerto Rico Trench. In : Géodynamique des Caraïbes, Symposium, Paris 5-8 février 1985, Ed. Technip, Paris, p. 287-304.

HORSFIELD, W.T., 1975. Quaternary vertical movements in the Greater Antilles. Géol. Soc. of Amer. Bull., vol. 86, p. 933-938.

HOULGATTE, E., STEPHAN, J.F. et MASCLE, A., 1984. Structure et signification géodynamique du Passage d'Anegada (NE de la plaque Caraïbe). 10ème Réun. Ann. Sciences de la Terre, Bordeaux, 1984, Soc. Géol. France, edit., Paris, p. 290.

IRVING, E.M. 1971. La evolucion estructural de los Andes mas septentrionales de Colombia. Bol. Geol. Ingeominas, 19, n° 2, p. 1-89.

JULIVERT, M., 1973 .- Les traits structuraux et l'évolution des Andes colombiennes. Revue de Géographie physique et de Géologie dynamique, Paris, (2), vol. XV, fasc. 1-2, p. 143-156.

KAFKA, A.L. 1980. Caribbean tectonic processes : seismic surface wave source and path property analysis (Ph. D thesis). State Univ. of New-York, 276 p.

KAFKA, A., WEIDNER, D.J., 1981 .- Earthquake focal mechanisms and tectonic processes along the Southern boundary of the Caribbean plate. J. Geophys. Res. 86, 84, p. 2877-2888.

KELLOG, J.N., BONINI, W.E., 1982. Subduction of the Caribbean plate and basement uplifts in the overriding South American plate. Tectonics, vol. 1, n° 3, p. 251-276.

LADD, J.W. and WATKINS, J.S., 1978 .- Active margin structures within the north slope of the Muertos Trench. Geol. en Mijnbouw, vol. 57, p. 255-260.

LADD, J.W., SHIH, T.C. and TSAI, C.J., 1981. Cenozoic tectonics of central Hispaniola and adjacent Caribbean Sea. Amer. Assoc. Petrol. Geol. Bull., vol. 65, p. 466-489.

LADD, J.W. and WATKINS, J.S., 1979. tectonic development of trench-arc complexes on the northern and southern margins of the Venezuela Basin, in Geological investigations of continental margins. A.A.P.G. Mem., 29, p. 363-371.

LAMB, J.L. and SULEK, J.A., 1968. Miocene turbidites in the Carapita formation of Eastern Venezuela. IVème Conf. Geol. Caraïbes, Port of Spain, Trinidad (1985), p. 111-119.

LONSDALE, P. and KLITGORD, K.D., 1978 .- Structure and tectonic history of the Eastern Panama Basin. Geol. Soc. Am. Bull., 85, p. 1697-1712.

LOWRIE, A., 1978 .- Buried trench south of the Gulf of Panama. Geology, 6, p. 434-436.

LOWRIE, A., STEWART, J., STEWART, R.H., Van ANDEL, T.J. and McRANEY, L., 1982. Location of the eastern boundary of the Cocos plate during the Miocene. Mar. Geol., 45, p. 261-279.

LU, R.S. and McMILLEN, K., 1983. Multichannel seismic survey of the Colombia basin and adjacent margins. Am. Assoc. Petrol. Geol. Mem. 34, p. 395-412.

Mc CANN, W.R. and SYKES, L.R., 1984. Subduction of aseismic ridges beneath the Caribbean plate : implications for the tectonics and seismic potential of the northeastern Caribbean. J. Geophys. Res., vol. 89, n° B6, p. 4493-4519.

Mac DONALD, K.C, HOLCOMBE, T.L., 1978. Inversion of magnetic anomalies and sea floor spreading in the Cayman trough. Earth and Planetary Science Letters, 40, p. 407-414.

Mac DONALD, W.D. and MELSON, W.G., 1969. A late Cenozoic volcanic province in Hispaniola. Carib. J. Sci. (Mayagüez), 9, p. 81-91.

MACELLARI, C.E., 1982. Tectonica compresional en el sur de los Andes venezolanos. V. Congr. Latinamer. Geol. Argentine, t. I, p. 403-418.

MATTHEWS, J.E. and HOLCOMBE, T.L., 1974. Possible Caribbean underthrusting of the Greater Antilles along the Muertos trough. Trans. 7th Caribbean Geol. Conf., Guadeloupe, (1974), p. 235-242.

MATTHEWS, J.E. and HOLCOMBE, T.L., 1976. Regional geological / geophysical study of the Caribbean Sea (Navy Ocean Area NA-9) - 1 : Geophysical Maps of the Eastern Caribbean, Washington D.C., Naval Oceanogr. Office, 43 p.

MINSTER, J.B. and JORDAN, T.H., 1978. Present day plate motions. J. Geophys. Res. 83, n° B11, p. 5331-5354.

MOLNAR, P. and SYKES, L.R., 1969. Tectonics of the Caribbean and middle America regions from focal mechanisms and seismicity. Geol. Soc. of Amer. Bull., vol. 80, p. 1639-1684.

NAGLE, 1972 .- Rocks from sea-mounts and escarpments on the Aves Ridge. VI. Conf. Geol. Caraîbes, Margarita, Vénézuéla (1971), p. 400-413.

NESTEROFF, W.D., 1976. Observations des phénomènes de subduction actuelle dans la fosse océanique de Porto Rico. 6ème Réun. Ann. Sciences de la Terre, Paris XII-Orsay (1976), p. 290.

NESTEROFF, W.D., 1982. Subsidence et sédimentation actuelles dans la fosse de Porto Rico. Oceanics, vol. 8, fasc. 8, p. 667-681.

PENNINGTON, W.D., 1981. Subduction of the Eastern Panama Basin and seismotectonics of northwestern South America. J. Geol. Res., 86, n° B11, p. 10753-10770.

PINET, B., LAJAT, D., LE QUELLEC, P., BOUYSSE, Ph., 1985. Structure of Aves Ridge and Grenada Basin from multichannel seismic data. In : Geodynamique des Caraîbes, Symposium, Paris, 5-8 Février 1985, Ed. Technip, Paris, p. 53-64.

RADELLI, L., 1967. Géologie des Andes colombiennes. Thèse, Fac. Sci., Univ. Grenoble, 455 p.

RICHARD, M., 1983. Pétrologie et signification géodynamique des ensembles magmatiques de la ride d'Aves (secteur Caraîbe) et de l'Est de Taîwan (Sud-Ouest Pacifique). Mémoire D.E.A., Brest.

ROD, E., 1956. Strike-slip faults of Northern Venezuela. Am. Ass. Petrol. Géol., Bull., 40, p. 457-476.

ROEMER, L., BRYANT, W. and FAHLQUIST, D., 1976. A geophysical investigation of the Beata ridge. VII Conf. Geol. Caraîbes, Saint-François, Guadeloupe (1974), p. 115-125.

ROSSI, T., 1985. Contribution à l'étude géologique de la frontière Sud-Est de la plaque Caraîbes : La Serrania del Interior Oriental (Vénézuéla) sur le transect Cariaco-Maturin. Synthèses paléogéographique et géodynamique. Thèse de Doctorat, Univ. Brest.

SAUNDERS, J.B., EDGAR, N.T., DONNELLY, T.W. and HAY, W.W., 1973. Cruise synthesis J.O.I.D.E.S. Leg 15. In : Rep. D.S.D.P., vol. 15, p. 1071-1111.

SCHELL, B.A. and TARR, A.C., 1978. Plate tectonics of the northeastern Caribbean Sea region. Geol. en Mijnbouw, vol. 57, p. 319-324.

SCHUBERT, C., LAREDO, M., 1979. Late Pleistocene and Holocene faulting in Lake Valencia basin, North-Central Venezuela. Geology, 7, p. 289-292.

SCHUBERT, C., 1982. Neotectonics of Bocono Fault, Western Venezuela. Tectonophysics, 85, p. 205-220.

SILVER, E.A., CASE, J.E. and Mac GILLAVRY, H.J., 1975. Geophysical study of the Venezuelan bordeland. Geol. Soc. Amer. Bull., 86, p. 213-226.

STEPHAN, J.F., BECK, C.M., BELLIZZIA, A. et BLANCHET, R., 1980. La chaîne Caraïbe du Pacifique à l'Atlantique. 26ème Cong. Geol. Intern., Coll. Geol. des Chaînes alpines issues de la Téthys, Paris, 7-17 juil., p. 38-59.

STEPHAN, J.F., 1982. Evolution géodynamique du domaine Caraïbe ; Andes et chaîne Caraïbe sur la Transversale de Barquisimeto (Vénézuéla). Thèse d'Etat, Paris, 512 p.

STEPHAN, J.F. 1985. Andes et Chaîne caraïbe sur la Tranversale de Barquisimeto (Vénézuéla). Evolution géodynamique. In : Géodynamique des Caraïbes, Symposium, Paris, 5-8 février 1985, Ed. Technip, Paris, p. 505-529.

STEPHAN J.F., BLANCHET, R., MERCIER DE LEPINAY, B., 1985. Les festons nord et sud Caraïbes (Hispaniola - Porto-Rico ; Panama et Colombie - Vénézuéla) : des pseudo-subductions induites par le raccourcissement Est-Ouest du bâti continental péri-caraïbe. In : Géodynamique des Caraïbes, Symposium, Paris, 5-8 février 1985, Ed. Technip, Paris, p. 35-51.

SYKES, L.R., McCANN, W.R. and KAFKA , A.L., 1982. Motion of Caribbean plate during last 7 million years and implications for earlier cenozoic movements. Jour. Geophys. Research, vol. 87, n° B-13, p. 10656-10676.

TALWANI, M., WINDISCH, C., STOFFA, P., BUHL, P. and HOUTZ, R., 1977. Multichannel seismic study in the venezuelan basin and the curacao ridge. In : Island arc, Deep sea trenches and back-arc basins, M. Ewing series 1, American Geophysical Union, p. 83-98.

TARDY, M., 1980. Contribution à l'étude géologique de la Sierra Madre Orientale du Mexique. Thèse d'Etat, Univ. Pierre et Marie Curie de Paris, 459 p.

VASQUEZ, E.E. and DICKEY, P.A., 1972. Major faulting in north-western Venezuela and its relation to global tectonics. IVème Cong. Géol. Caraïbes, Margarita, Vénéz. (1971), p. 191-202.

WADGE, G. and BURKE, K., 1983. Neogene Caribbean plate rotation and associated central american tectonic evolution. Tectonics, vol. 2, n° 6, p. 633-643.

WADGE, G. and WOODEN, J.L., 1982. Late Cenozoic alkaline volcanism in the northwestern Caribbean : tectonic setting and Sr isotopic characteristics. Earth Planet. Sci. Letters, vol. 57, p. 35-46.

WESTERCAMP, D., TAZIEFF, H., 1980. Martinique, Guadeloupe, Saint-Martin, La Désirade. Guide géologique, Masson, Paris, 135 p.

WILCOX, R.E., HARDING, T.P., SEELY, D.R., 1973. Basic wrench tectonics. Amer. Assoc. Petr. Geol. Bull., 57, 1, p. 74-96.

WOZNIAK, J. and WOZNIAK, M.H., 1979. Geologia de la parte Nor-Central de la Sierra de Falcon, hojas Coro y Cabure. Ministerio de Energia y Minas, Caracas, rapport inédit.

V. THE PACIFIC

THE ORIGIN OF THE PACIFIC ON AN EXPANDING EARTH

ARTHUR RAYMOND CRAWFORD

Department of Geology, University of Canterbury, Christchurch, NZ

ABSTRACT

The Pacific Ocean occupies a depression in the Earth's crust which is the largest of geomorphic units. Its sheer size suggests an old origin, in contrast with the Atlantic and Indian Oceans. On the hypothesis of an expanding Earth, originally covered completely by continental crust, the Pacific probably originated in the late Precambrian or early Palaeozoic, contemporaneously and comparably causally with the (global) 'Pan-African Event' (750-450 Ma). It formed catastrophically in common with fractures which outlined the broad shapes of the future continents. Continued expansion of the Earth was taken up largely by expansion of the Pacific until about -200 Ma when a greater rate of expansion forced dispersal of the continents.

The catastrophe was fundamentally endogenic and inevitable but may nevertheless have been triggered, and perhaps briefly anticipated, by impact of an extra-terrestrial body.

INTRODUCTION

The Pacific Ocean occupies a depression in the Earth's crust which is the largest of geomorphic units. However defined (for the western boundary is difficult to define) the Pacific covers at present between 35% and 30% of the Earth's surface. So great a proportion implies that the origins and histories of not only the Pacific and the Earth, but all the continental crust, must surely be closely related. This in itself suggests that the Pacific might be old. As the other oceans are now generally accepted to be young (of the order of 200 million years old) the Pacific is unique. Its origin and history are therefore of cardinal importance.

Yet the problem is scarcely recognized as one, let alone solved. In the context of an expanding Earth only Carey (1983) and Owen

(1983) have dealt with it at any length. Carey emphasizes the uniqueness of the Pacific and its greater age than those of the other oceans, but regards it nevertheless as no older than the Mesozoic. Owen on the other hand regards it as probably of late Precambrian-early Palaeozoic age, from a study of the sea-floor spreading data. I agree with Owen.

THE EARLY HISTORY OF THE EARTH

However the Earth formed, at some stage there developed a primitive crust and on that a hydrosphere. Whatever the particular processes it is to be expected that they would be essentially similar over all the Earth. There would therefore have been either a complete cover of water on the primitive crust if it lacked significant relief, or everywhere a similar pattern of pools of water of regular or irregular shape, probably in communication with one another, over all the surface. It is unlikely that water would be concentrated in one great body on one side of the Earth, for the processes which led to the falling of the first rains would be circum-global and everywhere broadly comparable in intensity. Although the primitive crust might reasonably be expected to be heterogeneous in detail, a heterogeneity so marked that it produced a depression of oceanic extent is unlikely on a rotating body.

It is of course possible that impact of a particularly large extra-terrestrial body might have produced a primitive Pacific basin. An impact in the early Precambrian is conceivable, especially as it is argued from lunar evidence that the Earth must have been violently and repeatedly impacted early in its history. Wetherill (1985) has postulated a very large single impact as an explanation, among other things, of the origin of the Moon; but it is plain that he is postulating one so early in the Earth's history that no persistent geomorphic effects would have been produced and his interesting suggestion is irrelevant to the problem of the origin of the Pacific (or indeed any of the existing geotectonic features).

THE CENOZOIC EARTH

It is now very widely accepted that the present continents had different positions in the past in relation to one another. Some workers think in terms of a Pangaea, others a juxtaposed Gondwana-cum-Laurasia. Many would also accept that the combined area of continental crust in the recent past (say 200 Ma ago) was substantially what it is today, so that when either of these assemblies is made there is left on an Earth of the present size a vast Panthalassa, i.e. a very greatly enlarged predecessor of the Pacific.

If so, the problem of the origin of the Pacific is made even more difficult, for the size of that predecessor needs to be such that it would have covered considerably more than half the surface of the Earth. It seems to me inherently improbable that so vast an ocean ever existed.

At this point it is important to remember that much evidence exists which suggests that nearly all Gondwana (in particular) retained its integrity for a very long time, longer than the period of time during which it dispersed. Gondwana existed from at least 1000 million years ago and began to break up only in the late Mesozoic. While on plate tectonics theory there are no limits (except those of time) to the number of crustal separations or accretions which may occur, that Gondwana (at least) existed as a unity for about 800 million years, and perhaps for much longer, suggests strongly that unity of continental crust is a terrestrial characteristic, and that dispersal was a unique event. (Some workers, such as Goodwin (1976) argue for such super continental unity from the beginning of the Archaean until break up, which would merely add emphasis to my suggestion).

Plate tectonics (or some varient) may well be the processes characteristic of today and the last 200 million years, but to argue that they applied during all or most of geological time when they were glaringly absent, or ineffective, over a period of at least 800 million years over a vast proportion of the Earth's surface seems absurd. I suggested (Crawford, 1982), and re-iterate, that they appeared for the first time about 200 million years ago.

THE PRESENT CONTINENTAL CRUST WOULD COVER COMPLETELY A SMALLER EARTH

It was shown by Hilgenberg (1933) that the entire Earth can be completely covered by an assembly of the present continents if its diameter is reduced to about 60% of that of today. This requires some distortion and fracture, for the curvature is different; but that the fit is reasonably good was a most significant discovery. Such a reconstruction was also demonstrated by Brösske (1962), using an Earth of 55% diameter. Creer (1965, 1967) did essentially the same, as did Barnett (1969); and Kirillov, Neiman, Milanovskii, Blinov and Vogel have each developed the concept (references in Carey, 1976). Vogel constructed transparent concentric globes to illustrate vividly the patterns (1983 and pers. comm.). Quite apart from other evidence, this alone gives strong support to a hypothesis of Earth expansion. To ignore it and its implications is today as unwise as it was to ignore in the nineteen - twenties the observation that the western continental margin of Africa fits remarkably well the eastern continental margin of South America.

On this hypothesis, once fracturing of a globally continuous crust took place, the continents then born became separated from one another by a dispersion due directly to increase in the radius of the Earth. This dispersion is totally distinct in style from any proposed by those who advocate conventional forms of continental drift or displacement on an Earth of constant dimensions. It involves strictly no movement of any continental crust 'horizontally' (with all that that may imply), but merely a 'vertical' (strictly, a centrifugal) movement, with creation of an ever - widening area of oceanic type crust as the continents become more and more widely separated.

AT WHAT TIME DID THE FRACTURING TAKE PLACE?

To estimate the time that the fracturing took place we need to work backwards. Let us, for the moment, make a first assumption that the area of continental crust on the Earth has indeed been constant. If so, then a Pangaea has always had the same area, whatever the dimensions of the Earth.

At present the total surface area of the Earth, if we use a figure of 6370 km as an average radius, is 510 x 10^6 km^2. If we regard the shelves as covering 5% of the land area, the total area of continental crust becomes 156 x 10^6 km^2. This, on our first assumption, is the area of Pangaea whatever its age. Now, the rate of expansion of the Earth according to Owen (1983), from careful analysis of the sea-floor spreading data, is such that in Jurassic time the Earth's radius was 80% that of today, i.e. 5090 km, with a surface area of 325 x 10^6 km^2. Its Pangaea, therefore, would have covered 48% of the surface, and a Jurassic 'Pacific', or Panthalassa (for there would have been no other oceans in existence) 52%. Its area would have been about 169 x 10^6 km^2.

The size of the Cambrian Earth is more difficult to estimate; we have no sea floor spreading data for the period before about 200 million years ago. Owen suggests a figure of 55% for its diameter. If we accept this estimate, the radius becomes 3500 km and the surface area 154 x 10^6 km^2. This is pretty well the area of a Cambrian Pangaea, so that there would have been no room on a Cambrian Earth for any predecessors of the Pacific; the continental crustal coverage would have been complete.

(This is not to say that there would have been no seas; epicontinental seas even of oceanic extent may well have existed; indeed, stratigraphic evidence indicates plainly that they did).

It is evident then, that the Pacific must have formed no earlier than the Ordovician if our assumptions are correct. With an estimated radius of 57% of that of today, an Ordovician Earth would have a total surface area of 165.6 x 10^6 km^2. A Pangaea would occupy 97% of its surface. Such a new-born Pacific would have had an area of slightly less than 5 x 10^6 km^2. That is about twice the area of the present Mediterranean, and nearly one-third the size of the present Arctic Ocean.

These estimates are of course crude indeed. If the existing continental shelves are regarded as covering more than the 5% of the land area, then the origin of the Pacific may have to be later, but even at 7% a Pangaea covers only 159.4×10^6 km^2, which can still be accomodated on an Ordovician Earth; at 10%, however, the origin would have to be no earlier than the Silurian. On the other hand, if the area of continental crust of today is rather greater than that of early Palaeozoic time (as some workers argue) the reverse would apply. For example, even if the shelves are so extensive as to be equivalent to 10% of the present land area, a growth of Pangaea by 20% in Phanerozoic time implies that a Cambrian Pacific could have been as large as 23×10^6 km^2, half as large again as the present Arctic Ocean. Its origin may have been in the late Precambrian, though only if no significant expansion of the Earth occurred until the Cambrian.

THE SIGNIFICANCE OF THE TIME OF ORIGIN

Let us for the moment accept an origin in Ordovician time. What is the significance? One remarkable feature of the Earth is the curious relationship of isotopic ages to the time-scale. A particularly prominent group of ages, especially mineral ages apparently associated with a thermal rather than an orogenic episode, is concentrated between about 750 and 450 million years. This concentration was first noticed in Africa and has become known as the 'Pan-African Event'. It has however been observed over much of the Earth, though restricted to certain zones. These, very curiously too, coincide with the margins (and some other zones) of the present continents, i.e. with those zones along which fractures developed into the great rifts from which the new oceans grew. I suggested (Crawford, 1982) that if we think in terms of a complete cover of continental crust these zones represent fractures resulting directly from Earth expansion at a time when the degree of such expansion was too much for the crustal cover to retain its continuity. Indeed, these fractures include many which became the broad continental outlines of the future.

CATASTROPHIC ORIGIN: ENDOGENIC OR EXOGENIC?

On the argument given above, if we assume that the Earth has expanded, the Pacific formed catastrophically. The catastrophe may have been wholly the product of the Earth's behaviour. Exponentially increasing expansion would produce the catastrophe itself. In this respect the experiments of Walter H. Bucher (Bucher 1924) are worth recalling. (These are illustrated and described also in his well-known and unique book The deformation of the Earth's Crust, first published by Princeton University Press in 1933 and reprinted by Hafner Publishing Company, New York, in 1957). (See Fig. 1).

All the experiments relate to the freezing of thin glass spheres filled with water. The catastrophic failure as the water expands to form ice is expressed by a system of cracks, which commonly spread from a centre of a short 'bar'. I have repeated the experiments with similar results. They are well known to engineers. If we may (with the usual risks) then compare the behaviour of such spheres with that of the Earth, in which the continental crust is equivalent to the thin glass, the centre of the small 'bar' can be regarded as the original, catastrophically produced Pacific, a small area where the previously globally continuous continental crust now exposes a gap to the mantle. The cracks radiating from this area would lie within the continental crust. It is noteworthy that those major fractures of irregular pattern seen in the experiments are very comparable in number to the great plate boundaries, and may be thought of as outlines of what would become the continents.

On such a model it is conceivable that the Pacific once formed as a tiny area could gradually expand in size without any other manifestation of Earth expansion elsewhere until the rate of increase of expansion became such that new oceans formed by widening of the irregular fractures. This, I argue, began to occur about

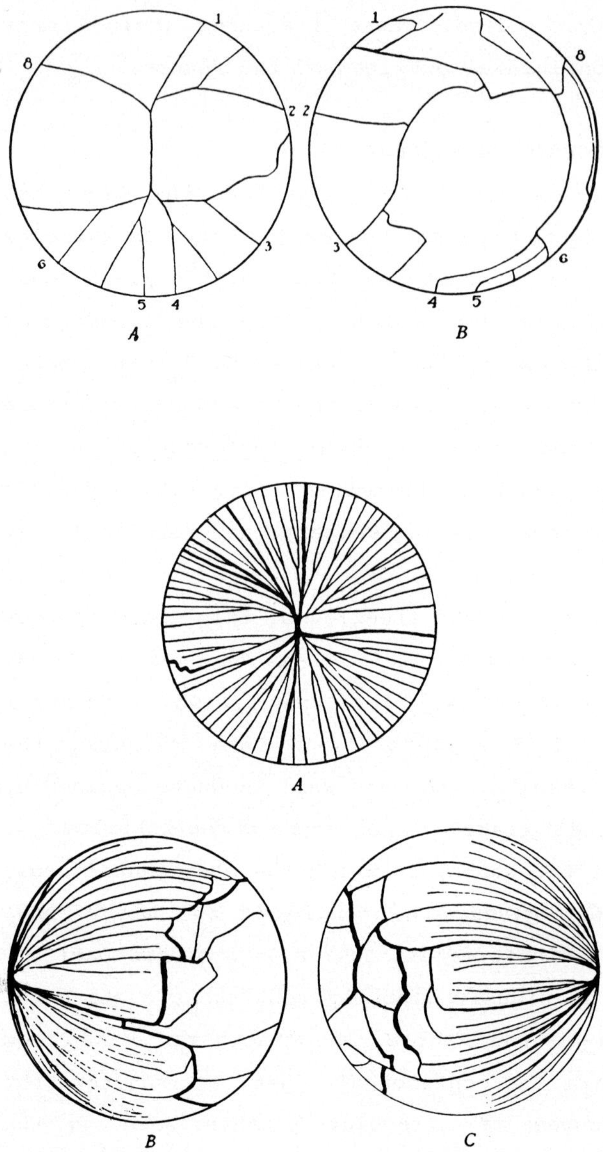

Fig. 1. Two examples of the manner of tension fracturing developed in thin glass spheres suddenly expanded by the freezing of enclosed water. Upper pair show sparse fracture pattern with a central bar in view A. Lower three (frozen more rapidly) show a radial group of narrow radial cracks in one hemisphere and a sparser pattern of wider cracks in the other. (Reproduced from Bucher, 1924).

200 million years ago. Thus for a period of perhaps twice that, and certainly as much as that, all increase in the global surface area was that of increase of the area of the Pacific.

On this hypothesis the origin of the Pacific would be catastrophic and endogenic. (In the direct sense; let us not be confused by the real possibility that the Earth, though expanding by internal processes, may ultimately be caused to be doing so by some persistent process originating outside it of which we at present know nothing). Evidence from engineering suggests that even on an apparently perfect sphere the slightest local weakness provokes catastrophic failure at that one point (in the case of a glass sphere the mere hint of perspiration from the touch of a finger is adequate enough). On the Earth it would be inevitable that some point on the globally continuous crust would be the weakest.

Yet it is equally possible that a catastrophic origin might be exogenic. There is little doubt that during the entire history of the Earth it has been impacted occasionally by extra-terrestrial bodies, even if rather less frequently, and more irregularly in time, than is supposed by a voluble group. This is quite distinct from the presumed very heavy bombardment early in its history simultaneously with that of the Moon. Even quite a small body could have triggered the catastrophic failure which would otherwise have occurred a little later. I question, though, whether such an impact, even of a larger body, could have created the Pacific in the absence of those conditions which were the direct consequence of Earth expansion.

The origin, then was essentially endogenic.

PARTICULAR ASPECTS OF THE PRESENT PACIFIC

In several ways the present Pacific is unique, apart from its size. It is conspicuously asymmetrical in having a spreading ridge in the south and east, so that the major geotectonic units are very different in size. It is also conspicuously asymmetrical in having a well-defined eastern boundary and a broad western one characterized by a complex mosaic of continental and oceanic crust, with several disjunctive basins. Several workers in the past preferred to regard the true Pacific as beginning on the eastern limit of this zone, along an 'Andesite Line'. Deep trenches, characteristic of much of the western, northern and eastern margins, are absent along the Antarctic margin, as they are from most of the Atlantic and Indian Oceans. The Pacific has innumerable islands, often in rectilinear or gently curving groups. These are almost all in the western part. The eastern Pacific is cut by immensely long straight fractures which in the north trend slightly south of west and in the south are predominantly WNW'ly.

All these features have significance.

Much of the asymmetry may, as Carey (1976) pointed out, relate to the direction of the Earth's rotation. The existence of innumerable islands (and still more seamounts) implies great and persistent volcanicity, which accords also with the intense volcanicity along almost the entire Pacific margin at the present time.

It is quite possible that the margin of the Pacific, as well as much of its floor, has always been volcanically active. Such activity is exactly what is to be expected from an ocean formed not by rifting and sea-floor spreading but catastrophically. In so far as its original growth related to withdrawal of continental crust and revelation of the mantle, it is to be expected that the entire margin would be volcanically active, as might the floor, and that this would continue. The development of a Mid-oceanic Ridge type fracture later in its history is not incompatible. That the east Pacific Rise is asymmetrically placed and is spreading exceedingly actively is in itself remarkable and indicates uniqueness.

A SPECULATION

Looking again at Bucher's experiments, we see in one example a division of the fracture pattern into two, with one hemisphere having numerous radial cracks, essentially parts of great circles, and one with 'plate boundary type' irregular cracks, far fewer in number. Could this have happened on the Earth? If so, then we have to accept that the hemisphere in which the infant Pacific was the centre, the fracture pattern would have been radial, and that only in the opposite hemisphere were the 'plate boundary type' cracks. As then the Pacific increased gradually in size, the non-Pacific, i.e. Pangaea, would have had a margin fractured essentially by parts of great circles, and a large central area, formerly but no longer of hemispheric size, dominated by the irregular cracks. One would then expect that the present condition of the Earth might show relics, at least, of part-great-circle cracks in those continents marginal to the present Pacific.

Do any exist? Perhaps so. For example, in Australia a very significant straight lineament is that extending WSW'ly along the Darling River and into South Australia. Though oddly elusive on the ground it is one along which lie the major base metal mineral bodies of Broken Hill and the Wallaroo-Moonta area of South Australia, and is at least 100 km long. In Indo-China there are several very important very long east-west faults which could be relics. In South America there is considerable evidence (Gansser, 1973) that the entire continent is fractured along an east-west zone followed by the Amazon.

As for the great straight fractures of the eastern Pacific, might not they too be related in some way and be in part inherited features, rather than just modern products consequent upon plate movement on a sphere?

CONCLUSION

On an expanding Earth the Pacific formed catastrophically as a small opening to the mantle in the late Precambrian or early Phanerozoic time as a direct consequence of increase in the area of the Earth.

ACKNOWLEDGEMENTS

I thank Professor Derek Ager, Dr John Bradshaw, Dr William Compston and Dr Stephen Weaver for valuable comment.

REFERENCES

Barnett, C.H., 1962. A suggested reconstruction of the land masses of the earth as a complete crust. Nature, 195: 447-448.

Brösske, L., 1962. Wächst die Erde mit Katastrophen? Dusseldorf

Bucher, W., 1924. The pattern of the Earth's mobile belts. J. Geol. 32: 265-290.

Carey, S.W., 1976. The Expanding Earth. Amsterdam, Elsevier.

Carey, S.W., 1983. Tethys, and her Forebears. In: Carey, S.W., (Editor), The Expanding Earth, A Symposium. Hobart, Tasmania, Geology Department, University of Tasmania, pp. 169-187.

Crawford, A.R., 1982. The Pangean paradox: where is it? J. Petr. Geol., 5(2): 149-160.

Creer, K.M., 1965. An expanding Earth? Nature, 205: 539-544.

Creer, K.M., 1967. A Synthesis of world-wide palaeomagnetic data. In: S.K. Runcorn (Editor), Mantles of the earth and terrestrial planets. New York. Interscience, pp. 351-382.

Gansser, A., 1973. Facts and theories on the Andes. J. Geol. Soc. Lond, 129: pp. 93-131.

Goodwin, A.M., 1976. Giant Impacting and the Development of Continental Crust. In: B.F. Windley (Editor), The Early History of the Earth, New York, Wiley, pp. 77-95.

Hilgenberg, O.C., 1933. Vom wachsenden Erdball. Berlin.

Owen, H.C., 1983. Ocean-Floor Spreading Evidence of Global Expansion. In: S.W. Carey (Editor), The Expanding Earth, A Symposium. Hobart, Tasmania. Geology Department, University of Tasmania, pp. 31-58.

Vogel, K., 1983. Global Models and Earth Expansion. In: S.W. Carey (Editor), The Expanding Earth, A Symposium. Hobart, Tasmania. Geology Department, University of Tasmania, pp. 17-27.

Wetherill, G.W., 1985. Occurrence of Giant Impacts During the Growth of the Terrestrial Planets. Science, 228: 877-879.

FACTS, IDEAS AND OPEN PROBLEMS ON TRENCH-ARC-BACKARC SYSTEMS

SEIYA UYEDA

Earthquake Research Institute, University of Tokyo
Tokyo 113 (JAPAN)

ABSTRACT

In the framework of plate tectonics, loci of trench-arc-backarc (T-A-BA) systems are identified with those of subduction, and various features characteristic to T-A-BA systems are causally attributed to subduction. While some aspects of such features as arc seismicity, seismic wave velocity attenuation structures and trench topography are rationally explained by subduction models, two fundamental features are difficult to explain. They are (1) the thermal state represented by high heat flow and arc volcanism in arc-backarc regions and (2) the extensional opening of backarc basins. Opening of backarc basins is closely related to another very basic problem of (3) why are arcs arcuate in shape? This paper reviews facts and ideas on some of these problems. Frictional heating between subducting slab and mantle wedge above the slab is probably of little importance in generating high heat flow and magma in arc-backarc regions. Convective asthenospheric motion mechanically induced by subducting slab may play a key role as it carries hot deeper mantle materials upwards. This motion, however, would not generate shear stress strong enough to cause backarc spreading. Backarc spreading would probably be a passive process governed chiefly by plate motions, i.e. by mutual retreat of overriding plate and trench line. Sphericity of the earth, as well as collision of buoyant features, may be responsible for making subduction zones arcuate in shape, and possibly play some role in the passive backarc spreading. Departure from simple 2-D treatment is essential for better understanding of subduction, i.e. it is important to look at individual subduction zones in more detail (comparative subductology), to look into the zones where collision is occurring (Vogt effects) and to consider 3-D aspects of shell tectonics. It is noted, however, that the more our knowledge and ideas advance, the more distant we are from ultimate solutions.

INTRODUCTION

In plate tectonics, subduction of oceanic lithosphere is the basic model for the tectonics of trench-arc-backarc (denoted as T-A-BA) systems. However, the subduction models can not readily explain some of the observed characteristics of T-A-BA systems (e.g. Sugimura and Uyeda, 1973; Uyeda, 1977). For example, Uyeda (1977), pointed out, in addition to the lack of convincing explanation for the diversity in the degree of convexity in shape, three basic inter-related unsolved problems.

1) Why is the upper mantle "hot" in the inner zone of the arc? (Origin of volcanism and high heat flow.)
2) Why is the stress "tensional" in the backarc areas? (Origin of backarc basins).
3) Why do the arcs rise? (Origin of arc mountain belts).

To understand these problems, it was suggested that solution of the following more specific problems might be helpful.

4) Origin of the trench outer swell and its diversity.
5) Shallow outer-arc seismicity and its diversity.
6) Real nature of the Wadati-Benioff zone.
7) Constance of dip angle of the Wadati-Benioff zones and its variation.
8) Time-space distribution of arc volcanic rocks.
9) Source material and depth of magma production. Does descending oceanic crust melt?
10) Possible role of water and CO_2 in magma production.
11) Is "frictional heating" important?
12) Possible effects of higher subduction rate.
13) Is there outer arc volcanism, or such a thing as initial "geosynclinal volcanism"?
14) Is there such a thing as a "cannibalistic" trench?
15) Processes in the trench-arc gap.
16) Is backarc spreading the same as the conventional sea-floor spreading?
17) Why some arcs develop backarc basins and others do not?
18) What does a collision of ridge and trench do?

Since 1977, considerable advance has been made so that at least partial solutions to some of the above listed problems have been found. Especially, the problems related with "diversity" have been extensively studied by what the present author calls "comparative subductology" (Uyeda and Kanamori, 1979; Uyeda, 1979; 1981; 1982; 1983; 1984). Many aspects of diversity are now interpreted in terms of diversity in the degree of mechanical coupling between overriding and subducting plates. However, the mechanism(s) controlling the degree of mechanical coupling is still a matter of debate (e.g. references cited above and Ruff and Kanamori, 1983; Peterson and Seno, 1984). Futhermore, many of the problems related to arc volcanism are still far from solutions although much work has been done (e.g. Wyllie, 1984). In this paper, we will review the present state of art in some large scale aspects of these problems.

CHARACTERISTIC FEATURES OF T-A-BA SYSTEMS

In this section, major characteristic features of T-A-BA systems will be summerized, taking the salient facts for the Japanese T-A-BA system as an example (e.g. Yoshii, 1979a, b; Hasegawa, Umino and Takagi, 1978; Honda, 1985). Fig. 1 shows the Quaternary volcanoes, volcanic front, epicenters of upper mantle earthquakes with focal depth of 40-60 km and aseismic front and Fig. 2 shows the heat flow distribution. Fig. 3 is the east-west cross-section of the Northeast Japan arc showing the earthquake mechanisms.

As seen in Fig. 3, earthquakes in island arc regions can be classified into five types and interpreted in terms of plate subduction as follows:

i) Normal fault-type shallow earthquakes due to downward bending of the oceanic lithosphere (Stauder, 1968).

ii) Large normal fault-type earthquakes due to tearing of the oceanic lithosphere by its own weight (Kanamori, 1971a).

iii) Thrust-type earthquakes due to subduction of the oceanic lithosphere. Most of great circum-Pacific inter-plate earthquakes belong to this category (Kanamori, 1971b).

iv) Intermediate and deep focus intra-plate earthquakes forming double Wadati-Benioff zone (Hasegawa, Umino and Takagi, 1978). Probably accompanying the unbending of the oceanic lithosphere, upper and lower zones show down-dip compression and tension mechanisms respectively.

v) Intra-plate shallow earthquakes under the arc and backarc regions. Mostly compressional or strike-slip type in Japan, but tensional in some arcs where backarc spreading is in progress (Uyeda and Kanamori, 1979; Eguchi and Uyeda, 1983; Eguchi, 1984a, b).

In the eastern margin of the Japan Sea, compressional events are interpreted as caused by initiation of subduction of the Japan Sea plate (Nakamura, 1983; Kobayashi, 1983; Tamaki and Honza, 1985).

These features are largely common to other T-A-BA systems and it can be said that as far as seismic activities are concerned, the new global tectonics have been quite successful, although there still remain some major problems, such as the physical mechanisms of deep earthquakes, and their double zone characteristics and the depth of penetration of subducting slabs (Creager and Jordan, 1984).

As to the thermal features of T-A-BA systems, the situation is different because subduction of cold oceanic lithosphere

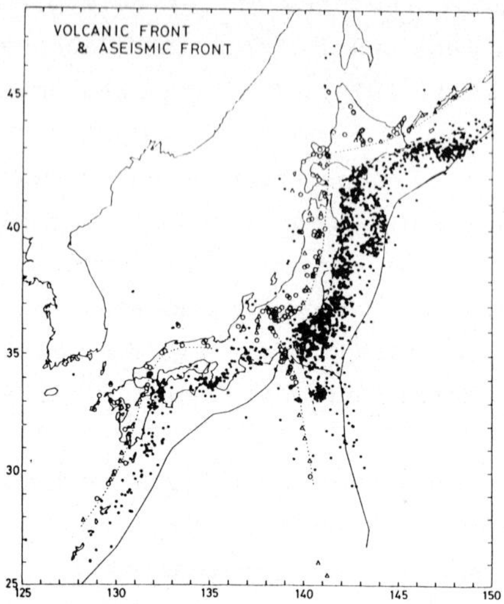

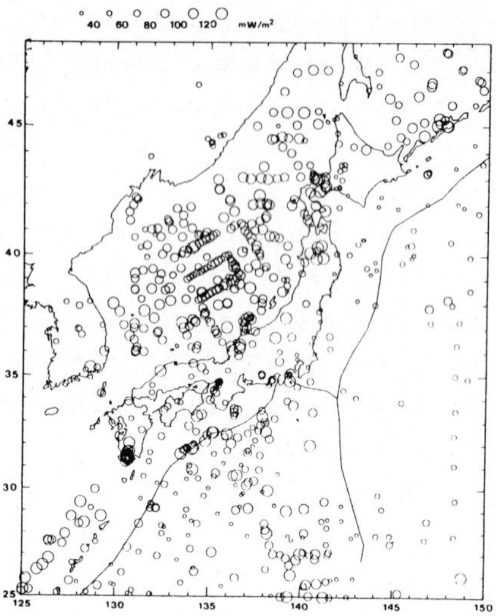

Fig. 1

Fig. 2

Fig. 1. Quaternary volcanoes (open marks) and epicenters of earthquakes with depth of 40 - 60km (black dots). Dotted and dashed curves indicate the volcanic front and the aseismic front respectively (Yoshii, 1979b).

Fig. 2. Distribution of heat flow in the Japanese region.

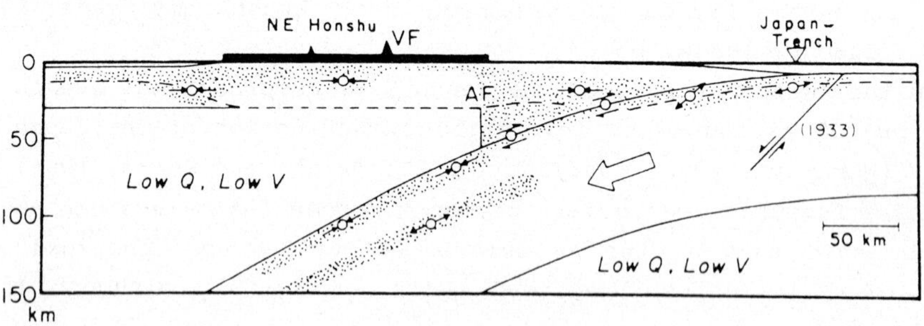

Fig. 3. Schematic representation of the focal mechanism beneath Northeast Japan. Dashed curves indicate the M-discontinuity. The stippled areas indicate seismically active zones. VF - volcanic front: AF - aseismic front (Yoshii, 1979a).

intuitively does not seem to result in either high heat flow in backarc regions or arc volcanism. Low Pn velocity and low Q under the arc and backarc regions (e.g. Utsu, 1971) and the existence of the aseismic front (Yoshii, 1975) also testify to the anomalously high temperature in the mantle wedge above the slab. This is the problem (1) mentioned in the Introduction. Closely related to this is of course the problem (2), namely the origin of backarc basins. Some backarc basins are believed to have been generated by backarc spreading (e.g. Karig, 1971; 1974) and dated magnetic lineations in many backarc basins testify to its validity (e.g. Weissel, 1981; see also references cited in Section 5). Moreover, past and present day occurrence of active backarc spreading has been demonstrated for the Mariana Trough, for example, by DSDP drilling results (e.g. Hussong and Uyeda, 1981). But why does the backarc spreading process ever take place behind arcs where oceanic lithosphere is thrusting down? As to the problem (3), namely the origin of arc mountain belts, since the advent of plate tectonics, subduction has been assumed to cause mountain building (e.g. Dewey and Bird, 1970; Matsuda and Uyeda, 1971). The Cordilleran or the Pacific-type orogeny has been distinguished from the Alpine-Himalayan orogeny which requires continental collisions. However, since the late 1970's, so called accretion tectonics have been postulated as the basic process for mountain building (e.g. Coney, Jones and Monger, 1980; Ben-Avraham et al., 1981; Nur and Ben-Avraham, 1982). It is said that numerous buoyant features on the ocean floor are accreted to continents during subduction and become mountains, and simple subduction of smooth oceanic plate can not give rise to sufficient deformations for building mountain ranges. Key to this question may be whether or not the Andean mountains consist of accreted terranes. At present, it is an open question.

Thus, the problems of T-A-BA systems are still largely unsolved, although some of them appear fairly well resolved.

HISTORY OF SOME BASIC IDEAS

Important tectonic features of T-A-BA systems were often generalized in a two dimensional cartoon as shown in Fig. 4, and many efforts have been made to explain them by simple subduction models, although such attempts are probably infertile as was pointed out by Uyeda (1982; 1983; 1984).

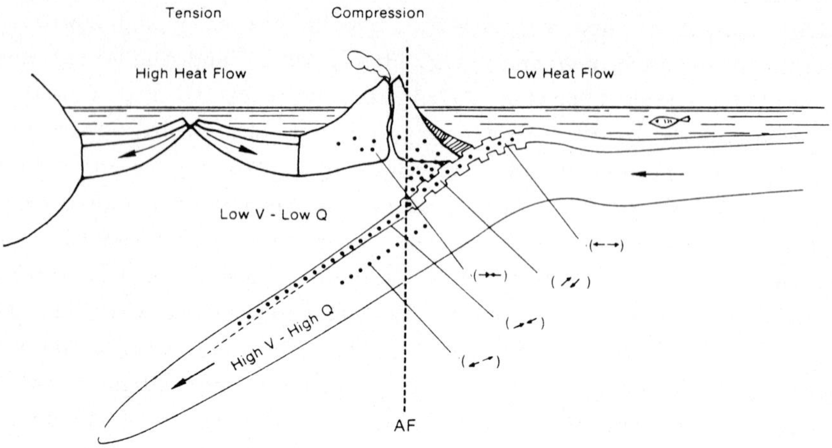

Fig. 4. Schematic cross-section of idealized subduction zone (not to scale). Dots - earthquake foci; heavy arrows - plate motions; V - seismic wave velocity; Q - seismic quality factor; AF - aseismic front. Arrows in parentheses indicate earthquake source mechanism.

Frictional Heat Models

Both Oxburgh and Turcotte (1968) and McKenzie and Sclater (1968) pointed out that simple subduction models cannot explain the backarc high heat flow and arc magmatism unless some special heat generating mechanism, such as frictional heating between the subducting slab and the mantle, exists. McKenzie and Sclater (1968) further noticed that even if enough heat was generated, it would take enormous time ($\sim 10^9$ years) for the heat to be conducted to the surface unless some extremely efficient heat transfer mechanism, such as penetrative convection, operates in the mantle wedge.

Toksöz and his associates studied numerically the thermal regime of subduction systems, taking all the conceivable thermal factors, such as adiabatic compression, phase changes, radioactivity and friction into consideration (Minear and Toksöz, 1970; Toksöz, Minear and Jullian, 1971; Toksöz, Sleep and Smith, 1973). Phenomena associated with the low temperature slab, such as some aspects of travel time anomaly, stress field of intermediate and deep focus earthquakes, were explained rather well by these studies. Magma production at the upper suface of the slab, however, was estimated to require frictional heat generation as high as 5-20 HFU at the upper surface of the slab (HFU is the heat flow unit: 10^{-6} cal/cm^2sec or 42 mw/m^2). Even with this high rate of heat generation at the slab surface, the surface high heat was not explained at all. Hasebe, Fujii and Uyeda (1970) took a different approach. They tried to find numerically how much heat has to be generated and how efficient the heat transfer has to be, in order to explain the observed surface heat flow anomaly of the Northeast Japan subduction zone. Their results also indicated that high frictional heat ($\sim$ 6 HFU) must be generated and the effective thermal conductivity of the mantle wedge must be almost 10 times as high as the ordinary lattice conductivity. The reason why heat production much higher than the observed heat flow anomaly (1 - 3 HFU) is needed is that the slab is a great heat sink. The 10 times more efficient heat transfer was considered possible only when significant amount of thermal diapirs ascend through the mantle wedge, so much that the Karig type opening of backarc basins might ensue (Matsuda and Uyeda, 1971).

Frictional heat models, however, have a fundamental weakness. Whatever its detailed mechanism, frictional heat w is expressed as $w = \sigma \times v$, where σ is the shear stress and v is the relative velocity. Simple calculation shows that in order to generate several HFU of heat, σ has to be of the order of a few kilobars. Such a high stress can not be sustained at high temperatures, so that the whole system is a self-limiting process which can not result in any significant melting (e.g. Uyeda, 1977; Yuen et al., 1978).

Wedge Flow Models

Contrary to frictional heat models, in wedge flow models the mantle wedge is treated as consisting of viscous liquid (Holmes, 1965; McKenzie, 1969; Sleep and Toksöz, 1971, Toksöz and Hsui, 1978; Hsui and Toksöz, 1979; Andrews and Sleep, 1974; Bodri and Bodri, 1978; Honda and Uyeda, 1981; Jurdy and Stefanick, 1983; Honda, 1985). In all these models, subduction of the slab is given kinematically, and the flow induced in the mantle and the temperature and stress fields are obtained either analytically or numerically.

In some models, temperature and pressure dependent viscosity was also considered. Fig. 5 is an example of interpreted diagram of such calculations. In the backarc area, where hot materials well up, about 1 km high ridge is formed by buoyancy, so that, adding to the shear stress due to the flow, a total of about 80 bar tensional stress is expected. This stress would contribute to backarc spreading once it started but it is questionable if the flow can initiate the spreading by overcoming the inter-plate compressional stress (Bird, 1978; Froidevaux, Uyeda and Uyeshima, 1986) and the strength of the overriding plate.

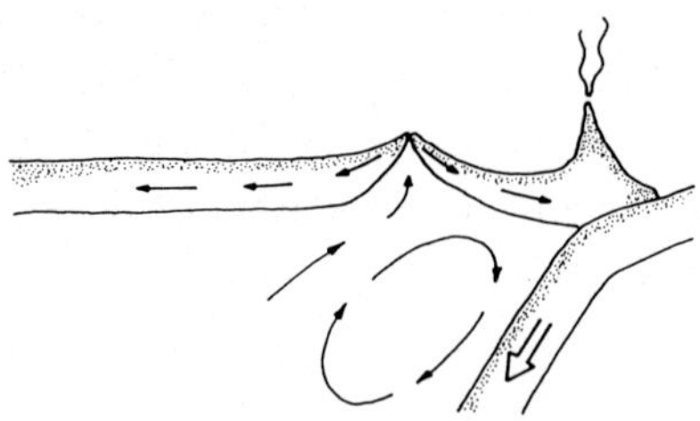

Fig. 5

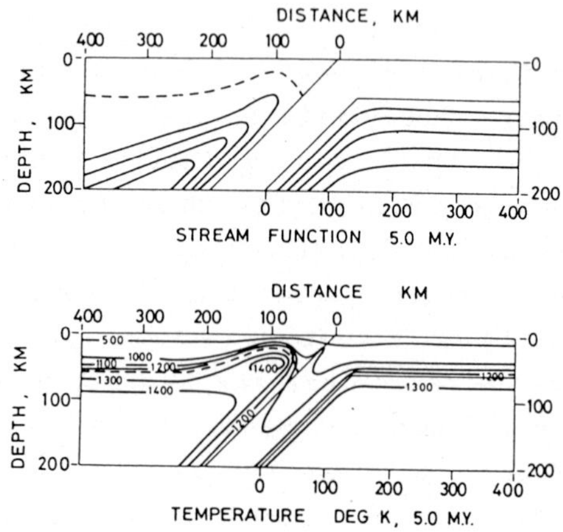

Fig. 6

Fig. 5. Schematic sketch of inter-arc spreading based on a wedge flow model (Toksöz and Hsui, 1978).

Fig. 6. Streamlines and isotherms at 5.0 m.y. from start of calculation on a wedge flow model (Bodri and Bodri, 1978).

In order to produce arc magma by this wedge flow mechanism, a strong ablation of the overriding plate by the high temperature flow at the corner of the wedge was invoked. Both Andrews and Sleep (1974) and Bodri and Bodri (1978) considered this problem by introducing the concept of yield stress by which the lower portion of the overriding plate near the wedge corner is forced to yield to the stress higher than say 1 kb and ablated (Fig. 6). The ablated and thinned portion of the arc lithosphere is the likely place where magma is segregated to ascend through fissures. This general picture seems to fit quite well with the thinned lithosphere beneath Japan. However, this process of ablation and magma ascent appears to be highly non-linear and impossible to be investigated in detail by the coarse mesh calculations conducted so far.

As to the Northeast Japan arc, Honda (1985) attempted to build a model that satisfies two major constraints. One was the observed heat flow profile and the other the temperature ($\sim 1400^{\circ}C$) in the mantle wedge as required by petrologic studies of arc basalts (Tatsumi et al., 1983). These authors estimated the chemical compositions of primary magmas for the Quaternary basalts in the Northeast Japan arc and, from the melting relations of the system concerned, deduced the pressure-temperature conditions when the magmas coexisted last with the upper mantle to be 10 - 20 kbar (35 - 70 km in depth) and around $1320^{\circ}C$. They further inferred that the melt was formed at greater depth in the mantle wedge at about $1400^{\circ}C$. According to Honda (1985), as Fig. 7 shows, one needs to assume quite a high temperature in the upper mantle under the Japan Sea to give rise to $1400^{\circ}C$ beneath the volcanic arc and to assume a rigid triangular region at the wedge corner to fit the relatively low heat flow in the trench-arc gap zone. Moreover, the high heat flow peak ($\gtrsim 200$ mW/m^2) on the arc has to be maintained by magma rise, which was estimated to be about 0.05 cm/yr in flux per unit area. This flux is one to two orders higher than the Cenozoic flux of erupted volcanic materials in Japan (Sugimura et al., 1963).

ARC VOLCANISM

As viewed in the preceding section, frictional heat models may reasonably be ruled out as a source of heat for high heat flow and arc magma. Instead, the magma is more likely to be produced as a result of lifting of high temperature material by the wedge flow. This view is supported by the observation that in the South American Cordillera, no active volcanism is found where the Wadati-Benioff zone is shallowly dipped and the mantle wedge does not exist (Sacks, 1977; Isacks and Barazangi, 1977).

One problem in arc volcanism is whether the subducted slab (sediments and crust) is melted and involved in arc volcanism or not. Since Coats (1962)'s pioneering work, controversies have been going around on this problem among petrologists and geochemists (e.g. Marsh, 1979), but, to the author's knowledge, there is still no unequivocally convincing evidence for or against slab melting. It is not clear to the present author if the chemical and isotopic data so far presented for slab melting, including those on Be^{10} (Brown et al., 1982), really prove it or they could also be interpreted as caused by transfer of water, containing various

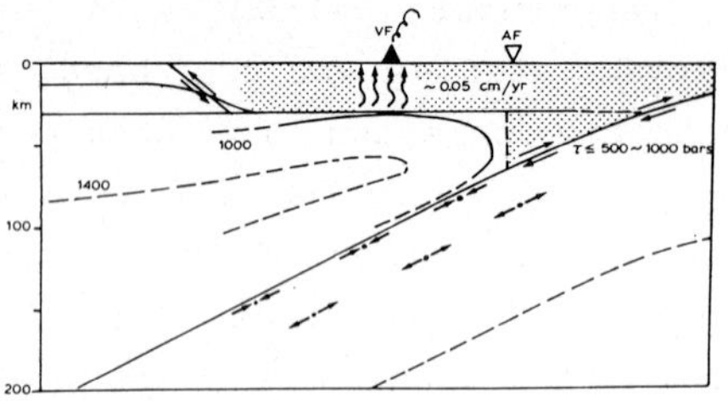

Fig. 7

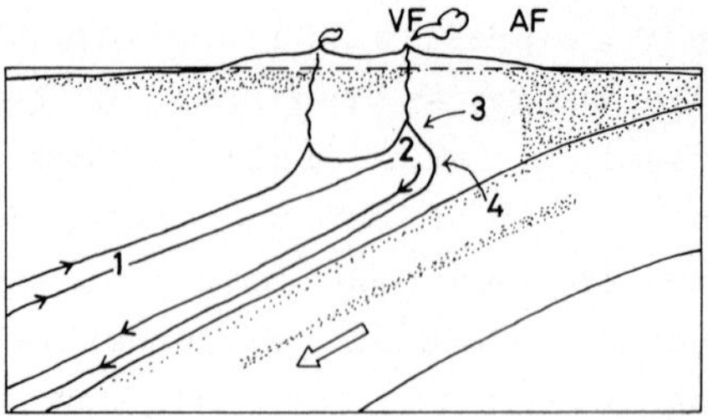

Fig. 8

Fig. 7. Schematic representation of the results of a wedge flow calculation, showing the isotherms in the mantle wedge. Stippled area shows the rigid part of overriding plate. Arrows indicate the focal mechanism (Honda, 1985).

Fig. 8. A wedge flow model for arc volcanism.

1 - Ascending wedge flow. Melt fraction increases because of lowering solidus by decompression.
2 - Melt fraction reaches maximum at the shallowest point of the flow.
3 - High stress-temperature ablation brings the flow locally up and the melt segregates for eruption to form the volcanic front.
4 - Descending flow. Any melt fraction remaining in the flow tends to solidify by compression.

Stippled areas show seismically active zones.

elements and isotopes in solution, from the slab alone. It is most likely that during subduction, sediments and oceanic crust of the slab undergo series of dehydration processes (e.g. Anderson, Uyeda, and Miyashiro, 1976) and the released water goes up into the mantle wedge, contributing in lowering its solidus temperature and giving it chemical and isotopic signatures of the slab. Thermal calculations of subduction models generally disfavor melting of slab, although Hsui, Marsh and Toksöz (1983) claim that melting can be demonstrated if sufficiently fine mesh is used in model calcualtion. Thus, the problem of slab melting has not yet been solved. However, now that the primary arc magma has been shown to have been in equilibrium with the mantle wedge (e.g. Tatsumi et al., 1983; Wyllie, 1984), minor involvement of materials from subducted slab, if any, would not constitute a problem of fundamental importance in arc magmatism, although slab melting itself is an interesting possibility in the thermal process of subduction zones.

The most intriguing feature of arc volcanism, in the author's view, is the volcanic front. It is almost always well defined and lies on the arc axis where the depth of the Wadati-Benioff zone is about 100-150 km for a variety of the dip of the Wadati-Benioff zone and the rate and obliquity of subduction. Yet, as Marsh (1979) pointed out, a long thin volcanic front such as the Aleutian volcanic front is commonly segmented and the pattern of segmentation reflects the fragmentation of subducting slab inferred from fracture zones of the seafloor approaching the trench. Based on these and other observations and some experimental results on Rayleigh-Taylor type instability, Marsh (1979) speculated that magmas ascend with regular spacing from a thin ribbon-like source layer of melt on the Wadati-Benioff zone. This is an interesting idea but it is not clear how the ribbon-like layer is formed at the desired depth of the Wadati-Benioff zone.

In the author's view (Fig. 8), as mentioned already, arc magma is produced essentially by the rise of hot asthenospheric materials from the depth due to the wedge flow, rather than by the rise of diapirs generated at greater depth (Marsh, 1979; Tatsumi et al., 1983). The rising asthenospheric materials may already contain a small fraction of initial melt, possibly the melting being aided by the water released from the slab, but the degree of melting should increase effectively during the virtually adiabatic ascent of the flow because of the lowering of the solidus temperature by decompression. The degree of melting or the melt fraction attains its maximum value precisely where the flow reaches the shallowest point near the uppermost wedge corner.

The downgoing flow would leave the melt at this depth because the melt has a lower density than the unmelted bulk of the flowing mantle materials. Therefore, the magma would be accumulated there until it starts to ascend to form the volcanic front directly above. When sufficient magma concentration is attained before reaching the shallowest point of the flow, a secondary or even third volcanic chain may be produced as in the Northeast Japan arc. No magma would be produced as soon as the flow starts downward motion because of the increase in pressure. Any fraction of melt remaining in the downgoing flow would start to solidify. This is why there are no volcanoes oceanward of the volcanic front. Thus, in this model, the geometry of the volcanic arc is controlled by the shallow magma zone. The narrowness of the arc volcanic belt is most likely caused by the local non-linear effects, involving the high temperature-high stress ablation mentioned in the previous section. By these effects the flow containing the magma sources can ascend locally to the depth of 30 or so km. The detail, of course, is yet to be investigated. The main problem in this view is to know why the shallowest point of the wedge flow is always located above the 100-150 km deep portion of the Wadati-Benioff zone. The whole problem, thus, is still open, but, perhaps, some useful clue to the solution may be found through detailed comparative studies of volcanic activity in different subduction zones in conjunction with numerical models tailored to individual cases.

BACKARC BASINS

Comparative studies have been more successful in dealing with the problem of backarc basin formation (Uyeda and Kanamori, 1979; Uyeda, 1984; Dewey, 1980; Taylor and Karner, 1983). Backarc basins may be classified according to models of their origin as in Table 1 (Uyeda, 1986).

First, class (A) basins were formed by entrapment of pre-existing oceans (Ben-Avraham and Uyeda, 1983), examples being the Aleutian Basin (Cooper, Marlow and Scholl, 1976) and the West Philippine Basin (Uyeda and Ben-Avraham, 1972; Hilde and Lee, 1984). Their entrapment origin has been verified by the magnetic lineation isochrons older than the associated island arcs. The Banda, Celebes and Sulu Seas may also be of entrapment origin (Hilde, Uyeda and Kroenke, 1977; MaCabe, Lee and Hilde, 1985).

Class (B) basins, that were formed in backarc environment, are divided into two subclasses (B-1) and (B-2) according to their possible mechanisms. The class (B-1) mechanism includes subsidence due to so-called oceanization or basification process (Beloussov and Ruditch, 1968) or due to gabbro-eclogite transformation in the lower crust (Artyushkov, Shlesinger and Yanshin, 1980). The latter mechanism was proposed mainly to explain rapid subsidence of intra-continental crust without such stretching as required by the contrasting and more popular view on subsidence (e.g. McKenzie, 1978). In the present author's view, no real backarc basin has been shown to belong to the class (B-1), whereas the stretching model, regardless of the origin of stretching, should apply to the initial stage of any backarc spreading. The present Okinawa Trough is probably one of such typical cases (Letouzey and Kimura, 1985; Kimura, 1985; Japanese DELP Research Group on Backarc Basins, 1986).

TABLE 1
ORIGIN OF BACKARC BASINS

ORIGIN OF BACKARC BASINS
(A) Entrapped ocean
(B) Basin formed in backarc environment
(B-1) By oceanization-basification
(B-2) By backarc spreading
(B-2a) Active spreading
Diapir rise
Wedge flow
"Hot Region"
(B-2b) Passive spreading
Landward retreat of landward plate
Slab "anchored" to stationary mantle
Retreat due to collision of buoyant feature
Trench retreat
Slab "anchored" to oceanward mantle flow
Gravitationally collapsing slab
Spherical shell buckling

The class (B-2), namely backarc spreading origin has been accepted more widely because it has been verified by dated magnetic lineation isochrons for a number of backarc basins, including the Scotia Sea (Barker, 1970), Lau Basin (Weissel, 1977), Shikoku Basin (e.g. Kobayashi and Nakada, 1978), Parece Vela Basin (Mrozowski and Hayes, 1979), Bismarck Sea (Taylor, 1979), and the Fiji Plateau (Malahoff, Feden and Fleming, 1982). The results of Deep Sea Drilling Project have provided sea-floor ages supportive of the spreading hypothesis notably in the Philippine Sea (Kroenke and Scott et al., 1980; Klein and Kobayashi et al., 1980; Hussong and Uyeda et al., 1981). Class (B-2) models are further subdivided into (B-2a) active spreading and (B-2b) passive spreading models.

Two typical (B-2a) type models, namely the diapir upwelling model (e.g. Hasebe, Fujii and Uyeda, 1970; Karig, 1971) and the wedge flow model (e.g. Sleep and Toksöz, 1971; Toksöz and Hsui, 1978; Jurdy and Stefanick, 1983; Ito, Masuda and Kinoshita, 1983) were explained in previous sections. Models relying on frictional heat for the production of diapirs are probably unrealistic, whereas the wedge flow appears to contribute to the spreading, although the shear stress generated by this mechanism would probably be not strong enough to initiate the spreading. In fact, if these models were powerful enough, backarc spreading would occur always at every subducting zone, which is not the case.

A novel variant of the active spreading model has been put forward recently by Miyashiro (1986). This model postulates that the series of backarc spreading in the western Pacific was due to the northward migration of one or two upper mantle "hot regions". This model, although provocative, appears to be difficult to verify at the present stage.

Alternative to active spreading (B-2a) models described above, passive backarc spreading (B-2b) models have been proposed in the last several years. Geometrically speaking, backarc spreading is a progressive separation of the landward or overriding plate and the arc plate attached to the trench hingline. In Fig. 9, h is the trench hingline of which absolute velocity is V_R, other quantities being self-evident. Backarc spreading will occur if and only if $V_o + V_R$ has a divergent component. If V_o and V_R are determined by factors other than those tensional forces considered for active spreading models, then backarc spreading can be regarded as a passive response rather than an active process. This way, backarc spreading is nothing but a pull-apart basin forming process (Lallemand and Jolivert, 1986) similar to the ordinary sea-floor spreading. In fact, the heat flow vs. age relation ($1/\sqrt{t}$ law) appears to hold for backarc basins, too (e.g. Sclater, Jaupart and Galson, 1980). Thus, the high heat flow in backarc basins is a natural consequence of sea-floor spreading. This general view was proposed by, among others, Chase (1978), Uyeda and Kanamori (1979), Dewey (1980), and Zonenshain and Savostin (1981).

If the subducted slab is "anchored" to the asthenosphere, V_R must be close to V_A, the absolute velocity of the asthenospheric flow. If V_A is small, backarc spreading occurs when V_O has a landward component (Landward retreat of landward plate). Chase (1978), Uyeda and Kanamori (1979) and Zonenshain and Savostin (1981) showed that the present absolute motions of landward plates of some actively spreading backarc basins do have landward velocity components and the landward plates of subduction zones without backarc spreading have oceanward velocity components. The typical

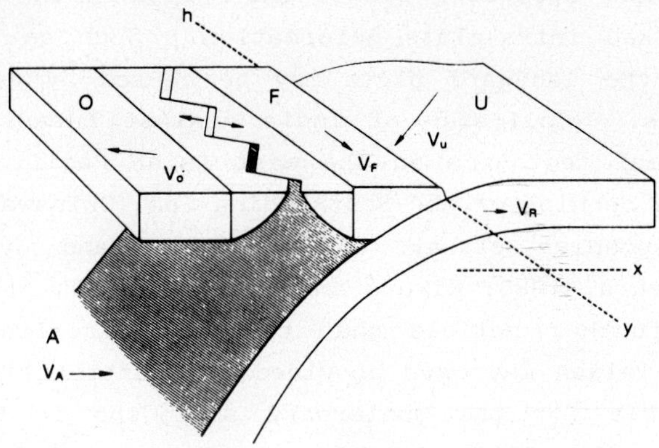

Fig. 9. Schematic diagram of subduction geometry (Uyeda, 1983).
O - overriding plate with velocity V_o;
U - underthrusting plate with velocity V_u;
F - arc plate with velocity V_F;
A - asthenosphere with velocity V_A;
V_R - roll-back velocity
h - trench hingeline

cases are the Mariana T-A-BA system and the Peru-Chile system respectively. In the Peru-Chile system, the South American plate is advancing trenchward so that the trench hingeline is forced to retreat, but no backarc spreading takes place. However, it appears that in some cases (e.g. South Sandwich Trench), trench hingeline retreats fast enough, so that $V_R + V_o$ has a divergent component, regardless of the sign of small V_o. Trench rereat may be caused either by a slab "anchored" to oceanward asthenospheric flow or by a gravitationally collapsing slab (Elsasser, 1971). Seno and Maruyama (1984) favored models for the Philippine Sea based on the trench retreating hypothesis. In the case of the South Sandwich Trench, the slab is probably "anchored" to eastward asthenospheric flow under the Drake Passage (Alvarez, 1982). As to possible motions of various trench hinglines, readers are referred to Carlson and Melia (1984).

Up to this point, problems have been treated essentially in two dimensional schemes. One important feature of backarc spreading is that it usually takes place along not the entire length of a convergent plate boundary but along a fraction of it. One of the reasons for this may be that the retreat of the landward plate causing the backarc spreading is not the motion of the whole plate but more localized intra-plate deformation. Such an intra-plate deformation of the landward plate may be caused by collision of buoyant features. Collision of India against Eurasia may have caused extensional tectonics in the east Asian Pacific margin to give rise to the opening of the South China Sea, Okinawa Trough and Japan Sea (Tapponnier et al., 1982; Eguchi and Uyeda, 1983; Letouzey and Kimura, 1985; Kimura and Tamaki, 1986). Letouzey and Kimura (1985) further noticed that the nearby collision of the Luzon arc with Taiwan may have provided an additional extensional stress responsible for the Quaternary spreading in the Okinawa Trough (Lee et al., 1980). McCabe and Uyeda (1982) and McCabe (1984) suggested a similar mechanism for the southern Mariana region: collision of the Caroline Ridge with the Yap Trench caused the sharp bend of the southern Mariana arc and helped the spreading of the southern Mariana trough, while Vogt (1973) and Vogt et al. (1976) postulated that collision of buoyant ridges would cause inhibition of backarc spreading only (Fig. 10).

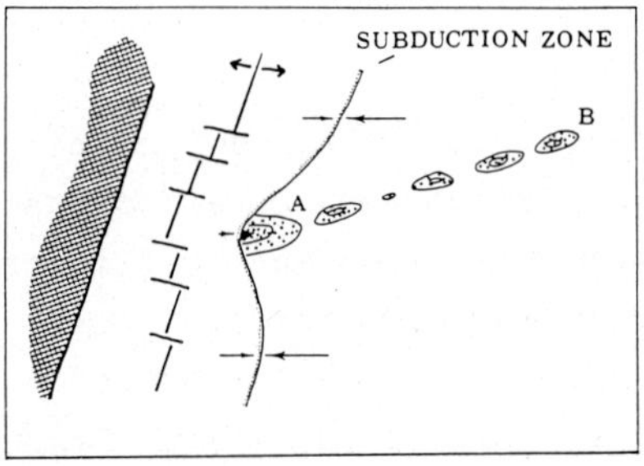

Fig. 10

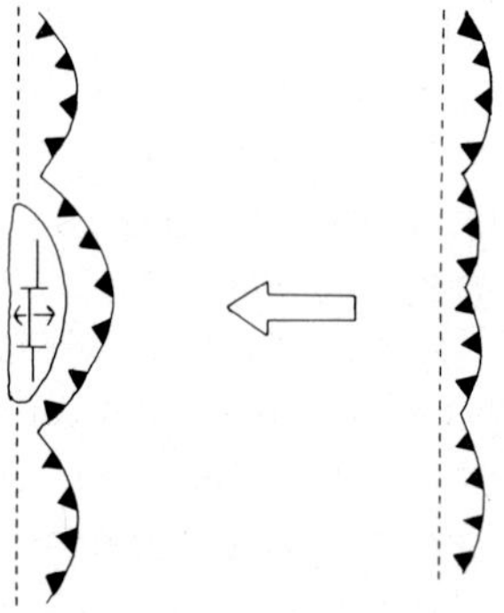

Fig. 11

Fig. 10. Schematic view of collision of a "semi-unsubductable" aseismic ridge with a subduction zone hindering backarc spreading and creating a cusp (Vogt, 1973).

Fig. 11. Backarc opening by buckling hypothesis. (Yamaoka, Fukao and Kumazawa, 1986a).

The last but perhaps not the least important aspect in non 2-D tectonics is the sphericity of lithosphere. Bayly (1982) proposed that backarc spreading occurs as a result of retreat of trench line accompanying the buckling of subducted spherical shell lithosphere. Recently, Yamaoka, Fukao and Kumazawa (1986a) and Yamaoka (1986), through both experimental and theoretical analysis, showed that during the process of subduction on a spherical earth, the mode of buckling abruptly changes, so that an arc gets longer and a backarc basin is formed (Fig. 11). This certainly seems a viable possibility particularly because in this mechanism the abrupt nature of backarc spreading and limited size of backarc basins can be explained plausibly. Recent paleomagnetic studies in Japan have indicated that both Southwest Japan arc and Northeast Japan arc rotated about 50° in clockwise and counterclockwise directions concurrently in a period between 21 - 11 Ma and the Japan Sea might have been opened by this process (e.g. Otofuji, Matsuda and Nohda, 1985). The age of the Japan Sea has never been really constrained by other means, but recent results of magnetic anomaly pattern analysis (Isezaki, 1986) tend to support the above age, while heat flow, basement depth and sediment stratigraphy of the Japan Basin suggest 30 - 15 Ma age (Tamaki, 1985). The age of the Japan Sea, thus, is still an open problem.

It must be noticed, moreover, that the arcuateness of a trench-arc system itself does not necessarily mean that its backarc basin was formed by backarc spreading, as in the case of the Aleutian arc-Bering Sea, because there has been no backarc spreading behind this well developed arc (Cooper, Marlow and Scholl, 1976), except for the Comandorian Basin in the western corner region where a Cenozoic spreading took place (Shor and Fornari, 1976; Bogdanov and Neprochenov, 1984), probably as a result of the collision of the Emperor Seamount chain. On the other hand, many backarc basins, such as the Japan Sea, Kurile Basin, Parece Vela-Shikoku Basin do not seem to be closed completely even if the arcs are straightened. Furthermore, the proposed mechanism has to show why suggested buckling takes place in some subduction zones and not in others. This mechanism may well be a part of the whole story, but how much part it plays is still an open question.

WHY ARE THE ARCS ARCUATE?

Two topics mentioned in the last part of the preceding section, namely the effects of collision of buoyant features and earth's sphericity are probably the most important factors in determining the arcuate form of trench-arc systems. Not only the trench-arc sytems, earth's orogenic belts in general have long been noticed for their curvatures. Fig. 12 shows the Mesozoic-Cenozoic mountain and island arc systems on an oblique Mercator Projection (Wilson, 1954). Both circum-Pacific and Alpine-Himalayan systems appear to lie broadly on great circles, but consist of series of arcs and cusps. The problem of why orogenic zones, in particular, the subduction zones, have circular or arcuate shape jointed by cusps has been discussed from time to time but, in spite of its obvious importance, conclusive solution has not yet been reached.

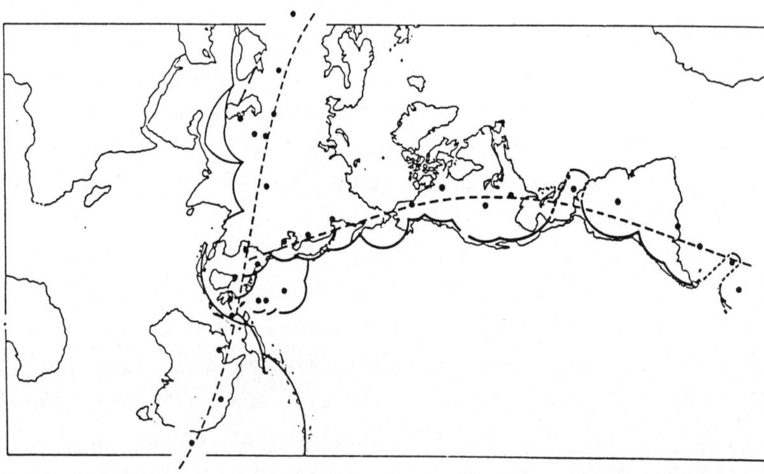

Fig. 12. The mountain and island arcs of the recent orogenetic system projected on oblique Mercator Projection (pole at 35.5°N, 2°E). Thick lines - mountain and island arcs; Dots - centers of arcs; Dashed lines - great circles through centers (Wilson, 1954).

In the 1950's, Carey (1955; 1958) postulated the orocline hypothesis. An orocline, which had a central importance among several other first order features such as rhombochasm, megashear etc in his theory, was defined as an orogenic belt with a curved trend imparted after its formation. Carey assumed that orogenic belts usually are formed in straight belts but bent as a result of

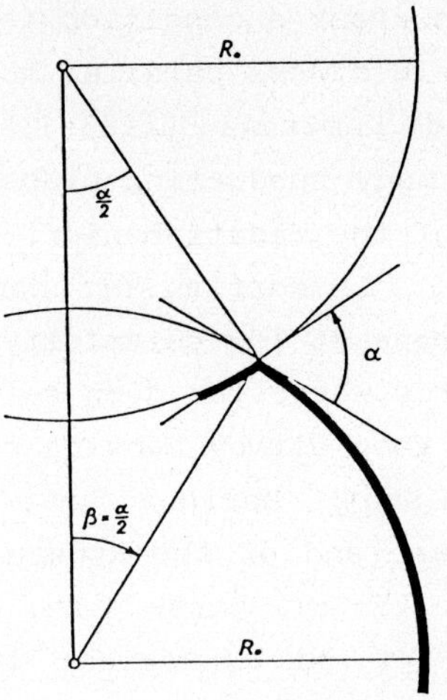

Fig. 13

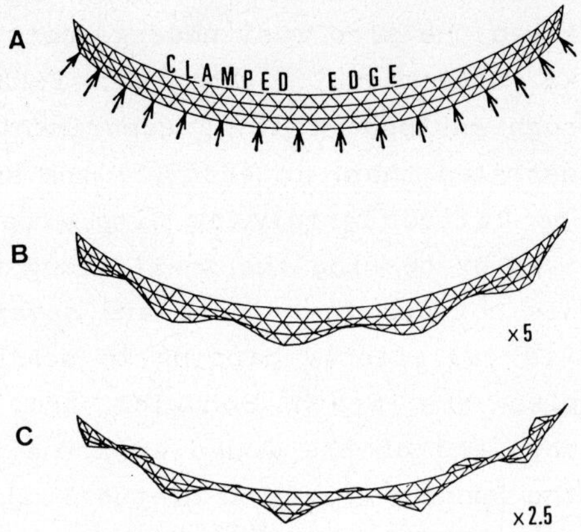

Fig. 14

Fig. 13. The ruptured ping-pong ball effect. = dip angle; ß = radius of the arc; R_O = earth's radius (Strobach, 1973).

Fig. 14. Result of finite element method calculation of bukcling deformation of a spherical shell (Yamaoka and Fukao, 1986)
A: Model of subducting lithosphere. Arrows show the uniform load on the edge.
B, C: Perspective views of subducted shell after buckling in the states of small (B) and large (C) deflections. Each figure is exaggerated in the radial direction by 5 or 2.5 times.

rotational motions of continental drift. Typical examples were the Alaskan Orocline and Peru-Bolivia Orocline of the Cordilleras. Carey's assumption that the bend was secondary may be quite valid (e.g. Heki, Hamano and Kono, 1983; Kono, Heki and Hamano, 1985). However, the mechanism which caused bending was not clear in Carey's theory.

As soon as the advent of the concept of plate subduction, Frank (1968) pointed out that a subducting spherical lithospheric plate, if flexible but inextensible, may be bent inwards through an angle α on and only on a circle whose radius of curvature is $\alpha/2$ (ruptured ping-pong ball effect; Fig. 13) and several studies followed along this line (e.g. Strobach, 1973; Le Pichon, Francheteau and Bonnin, 1973; DeFazio, 1974; Aoki, 1974; Laravie, 1975; Tovish and Schubert, 1978; Bayly, 1982). It was found that Frank's curvature-dip relation holds in some cases but not in others. Strobach (1973) showed that when Frank's condition is not fulfilled the slab must undergo considerable stress parallel to its strike directions. Yamaoka, Fukao and Kumazawa (1986b) made thorough analogue fitting experiments on many subducting slabs and demonstrated that, indeed, (1) the shape of the Wadati-Benioff zone can be fitted largely by simple bending, (2) most misfit can be resolved by tearing the shell along the trace of low seismicity.

They noticed that there are several arcs where the deep seismic zone is too steeply dipping to achieve a good fit by bending only. Examples are South Sandwich arc, the South Mariana arc, the northern end of the Tonga arc, the southern end of the Ryukyu arc and the Banda Sea region of the Sunda arc. In all these arcs, some seismic gaps were observed and tearing of the spherical shell explains both the seismic gap and the too steep dip of the slab. Except for the Banda arc, all these arcs have active backarc spreading. Based on these observations, these authors claim some causal relationship between backarc spreading and steepening/-tearing of the slab. It is, however, also true that in all these cases, sharp bends of arcs appear to be caused by indentation due to collision of buoyant features, and backarc spreading might have been invoked by the collision/lateral backarc spreading mentioned in the preceding section.

Yamaoka and Fukao (1986) showed also, through numerical and laboratory experiments, that cusps are intrinsic results of the buckling process (Fig. 14). Again, it appears also true that cusps are well developed where buoyant features are indenting the overriding plates as Vogt (1973) and Vogt et al. (1976) demonstrated.

Thus, the competing two theories on arc-cusp formation at subduction zones are at present almost equally persuasive. It may be that the two theories are not mutually exclusive and, as pointed out by Yamaoka and Fukao (1986), collision of buoyant features reduces the stress needed for buckling and induces cusp formation.

It is, however, to be noted that the orocline such as the Peru-Bolivia Orocline can be explained by neither mechanism. The only explanation, to the author's knowledge, that might be applied to the South American trench-arc curvature, is that of Peltzer (1981). He deduced from theoretical consideration on the driving forces that subduction plate boundaries should take a form of segments of helices about the axis of relative motions. Peltzer (1981)'s inference is based on the assumption of local balance of slab pull and slab resistance at each segment of consuming plate boundaries. Even if this theory is right, it explains the present plate motions but does not necessarily explain how oroclines were formed.

CONCLUSION

As stated in the Introduction, although much progress has been made in understanding the origin of arcs and subduction tectonics, we are still far from the ultimate solutions of a number of fundamental problems.

ACKNOWLEDGEMENT

The author thanks Ms. Nobuyo Matsushima for her assistance in preparing the manuscript.

REFERENCES

Alvarez, W., 1982. Geological evidence for the geographical pattern of mantle return flow and driving mechanism of plate tectonics. J. Geophys. Res., 87: 6697-6710.

Anderson, R.N., Uyeda, S. and Miyashiro, A., 1976. Geophysical and geochemical constraints at converging plate boundaries - Pt. I, Dehydration in the downgoing slab. Geophys. J. Roy. astr. Soc., 44: 333-357.

Andrews, D.J. and Sleep, N.H., 1974. Numerical modeling of tectonic flow behind island arcs. Geophys. J. Roy. astr. Soc., 38: 237-251.

Aoki, H., 1974. Plate tectonics of arc-junction at Central Japan. J. Phys. Earth, 22: 141-161.

Arthyshkov, E.V., Schlesinger, A.E. and Yanshin, A.L., 1980. The origin of vertical crustal movements within lithospheric plates. In: Dynamics of Plate Interiors. A.W. Bally, P.L. Bender, T.R. McGetchin, R.I. Walcott (Editors), Amer. Geophys. Un/Geol. Soc. Amer. Geodynamics Series, 1: 37-51.

Barker, P.F., 1970. A spreading centre in the East Scotia Sea. Earth Planet. Sci. Lett., 15: 123-132.

Bayly, B., 1982. Geometry of subducted plates and island arcs viewed as a buckling problem. Geology, 10: 629-632.

Beloussov, V.V. and Ruditch, E.M., 1961. Island arcs in the development of the earth's structure (especially in the region of Japan and Sea of Okhotsk). J. Geol., 69: 647-658.

Ben-Avraham, Z., Nur, A., Jones, D. and Cox, A., 1981. Continental accretion: From oceanic plateaus to allochtonous terranes. Science, 213: 47-54.

Ben-Avraham, Z. and Uyeda, S., 1982. Entrapment origin of marginal seas. In: T.W.C. Hilde and S. Uyeda (Editors), Geodynamics of the Western Pacific-Indonesian Region. Amer. Geophys. Un./Geol. Soc. Amer. Geodynamics Series, 11: 91-104.

Bird, P., 1978. Stress and temperature in subduction shear zones: Tonga and Mariana. Geophys. J. R. astr. Soc., 55: 411-434.

Bogdanov, N.A. and Neprochenov, and Yu. P., 1984. Geology of the Bering Sea deep basins, In: Proc. 27th Intern. Geol. Congr., 23:

Origin and History of Marginal and Inland Seas, VNU Science Press, 1-17.
Bodri, L. and Bodri, 1978. Numerical invesitgation of tectonic flow in island-arc areas. Tectonophys., 50: 163-175.
Brown, L., Klein, J., Middleton, R., Sacks, I.S. and Tera, F., 1982. ^{10}Be in island-arc volcanoes and implications for subduction. Nature, 299: 718-720.
Carey, W., 1955. The orocline concept in geotectonics. Proc. Roy. Soc. Tasmania, 89: 255-288.
Carey, W., 1958. A tectonic approach to continental drift, In: Continental Drift - A Symposium, W. Carey (Editor), 177-355, Univ. Tasmania.
Carlson, R.L. and Melia, P.J., 1984. Subduction hinge migration. Tectonophys., 102: 399-411.
Chase, C., 1978. Extension behind island arcs and motions relative to hot-spots. J. Geophys. Res., 83: 5385-5387.
Coats, R., 1962. Magma type and crustal structure in the Aleutian arc. In: G. MacDonald and H. Kuno (Editors), The Crust of the Pacific Basin. Amer. Geophys. Un. Mon. 6: 92-109.
Coney, P.J., Jones, D.L. and Monger, W.H., 1980. Cordillera suspect terranes, Nature, 288: 329-333.
Cooper, A.K., Marlow, M.S., and Scholl. D.W., 1976. Mesozoic magnetic lineations in the Bering Sea marginal basin. J. Geophys. Res., 81: 1916-1934.
Creager, K.C. and Jordan, T.H., 1984. Slab penetration into the lower mantle. J. Geophys. Res., 89: 3031-3049.
DeFazio, T.L., 1974. Island-arc and underthrust-plate geometry. Tectonophys., 23: 149-154.
Dewey, J.L., 1980. Episodicity, sequence and style at convergent plate boundaries. In: D.W. Strangway, (Editor), The Continental Crust and its Mineral Resources. Geol. Assoc. Canada, Spec. Pap., 20, 553-574.
Dewey, J.F. and Bird, J.M., 1970. Mountain belts and the new global tectonics. J. Geophys. Res., 75: 2625-2674.
Eguchi, T. and Uyeda, S., 1983. Seismotectonics of the Okinawa Trough and Ryukyu Arc. Memoir No. 5, Geol. Soc. China, 189-210.
Eguchi, T., 1984a. Seismotectonics of the Fiji Plateau and Lau Basin. Tectonophys., 102: 17-32.
Eguchi, T., 1984b. Seismotectonics around the Mariana Trough. Tectonophys., 102: 33-52.
Elsasser, W.M., 1971. Sea-floor spreading as thermal convection. J. Geophys. Res., 76: 1101-1112.
Frank, F.C., 1968. Curvature of island arcs. Nature, 220: 363.
Froidevaux, C., Uyeda, S. and Uyeshima, M., 1986. Island arc tectonics. Tectonophys. (in press).
Hasebe, K., Fujii, N. and Uyeda, S., 1979. Thermal processes under island arcs. Tectonophys., 10: 335-355.
Hasegawa, A., Umino, N. and Takagi, A., 1978. Double-planed deep seismic zone and upper-mantle structure in the Northeastern Japan Arc. Geophys. J. Roy. astr. Soc. 54: 281-296.
Heki, K., Hamano, Y. and Kono, M., 1983. Rotation of the Peruvian Block from paleomagnetic studies of the Central Andes. Nature, 305: 514-516.
Hilde, T.W.C. and Lee, C.S., 1984. Origin and evolution of the West Philippine Basin: a new interpretation. Tectonophys., 102: 85-104.
Hilde, T.W.C., Uyeda, S. and Kroenke, L., 1967. Evolution of the Western Pacific and its margin. Tectonophys. 38: 145-165.
Holmes, A., 1965. Principles of Physical Geology, 1288pp., Nelson, London.
Honda, S., 1985. Thermal structure beneath Tohoku, Northeast Japan

- a case study for understanding the detailed thermal structure of the subduction zone. Tectonophys., 112: 69-102.
Honda, S. and Uyeda, S., 1983. Thermal process in subduction zones. In: D. Shimozuru and I. Yokoyama (Editors), Arc Volcanism: Physics and Tectonics. Terra Sci. Pub. Co., Tokyo, 117-140.
Hsui, A.T., Marsh, B.D. and Toksöz, M.N., 1983. On melting of the subducted ocean crust: effects of subduction induced mantle flow. Tectonophys., 99: 207-220.
Hsui, A.T. and Toksöz, M.N., 1979. The evolution of thermal structures beneath a subduction zone. Tectonophys., 60: 43-60.
Hussong, D. and Uyeda, S., 1981. Tectonic processes and the history of the Mariana arc: a synthesis of the results of Deep Sea Drilling Project Leg. 60, Init. Rep. DSDP, U.S. Gov. Print. Office, 60: 909-929.
Hussong, D. and Uyeda, S. et al., 1981. Init. Rep. DSDP, 60: Washington D.C., U.S. Govt. Print. Office.
Isacks, B.L. and Barazangi, M., 1977. Geometry of Benioff zones: Lateral segmentation and downward bending of the subducted lithosphere. In: M. Talwani and W.C. Pitman III (Editors), Island Arcs, Deep Sea Trenches and Back-arc Basins. Maurice Ewing Series 1: 99-114, Amer. Geophys. Un.
Isezaki, N., 1986. A magnetic anomaly map of the Japan Sea. J. Geomag. Geoelectr. (in press).
Ito, H., Masuda, Y. and Kinoshita, O., 1983. Mantle vortex induced by downgoing slab: Experimental simulation and its application to trench-arc systems. Bull. Univ. Osaka Prefecture, ser. A, 32: 47-63.
Japanese DELP Research Group on Back-arc Basins, 1986. Report on DELP 1984 Cruises in the Middle Okinawa Trough, Bull. Earthq. Res. Inst., (in press).
Jurdy, D.M. and Stefanick, M., 1983. Flow models for back-arc spreading. Tectonophys., 99: 191-206.
Kanamori, H., 1971a. Seismological evidence for a lithospheric normal faulting - The Sanriku earthquake of 1933. Phys. Earth Planet. Inter., 4: 289-300.
Kanamori, H., 1071b. Great earthquakes at island arcs and the lithosphere. Tectonophys., 12: 187-198.
Karig, D.E., 1971. Origin and development of marginal basins in the western Pacific. J. Geophys. Res., 76: 2542-2561.
Karig, D.E., 1974. Evolution of arc systems in the western Pacific. Ann. Rev. Earth Planet. Sci., 2: 51-75.
Kimura, M., 1985. Back-arc rifting in the Okinawa Trough. Marine and Petroleum Geol. 2: 222-240.
Kimura, G. and Tamaki, K., 1985. Collision, rotation and back-arc spreading: the case of the Okhotsk and Japan Seas. Tectonics (in press).
Klein, G., Kobayashi, K. et al., 1980. Init. Rep. DSDP, 58: Washington, D.C. U.S. Govt. Print. Office.
Kobayashi, Y., 1983. Initiation of "subduction" of plates. Chikyuu (Earth Monthly), 3: 510-518 (in Japanese).
Kobayashi, K. and Nakada, M., 1978. Magnetic anomalies and tectonic evolution of the Shikoku Inter-arc Basin. J. Phys. Earth., 26: Suppl. s391-s402.
Kono, M., Heki, K. and Hamano, Y., 1985. Paleomagnetic study of the central Andes: counterclockwise rotation of the Peruvian Block. J. Geodyn., 2: 193-209.
Kroenke, L.W., and Scott. R.B., et al., 1980. Init. Rep. DSDP, 59: Washington D.C., U.S. Govt. Pirnt. Office.
Lallemand, S. and Jolivert, L., 1986. Japan Sea: a pull-apart basin? Earth Planet Sci. Lett., 76: 359-389.

Laravie, J.A., 1975. Geometry and lateral strain of subducted plates in island arcs. Geology, 3: 484-486.

Lee, C.S., Shor, G., Bibee, L.D., Lu, R.S. and Hilde, T.W.C., 1980. Okinawa Trough: origin of a back-arc basin. Mar. Geol., 35: 219-241.

Le Pichon, X., Francheteau, J. and Bonnin, J., 1973. Plate Tectonics, pp. 229, Elsevier, Amsterdam.

Letouzey, J. and Kimura, M., 1985. Okinawa Trough genesis: structure and evolution of a backarc basin developed in a continent. Marine and Petroleum Geol. 2: 111-130.

Malahoff, A., Feden, R.H. and Fleming H.F., 1982. Magnetic anomalies and tectonic fabric of marginal basins north of New Zealand. J. Geophys. Res., 87: 4109-4125.

Marsh, B.D., 1979. Island arc development: some observations, experiments, and speculations. J. Geol., 687-713.

Matsuda, T. and Uyeda, S., 1971. On the Pacific-type orogeny and its model - extension of the paired belts concept and possible origin of marginal seas. Tectonophys., 11: 5-27.

McCabe, R., 1984. Implication of paleomagnetic data on the collision related bending of island arcs. Tectonics, 3: 409-428.

McCabe, R., Lee, C.H. and Hilde, T.W.C., 1985. The Sulu-Celebes-Banda Basin - a trapped piece of ocean crust. EOS, 66: 1078 (abstract).

McCabe, R. and Uyeda, S., 1982. Hypothetical model for bending of the Mariana Arc. In: D.E. Hayes (Editor), The Tectonic and Geologic Evolution of Asian Seas and Islands. Pt. 2, Geophys. Monogr., 27: Amer. Geophys. Un., 281-293.

McKenzie, D.P., 1969. Speculations on the consequences and causes of plate motions. Geophys. J. Roy. astr. Soc., 18: 1-32.

McKenzie, D.P., 1978. Some remarks on the development of sedimentary basins. Earth Planet. Sci. Lett., 40: 25-32.

McKenzie, D.P. and Sclater, J.C., 1968. Heat flow inside the island arcs of the northwestern Pacific. J. Geophys. Res., 73: 3137-3179.

Minear, J.W. and Toksöz, M.N., 1970. Thermal regime of a downgoing slab and new global tectonics. J. Geophys. Res., 75: 1397-1419.

Miyashiro, A., 1986. Hot regions and the origin of marginal basins in the western Pacific. Tectonophys. (in press).

Mrozowski, C.L. and Hayes, D.E., 1979. The evolution of the Parece Vela Basin, eastern Philippine Sea. Earth Planet. Sci. Lett., 46: 49-67.

Nakamura, K., 1983. Possible nascent trench along the eastern Japan Sea as the convergent boundary between Eurasian and North American plates. Bull. Earthq. Res. Inst., 58: 711-722, (in Japanese with English abstract).

Nur, A. and Ben-Avraham, Z., 1982. Oceanic plateaus, the fragmentation of continents, and mountain building. J. Geophys. Res., 87: 3644-3661.

Otofuji, Y., Matsuda, Y. and Nohda, S., 1985. Opening mode of the Japan Sea inferred from the plaeomagnetism of the Japan arc. Nature, 317: 603-604.

Oxburgh, E.R. and Turcotte, D.L., 1968. Problem of high heat flow and volcanism associated with zones of descending mantle convective flow. Nature, 218: 1041-1043.

Peltzer, G., 1981. On the geometry of convergent plate boundaries. Earth Planet. Sci. Lett., 55: 463-472.

Peterson, E. and Seno, T., 1984. Factors affecting seismic moment release rates in subduction zones. J. Geophys. Res., 89: 10,233-10,248.

Ruff, L. and Kanamori, H., 1983. Seismic coupling and uncoupling at subdution zones. Tectonophys., 99: 99-117.

Sacks, I.S., 1977. Interrelationships between volcanism, seismicity, and anelasticity in western South America. Tectonophys., 37: 131-138.
Sclater, J.G., Jaupart, C. and Galson, D., 1980. The heat flow through oceanic and continental crust and the heat loss of the earth. Rev. Geophys. Space Phys. 18: 269-311.
Seno, T. and Maruyama, S., 1984. Paleogeographic reconstruction and origin of the Philippine Sea. Tectonophys., 102: 53-84.
Shor, G.G. and Fornari, D.L., 1976. Seismic refraction measurements in the Kamtchatka Basin, western Bering Sea. J. Geophys. Res., 81: 5260-5266.
Sleep, N. and Toksöz, M.N., 1971. Evolution of marginal basins. Nature, 233: 548-550.
Stauder, W., 1968. Tensional character of earthquake foci beneath the Aleutian Trench with relation to sea-floor spreading. J. Geophys. Res., 73: 7693-7701.
Strobach, K., 1973. Curvature of island arcs and plate tectonics. J. Geophys., 39: 819-831.
Sugimura, A., Matsuda, T., Chinzei, K. and Nakamura, K., 1963. Quantitative distribution of late Cenozoic volcanic materials in Japan. Bull. Volcanol., 26: 125-140.
Sugimura, A. and Uyeda, S., 1973. Island Arcs - Japan and its Environs, Elsevier, Amsterdam, pp.247.
Tamaki, K., 1985. Age of the Japan Sea. Chigaku Zasshi, 94: 14-29 (in Japanese with English abstract).
Tamaki, K. and Honza, E., 1985. Incipient subduction and obduction along the eastern margin of the Japan Sea. Tectonophys., 119: 381-406.
Tapponnier, P., Peltzer, G., Le Dain, A.Y. and Armijo, R., 1982. Propagating extrusion tectonics in Asia: new insights from simple experiments with plasticine. Geology, 10: 609-688.
Tatsumi, Y., Sakuyama, M., Fukuyama, H. and Kushiro, I., 1983. Generation of arc basalt magmas and thermal structure of the mantle wedge in subduction zones. J. Geophys. Res., 88: 5815-5825.
Taylor, B., 1979. Bismark Sea: evolution of a back-arc basin. Geology, 7: 171-174.
Taylor, B. and Karner, G.D., 1983. On the evolution of marginal basins. Rev. Geophys. Space. Phys., 21: 1727-1741.
Toksöz, M.N. and Hsui, A.T., 1978. Numerical studies of back-arc convection and the formation of marginal basins. Tectonophys., 50: 177-196.
Toksöz, M.N., Minear, J.W. and Jullian, B.R., Temperature field and geophysical effects of a downgoing slab. J. Geophys. Res. 76, 5: 1113-1138.
Toksöz, M.N., Sleep N.H. and Smith, A.T., 1973. Evolution of the downgoing lithosphere and mechanisms of deep focus earthquakes. Geophys. J. Roy. astr. Soc., 35: 285-310.
Tovish, A. and Schubert, G., 1978. Islsnd arc curvature, velocity of convergence and angle of subduction. Geophys. Res. Lett., 5: 329-332.
Utsu, T., 1971. Seismological evidence for anomalous structure of island arcs with special reference to the Japanese region. Rev. Geophys. Space Phys., 9: 839-890.
Uyeda, S., 1977. Some basic problems in the trench-arc-back arc systems, In: M. Talwani and W.C. Pitman III, (Editors), Island Arcs, Deep Sea Trenches, and Back arc Basins. Maurice Ewing Ser., 1, Am. Geophys. Un.: 1-14.
Uyeda, S., 1979. Subduction zones: facts, ideas and speculations. Oceanus, 22: 52-62.

Uyeda, S., 1981. Subduction zones and back arc basins - a review. Geol. Rundsch., 70: 552-569.
Uyeda, S., 1982. Subduction zones: an introduction to comparative subductology. Tectonophys., 81: 133-159.
Uyeda, S., 1983. Comparative subductology. Episodes, Vol. 1983, 2: 19-24.
Uyeda, S., 1984. Subduction Zones: Their diversity, mechanism and human impacts. GeoJournal, 8.4: 381-406.
Uyeda. S., 1986. Origin of back-arc basins. In: Proceedings of USSR-Japan Symposium on Geology and Geophysics of East Asian Marginal Seas. Khabarovsk.
Uyeda, S. and Ben-Avraham, Z., 1972. Origin and development of the Philippine Sea. Nature, Phys. Sci., 240: 176-178.
Uyeda, S. and Kanamori, K., 1979. Back-arc opening and the mode of subduction. J. Geophys. Res., 84, 1049-1061.
Vogt, P.R., 1973. Subduction and aseismic ridges. Nature, 241: 189-191.
Vogt, P.R., Lowrie, A., Bracey, D.R. and Hey, R.N., 1976. Subduction of aseismic ridges: effects on shape, seismicity, and other characteristics of consuming plate boundaries. Geol. Soc. Amer. Sp. Paper, 172, pp. 59.
Weissel, J.K., 1977. Evolution of the Lau Basin by the growth of small plates. In: M. Talwani and W.C. Pitman III (Editors), Island Arcs, Deep Sea Trenches and Back-arc Basins. Maurice Ewing Ser., 1, Amer. Geophys. Un.: 429-436.
Weissel, J.K., 1981. Magnetic lineations in marginal basins of the western Pacific. Phil. Trans. R. Soc. Lond., A300: 223-247.
Wilson, J.T., 1954. The development and structure of the crust. In: G.P. Kuiper (Editor), The Earth as a Planet. Univ. Chicago Press. 138-214.
Wyllie, P.J., 1984. Constraints imposed by experimental petrology on possible and impossible magma sources and products. Phil. Trans. R. Soc. Lod. A 310: 439-456.
Yamaoka, K., 1986. Spherical shell tectonics: on the buckling of the lithosphere at subduction zones. Tectonophys. (in press).
Yamaoka, K. and Fukao, Y., 1986. Why do island arcs form cusps at their junctions?, Nature (in press).
Yamaoka, K., Fukao, Y. and Kumazawa, M., 1986a. Spherical shell tectonics and the evolution of subduction zones. In: C. Froidevaux and K. Fuchs (Editors), The Composition, Structure and Dynamics of the Lithosphere Asthenosphere System. Amer. Geophys. Un., (in press).
Yamaoka, K., Fukao, Y. and Kumazawa, M. 1986b. Spherical shell tectonics: effects of sphericity and inextensibility on the geometry of the descending lithosphere. Rev. Geophys.Space Phys. (in press).
Yoshii, T., 1975. Proposal of the "Aseismic Front", Zisin, 28: 365-367 (in Japanese with English abstract).
Yoshii, T., 1979a. A detailed cross-section of the deep sesimic zone beneath northeastern Honshu, Japan. Tectonophys., 55: 349-360.
Yoshii, T., 1979b. Compilation of geophysical data around the Japanese Island II, Bull. Earthq. Res. Inst., 54: 75-117 (in Japanese with English abstract).
Yuen, D.A., Fleitout, L., Schubert, G. and Froidevaux, C., 1978. Shear deformation zones along major transform faults and subducting slabs. Geophys. J. R. astr. Soc., 54: 93-119.
Zonenshain, L.P. and Savostin, L.A., 1981. Movement of lithospheric plates relative to subduction zones: formation of marginal seas and active continental margins. Tectonophys., 74: 57-87.

SUBDUCTION IN THE JAPAN TRENCH : THE KAIKO RESULTS

J.P. CADET[1], K. KOBAYASHI[2], S. LALLEMAND[1] and L. JOLIVET[3]

[1]Laboratoire de Géologie Dynamique, Département des Sciences de la Terre, Université d'Orléans, B.P. 6759, 46067 ORLEANS cedex 2, France, UA CNRS 215

[2]Ocean Research Institute, University of Tokyo, 1-15-1 Minamidai, Nakano-ku, Tokyo 164, Japan

[3]Laboratoire de Géologie, Ecole Normale Supérieure, 24 Rue Lhomond, 75231 PARIS cedex 05, France, UA CNRS 215

ABSTRACT

This paper presents the results of a detailed survey combining Seabeam mapping, gravity, geomagnetic measurements and single channel seismic reflection observations in the Japan trench and the juncture with the Kuril trench during the French-Japanese Kaiko project (Leg 3) aboard the R/V Jean Charcot (1984) as well as deep scientific dives using the Nautile aboard the R/V Nadir (1985).

Combined with the previous results (reprocessed multichannel records...), the data indicate new structural interpretations which lead to the conclusion that along the northern Japan trench there is little evidence of accretion but, instead, a tectonic erosion of the overriding plate. The tectonic pattern, on the oceanic side of the trench is controlled by the creation of new normal faults parallel to the Japan trench axis, which is a direct consequence of the downward flexure of the Pacific plate. In addition to these new faults, ancient normal faults trending parallel to the N 65° oceanic magnetic anomalies and oblique to the Japan trench axis are reactivated, so that two directions of normal faulting are observed seaward of the Japan trench. Only one direction of faulting is observed seaward of the Kuril trench because of the parallelism between the trench axis and the magnetic anomalies. The convergent front of the Kuril trench is offset left-laterally by 20 km relative to those of the Japan trench. This transform fault and the lower slope of the southernmost Kuril trench are represented by very steep scarps more than 2 km high, affected by active erosion with very recent large scale slumpings toward the trench. Slightly south of the juncture, the Erimo seamount riding on the Pacific plate, is now entering the subduction zone. One other seamount located in the southern part of the Japan trench, the Daiichi Kashima seamount is split into two halves by a set of high angle normal faults. The contact between the down-faulted block of the seamount and the landward wall of the trench is outlined by a 30 m thick zone of crushed Lower Cretaceous limestone which forms the limit between the oceanic and the continental slopes.

INTRODUCTION

The southernmost and the northernmost Japan trench areas with its northern junction with the Kuril trench were surveyed in detail during the leg 3 of the French-Japanese Kaiko project onboard R/V Jean Charcot in 1984 (see fig. 1 and 2 for location), (Cadet et al., 1985) and then onboard R/V Nadir in 1985 with the new French 6000 m-submersible Nautile.

The R/V Charcot detailed survey covers about 25 000 square kilometers along the trenches with accurate Seabeam bathymetry and simultaneous recording of the magnetic and gravity fields closely spaced. Single channel seismic records were also obtained. All these data were recorded continuously along the tracks. Navigation in a Loran-C net provided continuous position information nominally precise to about 100 m, and onboard computer and plotting facilities enabled us to make real-time interpretations.

Ten dives with the "Nautile" were devoted to selected areas : Japan trench landward slope (2 dives), Kuril trench landward slope (2), Erimo seamount (3), Daiichi Kashima seamount (3), with an average of 4 to 5 hours on the bottom between 5000 and 6000 m (fig. 3 to 5). Numerous sediment cores, rocks, in situ temperature measurements, bivalves and water samplings were made in addition to continuous video-recording and selected photographs along more than 22 kilometers of precisely navigated tracks.

NEW INSIGHTS ABOUT TECTONIC EROSION AT THE FRONT OF THE NORTHERN JAPAN TRENCH LANDWARD SLOPE

The Japan trench is one of the most intensively studied subduction zones in the world, that is to say many multichannel seismic profiles were recorded in the frame of the pre-site survey of DSDP cruises and the following DSDP holes (fig. 1) results (Legs 56,57 and 87).

Fig. 1 : Location of the Kaiko I survey (leg 3, Box 1) off Northeast Japan in addition to the previous multichannel seismic lines : JNOC 1-2 (Japan National Oil Corporation, 1976), ORI 78-3 and 78-4 (Ocean Research Institute and Japan Petroleum Exploration Company, 1978) and P. 849 (Shell Oil Company, Beck et al., 1976). DSDP sites (legs 56-57 and 87) were also plotted on this map. Isobaths are derived from the bathymetric chart of the adjacent seas of Nippon n°6301 (Hydrographic Department, Maritime Safety Agency, Japan, 1966). Location of fig. 7.

Fig. 2 : Location of the surveyed area (box 2) during Kaiko I, leg 3. The Japanese multichannel seismic profile : MSA-M1 is represented.

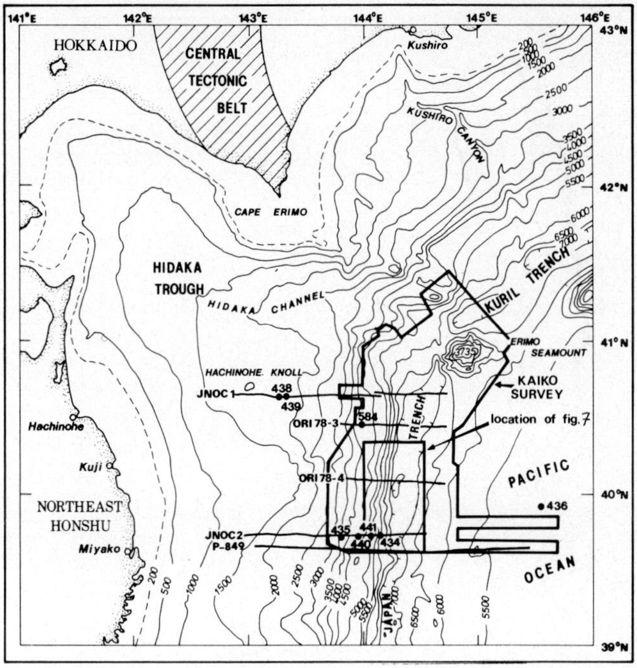
141°E
142°E
143°E
144°E
145°E
146°E
43°N
42°N
41°N
40°N
39°N
HOKKAIDO
CENTRAL
TECTONIC
BELT
Kushiro
KUSHIRO CANYON
CAPE ERIMO
HIDAKA
TROUGH
HIDAKA CHANNEL
KURIL TRENCH
ERIMO
SEAMOUNT
KAIKO
SURVEY
HACHINOHE KNOLL
JNOC 1
438
439
ORI 78-3
584
location of fig.7
TRENCH
Hachinohe
Kuji
ORI 78-4
PACIFIC
436
NORTHEAST
HONSHU
JNOC2
P-849
435
441
440
434
Miyako
OCEAN
JAPAN

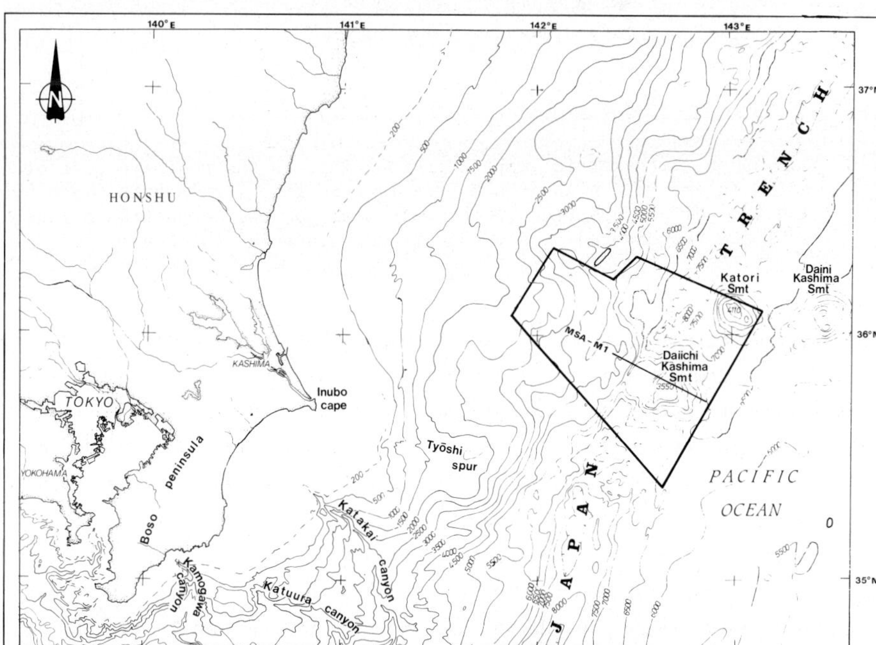
140°E
141°E
142°E
143°E
37°N
36°N
35°N
N
HONSHU
TRENCH
Katori
Smt
Daini
Kashima
Smt
MSA-M1
Daiichi
Kashima
Smt
KASHIMA
Inubo
cape
TOKYO
YOKOHAMA
Boso peninsula
Tyōshi
spur
Katakai canyon
Kamogawa canyon
Katuura canyon
JAPAN
PACIFIC
OCEAN

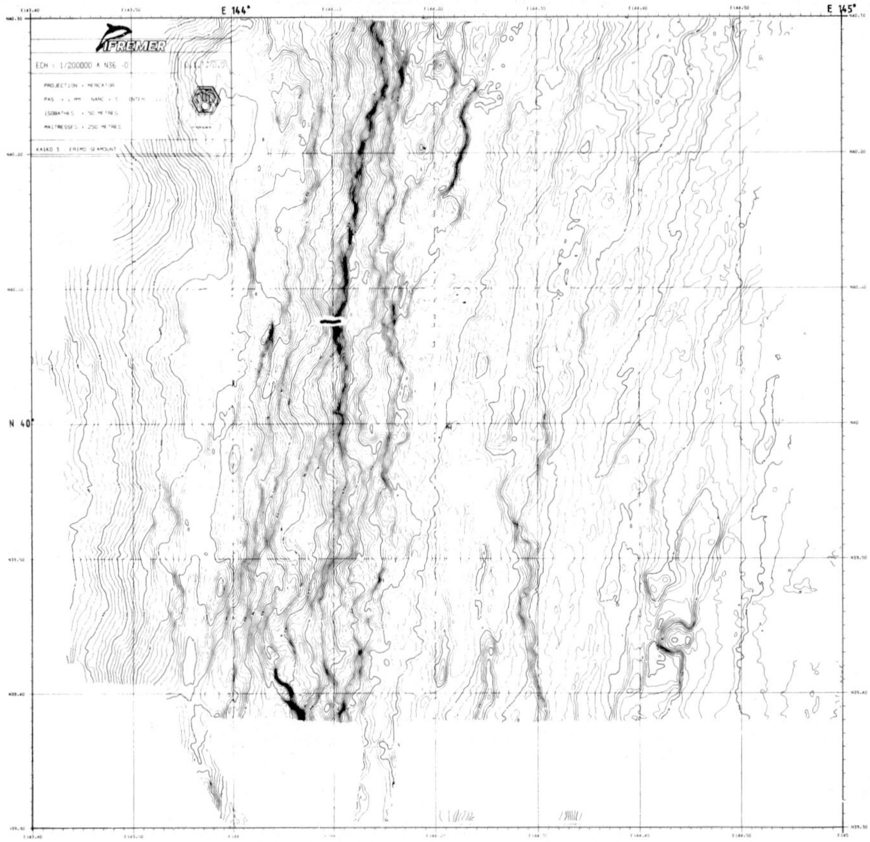

Fig. 3 : Seabeam bathymetric map corresponding to the southern part of the Kaiko survey : JAPAN TRENCH (box 1) (see fig. 1). Isobaths each 50 m. The two dives devoted to the main scarp of the Japan trench landward wall are represented by a single track because of the scale.

The Neogene tectonic history of the Japan trench landward slope (von Huene et al., 1982) is dominated by massive subsidence occasionally associated with arc volcanism and subduction until about 5 m.y. ago, and one of uplift for the last 3 m.y. The rate of vertical motion near shore is much less and perhaps negligeable. The history is based on the lithologic and paleontological sequences recovered both from drill holes near shore and also near the edge of the deep sea terrace. The temporal boundaries located by means of the drill holes were followed in seismic records to show the seaward increase in subsidence corresponding to the increase in depth of the subaerial unconfirmity cutting the Cretaceous. The subsidence was like a large flap with a quasi-stable hinge point

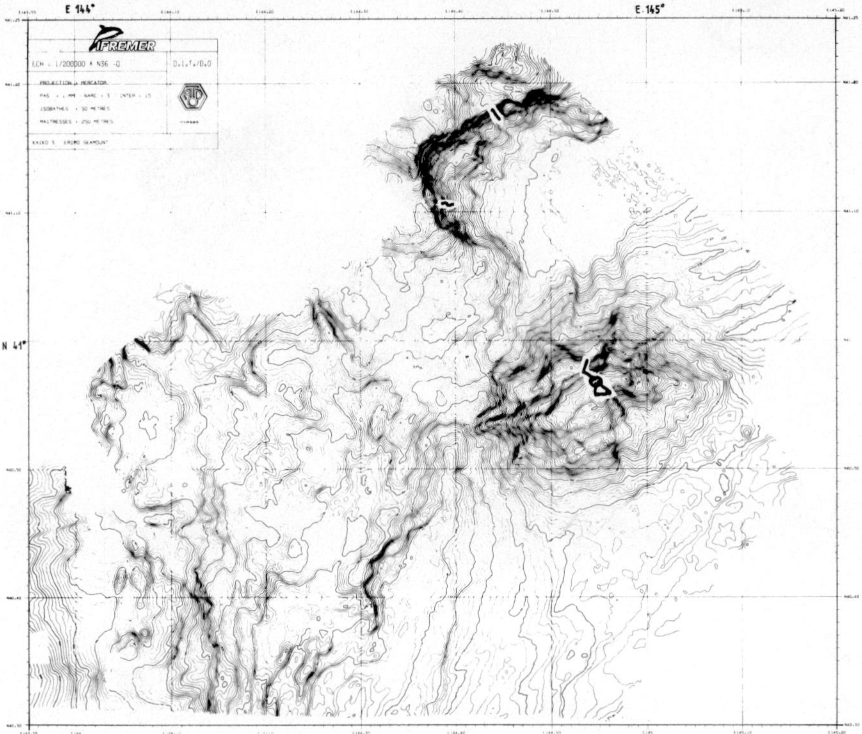

Fig. 4 : Seabeam bathymetric map corresponding to the northern part of the Kaiko survey : TRENCH JUNCTION (box 1) (see fig. 1). Isobaths each 50 m. 3 dives took place on the northern flank and at the top of the Erimo Seamount (2 for geophysical matter) and 2 dives were devoted to the landward wall of the Kuril trench.

near the present shoreline and an edge that subsided more rapidly near the trench. Subsidence resulted in the progressive seaward shift of the Neogene depocenter. Most of the sediments sampled were a monotonous sequence of silt and claystone of hemipelagic origin with rare interbeds of sand turbidites. A progressive microfracturing of the cored sediments, whose origin is thought to result from hydrofracturing as a consequence of overpressured pore fluids increased progressively toward the trench (Carson et al. 1982).

An excellent seismic profile (ORI 78-4, see fig. 1 for location) reprocessed by von Huene (Fig. 6, von Huene et al, in press) indicates that most of the oceanic sedimentary cover enters without obvious accretion down the subduction zone. It is also possible to follow the horst and graben structure of the oceanic plate under the margin. The front of the margin is composed of a thick

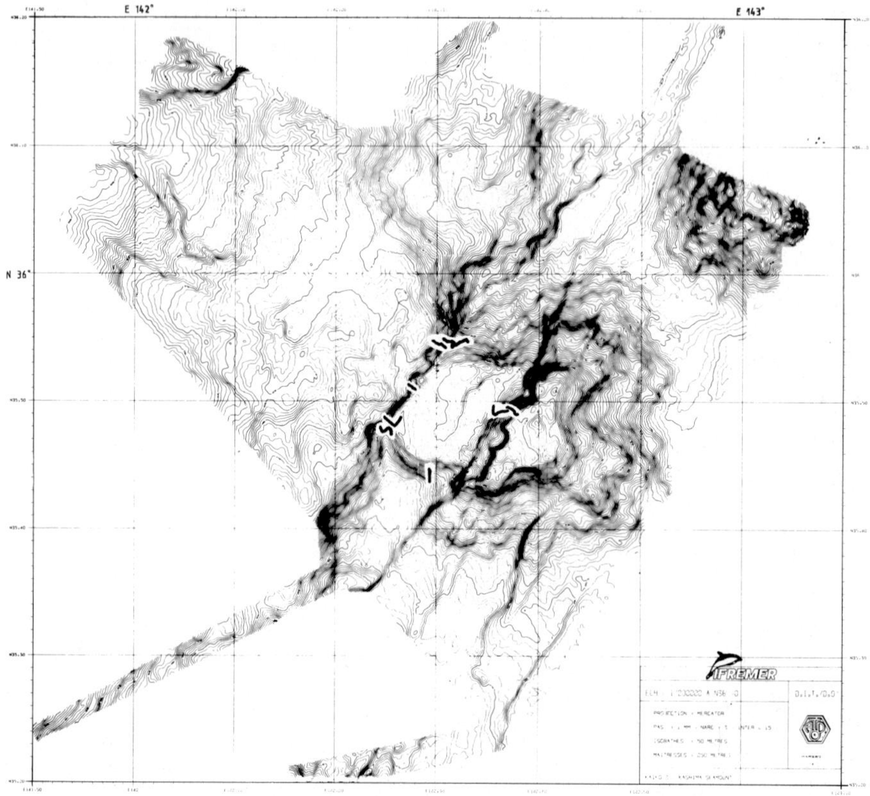

Fig. 5 : Seabeam bathymetric map corresponding to the box 2 of the Kaiko survey : DAIICHI-KASHIMA Seamount area (see fig. 2). Isobaths each 50 m. 5 dives took place during Leg 2 (Pautot et al., in press) and 3 dives occured during Leg 3 (the southernmost ones).

sequence of probable Cretaceous or Paleogene strata whose reflectors are cut at the contact with the oceanic plate. It indicates that abrasion by the oceanic subducting crust occurs below the margin. A Neogene sedimentary cover unconformably overlies the older consolidated sediments which outcrop at the favor of a huge slump scar.

The general morphological subdivisions developed from seismic records and single bathymetric profiles are characterized in greater detail and, on the landward slope of the trench, an upper slope, a mid-slope terrace and a lower slope are easily distinguished.

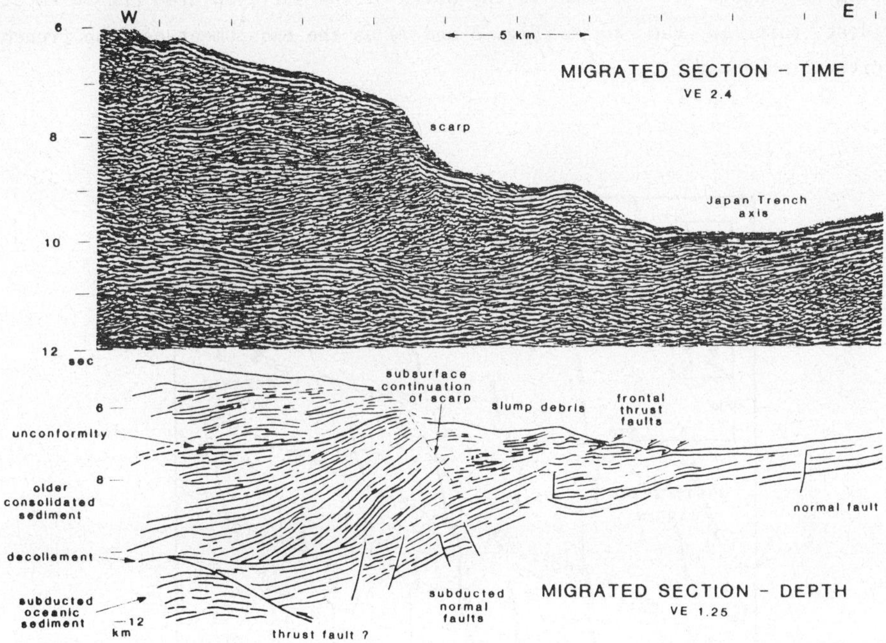

Fig. 6 : Reprocessed seismic record section ORI 78-4. Location on fig. 1 (Von Huene et al., in preparation).

The adjacent terrane immediate down-slope from the terrace is the least disrupted part of the lower slope. Here the Seabeam contours trend generally North, show a 9° dip, and are locally deflected by shallow re-entrants and small highs (Fig.3). Several short "canyon-like" features trend down-dip, suggesting down-slope transport of sediment. Notable is the lack of a strong structural grain in comparison with the adjacent areas.

A dominant morphological feature of the Seabeam map is a sinuous escarpment at about 6000 m depth that trends subparallel to the trench axis (fig. 3 and 7). The strong continuity of the escarpment is broken toward the southern part of the map although clearly defined in 4 of the 5 multichannel seismic records crossing the slope . The scarp has a remarkably uniform width of about 2 km, an average slope of 20° and a length of about 120 km. Locally, it is 1 km high.

Between the scarp and the trench axis is an area of generally disorganized topography (Fig. 3). Closed depressions, hills and lobate bodies suggest slumps. A locally strong but discontinuous linear topography is developed near the trench axis and parallels it ; a steep front commonly bounds the axis. This morphology is about 10 km wide in the south of the surveyed area and 38 km at its widest part in the north (Fig. 3 and 7) as the escarpment and the trench axis diverge northward.

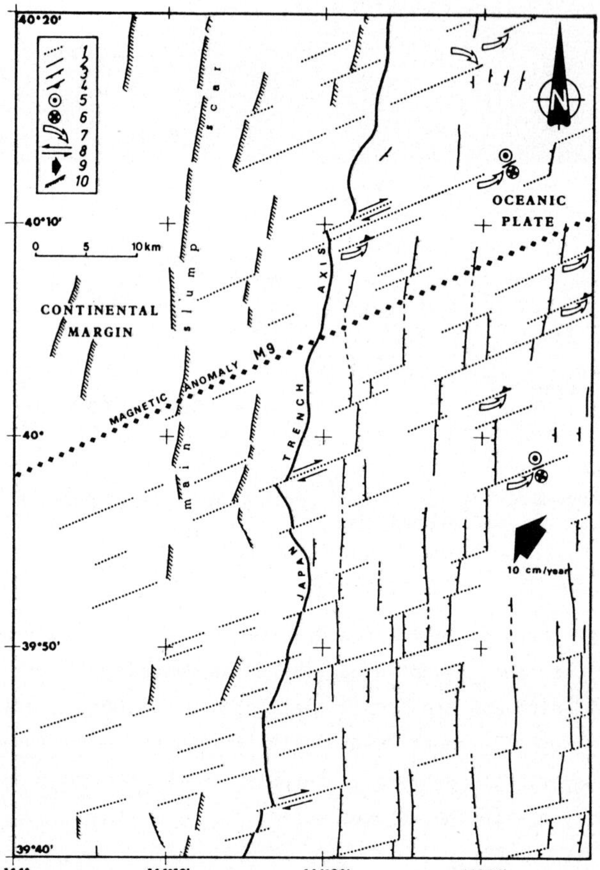

Fig. 7 : Structural map (Lallemand et al, in press). See location on the fig. 1. 1. Structural alignments or "lineaments" subparallel to oceanic magnetic lineations. 2 - Vertical faults. 3 - Normal faults. 4 - Apparent reverse faults on seismic profiles. 5 - Upfaulted block. 6 - Downfaulted block. 7 - Intersection point between N 65° "lineaments" and E-W seismic profile. 8 - Fault with at least an important right-lateral component. 9 - Pacific plate motion relative to the Japanese plate (N 55°W). 10 - Escarpments facing the trench axis.

The N-S trend prevails on the oceanic side and corresponds to the horst and graben structure due to the bending of the oceanic plate before entering the subduction zone. The second trend on the oceanic plate is parallel to the magnetic anomalies (N 65°) and, thus, corresponds to the original structure of the plate (Cadet et al., 1985 and in press). The main conclusion is that both directions prevail on the continental side, especially in the southern half of the surveyed area. Some N 65° lineaments are continuous from the oceanic side to the continental side through the trench axis. These features are summarized on the structural map (fig. 7, see fig. 1 for location) (Lallemand et al., 1986 and in press).

Different arguments (Seabeam map analysis, seismic profiles...) show that the N 65° oceanic lineaments correspond to old normal faults inherited from the original structure of the crust, reactivated into vertical (or reverse) faults with a major right-lateral component after the bending of the oceanic lithosphere because they offset the recent N-S fault set. The trench axis is also offset right-laterally along these faults. The apparent offset (horizontal and vertical) calculated from the Seabeam map on the continental side (probably close to the real offset because of the recency of the fault activity) is comparable to the oceanic one close to the trench axis and decreases landward in amplitude until zero at approximately 25 km from the trench axis. Then, it may affect a maximum thickness of 8 km of continental sediments, if we consider that the oceanic plate is at the origin of the offset. This fact is confirmed by focal mechanisms of earthquakes occuring either in the subducting or the overriding plate (Lallemand et al., 1986). Thus, it seems clear that the intermittent fault motion of the N 65° oceanic faults continues under the continental margin and induces from time to time major block movements of the front of the margin. Consequently, the N-S oceanic fault set subparallel to the trend of the margin will induce ever more block movements toward the trench. In other words, the oceanic fault (N 65° and N-S) motion below the continental margin induces gravity sliding at the base of the landward slope guided by the trend of fault bounding of the oceanic blocks (Lallemand et al., 1986).

Two dives between 5969 m and 5222 m were devoted to the landward wall of the northern portion of the Japan trench at the top of the large faulted scarp detected by seismic reflection records and an E-W cross-section normal to the trend of trench axis was surveyed with a total length of about 3 km. Overall morphology is characterized in the lower part of the wall by steep slopes or cliffs related to N 30° faults (trend of the trench axis) N 60° (trend of the oceanic magnetic anomalies) and N 8° faults, separated by more gentle benches

Fig. 8 : This photograph illustrates the typical landslide morphology of the Japan trench landward lower slope (see fig. 3 for location) (depth : 5 800 m).

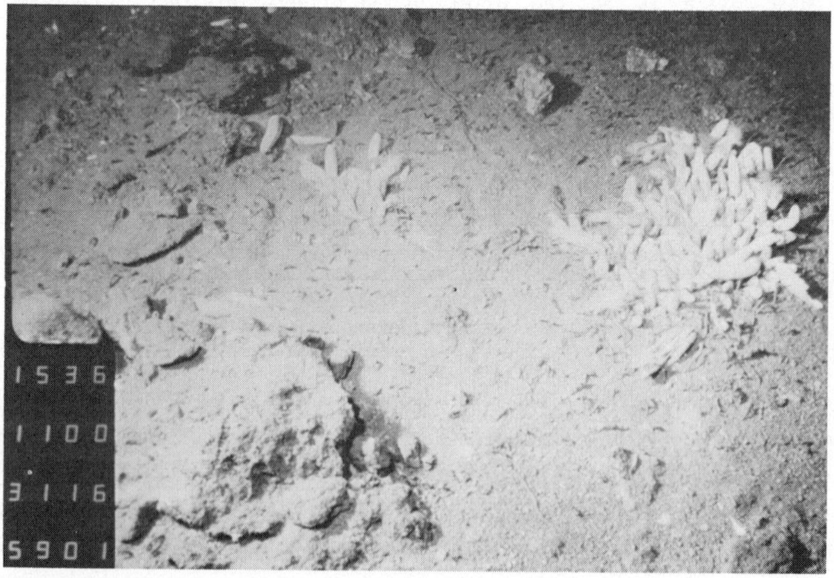

Fig. 9 : One of the largest clam colonies with several tens of living shells (depth = 5900 m, see fig. 3 for location).

and terraces, with typical evidences of scars, slumps and gravity slides (fig. 8). Outcrops are continuous and consist of landward dipping Neogene mudstone. Microfractures are conspicuous with strikes of N 30°, 80° and 140°. Toward the top of the section, overall morphology becomes smoother than the deeper parts and moderately steep slopes are coated with 4-5 cm thick Mn-dioxide black crusts, often broken. "Clam" colonies were frequently observed from 5969 m and, as a first approximation, both abundance and size of clam colonies decrease upward in the observed track. Small clam colonies are composed of 5 to 12 vertically living shells (Calyptogena n. sp.) and the larges ones are composed of 50 to 60 shells with an average length of 15 cm (fig. 9). Round colonies are located on benches without any apparent relations with the tectonic but others are clearly aligned along fracture. A few isolated clams were observed far from any colonies.

Thus, mass sediment sliding occurs on the landward slope of the trench, fills the axial zone of the trench and reduces it to a succession of short and narrow basins which are offset on the oceanic side by faults trending obliquely to the trench axis. The topographic Japan trench is only the frontal boundary of the sedimentary gravity sliding. These slide deposits are then probably subducted together with oceanic crust and its cover.

The Japan trench could be a model for tectonic erosion linked with subduction : the continental slope is being destroyed by superficial collapse and the resulting material may ultimately be carried down the subduction zone. The continental plate has thus retreated. In addition, it is likely that erosion of the continental margin from below by the subducting oceanic plate occurs (Langseth et al., 1981).

THE JAPAN-KURIL TRENCHES JUNCTION

The second major objective of the Kaiko program leg III was to explore the Japan and Kuril trench juncture near Erimo seamount. This already complex juncture is further complicated by a southward projection of the Central Tectonic Belt on the island of Hokkaido that passes beyond Cape Erimo into the Pacific (Fig. 1). The belt may have been a major plate boundary either between the American and Eurasian plates or between the Okhotsk and Amurian plates in Neogene time (Chapman and Solomon, 1976 ; Jolivet and Cadet, 1984 ; Kimura and Tamaki, in press). The role of Erimo seamount at this juncture is debated from several points of view, because neither the bathymetry nor the seismicity indicates a clear structural relation with other major tectonic features.

The Seabeam topography (fig. 4) shows clearly, first a NW-SE left-lateral offset (transform fault ?) by about 20 km of the convergent front in the Kuril trench relative to the Japan trench. Second, the trench axis trends N 25°E south of the previous tectonic boundary and N 55°E north of it. Third, the Erimo seamount is located immediately south of this tectonic boundary, but an already subducted seamount (less important than Erimo), is observed, on the basis of magnetic anomalies and seismic profiles, at the upper corner of the trench-juncture. Are these results sufficient to say that the sharp curvature of the trench axes is due to the collision of a possible chain of seamounts preceding Erimo seamount, according to the conceptions of Vogt et al. (1976) ?

Keeping in mind that our survey includes only the frontal zone of the continental plate and also only the southernmost part of the Kuril trench, it seems very difficult to clarify this problem without taking into account the adjacent areas, especially northward and eastward of the present survey. It seems that the oceanic plate is not concerned by this curvature because the normal faults are parallel to both trenches with an intermediate zone south of the Erimo seamount, where we can observe simultaneously both directions : N 10°E and N 60°E. The seismicity in this area does not confirm an active collision between an already subducted chain of seamount and the margin (Lallemand et al., submitted).

Another possibility is to relate the "transform fault" to a failure line of a present boundary between the Japanese microplate and the Okhotsk plate (Seno, 1985), but there is no evidence of intracontinental seismic activity in the assumed boundary zone (between Erimo Cape and Erimo Seamount). Concerning the curvature of trenches near Erimo seamount, it may be due to a different tectonic behaviour of plates on either side of the Neogene plate boundary between America and Eurasia (Chapman et Solomon, 1976).

Two dives were devoted to investigate the very steep, 2 000 m high, landward walls of the westernmost portion of the Kuril trench. One dive is situated at its northern (south-facing) wall and the other at its western slope (fig. 4). In about 30 % of the surveyed track, the slopes are composed of vertical cliffs even with overhanging topography, with talus of gravels, sands and fallen rocks (fig. 10). Outcrops of layered grey-greenish diatomaceous mudstone with white volcanogenic tuff have a very fresh surface indicating that the slope was formed very recently by active slidings and that erosion is very active there. The cliffs are built up by two sets of subvertical faults ; one striking N 60° and the other N 330°. The former determines general direction of main cliffs parallel to trench axis. The latter determines the transverse cliffs. Observations throughout the dive at the western wall of the Kuril trench are not in disagreement with the previous proposition of the left-lateral strike-slip origin of this wall. Clam (Calyptogena) colonies were observed at various depths. They are generally small and scattered or isolated. Rare gastropodes were found with clams. Temperature of sediments inside a colony at 5 130 m was 1.120°C slightly higher than those at the edge (1.111°C, 1.114°C) and outside of it (1.111°C).

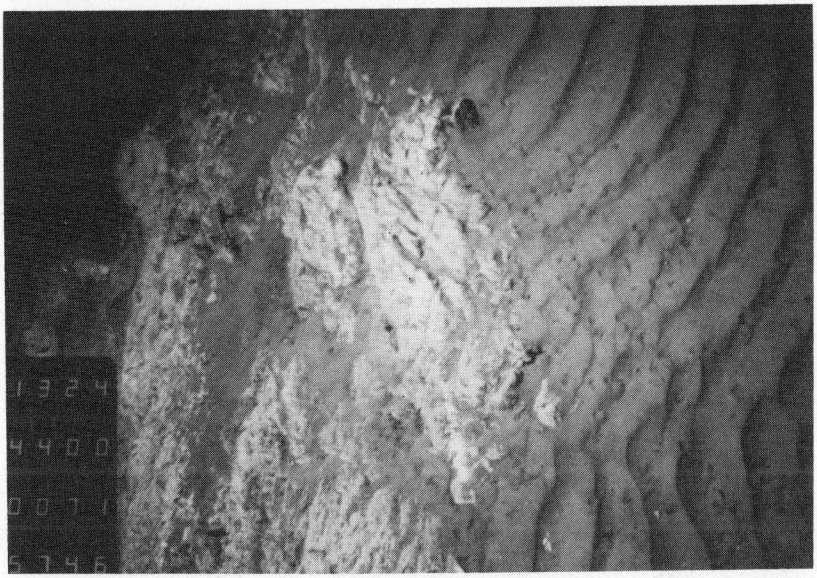

Fig. 10 : An abrupt outcrop is clearly observed on the left side of the photograph, while current ripple marks affecting recent sediments are observed on the right side. Kuril trench landward wall (see fig. 4 for location, depth : 5750 m.

Thus the transition with the Kuril trench corresponds to a left lateral transform fault which offsets the continental slope and leads to a wide flat plain in the axial zone strongly contrasting with the small basins of the northern Japan trench. The Erimo seamount, located slightly south of the junction, may belong to a chain, already subducted, responsible for the trench curvature by collision. This trench transition may also be related to the intracontinental boundary between American and Eurasian plates. Nevertheless, no definitive conclusion can be given at present.

HOW A SEAMOUNT SUBDUCTS IN THE JAPAN TRENCH ? THE CASE OF THE DAIICHI KASHIMA SEAMOUNT

Mogi and Nishizawa (1980) first found the scarp assimilated to a normal fault separating Daiichi Kashima Seamount into two blocks based upon their compilation of bathymetric data.

The Research Group for Daiichi Kashima Seamount (Tokai University, 1985) argued that the present topography of the seamount was originally formed in early Cretaceous and was not due to the down-faulting. They claimed, based upon their paleontological study of some dredge samples, that a reef limestone from the western half is of Barremian age which is older than an Albian limestone collected from the eastern half.

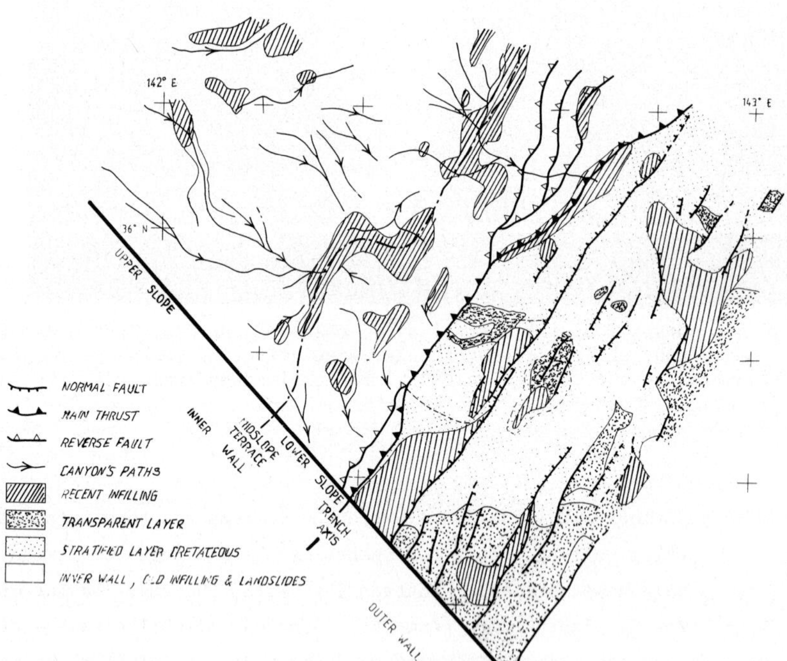

Fig. 11 : Distribution of sedimentary facies in an area surrounding the Daiichi Kashima Seamount. Canyon paths are represented on the Japan Trench landward slope.

Map (Fig. 5) represents a redrawn topographic map of the seamount area at a contour interval of 50 m (Kobayashi et al., in press). The first continuous delineation of a nearly straight 100 km long normal set of faults trending N 30°E which cuts the Daiichi Kashima Seamount into halves is most remarkable (Fig. 11). Diameter of the exposed seamount is about 50 km but the faults extend further along the axial zone of the Japan trench. The deepest part of the Japan trench (approximately 8 km deep) is along the northern extension indicating that the greatest depression of the trench was also achieved along the faults. The greatest vertical displacement along the fault planes at the crest of the seamount is 1 600 m accomodated by many smaller offset displacements, gradually decreasing toward both ends of the fault which extend far from the flank of the seamount.

Topographic and seismic reflection results obtained during the Kaiko survey provide definitive support for the splitting of the Daiichi Kashima Seamount by young normal faults. Such normal faults are observed across the oceanward slope of the Japan trench but not east of the marginal swell about 100 km from the trench axis in the abyssal basin of the Pacific ocean (Cadet et al., in press). The present faults are not a relict of normal faults formed at the spreading axis, because they are oblique to the magnetic lineations in this region.

Thus the Kashima seamount "falls" into the trench before being subducted ; this phenomenon belongs to the normal faulting of the oceanic plate when entering into the trench. The set of faults which cuts the seamount also cuts the oceanic crust and nothing supports the idea of a simple collapse of the seamount as sometimes stated. The normal faulting increases when the Kashima seamount approaches the trench. It is therefore situated in a transverse syncline of the oceanic basement of the trench ; which consequently dips axially toward the down-faulted part of the seamount.

The down-faulted part of the Kashima seamount just collides with the landward slope and passes underneath as seen on the seismic records ; the landward slope is uplifted and tilted in its lower part ; only a gentle deflection is related to that collision that might suggest that the seamount is downfaulted another time just below the subduction front.

Finally the subduction of the Kashima seamount seems to be as easy as for the normal oceanic crust : the main factor seems to be the gravity as suggested by the classical normal faulting of the oceanic plate before getting in the trench. It is reinforced by the fact that the seamount is more downfaulted than the adjacent area when going into the trench. The conclusion has certainly to be tested in other trenches, but suggests that seamounts have no special destiny in the subduction processes.

Three dives in Kaiko leg III were devoted to the southwestern part of the Daiichi Kashima Seamount (fig. 5). The first one has a nearly N-S upslope track located on the southern outer slope of the lower body of the faulted seamount. The basalts, and the boundary between basalts and overlying limestone were observed and the samples although partially phosphatized and sometimes encrusted with ferromanganese oxides, suggests its lithological and paleontological (Aptian to Albian Orbitolinids, Rudists), similarity with those collected from the upper (eastern) body of the Daiichi Kashima Seamount (Pautot et al, in press) as well as the previous dredge hauls. The total thickness of sedimentary caps (mostly shallow-water reef limestone) is approximately 400 m in the present track which is roughly the same as or slightly thicker than that on the upper body determined by submersible observations and seismic records (Oshima et al., 1985).

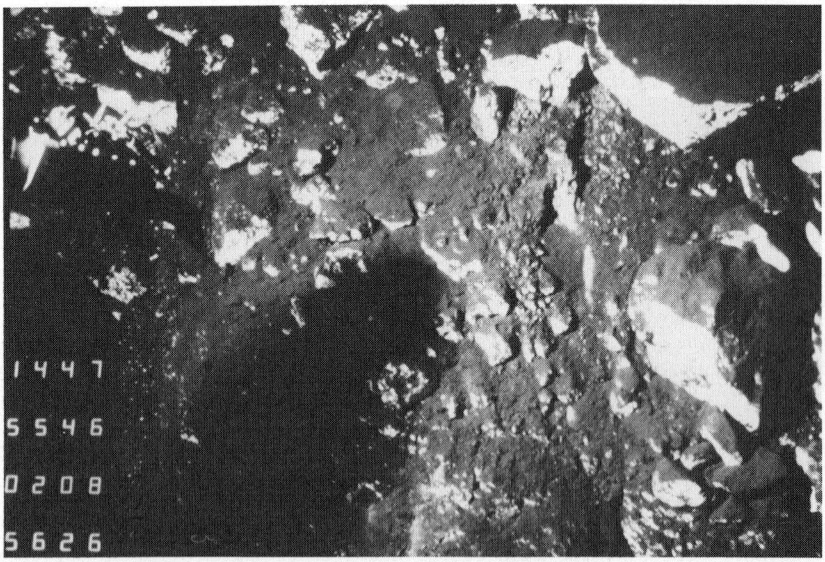

Fig. 12 : Crushed limestones marking the tectonic boundary between the down-faulted part of the Daiichi-Kashima seamount and the continental margin (depth : 5626 m, see fig. 5 for location).

The last two dives were devoted to the investigation of the plate boundary between the subducting southwestern flank of the seamount and the landward slope. A zone of deformed limestone was clearly observed along the apparent plate boundary near the trench axis. It consists of very angular flat boulders of limestone (fig. 12) up to a few meters long. On the landward slope a 30 m high talus of highly brecciated limestone underlying thin limestone bed, which might be an accreted material was observed in one of the two tracks. Thus for the first time the oceanic sea-floor has been found to be covered by the continental slope, along this deformation zone, demonstrating active subduction.

Above this deformation zone, the Japan continental slope is very steep and characterized by almost vertical cliffs of siltstone and mudstone and supposed to be formed by a succession of several stepwise faults. Rare bivalve communities have been found near the contact.

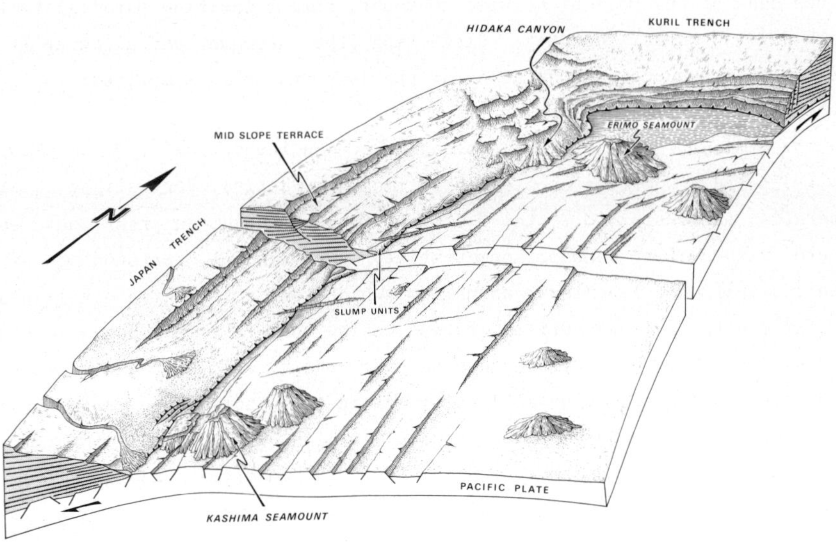

Fig. 13 : Perspective diagram illustrating the main results of the Kaiko project (leg 3).

CONCLUSION

The combined survey, using especially a Seabeam sounder and a deep-sea submersible during the Kaiko project, provided us with a great amount of geophysical, geological, visual and sampling data. The closely spaced seismic profiles in addition to bottom data allowed us to have a three-dimensional view of this area (fig. 13). The scientific community is now analysing these data in order to finalize the discoveries. But, the preliminary results can be summarized as follows :

1. There is no indication of active accretion of oceanic sediments to the landward wall of the Japan trench.
2. On the contrary, the continental landsliding material and the oceanic seamounts are carried along by the Pacific plate when entering the subduction zone.
3. Tectonic erosion associated with megaslumps is a characteristic of the landward wall of the Japan trench and mainly explains its retreat.
4. The mechanism of this tectogenesis of the lower slope of the Japan trench consists in the reactivation of oceanic N-S and N 65° faults below the continental margin (Lallemand et al., 1986)
5. In the case of the Daiichi Kashima seamount, normal faulting parallel to the trench axis occured recently, dissecting the seamount and allowing it to enter in the subduction zone, even if the landward wall is uplifted in front of the seamount.
6. Bivalve communities are found along the continental wall, in decreasing density upward, wherever erosion is sufficiently active. Very slight in situ temperature increases (0.2 to 0.4°C, Le Pichon et al., in press) have been measured within the densest communities. Analysis of surrounding water reveals significant chemical anomalies which confirm that the bivalves are associated with overpressurized water oozing out from the zone of contact between plates. The energy necessary for their life probably comes from methane and hydrogen sulphide contained in the water. The overpressurized water level enables the oceanic plate to slide without too much friction below the continental slope. These bivalve communities can be used as tectonic markers of this gliding plane (Le Pichon, in press).

REFERENCES

Cadet, J.P., Kobayashi, K., Aubouin, J., Boulègue, J., Dubois, J., von Huene, R., Jolivet, L., Kanazawa, J., Koisumi, K., Lallemand, S., Nakamura, Y., Pautot, G., Suyehiro, K., Tani, S., Tokuyama, H. and Yamasaki, T., 1985. De la fosse du Japon à la fosse des Kouriles : Premiers résultats de la campagne océanographique franco-japonaise Kaiko (Leg III). C.R.Acad. Sci. Paris, t. 301, 5 : 287-296.

Cadet, J.P., Kobayashi, K., Aubouin, J., Boulègue, J., Deplus, C., Dubois, J., Jolivet, L., Kanazawa, T., Kasahara, J., Koizumi, K., Lallemand, S., Nakamura, Y., Pautot, G., Suyehiro, K., Tani, S., Tokuyama, H., von Huene, R., and Yamazaki, T., in press. The Japan Trench and its juncture with the Kuril Trench, Cruise results of the Kaiko Project, Leg 3. Earth Planet. Sc. Lett.

Cadet, J.P., Kobayashi, K., Aubouin J., Boulègue J., Dubois, J., Hotta, H., Ishii, T., Jolivet, L., Konishi, K., Lallemand, S., Niitsuma, N. and Shimamura, H., in Press. Deep scientific dives in the Japan Trench and its adjacent areas. Earth Planet. Sc. Lett.

Carson, B., von Huene, R. and Arthur, M., 1982. Small-scale deformation structures and physical properties related to convergence in Japan trench slope sediments. Tectonics, 1 : 277-302.

Chapman, M.E. and Solomon, S.C., 1976. North American-Eurasian plate boundary in northeast Asia, J. Geophys. Res., 81 : 921-930.

Jolivet, L. and Cadet, J.P., 1984. Mouvements décrochants, structuration et limite de plaques dans l'île d'Hokkaido (Japon septentrional). Annales de la Société Géologique du Nord, C III : 345-352.

Kimura, G. and Tamaki, K., 1986. Collision, rotation and back-arc spreading : the case of the Okhotsk and Japan Seas. Proceedings of the Geodynamic Symposium on "Collision Tectonics", Texas A & M University. Tectonophysics.

Kobayashi, K., Cadet, J.P., Aubouin, J., Boulègue, J., Dubois, J., von Huene, R., Jolivet, L., Kanazawa, T., Kasahara, J., Koizumi, K., Lallemand, S., Nakamura, Y., Pautot, G., Suyehiro, K., Tani, S., Tokuyama, H., Yamazaki, T., in press. Normal faulting of Daiichi Kashima seamount, Japan Trench, revealed in topography surveyed during the Kaiko 1 cruise, leg 3. Earth Planet. Sc. Lett.

Lallemand, S., Cadet, J.P. and Jolivet, L., in press. Mecanisme de tectogénèse à la base du mur interne de la fosse du Japon (au large de Sanriku, Japon NE) : rejeu des failles océaniques sous la marge (in french with english abstract). C.R. Acad. Sc. Paris.

Lallemand, S., Cadet, J.P. and Jolivet L., in press. Geodynamic evolution of the northeast Japan continental margins from tertiary time to present : new insights. Bull. Soc. Géol. France.

Langseth, M.G., von Huene, R., Nasu, N. and Okada, H., 1981. Subsidence of the Japan trench forearc region of northern Honshu, Oceanologica acta, Proc. 26th Int. Geol. Congr., Geol. Cont. Marg. Symp. Paris, 173-179.

Le Pichon, X., Iiyama, T., Boulègue, J., Charvet, J., Faure, M., Kano, K., Lallemant, S., Okada, H., Rangin, C., Taira, A., Urabe, T. and Uyeda, S., in press. Nankai trough and Zenisu ridge : deep-sea submersible survey. Earth Planet. Sc. Lett.

Mogi, A. and Nishizawa, K., 1980. Breakdown of a seamount on the slope of the Japan trench. Proc. Jpn Acad., 56 : 257-259.

Pautot, G., Nakamura, K., Angelier, J., Bourgois, J., Fujioka, K., Huchon, P., Kanazawa, T., Nakamura, Y., Ogawa, Y., Seguret, M. and Takeuki, A., in preparation. Deep-sea visual observations of deformation of the toe and a seamount subduction, preliminary results of 1985 Kaiko cruise, leg 2. Earth Planet. Sc. Lett.

Oshima, S., Ogino, T., Katsura, T., Ikeda, K., Uchida, M., Nagano, M., Hayashida, M., Muneda, K., Kasuga, S. and Tani, S., 1985. Subduction of the Daiichi Kashima seamount into the landward slope of the Japan trench. Rep. Hydr. Res., 20 : 25-46.

Research Group for Daiichi-Kashima seamount (Tokai University), 1985. Daiichi-Kashima Seamount. Tokay University Press.

Seno, T., 1985. Is northern Honshu a microplate ? Tectonophysics, 115 : 177-196.

Vogt, P.R., Lowrie, A., Bracey, D.R. and Hey, R.N., 1976. Subduction of aseismic oceanic ridges : effects on shape, seismicity and other characteristics of consuming plate boundaries. Geol. Soc. Amer., spec. paper, 172 : 59 p.

Von Huene, R., Langseth, M., Nasu, N. and Okada, H., 1982. A summary of cenozoic tectonic history along the IPOD Japan trench transect, Geol. Soc. Am. Bull., 93 : 829-846.

Von Huene, R., Nasu, N., Culotta, R. and Aoki, Y., in preparation.

Kaiko project is a cooperative french-japanese program organized under the patronage of IFREMER and CNRS for France and Monbusho for Japan to study the Japanese subduction trenches in two phases (July-August 1984 with the Jean Charcot, and July-August 1985 with the Nautile).

Participating scientists to leg III (Japan trench) were for the first phase (1984) :
Jean-Paul CADET and Kazuo KOBAYASHI (co-chiefs), Jean AUBOUIN, Jacques BOULEGUE, Jacques DUBOIS, Roland von HUENE, Laurent JOLIVET, Toshihiko KANAZAWA, Junzo KASAHARA, Kin-Ichiro KOIZUMI, Serge LALLEMAND, Yasuo NAKAMURA, Guy PAUTOT, Kiyoshi SUYEHIRO, Shin TANI, Hidekazu TOKUYAMA and Toshitsugu YAMAZAKI

and for the second phase (1985) :
Jean-Paul CADET and Kazuo KOBAYASHI (co-chiefs), Jean AUBOUIN, Jacques BOULEGUE, Jacques DUBOIS, Hiroshi HOTTA, Teruaki ISHII, Laurent JOLIVET, Kenji KONISHI, Serge LALLEMAND, Nobuaki NIITSUMA and Hideki SHIMAMURA.

GEOCHEMISTRY OF VOLCANIC ROCKS FROM THE MARIANA, YAP AND PALAU TRENCHES BEARING ON THE TECTONO-MAGMATIC EVOLUTION OF THE MARIANA TRENCH-ARC-BACKARC SYSTEM

L. BECCALUVA[1], G. SERRI[2] and J. DOSTAL[3]

[1]Dipartimento di Scienze della Terra, Via Mezzocannone 8, Napoli - Italy

[2]Dipartimento di Scienze della Terra, Via Santa Maria 53, Pisa - Italy

[3]Department of Geology, St. Mary University, Halifax, Nova Scotia - Canada

ABSTRACT

New trace element analyses, including REE, on volcanic rocks from Mariana, Yap and Palau trenches strongly support the conclusions of Beccaluva et al. (1980) that the rocks dredged from the inner and outer walls of these trenches respectively derived from island arc/interarc and ocean island/mid-ocean ridge mantle sources. Tholeiitic rocks recovered from the inner wall of the Mariana Trench at station 1438 and 1404 are of great petrogenetic interest in that they have mineralogical and geochemical features respectively intermediate between island arc tholeiites/boninites and island arc tholeiites/MORB.

The genesis of 1404 tholeiites can be best explained considering that they derived by higher degrees of melting than, but from a mantle source largely comparable to, normal MORB. Incompatible element ratios indicate that the source was respectively slightly and significantly enriched in LREE + Ba and K + Rb, very likely via hydrous fluids derived from the subducted oceanic plate.

Orthopyroxene-phyric low-Ti tholeiitic series (station 1438), ranging in composition from magnesian quartz tholeiites to bronzite andesites, is mineralogically and chemically distinct from normal island arc tholeiites as well as from boninites. They very much resemble low-Ti basalts or LREE-depleted boninite-like lavas from low-Ti ophiolites.

Higher CaO/Al_2O_3, much lower or absent LREE enrichment, much higher Ti/Zr, Tb/Zr and Y/Zr ratios respectively greater or equal or much lower than primitive earth mantle of the low-Ti basalts with respect to typical boninites point to different mantla sources and to distinct metasomatizing fluids from the underlying subduction zone. Nevertheless the genesis of parental magmas of both series requires high degrees of melting (>15-20%) of depleted mantle source, i.e. cpx-poor harzburgites and cpx-poor lherzolites for boninites and low-Ti basalts respectively , under hydrous conditions, at pressure where orthopyroxene melts incongruently (<10-15 Kb).

Data so far available restrict the genesis of low-Ti basalts and boninites to supra-subduction intraoceanic settings.

INTRODUCTION

The petrogenesis of island arc magmas has been extensively studied and debated in last decade. Numerous petrogenetic models for their genesis have been invoked. Currently, the most popular of these involve both the subducted slab and overlying metasomatized mantle wedge (Ringwood, 1973, 1974; Green, 1976; Hawkesworth et al., 1977; Kay, 1977, 1980).

Whatever the mode of arc magma petrogenesis, is should be borne in mind that until recently, virtually everything that was known about arc-related magmatism was deduced from samples representative of only a very small volumetric fraction of the arcs. The subaerial parts of the island arc volcanoes in oceanic arcs probably represent only one hundredth to one thousandth of the total volume of the volcanic edifice. At the present time, nearly nothing is known on the nature and the composition of the products of the earliest submarine stages of arc volcanism, or of the volcanics (if any) erupted between adjacent arc volcanoes. Similarly, with the exception of data from the Mariana forearc (Bogdanov et al., 1977; Dietrich et al., 1978; Beccaluva et al., 1980; Natland and Tarney, 1981; Bloomer, 1982, 1983) little is known of the arc-related rocks which form the basement of the forearc reqions of oceanic island arcs. Although a great deal of data on volcanic rocks erupted in back-arc basins has been recently published (Ridley et al., 1974; Gill, 1976; Hawkins, 1977; Saunders and Tarney, 1978; Kay and Hubbard, 1978; Wood et al., 1980; Natland and Tarney, 1981) none, except those of Weaver et al. (1979) and of Tarney et al. (1981), concerns the composition of lavas erupted during the initial stages of arc rifting and back-arc opening.

There is now a wide consensus that most of the well-preserved and well-studied Phanerozoic ophiolites such as Oman, Troodos, Vourinos, Betts Cove, etc. were formed in subduction-related settings, i.e. back-arc, interarc basins and intraoceanic island arcs (Beccaluva et al., 1980; Serri, 1981; Alabaster et al., 1982; Coish et al., 1982; Cameron et al., 1983; Beccaluva et al., 1984). As they share with the Mariana arc-backarc system a surprising variety of lavas including mid-ocean ridge basalts (MORB), island arc tholeiites (IAT) and boninites , a careful evaluation of the magma-types occurring in these ophiolites can give important information for unravelling the complexities in the genesis of subduction-related magmas from intraoceanic settings.

In this paper we present geochemical data on volcanic rocks dredged during the 1976 Dimitry Mendeleev's cruise, from the inner and outer walls of Yap, Mariana and Palau trenches in order to put constraints on their petrogenetic affinities . We also rewiev available geological data on the West Philippine-Mariana region aimed at providing an up to date reconstruction of its tectono-magmatic evolution.

GEOLOGICAL AND PETROCHEMICAL OUTLINES

The Mariana and Yap trenches are the south-eastern boundary of the most developed and best studied island arc-backarc systems of the world. The most recent syntheses of its tectonic and magmatic evolution are given by Scott and Kroenke (1980), Natland and Tarney (1981), Crawford et al. (1981) and Bloomer (1983). The main results of our former paper (Beccaluva et al.,1980), later confirmed by D.S.D.P. Leg 60 data (Natland and Tarney, 1981) and dredging of the inner Mariana wall (Bloomer, 1983),were the recognition that almost all the rocks recovered from the island arc slope of Mariana, Yap and Palau trenches have petrochemical similarities with subduction-related products. On the contrary, the volcanic rocks dredged from the Pacific side of the trenches all show a clear affinity with mid-ocean ridge or ocean island magmas. This is now considered as related to important removal of forearc material (by tectonic erosion and subduction) largely during the Eocene-Oligocene subduction of the Pacific plate (Hussong et al., 1978; Beccaluva et al., 1980; Natland and Tarney, 1981; Bloomer, 1983).

The volcanic rocks analysed in this paper were selected from a large set of samples previously studied by Beccaluva et al. (1980). Their precise dredge-site locations together with petrochemical data are given by Bogdanov et al. (1977), Beccaluva et al. (1980) and Peive (1980). Except the rocks from the Palau trench and the Yap trench dacite (sample 1429A), the studied samples suffered little or no mineralogical evidence of halmyrolisis. The K/Ar ages and geological setting of the majority of the samples are relatively well known (Bogdanov et al., 1977; Beccaluva et al., 1980).

Major and trace element analyses of 15 representative samples are reported in Table 1. All major and trace element determinations (except REE, Sc, Co, Hf and Th by INAA; see Dostal and Capedri, 1979) were carried out by XRF on powder pellets with a PW 1450 automated spectrometer (see Leoni and Saitta, 1976 and Beccaluva et al., 1980 for methods and precision). Rb and Ba are new XRF analyses. The remaining trace elements were re-checked by XRF; the new analyses are in good agreement with those previously published (Beccaluva et al., 1980) but revised values for P_2O_5 and Y (2 and 3 samples respectively) at particularly low concentrations are given (Table 1). The precision is estimated to be better than 10% for all the INAA and XRF determined trace elements.

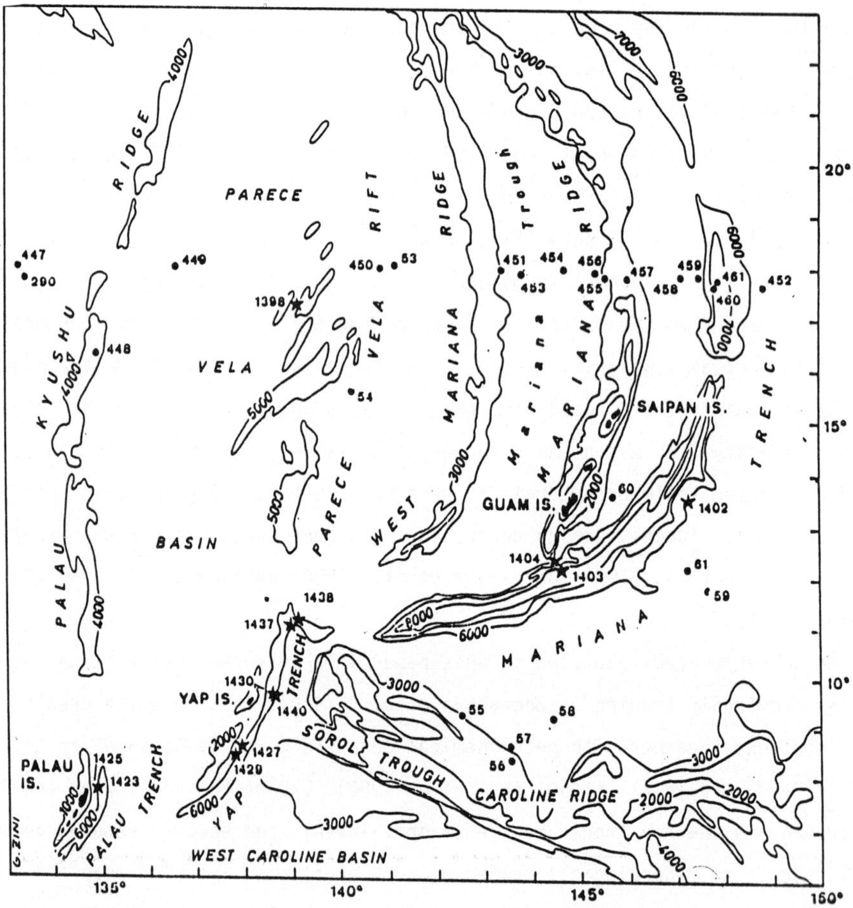

Fig. 1. Map of West Philippine Sea-Yap arc-Mariana arc area showing locations of Dimitry Mendeleev dredge sites (stars) and D.S.D.P. sites (filled circles). Bathymetric contours are in meters (afeter Beccaluva et al., 1980).

Off-shore station

The 1402 dredge haul from the outer wall of the Mariana Trench (depth 5400-5600 m) yielded variously evolved and altered alkalic and tholeitic lavas likely derived from a volcanic seamount of Cretaceous age (Dietrich et al., 1978; Beccaluva et al., 1980). For this paper two sample have been selected: one slightly altered alkalibasalt (sample 1402A) containing iddingsitised olivine phenocrysts set in an ophitic groundmass consisting of altered olivine, Ti-salite, plagioclase and interstitial alkalifeldspar, and one relatively altered tholeitic basalt (sample 1402B) carrying phenocrysts of plagioclase, clinopyroxene and possibly olivine in an originally glass-rich groundmass.

Primitive MOR-type tholeiites and pillow lavas likely formed about 7 m.y. ago at the Sorol Trough spreading centre (Beccaluva et al., 1980) were dredged one above the other (samples 1430 and 1440, depths 7250-7600 and 5800-6100 m respectively) from the outer wall of the Yap Trench (Fig. 1). The two selected samples are relatively fresh, except for altered olivine in specimen 1430B and sparse chlorophaeite after glass in sample 1440C. They are both slightly porphyritic with plagioclase (An 83) and olivine (Fo 88) as liquidus phases. The latter mineral constantly bears inclusions of tiny chromites.

Near-shore stations

A great variety of volcanic rocks belonging to different magma series were dredged from the inner walls of the Palau, Yap and Mariana Trenches (Bogdanov et al., 1977; Dietrich et al., 1978; Beccaluva et al., 1980; Peive, 1980; Bloomer, 1982).

Mariana Trench - All the dredge stations of the Dimitry Mendeleev cruise from the Mariana trench are located in the southern sector where the present day plate motion is dominantly strike-slip (Katsumata and Sykes, 1969). Here, differently from the northern section, the Benioff zone is poorly defined and the maximum focal depth of earthquakes barely reaches 100 Km. An oblique plate convergence is associated with strike-slip faulting along east-west fractures which define a transform fault between the backarc spreading ridge and the trench. This probably started about 7 m.y. ago and perhaps even earlier (Karig et al., 1978; Karig and Ranken, 1983). Here, as north of Guam, the limited subduction of the leading edge of the overriding plate resulted in an unusual topography of the inner slope characterised by a series of east-west trending ridges and troughs as well as the absence of the forearc basin and the frontal arc (Beccaluva et al., 1980). The arc structure itself is disrupted and locally the inner slope of the trench directly abuts the southern end of the Mariana Trough (Bloomer, 1982).

The seven samples selected for this paper come from two (1404 and 1438) dredge hauls. Station 1404 (depth 5405-5600 m) is located south of Guam upslope the Station 1403 (depth 8100-8400) where boninites have been rediscovered in the 1976 (Bogdanov et al., 1977; Peive, 1980). Here in addition to the boninites (Dietrich et al., 1978; Sharaskin et al., 1980; Hickey and Frey, 1982) there was the recovering of whole ophiolitic sequence consisting of various types of ultramafic and gabbroic rocks including abundant harzburgites, enstatite-bearing norites and scarce basaltic rocks, largely dolerites. Upslope station 1404 only tholeiitic dolerites and pillow lavas, mineralogically and chemically analogous to those of the previous station, were found (Bogdanov et al., 1977; Dietrich et al., 1978; Beccaluva et al., 1980).

Volcanic rocks belonging to the boninite series, ranging from olivine boninites to hyperstene dacites, were more recently dredged slighly upslope (depth 5093-3846 m) from the same locality. Here basaltic rocks similar to 1403/1404 tholeiites were also recovered (Bloomer, 1982).

K/Ar dating of doleritic basalts from Site 1404, carried out independently by Sharaskin et al. (1980) and Beccaluva et al. (1980), gave ages of 33 and 28 respectively. According to the tectono-magmatic evolution of the Mariana region (Scott and Kroenke, 1980; Natland and Tarney, 1981; Hussong and Uyeda, 1982; Crawford et al., 1981) and considering that the 1404 tholeiites are very likely cogenetic we interpret (Beccaluva et al., 1980) this suite of rocks as formed during the early stages of opening of the Parece Vela Basin (32-28 m.y.).

Three samples representative of the whole petrochemical variations of station 1404 have been selected for detailed analyses. In contrast with typical MORB (Bryan, 1983), 1404 tholeiites contain clinopyroxene as liquidus phase even in the most primitive samples, constantly associated with plagioclase, but generally crystallizing before it (Table 2). Olivine, now altered to chlorophaeite, can be recognized unambiguously only in one sample (1404 F), both as phenocrysts and dubitatively in the groundmass. It is to be noted that in the 1404 doleritic basalts (this work and Dietrich et al., 1978) pigeonite and very scarce interstitial quartz are present in the groundmass. This indicates an original slight SiO_2-oversaturation of the magmas. The zoning (decreasing Ca/ Ca+Mg+Fe from core to rim)and composition of the pyroxene phenocrysts determined in the glass-bearing samples (1404 F and C; Fig. 2; Beccaluva et al.,1980) suggests that this might also hold for all the analysed 1404 tholeiites. Station 1438 is located in the inner trench wall (forearc of the Mariana Trench, depth 6400-6700 m) close to the bowl-shaped depression more than 8000 m deep,which marks its southwestern termination at the intersection with the Yap Trench (Bogdanov et al., 1977; Beccaluva et al., 1980). The volcanic rocks recovered at Site 1438 have an unusual chemical and mineralogical composition compared to ocean floor, ocean island and island arc magmas.

Beccaluva et al. (1980) presented K/Ar dates for two 1438 rocks. A slightly Ne-normative, Ti-augite bearing basalt 1438 D yielded an age of 7.8 ± 1.3 m.y. A tholeiitic andesite with island arc affinity (1438 E) gave an age of 10.8 ± 0.4 m.y. which was considered to be probably higher than the eruption age, due to possible retention of radiogenic Ar in the abundant glass of this sample. Rocks from station 1438 represent a suite of highly incompatible element depleted arc tholeiites ranging in composition from magnesian quartz-tholeiites to andesites. They are in many respects (constantly opx-phyric, very low and high contents of incompatible and compatible elements respectively) transitional to boninite series (Beccaluva et al., 1980; Beccaluva and Serri, 1981; Crawford et al., in press). The 1438 lavas were interpreted by Beccaluva et al. (1980) as to represent magmas generated during the early stages of formation of the Mariana Trough and therefore roughly emplaced in the same tectoic setting of the associated mildly-alkalic basalts.

Yap and Palau Trenches - The physiography of these trenches is unusual in that arc-trench gap is narrower than any other arc-trench system of the world. In addition, although they have the sea-floor well below the 6000 m isobath (Fig. 1) seissmic activity has not been so far recorded. Apart from the 17th cruise of the Dimitry Mendeleev (Bogdanov et al., 1977; Peive, 1980) no other work on rocks collected from the Palau trench as been to our knowledge published. These rocks were studied in detail by Beccaluva et al. (1980) who recognised them as calcalkaline despite of their low-grade metamorphism.

Two relatively primitive, but unfortunately severely altered, samples have been selected for this work. The basalt 1423 B contains, as it is common for calcalkaline rocks, phenocrysts of clinopyroxenes, plagioclase and orthopyroxene. In the basaltic andesite 1423 D rare yellow-brown hornblende is an additional phenocryst phase (Beccaluva et al., 1980).

The inner wall of the Yap Trench an the Yap Island are relatively well studied. Nevertheless their origin is still debated (Shiraki, 1971; Divis, 1975; Hawkins and Batiza, 1977; Bogdanov et al., 1977; Beccaluva et al., 1980; Mc Cabe and Uyeda, 1983). Both on the island and in the trench highly metamorphic sheared rocks such as amphibolites, greenschists, marbles, etc. have been recovered (Shiraki, 1971; Hawkins and Batiza, 1977; Bogdanov et al., 1977; Peive, 1980). Volcanic rocks erupted after the deformation, metamorphism and emplacement of the Yap allochton are exposed on the island (Tomil Formation conglomerates, Miocene age; Divis, 1975) and in the trench (Bogdanov et al., 1977; Hawkins and Batiza, 1977; Peive, 1980). K/Ar ages of two fresh rocks from station 1427 (an island arc tholeiite, 1427 B, studied also in this paper and a two-pyroxene quartz-bearing gabbro, 1427 C) gave ages of 7.6 ± 1.0 and 10.9 ± 3.0 m.y. respectively (Beccaluva et al., 1980).

Of the two samples selected for this study, one (1429 A) is particularly altered. Because of an originally glassy matrix and phenocrystal plagioclase and pyroxene, it is tentatively classified as dacite. The fresh sample is a typical island arc tholeiite, with abundant plagioclase (about 30%, An 93-85), subordinate clinopyroxene and rare olivine as phenocrysts (Table 2) set a microcrystalline groundmass (Beccaluva et al., 1980).

GEOCHEMISTRY

Major element, mineral chemistry and petrography as well as some XRF determined trace elements such as Zr, Y and Nb were extensively discussed in Beccaluva et al. (1980). In this paper we will deal in more detail with the whole trace element distribution of the rocks from the Mariana, Yap and Palau Trenches with particular reference to incompatible elements including rare earths.

As many studied rocks underwent some submarine alteration and those from Palau Trench even low-grade metamorphism, most of the discussion is based upon elements such as Y, Ti, Zr, Hf, P, Nb, Th and REE, which are relatively immobile during halmyrolisis (Wood et al., 1980) and sub-seafloor hydrothermal alteration (Humphris and Thompson, 1978 a, b). Ba and Sr are also included in the discussion; however, since they are alteration sensitive particularly if calcite and epidote are generated, their real abundance in the original igneous rocks

Table - 1 Chemical composition of the rocks from the Mariana, Yap and Palau Trenches

	Mariana Trench									Yap Trench				Palau Trench	
	Off-shore		Near-shore							Off-shore		Near-shore		Near-shore	
	1042 A	1042 B	1438 C	1438 B	1438 E	1438 D	1404 A	1404 F	1404 C	1430 B	1440 C	1427 B	1429 A	1423 B	1423 D
SiO_2(Wt%)	45.77	46.18	47.98	58.41	61.36	49.32	48.50	48.06	48.43	47.87	48.91	47.48	63.90	49.15	54.17
TiO_2	2.12	1.93	0.44	0.67	0.72	1.61	0.89	1.34	2.09	1.14	1.04	0.59	0.68	0.68	0.82
Al_2O_3	16.31	15.46	14.81	14.67	14.28	17.61	14.90	13.99	14.11	15.31	16.33	18.33	14.50	16.34	15.82
Fe_2O_3	9.36	10.05	5.46	5.47	5.08	6.09	6.97	6.69	8.84	5.49	4.83	5.99	3.20	5.25	5.15
FeO	2.89	3.23	1.65	2.65	3.55	2.80	3.60	7.28	3.94	5.15	3.91	5.40	3.66	3.72	2.96
MnO	0.16	0.15	0.11	0.15	0.13	0.16	0.12	0.24	0.13	0.24	0.15	0.18	0.21	0.15	0.15
MgO	6.31	4.61	12.15	3.90	2.44	4.57	8.27	7.09	6.39	8.91	7.60	6.06	4.43	6.93	4.48
CaO	8.15	10.97	0.17	8.90	8.12	11.08	11.28	11.48	8.99	10.35	12.52	12.33	1.85	8.66	8.65
Na_2O	3.81	2.79	1.77	2.23	2.63	4.45	2.36	2.42	3.29	2.48	2.58	2.05	3.10	4.17	3.06
K_2O	1.06	1.11	0.84	0.61	0.55	0.25	0.64	0.38	0.72	0.09	0.20	0.33	0.90	0.28	0.12
P_2O_5	0.34	0.16	0.02*	0.05	0.08	0.24	0.07	0.07	0.15	0.10	0.08	0.10*	0.13	0.10	0.17
L.O.I	3.72	2.00	4.61	1.79	1.06	1.81	2.07	0.96	2.91	2.87	1.84	1.14	3.44	4.58	4.44
Mgvalue	0.53	0.43	0.79	0.50	0.38	0.53	0.63	0.52	0.52	0.64	0.66	0.54	0.58	0.62	0.52
Sc (ppm)	26	39	47		33	32	45			38	35	41	21	26	25
V	258	297	262	311	321	261	282	423	501	271	276	365	127	285	243
Cr	338	255	879	73	<15	230	352	259	144	425	517	90	<15	288	180
Co INAA	41	42	36		21	32	47			33	41	37	11	32	23
Ni	235	75	204	33	17	83	100	94	68	97	185	32	5	127	62
Zr	153	87	16	24	33	124	40	62	86	72	60	21	81	63	93
Hf INAA	3.5	2.3	0.6		1.2	2.7	1.1			1.8	1.6	0.9	1.5	1.6	2.3
Nb	24	10	<1	1	1	6	2	2.5	3.5	3.5	2.5	1.5	2.5	3.5	5
Y	26*	24	11*	21	27	27	19	33	45	27	24	14*	34	17	22
Th INAA	1.6	0.8	<0.3		0.3	0.4				0.1	0.1	0.3	0.6	1.1	1.5
Rb XRF	18	19	8	8	10	8	13	5	11	1	2	6	5	6	3
Sr	368**	213	36	54	69	289	67	64	77	93	102	302	141	222	314
Ba XRF	155	47	21	30	38	40	16	18	20	11	14	56	80	27	20
La INAA	13.1	7.41	0.38		1.09	4.51	1.51	2.14	2.94	2.24	1.88	1.93	3.61	7.92	8.74
Ce INAA	33	19.3	1.16		3.27	15.0	4.1	6.42	8.18	6.90	5.56	5.49	10.6	16.6	20.4
Nd INAA	17.9		1.33		3.80	11.7	4.2			6.43	5.92	4.64	8.99	9.7	11.8
Sm INAA	4.58	3.69	0.73		1.99	3.64		2.03	2.84	2.66	2.46	1.47	3.19	2.52	2.87
Eu INAA	1.59	1.30	0.30		0.77	1.39	0.6	0.81	1.08	0.99	0.92	0.55	1.09	0.88	0.99
Tb INAA	0.84	0.78	0.25		0.65	0.85	0.5	0.62	0.85	0.69	0.69	0.35	0.82	0.51	0.57
Yb INAA	2.46	2.19	1.05		2.92	2.71	2.2	2.58	3.67	2.67	2.40	1.56	3.58	1.92	2.29
Lu INAA	0.39	0.35	0.18		0.48	0.43	0.4	0.44	0.60	0.41	0.38	0.26	0.58	0.31	0.37

Mg-value is $Mg/Mg+Fe^{++}$ calculated with Fe_2O_3=0.15 FeO. The elements labelled with INAA and XRF are new analyses performed at Halihax and Pisa respectively. The other elements are from Beccaluva et al. (1980). *These values are more precise and supercede those of Beccaluva et al. (1980). **Value changed from 268 to 368 because of printing error in the original paper.

must be evaluated with care (Hart et al., 1974; Wood et al., 1980). K and Rb concentration in few fresh samples is used to infer the original composition of petrogenetically equivalent rocks.

Compatible elements

Owing to the small amounts and small size of phenocrysts, generally all studied samples (except 1427 B which certainly contains accumulated plagioclase) represent essentially liquid composition (Table 2). Only two samples (1438 C and 1440 C) have Cr and Ni values (879, 204 ppm and 517, 185 ppm respectively; Table 1) compatible with direct mantle peridotite melting (cf. Frey et al., 1978; Clague and Frey, 1982). Other volcanic rocks have suffered only minor mafic mineral fractionation (1402 A, 1430 B, 1404 A, 1404 F, 1423 B; Cr and Ni respectively of 338, 235 ppm; 425, 97 ppm; 352, 100 ppm; 259, 94 ppm; 288, 127 ppm; Table 1). The remaining rocks are relatively more evolved. Three still have a basaltic composition (1402 B, 1427 B and 1404 C) and the remaining four range from basaltic andesites (1423 D) through andesites (1438 B and E) to dacites (1429 A).Concomitantly with the decrease of Cr and Ni values, Co becomes more and more depleted with advancing fractionation. This is less obvious for Sc (Table 1). In the rocks with basaltic composition the rapid (Cr and Ni) and weak (Co) decrease of elements highly compatible with spinel and olivine uncorrelated with Sc, suggests that early stages of fractionation were dominated by olivine and spinel with a subordinate role of pyroxenes.

Rare earth and other incompatible elements

As REE have an intermediate field strength between Zr, Hf, Ti, Nb (high field-strength elements, HFSE) and Rb, K, Ba, Sr (low field-strength elements, LFSE; Saunders et al., 1980), they are among the less useful parameters to distinguish between magmas generated under different tectonic settings (Fig. 2). When coupled with HFS and LFS elements, REE become instead a powerful tool to better understand the petrogenesis of magmas (Fig. 3). The REE alone do not closely discriminate between volcanic rocks dredged from the inner (nearshore) and outer (offshore) walls of the Philippine Sea Trenches (Fig. 1). In the framework of the conclusions of Beccaluva et al. (1980), based on major and trace elements as well as mineral chemistry, REE patterns are easier to evaluate.

Off-shore stations - REE patterns of the rocks from 1402 have slight enrichment in light REE (LREE) with the La/Yb ratios of 3.4 to 5.4. Comparable

Table - 2 Modal composition of the porphyric rocks from the Mariana and Yap Trenches

	1402 A	1438 C	1438 B	1438 E	1404 A	1404 F	1404 C	1430 B	1440 C	1427 B
Phenocrysts										
ol	(10.1)	(0.9)	-	-	-	(2.2)	-	(1.5)	5.3	(2.0)
opx	-	4.1	2.0	0.5	-	-	-	-	-	-
cpx	-	7.5	2.7	0.8	2.1	3.5	2.4	-	-	4.5
plag	-	-	7.4	2.3	3.6	12.3	3.1	3.4	15.5	29.1
groundmass	89.9	75.0	74.5	83.7	94.3	76.6	87.2	95.1	86.2	60.1
vescicles	-	12.5	13.4	12.7	-	5.4	7.3	-	3.0	5.3
key minerals of the groundmass	alk-feldspar	-	-	-	quartz pigeonite	-	-	-	-	-

(), In all the samples, but 1440C, olivine is replaced by secondary minerals. Modal compositions performed with more than 3000 points in two standard-type thin sections for each sample.

rocks were also described for this station by Dietrich et al. (1978) who, in addition, have found strongly alkalic types with La 200 and 300 times the chondritic value. In our samples (1402 A, mildly alkalic basalt; 1402 B, tholeiitic basalt) heavy REE (HREE) are slightly but significantly fractionated with respect to chondrites or primitive earth mantle (cf. Clague and Frey, 1982; Jagoutz et al., 1979). This may reflect a small amount of residual garnet or, if melting occured in the spinel facies, a slight LREE enriched source. According to major element and mineral chemistry, LREE are more enriched in the alkalic than in the tholeiitic basalts. These rocks are compositionally indistinguishable (Fig. 2 and 3) from plume-type MORB (cf. Sun et al., 1979; Shibata et al., 1979; Le Roex et al., 1983) or ocean islands/aseismic ridges (cf. Humphris and Thompson, 1983; Roden et al., 1984). If strongly alkalic basalts of Dietrich et al. (1978) are taken into consideration, the whole 1402 rock-suite can be best interpreted as originally done by Beccaluva et al. (1980), as formed in intraplate setting.

Rocks from outer wall of the Yap Trench (1430 B and 1440 C) are normal MORB generated at the Sorol Trough spreading centre about 7 m.y. ago. They exhibit flat HREE and strongly and increasingly depleted LREE (Fig. 2). Accordingly with their near-primary liquid composition, they have an incompatible element pattern parallel but at lower values than average normal MORB (Fig. 3).

Near-shore stations - The rocks dredged from the Marian Trench (Site 1404, south of Guam), although they are in many respects within the range of variation of MORBs, show subtle but significant differences from them (constant presence of cpx phenocrysts also in the near-primary magmas, groundmass pigeonite and quartz in the microcrystalline samples, higher Ti/Zr ratios, Zr/Y values at the lower range of MORBs). The LREE of the three 1404 basaltic samples with different degree of evolution are less depleted if compared with near-primary normal MORB from Yap Trench (Sorol Trough).Nevertheless, intermediate REE (Sm, Eu and Tb) contents in the relatively primitive samples (1404 A and F) are well below those of Sorol Trough normal MORB (Fig. 2), but comparable to some extremely intermediate REE-depleted MORB which are however more primitive than 1404 rocks (cf. Shilling et al., 1983). Differently from these very rare MORB and Sorol Trough equivalents, 1404 tholeiites show distinctly higher La/Sm ratios. REE patterns tend to be similar to island arc tholeiites such as those from Tonga (Ewart et al., 1977) than to MORB. This supports the conclusions of Bec-

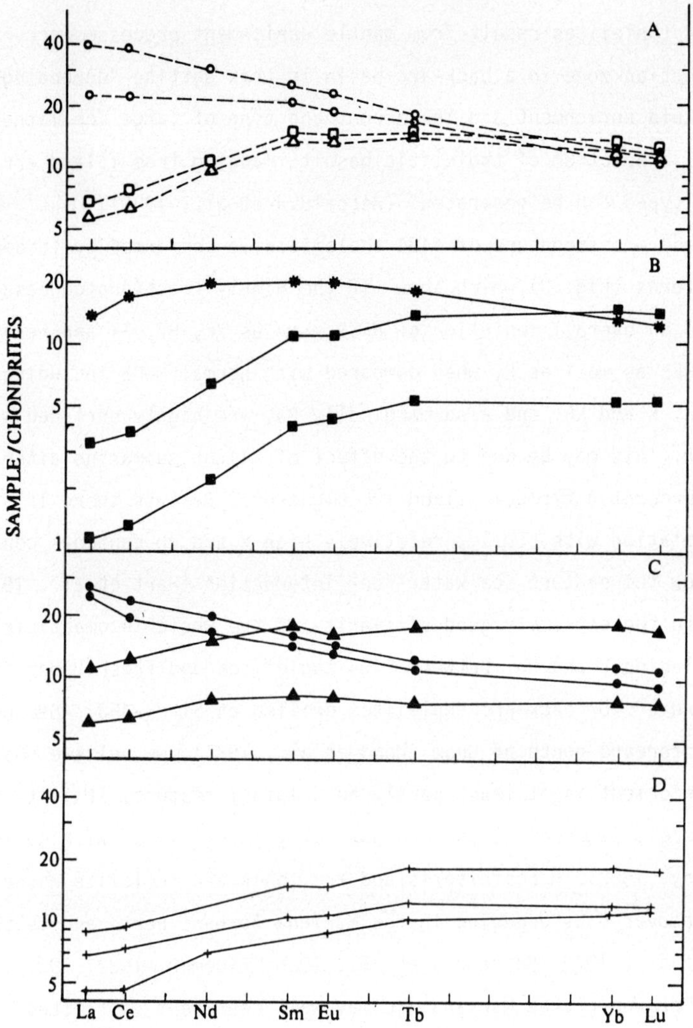

Fig. 2. REE abundances in volcanic rocks from the Philippine Sea normalized to chondrites(Frey et al., 1968). A - rocks from the off-shore slope stations of the Mariana Trench (samples 1402 A and B, open circles) and Yap Trench (samples 1430 B and 1440 C, open squares and triangles respectively). B - rocks from the near-shore slope stations of the Mariana Trench close to Yap (samples 1438 C and 1438 E, closed squares; 1438 D, asterisk). C - rocks from the near-shore slope stations of the Yap Trench (samples 1427 B and 1429 A, closed triangles) and the Palau Trench (samples 1423 B and 1423 D, dots). D - rocks from the near-shore slope slope stations of the Mariana Trench south of Guam (samples 1404 A, F and C, crosses).

caluva et al. (1980) that the transitional features (between normal MORB and IAT) of 1404 tholeiites result from mantle enrichment processes occurring above the subduction zone in a back-arc basin. In this setting "depending on the degree of fluid enrichment and the amount and type of large ion lithophile elements added, a spectrum of tholeiitic basalts ranging from island arc to mid-ocean ridge types can be generated" (Beccaluva et al., 1980).

The island arc signature of 1404 tholeiites is confirmed by incompatible element patterns (Fig. 3) which show (if the highly fractionated basalt 1404 C is excluded) an overall depletion of HFSE such as Zr, Hf, Ti and heavy and intermediate REE as well as P, when compared with normal MORB including those of Sorol Trough. K and Rb, and also marginally Ba, are highly enriched with respect to MORB. This may be due to the effect of slight submarine alteration or it would represent a typical island arc signature. Because there is a rough positive correlation with L.O.I., relatively high K and Rb contents could indeed be due to low temperature sea water-rock interaction (Hart et al., 1974). However, owing to the extremely good similarity of the whole incompatible element pattern including K and Rb, between 1404 tholeiites and fresh Upper Eocene-Lower Oligocene arc or back-arc tholeiites drilled at Sites 458, 459 and 460 in the Mariana forearc north of Guam (Wood et al., 1981), we believe that marked K and Rb enrichment is at least partly an original feature. It is to be noted that 458 tholeiites underlie and are partially interlayered with boninites (Hickey and Frey, 1982). A boninite/island arc, back-arc tholeiite association has also been recovered by dredging in the Mariana forearc north and south of Guam (Dietrich et al., 1978; Beccaluva et al., 1980; Bloomer, 1982, 1983). Therefore the forearc region of the Mariana arc may well represent a rifted off slice of the Kyushu-Palau Ridge (Natland and Tarney, 1981) including oceanic crust formed in the earliest stages of the Kyushu-Palau ridge rifting (embryonic Parece Vela Basin) as pointed out by Beccaluva et al. (1980) and Crawford et al. (1981).

Major and trace element composition of the selected basalts (1427 B) and dacite (1429 A) from the Yap Trench is typical of island arc tholeiitic suites (cf. also Beccaluva et al., 1980). They have very low abundances of Ti, Zr, P and Y. The contents of P, Zr, Y, Ba and REE increase while transition elements (see earlier section) and Sr decrease with fractionation. The basalts have low absolute abundances of REE than the dacites. They both have rather similar, re-

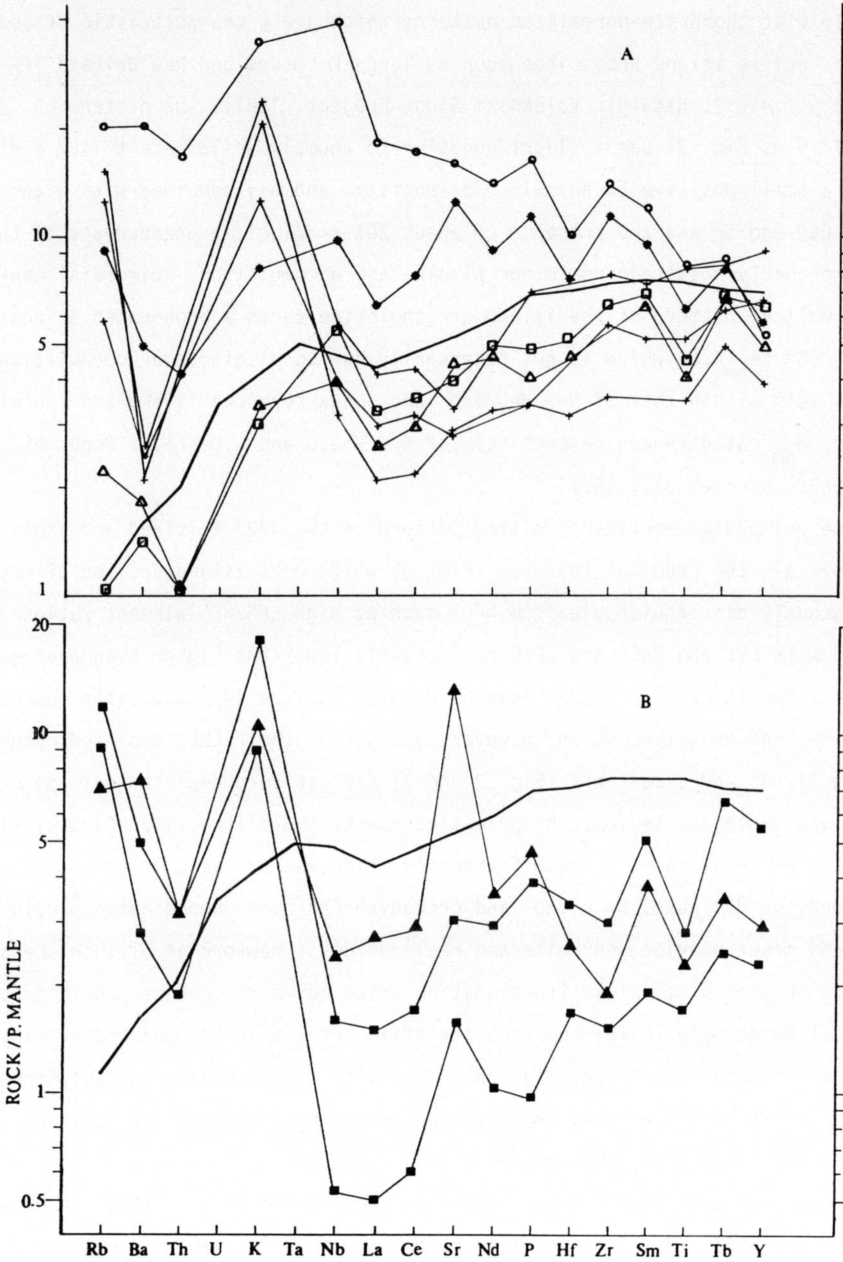

Fig. 3. Incompatible element abundances normalized to primordial mantle of Wood (1979) of selected volcanic rocks of the Mariana and Yap Trenches compared to normal MORB. A - inner and outer walls tholeiites and mildly alkalic basalts. B - inner wall normal (1427 B) and low-Ti (1438) island arc tholeiite series. Symbols as in Fig. 2.

latively flat chondrite-normalized patterns which are a characteristic of some Western Pacific island arc suites such as Tonga-Kermadec and New Britain (Ewart et al., 1977; Basaltic Volcanism Study Project, 1981). The pattern of dacite (1429 A, Fig. 2) has a slight negative Eu anomaly while basalt 1427 B displays a small positive Eu anomaly. The positive anomaly combined with high Al_2O_3, CaO and Sr and the presence of about 30% plagioclase phenocrysts in this sample probably resulted from minor plagioclase accumulation. Primordial mantle-normalized pattern of the island arc tholeiite shows a pronounced Sr positi anomaly (Sr/Ce_{pm} 4.5) which is not necessarily due to plagioclase accumulation because well within that of New Hebrides and Tonga-Kermadec island arc tholeiites (Sr/Ce_{pm} ratios ranges respectively of 6.5 - 3.0 and 6.0 - 1.5; Dupuy et al., 1983; Ewart et al., 1977).

The primordial mantle-normalized pattern of the 1427 B island arc tholeiites shows all the chemical features (Fig. 3) which make island arc tholeiites unambiguously distinguishable from MORB such as high LFS/HFS element ratios coupled with REE and HFSE,and LFSE respectively lower and higher than average MORB (cf. Perfit et al., 1980). This holds also for the 1438 magnesian quartz-tholeiites and andesites which, however, because of their LREE-depleted nature and low Ti, Zr and Y contents (Fig. 2 and 3) fall at very depleted extreme of island arc tholeiite series. The primitive quartz-tholeiites (1438 C) qualifies to be the parental magma of 1438 E andesite which was produced by about 70% of fractionation dominated by clino- and orthopyroxene over subordinated plagioclase and minor olivine, chromite and Fe-Ti oxides (Crawford et al., in press). In spite of some plagioclase fractionation which may have been responsible for the small Eu anomaly (0.76) the andesite still retains in the primordial mantle normalized diagram the Sr positive anomaly of its parent magma. In fact, the Sr/Ce_{pm} ratio decreases only from 2.6 to 1.6 during fractionation. This reflects minor plagioclase separation.

In contrast with most island arc tholeiites (Perfit et al., 1980; Arculus and Johnson, 1981), the 1438 volcanic rocks do not exhibit any Ba positive anomaly (Fig. 3). A smooth increase from Th to Ba up to Rb, and extremely pronounced K and Sr anomalies are, however, common features of boninites (Hickey and Frey, 1982 and reference therein), particularly those which are slightly LREE-depleted (Mariana Trench, Site 458) or little REE enriched (Mariana Trench, Station 1403; Bonin islands; Cap Vogel, sample 2984). However, it should be empha sized that boninites (except those from Site 458) have characteristically concave-upward REE pattern whereas that of the studied magnesian quartz-tholeiite is exactly the inverse.

In conclusion, these mineralogical and chemical features make the 1438 suite of rocks transitional between island arc tholeiite and boninite series. Rocks with analogous composition to 1438 tholeiites are commonly found in low-Ti ophiolites such as Troodos, Vourinos, Betts Cove, etc. (Sun and Nesbit, 1978; Beccaluva et al., 1980 b; Serri, 1981; Coish et al., 1982; Beccaluva et al., 1984; Cameron, 1985).

At station 1438 some slightly Na-alkaline basalts have also been recovered. The representative sample (1438 D) has chemical features (convex upward REE pattern, Fig. 2; enrichment of Rb, K and Ba with respect to normal MORB; pronounced Sr positive anomaly and relatively high K/Nb, Sr/Nb and Rb/Nb ratios, Fig. 3) and radiometric age (7.8 ± 1.3 m.y.) that can be best explained as formed in the initial stages of opening of the Mariana Trough (Beccaluva et al., 1980).

The rocks from the Palau Trench (Station 1423) consist of basalts and andesites. The major element composition, including low TiO_2, is typical of an island arc environment. The relationship between Ti, Zr, Y and Nb indicates that the rocks have calcalkaline affinity (Beccaluva et al., 1980). Likewise, the REE patterns of two selected samples (1423 B and D) show slight LREE enrichment similar to island arc calcalkaline suites. Total REE abundances increase with the fractionation without significant element/element variations. In fact, their chondrite-normalized patterns are roughly parallel to each other (Fig. 2). Because they are severely altered, these rocks are not reported in the incompatible element diagram of Fig. 3.

IMPLICATIONS FOR THE TECTONO-MAGMATIC EVOLUTION OF THE MARIANA TRENCH-ARC-BACKARC SYSTEM

New trace element analyses, including REE, on Mariana, Palau and Yap Trenches strongly support the conclusion of Beccaluva et al. (1980), which regards all the volcanic rocks dredged from the inner and outer wall of the Mariana, Yap and Palau Trenches respectively as being derived from island arc-interarc and ocean island-mid ocean ridge mantle sources. This implies a significant amount of tectonic erosion of the leading edge the the arc plate since the Eocene, as pointed out by Hussong et al. (1978), Beccaluva et al. (1980), Hussong and Uyeda (1981). Bloomer (1982, 1983) recently confirmed this hypothesis by numerous dredges from the inner wall of the Mariana Trench. Additional data (Bloomer, 1983; Karig and Ranken, 1983) have shown that most of the removal by subduction of the forearc material took place between the assumed beginning of the Pacific plate subduction underneath the Palau-Kyushu Ridge, i.e. 45 m.y. ago (Scott et al., 1980), and Oligocene time. Since then,limited accretion of oceanic material,essentially ocean island alkalibasalts and Mesozoic pelagic sediments, to the inner wall of the Mariana trench occured occasionally and intermittently (Bloomer, 1983; Karig and Ranken, 1983).

The timing of back-arc spreading and arc volcanism has been the object of a still standing controversy between those who regard the two processes as synchronous, i.e. the peak of arc activity coincides with the early stages of back-arc beasin opening (Karig, 1975; 1983) or diachronous, i.e. back-arc spreading iniatiates during a marked decrease or even absence of arc volcanism (Scott and Kroenke, 1980; Crawford et al., 1981; Sharaskin et al., 1981). Since our earlier papers on the Philippine-Mariana region (Beccaluva et al., 1980; Crawford et al., 1981), a large amount of new chronological data has become available. Their review, schematically reported in Fig. 4, allows to mitigate this debate with a more satisfactory explanation. The discrepancy between the two alternative models is partly due to the different emphasis given by the authors to the significance of the volcanogenic material found in the stratigraphic column of D.S.D.P. holes in the Philippine Sea. Since ash deposition can have an extremely wide dispersal and some of the volcanogenic material is found in turbites, it is obviously uncertain that any occurrence represents explosive activity taking place in the Mariana region and at the time paleontologically determined in the stratigraphic column. Nevertheless it is evident (Fig. 4) that the two periods of low level and/or absence of explosive volcanic activity in the Philippine Sea coincide with the beginning of the spreading in the Parece Vela and Mariana Trough basins. This is in favour of the hypothesis of diachronous spreading and arc volcanism rather than of the synchronous one. At any rate it is clear that the new results based on the record of volcanic ash and pyroclastic material considerably reduce the period of quiescence of arc volcanism as reported by Scott et al. (1980) and Crawford et al. (1981).

As put forward by Hussong and Uyeda (1981) it is not certain if the apparent decrease or lack of pyroclastic volcanic material at the beginning of back-arc spreading is really due to the absence of arc volcanism at that time. In fact, the inception of a new interarc basin causes, concomitantly with the rifting and splitting of the arc at or near the axis of volcanism, a major sundering and subsidence of the arc structure. In this case, even if arc volcanism was continuous, there would be very little chance to find the record of this submarine activity, due to the lack of dispersal of pyroclastic products.

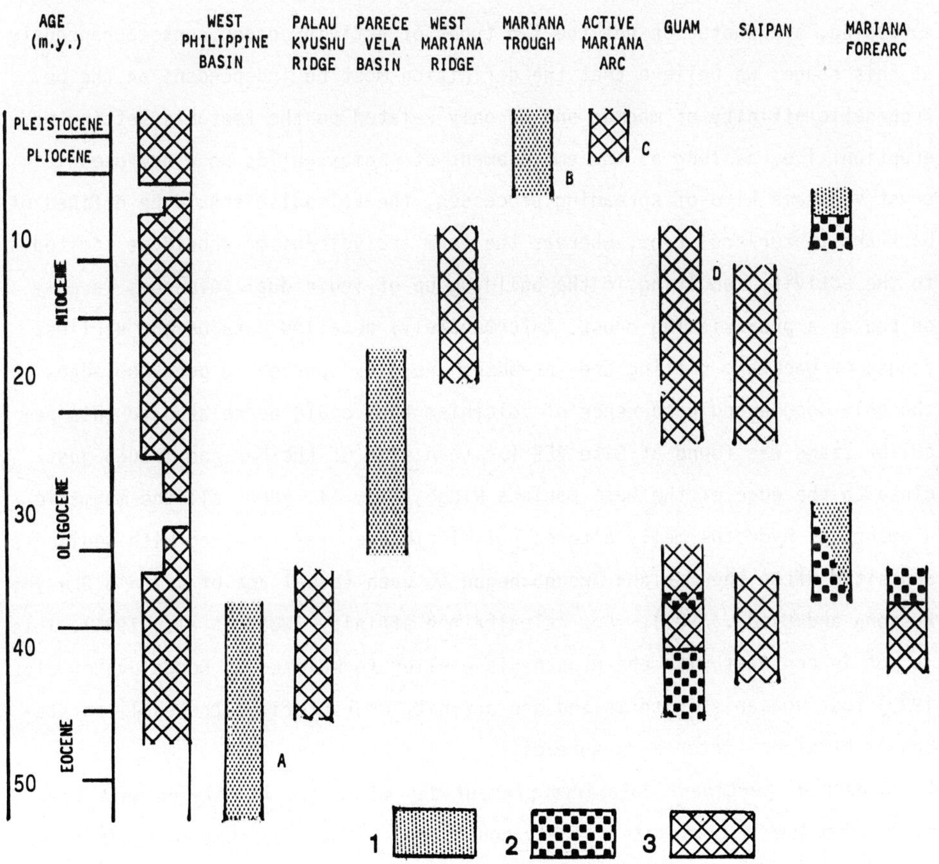

Fig. 4. Chronological evolution and petrogenetic affinity of the West Philippine-Mariana region volcanism (after Crawford et al., 1981, with modifications: A - Scott et al., 1980; B - Hussong and Sinton, 1983; C- Tracey et al., 1964; D - Meijer et al., 1983, Reagan and Meijer, 1984). 1 - back-arc basin volcanism 2 - boninitic volcanism ; 3 - island arc volcanism. On the left column also indicated arc volcanic activity (high or moderate) from volcanogenic levels in D.S.D.P. holes (Hussong and Uyeda, 1981).

The hypothesis of continuous arc volcanism is favoured by Hussong and Uyeda (1981) who state: "At this earliest rifting stage the axis of back-arc spreading and the axis of the arc volcanism are essentially in the same location. After a period of back-arc spreading, the centre of spreading migrates from the axis of arc volcanism, which must remain fixed relative to the descending oceanic slab".

In our opinion the point of the discussion is how do we define the arc and the back-arc (or inter-arc) volcanism at the earliest stage of rifting and spreading, and wheter or not the two types of activity occur contemporaneously at this stage. We believe that the definition must be independent on the petrogenetic affinity of magmas and be only related to the tectonic setting of eruption: i.e. as long as the emplacement of magmas builds up a new"oceanic" crust via some kind of spreading processes, the volcanism should be defined of back-arc (inter-arc) type, whereas the term"arc volcanism" should be limited to the activity resulting in the building up of individual volcanoes largely on top of a pre-existing crust. Unfortunately, detailed data on the earliest stages of back-arc rifting and spreading are very scarce. To our knowledge, the only documented occurrence of volcanism that could be related to this peculiar stage was found at Site 453 (western side of the Mariana Trough just close to the edge of the West Mariana Ridge, Fig. 1). Here, all the magmatic elements of hydrothermally altered polymictic breccias, covered with sediments deposited after the Mariana Trough began to open (basal age of 4.7 - 5.0 m.y.; Hussong and Uyeda, 1981), show calcalkaline affinity (Wood et al., 1981). This is not in contrast with the hypothesis earlier formulated by Beccaluva et al., 1980) that volcanism with island arc affinity characterizes the earliest stages of back-arc (inter-arc) spreading.

In absence of pertinent data from present day settings, we rely on what is known from low-Ti ophiolites widely considered to be generated in intraoceanic supra-subduction settings. The axis sequence of some of these ophiolites, i.e. the part formed by spreading processes, consists of volcanic rocks with island arc affinity either belonging to the island arc tholeiite series as Vourinos (Beccaluva et al., 1984) and Troodos (Robinson et al., 1983) or to the "boninite series", as Betts Cove (Coish et al., 1982) , Kan Taishir (Zonenshein and Kuzmin, 1978).and partly also Troodos (Arakapas Fault Belt;Cameron, 1985).

This clearly shows that the magma-types emplaced during the spreading processes which took place in supra-subduction settings were in some instances entirely of island arc affinity. If this was the case also in modern settings as it would appear from the results obtained at Site 453, the statement of Hussong and Uyeda (1981) "i.e. at this earliest rifting stage the axis of back-arc spreading and the axis of arc volcanism is essentially in the same location." would turn into: at the earliest stages of rifting and back-arc spreading the volcanism has essentially an island arc affinity. Namely the geochemistry of mantle sources of these magmas is still largely controlled by the subduction processes, whereas the mode of emplacement and probably the degree of fractionation by the tectonic setting.

Thus,in the light of data deduced from the study of low-Ti ophiolites showing that magmas emplaced in the earliest stage of back-arc spreading have frequently island arc affinity, we concur with the hypothesis (Hussong and Uyeda, 1981) that the apparent decrease of arc volcanism in the earliest stages of backarc spreading is largely due to the mode of emplacement. On the contrary, we maintain (Crawford et al., 1981) that the volcanism at the spreading centers of back-arc (interarc) basins rapidly (order of 1 m.y.) shifts toward MORB-type affinity related to extensive melting of ascending astenospheric diapirs. We consider that the apparent absence or strong decrease of arc volcanism in this period of back-arc spreading is due to the interference of early astenosphere mantle diapirs with the processes of arc magma generation. With the widening of the back-arc basin the subducted plate would necessarily move away from the astenospheric diapirs and, consequently, the arc magmas would again have easily access to the surface. Eventually the eruption of these magmas would built a new volcanic arc toward the trench on the crust previously formed in the early stages of back-arc opening, as occurred in the Pliocene to recent history of the active Mariana arc (Hussong and Uyeda, 1981).

Judging from the data obtained from Legs 59 and 60 (Fig. 4), the period of arc volcanism quiescence in the Tertiary evolution of the Mariana arc-backarc system, if present, was quite short, in the order of 2 m.y. The apparent diachronous arc and back-arc volcanism fits very well with the model of tectonic evolution of the Mariana region proposed by Hussong and Uyeda (1981) where the early stages of arc rifting and back-arc spreading are related to decoupling of the plates with the subduction shifting to"Mariana-type", whereas periods of vo luminous arc volcanism are connected to recoupling of the plates with a "Chilean-type subduction".

The Mariana arc-backarc system is the classical modern tectonic setting where boninite magmatism is well represented in several fore-arc localities (Bogdanov et al., 1977; Beccaluva et al., 1980; Natland and Tarney, 1981; Bloomer, 1982, 1983; Reagan and Meijer, 1984). In fact the Bonin Islands are the boninite-type locality (Beccaluva and Serri, 1981). Moreover Tertiary boninites have been found at Cape Vogel, Papua New Guinea (Cameron et al., 1983) and also recently in the Tonga (Sharaskin et al., 1983) and Bonin (Johnson and Nakamura, 1984) Trenches.Although they appear to be typically recovered from the fore-arc region of intraoceanic island arcs, their specific original setting is still under debate. They have been either considered to be generated during the early stages of intraoceanic subduction and consequent island arc formation (Meijer, 1980; Sharaskin et al., 1980; Bloomer, 1982) or during the first stages of arc rifting at the onset of back-arc (interarc) spreading (Beccaluva et al., 1980; Crawford et al., 1981; Coish et al., 1982; Hickey and Frey, 1982).

Recent chronological and stratigraphical data from Guam Island (Meijer et al., 1983; Reagan and Meijer, 1984), schematically reported in Fig. 4, have clearly documented that boninitic magmatism, in addition to Lower Oligocene, also took place in the Middle and Upper Eocene (oldest K/Ar age of 43.8 m.y.; Meijer et al., 1983). Similar K/Ar ages (38.6 - 43.0 m.y.; Tsunakawa, 1983) have been obtained on boninite series rocks from Bonin Islands. This shows that boninitic magmatism is not only confined to the initial stages of interarc rifting and spreading (Parece vela Basin and Mariana Trough, Fig. 4) but also occurred earlier in the Izu-Mariana arc setting (Kyushu-Palau Ridge).

In conclusion, although petrological and geological arguments consistently indicate a supra-subduction mantle source for the boninitic magma genesis, the precise timing and specific setting of their eruption in the arc-backarc system is still an open question. The data so far available suggest that boninitic magmatism may be generated during more than one stage of the tectonic evolution of the intra-oceanic island arc-backarc systems.

REFERENCES

Alabaster, T., Pearce, J.D. and Malpas, J., 1982. The volcanic stratigraphy and petrogenesis of the Oman ophiolite complex. Contrib. Mineral. Petrol., 81: 168-183.

Arculus, R.J. and Johnson, R.W., 1981. Island arc magma sources: a geochemical assessment of the role of slab-derived components and crustal contamination. Geochem. J., 15: 109-149.

Baker, P.E., 1978. The South Sandwich Islands: III. Petrology of the volcanic rocks. Brit. Abstract Surv. Sci. Rept., 93: 1-34.

Basaltic Volcanism Study Project, 1981. Basaltic Volcanism on the Terrestrial Planets. Pergamon Press Inc., New York, 1268 pp.

Beccaluva, L., Macciotta,G., Savelli,C., Serri, G. and Zeda, O., 1980. Geochemistry and K/Ar ages of volcanics dredged in the Philippine Sea (Mariana, Yap, Palau Trenches and Parece Vela Basins. In: D.E. Hayes (Editor), The tectonic and geologic evolution of Southeast Asian Seas and islands. Am. Geophys. Union, Geophys. Monogr. Ser., 23: 247-270.

Beccaluva, L., Piccardo, G.B. and Serri, G., 1980b. Petrology of Northern Apennine ophiolites and comparison with other Tethyan ophiolites. In: A. Panayiotou (Editor), Ophiolites, Proc. Int. Ophiolite Symp., Nicosia, Cyprus, pp. 314-331.

Beccaluva, L. and Serri, G., 1981. La serie boninitica: nuova associazione magmatica di margini di placche convergenti. Rend. Soc. Geol. It., 4: 305-314.

Beccaluva, L., Ohnenstetter, D., Ohnenstetter, M. and Paupy, A., 1984. Two magmatic series with island arc affinities within the Vourinos ophiolites. Contrib. Mineral. Petrol., 85: 253-271.

Bloomer, S., 1982. Structure and evolution of the Mariana Trench: petrologic and geochemical studies. Ph. D. Dissertation, Univ. of California, San Diego.

Bloomer,S., 1983. Distribution and origin of igneous rocks from the landward slopes of the Mariana Trench: implications for its structure and evolution. J. Geophys. Res., 88: 7411-7428.

Bogdanov, N. et al., 1977. Initial report of the geological study of the oceanic crust of the Philippine sea floor- Investigations by the International Working Group on the IGCP Poject "Ophiolites". Ofioliti, 2: 137-168.

Bougault, H., Cambon, P., Corre, O., Joron, J.L. and Treuil, M., 1979. Evidence for variability of magmatic processes and upper mantle heterogeneity in the axial region of the mid-atlantic ridge near 22° and 36° N. Tectonophysics, 55: 11-34.

Bougault, H., 1980. Contributions des élements de transition à la comprension de la génèse des basalts océaniques. Thèse, Université de Paris VI.

Bryan, W.B., 1983. Systematics of modal phenocrysts assemblages in submarine basalts: petrologic implications. Contrib. Mineral. Petrol., 83: 62 pp.

Cameron, W.E., Nisbet, E.G. and Dietrich, V.J., 1979. Boninites, komatiites and ophiolitic basalts. Nature, 28: 550-553.

Cameron, W.E., Mc Culloch, M.T. and Walker, D.A., 1983. Boninite petrogenesis: chemical and Nd-Sr isotopic constraints. Earth Plan. Sci. Lett., 65: 75-89.

Cameron, W.E., 1985. Petrology and origin of primitive lavas from the Troodos ophiolites, Cyprus. Contrib. Mineral. Petrol., 89: 239-255.

Coish, R.A., Hickey, R. and Frey, F.A., 1982. Rare earth element geochemistry of the Betts Cove ophiolites, Newfoundland: complexities in ophiolite formation. Geochim. Cosmochim. Acta, 46: 2117-2134

Clague, D.A. and Frey, F.A., 1982. Petrology and trace element geochemistry of the Honolulu volcanics, Oahu: implications for the oceanic mantle below Hawaii. J. Petrol., 23: 447-504.

Crawford, A.J., Beccaluva, L. and Serri, G., 1981. Tectono-magmatic evolution of the West Philippine-Mariana Region and the origin of boninites. Earth Planet. Sci. Lett., 54: 346-356.

Dick, H.J.B., Fisher, R.L. and Bryan, W.B., 1984. Mineralogic variability of the mid-ocean ridges. Earth Planet. Sci. Lett., 69: 88-106.

Dietrich, W., Emmermann, R., Oberhansli, R. and Puchelt, H., 1978. Geochemistry of basaltic and gabbroic rocks from the West Mariana Basin and the Mariana Trench. Earth Planet. Sci. Lett., 39: 127-144.

Divis, A., 1975. Structure and metamorphism of the Yap Island, Western Caroline Islands. Geol. Soc. Amer. Abstr. Prog., 7: 1053-1054.

Dixon, T.H. and Stern, R.J., 1983. Petrology, chemistry and isotopic composition of submarine volcanoes in the Southern Mariana Arc. Geol. Soc. Am. Bull., 94: 1159-1172.

Dostal, J. and Capedri, S., 1979. Rare earth elements in high-grade metamorphic rocks from the Western Alps. Lithos, 12: 41-49.

Dungan, M.A. and Rhodes, J.M., 1978. Residual glass and melt inclusions in basalts from D.S.D.P. Legs 45 and 46: evidence for magma mixing. Contrib. Mineral. Petrol., 67: 417-431.

Dupuy, C., Dostal, J., Marcelot, G., Bougault, H., Joron J.L. and Treuil, M., 1982. Geochemistry of basalts from Central and Southern New Hebrides arc: implication for their source rock composition. Earth Planet. Sci. Lett., 60: 207-225.

Ewart, A., Bryan, W.B. and Gill, J.B., 1973. Mineralogy and geochemistry of the younger volcanic islands of Tonga, S.W. Pacific. J. Petrol., 14: 429-465.

Ewart, A., Brothers, R.N. and Mateen, A., 1977. An outline of the geology and geochemistry and the possible petrogenetic evolution of the volcanic rocks of the Tonga-Kermadec-New Zealand island arc. J. Volcanol. Geotherm. Res., 2: 205-250.

Fornari, D. J.,, Weissel, J.K., Perfit, M.R. and Anderson, R.N., 1979. Petrochemistry of the Sorol and Ayu Troughs: implications for crustal accretion at the northern and western boundaries of the Caroline plate. Earth Planet. Sci. Lett., 45: 1-15.

Franzini, M., Leoni, L. and Saitta, M., 1975. Revisione di una metodologia analitica per fluorescenza - X, basata sulla correzione completa degli effetti di matrice. Rend. Soc. Mineral. Ital., 31: 365-378.

Frey, F.A., Haskin, M.A., Poetz, J.A. and Haskin, L.A., 1968. Rare earth abundances in some basic rocks. J. Geophys. Res., 73: 6085-6098.

Frey, F.A., Bryan, W.B. and Thompson, R.N., 1974. Atlantic ocean floor: geochemistry and petrology of basalts from Legs 2 and 3 of the Deep Sea Drilling Project. J. Geophys. Res., 79: 5507-5527.

Frey, F.A., Green, D.H. and Roy, S.D., 1978. Integrated models of basalt petrogenesis: a study of quartz tholeiites to olivine melilitites from Southern Australia utilizing geochemical and experimental petrological data. J. Petrol., 19: 463-513.

Fuji, T. and Bougault, H., 1983. Melting of a magnesium abyssal tholeiite and the origin of MORBs. Earth Planet. Sci. Lett., 62: 283-295.

Gill, J.B., 1976. Composition and the age of Lau Basin and Ridge volcanic rocks: implications for evolution of an interarc basin and remnant arc. Geol. Soc. Am. Bull., 87: 1384-1395.

Green, D.H., 1976. Experimental testing of "equilibrium" partial melting of peridotites under water saturated, high pressure conditions. Can. Mineral., 14: 255-268.
Hart, S.R., Glassley, W.E. and Karig., D.E., 1972. Basalts and sea floor spreading behind the Mariana island arc. Earth Planet. Sci. Lett., 15: 12-18.
Hart, S.R., Erlank, A.J. and Kable, E.J.D., 1974. Sea floor basalt alteration: some chemical and Sr isotopic effects. Contrib. Mineral. Petrol., 44: 219-230.
Hawkesworth, C.J., 1982. Isotope characteristics of magmas erupted along destructive plate margins. Thorpe, R.S., Wiley, J. and Sons.
Hawkesworth, C., O'Nions, R., Pankhurst, R., Hamilton, P. and Evensen, N., 1977. A geochemical study of island arc and back-arc tholeiites from the Scotia Sea. Earth Planet. Sci. Lett., 36: 253-262.
Hawkins, J.W., 1977. Petrologic and geochemical characteristics of marginal basins basalts. In: Island Arcs, Deep Sea Trenches and Back-Arc Basins. Talwani, M. and Pittman, W.C. (Editors), M. Ewing Ser. 1: 355-366.
Hawkins, J.W. and Batiza, R., 1977. Metamorphic rocks of the Yap arc-trench system. Earth Planet. Sci. Lett., 37: 216-229.
Hickey, R.L. and Frey, F.A., 1982. Geochemical characteristics of boninite series volcanics: implications for their source. Geochim. Cosmochim. Acta, 46: 2099-2115.
Howard, A.H. and Stolper, E., 1981. Experimental crystallization of boninites from the Yap Trench. EOS, Trans. Am. Geophys. Union, 62: 1091.
Humphris, S.E. and Thompson, G., 1978a. Trace element mobility during hydrothermal alteration of oceanic basalts. Geochim. Cosmochim. Acta, 42: 127-136.
Humphris, S.E. and Thompson, G., 1978b. Hydrothermal alteration of oceanic basalts by sea water. Geochim. Cosmochim. Acta, 42: 107-125.
Humphris, S.E. and Thompson, G., 1983. Geochemistry of rare earth elements in basalts from Walvis Ridge: implications for its origin and evolution. Earth Planet. Sci. Lett., 66: 229-242.
Hussong, D.M. et al., 1978. Leg 60 ends in Guam. Geotimes, 12: 19-22.
Hussong, D.M. and Uyeda, S., 1981. Tectonic processes and the history of the Mariana arc - A synthesis of the results of DSDP Leg 60. DSDP Init. Repts, 60: 909-929.
Hussong, D.M. and Sinton, J.B., 1983. Seismicity associated with back-arc crustal spreading in the central Mariana Trough. In: Hayes, D.E. (Editor), The tectonic and geological evolution of Southeast Asian seas and islands, part 2, Am. Geophys. Un. Mon., 27: 217-235.
Ingle, J.C., Jr., 1975. Summary of Late Paleogene-Neogene insular stratigraphy, paleobathymetry and correlations, Philippine Sea and Sea of Japan region. DSDP Init. Repts., 31: 837-855.
Jenner, G.A., 1981. Geochemistry of high-Mg andesites from Cape Vogel, Papua New Guinea. Chem. Geol., 33: 307-332.
Jenner, G.A. and Green, D.H., 1982. High-Mg andesites:petrology and some constraints on their petrogenesis. Mineral. Mag.
Jagoutz, E., Palme, H., Baddenhousen, H., Blum, K., Cendales, M., Dreibus, G., Spettel, B., Lorenz, V. and Wanke, H., 1979. The abundances of major, minor and trace elements in the earth's mantle as derived from primitive ultramafic nodules. Proc. Lunar Planet. Sci. Conf., 10: 2031-2050.
Karig, D.E., 1975. Basin genesis in the Philippine Sea. DSDP Init. Repts., 31: 857-880.
Karig, D.E., 1983. Temporal relationships between back-arc basin formation and

arc volcanism with special reference to the Philippine Sea. In: Hayes, D.E. (Editor), The Tectonic and geological evolution of Southeast Asian Seas and islands, Part 2, Am. Geophys. Union. Mon., 27: 318-325.

Karig, D.E., Anderson, R.N. and Bibee, L.D., 1978. Characteristics of back-arc spreading in the Mariana Trough.J. Geophys. Res., 83: 1213-1226.

Karig, D.E. and Ranken, B., 1983. Marine geology of the fore-arc region, Southern Mariana Island arc. In: Hayes, D.E. (Editor), The tectonic and geological evolution of Southeast Asian Seas and islands, Part 2, Am. Geophys. Union. Mon., 27: 266-293.

Katsumata, M. and Sykes, L.R., 1969. Seismicity and tectonics of the Western Pacific: Izu-Mariana-Caroline and Ryukyu-Taiwan regions. J. Geophys. Res., 74: 5923-5949.

Kay, R.W., 1977. Geochemical constraints on the origin of the Aleutian magmas. In: Talwani, M. and Pittman, W.C. (Editors), Island arcs, deep sea trenches and back-arc basins, Am. Geophys. Union, 229-242.

Kay, R.W., 1980. Volcanic arc magmas: implications of a melting-mixing model for element recycling in the crust-upper mantle system. J. Geol., 88: 498-522.

Kay, R.W. and Hubbard, N.J., 1978.Trace elements in ocean ridge basalts. Earth Planet. Sci. Lett., 38: 95-116.

Langmuir, C.H., Bender, J.F., Bence, A.E., Hanson, G.N. and Taylor, S.R., 1977. Petrogenesis of basalts from the Famous area: mid-Atlantic ridge. Earth Planet. Sci. lett., 36: 133-156.

Leoni, L. and Saitta, M., 1976. X-ray fluorescence analysis of 29 trace elements in rock and mineral standards. Rend. Soc. It. Min. Petrol., 32: 497-510.

Le Roex, A.P., Dick, H.J.B., Erlank, A.J., Reid, A.M., Frey, F.A. and Hart, S. R., 1983. Geochemistry, mineralogy and petrogenesis of lavas erupted along the southwest Indian Ridge between the Bouvet triple junction and 11° E. J. Petrol., 34: 267-318.

Mattey, D.P., Marsh, N.G. and Tarney, J., 1980. The geochemistry, mineralogy and petrology if igneous rocks from the West Philippine and Parece Vela Basins, and from the Palau-Kyushu and West Mariana Ridges, IPOD Leg 59. DSDP Init. Repts., 59: 753-800.

Mc Cabe, R. and Uyeda, S., 1983. Hypothetical model for the bending of the Mariana Arc. In: Hayes, D.E. (Editor), The tectonic and geological evolution of Southeast Asian Seas and Islands, Part 2, Am. Geophys. Union Mon., 27: 281-293.

Meijer, A., 1980. Primitive arc volcanism and a boninite series: examples from Western Pacific island arcs. In: Hayes, D.E. (Editor), The tectonic and geological evolution of Southeast Asian Seas and Islands, Part 1, Am. Geophys. Union Mon., 23: 271-282.

Meijer, A., Reagan, M., Ellis, H., Shafiqullah M., Sutter, J., Damon, P. and Kling, S., 1983. In: Hayes, D.E. (Editor), The tectonic and geological evolution of Southeast Asian Seas and Islands, Part 2, Am. Geophys. Union Mon., 27: 349-359.

Natland, J.H. and Tarney, J., 1981. Petrologic evolution of the Mariana arc and back-arc system - a synthesis of drilling results in the Southern Philippine Sea. DSDP Init. Repts., 60: 877-908.

Nicholls, I.A. and Ringwood, A.E., 1973. Effects of water on olivine stability in tholeiites and the production of silica-saturated magmas in the island arc environment. J. Geol., 81: 285-300.

Pearce, J.A., 1975. Basalt geochemistry used to investigate past tectonic environments on Cyprus. Tectonophysics, 25: 41-67.

Pearce, J.A. and Norry, M.J., 1979. Petrogenetic implications of Ti, Zr and Nb variations in volcanic rocks. Contrib. Mineral. Petrol., 69: 33-47.

Pearce, J.A., 1982. Trace element characteristics of lavas from destructive plate boundaries. In: Thorpe, R.S., Wiley, J. and Sons (Editors), Andesites, 525-548.

Peive, A.V., 1980. Geology of the Philippine sea floor. Publishing Office Nauka, Moscow, 258 pp.

Perfit, M.R., Gust, D.A., Bence, A.E., Arculus, R.J. and Taylor, S.R., 1980. Chemical characteristics of island arc basalts - Implications for mantle sources. Chem. Geol., 30: 227-256.

Reagan M.K. and Meijer, A., 1984. Geology and geochemistry of early arc-volcanic rocks from Guam. Geol. Soc. Am. Bull., 95: 701-713.

Ridley, W.I., Rhodes, J.M., Reid, A.M., Jakes, P., Shih, C. and Bass, M.N.,1974. Basalts from Leg 6 of the Deep Sea Drilling Project. J. Petrol., 15: 140-159.

Ringwood, A.E., 1974. The petrological evolution of island arc systems. J. Geol. Soc. London, 130: 183-204.

Robinson, P.T., Melson, W.G., O'Hearn T. and Schmincke, H.U., 1983. Volcanic glass compositions of the Troodos ophiolites, Cyprus. Geology, 11: 400-404.

Roden, M.F., Frey, F.A. and Clague, D.A., 1984. Geochemistry of tholeiitic and alkalic lavas from the Koolau Range, Oahu, Hawaii: implications for hawaiian volcanism. Earth Planet. Sci. Lett., 69: 141-158.

Saunders, A.D. and Tarney, J., 1979. The geochemistry of basalts from a back-arc spreading centre in the East Scotia Sea. Geochim. Cosmochim. Acta, 43, 555-572.

Saunders, A.D., Tarney, J., Marsh, N.G. and Wood, D.A., 1980. Ophiolites as ocean crust or marginal basin crust: a geochemical approach. In: Panayiotou, A. (Editor), Ophiolites, Int. Ophiol. Symp., Nicosia, Cyprus, 193-204.

Schilling, J.G., Zajac, M., Evans, R., Johnston,, T., White, W., Devine, J.D. and Kingsley, R. (1983): Petrologic and geochemical variations along the mid-Atlantic ridge from 29° N to 73° N. Am. J. Sci., 283: 510-586.

Scott, R. and Kroenke, L., 1980. Evolution and back-arc spreading and arc volcanism in the Philippine Sea. In: Hayes, D.E. (Editor), The tectonic and geological evolution of Southeast Asian Seas and Islands, Part 1, Am. Geophys. Union, Mon. 23: 283-291.

Scott, R.B., Kroencke, L. and Zakariadze, G., 1980. Evolution of the South Philippine Sea: Deep Sea Drilling Project Leg 59 results. DSDP Init. Repts., 59: 803-815.

Serri, G., 1981. The petrochemistry of ophiolite gabbroic complexes: a key for the classification of ophiolites into low-Ti and high-Ti types. Earth Planet. Sci. Lett., 52: 203-212.

Sharaskin, A.J., Dobretsov, N.L. and Sobolev, N.V., 1980. Marianites: the clinoenstatite-bearing pillow lavas associated with the ophiolite assemblages of Mariana Trench. In: Panayiotou, A. (Editor), Ophiolites, Int. Ophiol. Symp., Nicosia, Cyprus, 473-479.

Sharaskin, A.Y., Bogdanov, N.A. and Zakariadze, G., 1981. Geochemistry and timing of the marginal basin and arc magmatism in the Philippine Sea. Phil. Trans. R. Soc. London, A 300: 287-297.

Shibata, T., Thompson, G. and Frey, F.A., 1979. Tholeiitic and alkalibasalts from the mid-Atlantic Ridge at 43° N. Contrib. Mineral. Petrol., 70: 127-141.

Shiraki, K., 1971. Metamorphic basement rocks of Yap Island, Western Pacific: possible oceanic crust beneath an island arc. Earth Planet. Sci. Lett., 13: 167-174.

Smewing, J.D. and Potts, O.J., 1976. Rare earth abundances in basalts and metabasalts from the Troodos Massif, Cyprus. Contrib. Mineral. Petrol., 57: 245-258.

Sun, S.S., 1980. Lead isotope study of young volcanic rocks from mid-ocean ridges, ocean islands and island arcs. Phil. Trans. R. Soc. London, 297: 409-450.

Sun, S.S. and Nesbit, R.W., 1978. Chemical regularities and genetic significance of ohiolitic basalts. Geology, 6: 689-693.

Sun, S.S., Nesbit, R.W. and Sharaskin, A.Y., 1979. Geochemical characteristics of mid-ocean ridge basalts. Earth Planet. Sci. Lett., 44: 119-138.

Tarney, J., Saunders, A.D. and Weaver, S.D., 1977. Geochemistry of volcanic rocks from the island arcs and marginal basins of the Scotia Arc region. In: Ewing, M. (Editor), Island Arcs, Deep Sea Trenches and Back-Arc Basins, Am. Geophys. Union, Ser. 1, 367-377.

Tarney, J., Saunders, A.D., Mattey, D.P., Wood, D.A. and Marsh, N.G., 1981. Geochemical aspects of back-arc spreading in the Scotia Sea and Western Pacific. Phil. Trans. R. Soc. London, A 300: 263-285.

Thompson, R.N., Morrison, M.A., Hendry, G.L. and Parry, S.J., 1984. An assessment of the relative roles of crust and mantle in magma genesis: an elemental approach. Phil. Trans. R. Soc. London, 310: 549-590.

Tracey, J.I., et al., 1964. General geology of Guam. U.S. Geol. Surv. Prof. Pap., 403 A, 104 pp.

Walker,D.A. and Cameron,W.E.,1983.Boninite primary magmas:evidence from the Cape Vogel Peninsula.Contrib.Mineral.Petrol.,83:150-158.

Weaver, S.D.,Saunders,A.D.,Pankhurst,R.J. and Tarney,J.,1979.A geochemical study of magmatism associated with the initial stages of backarc spreading.Contrib. Mineral.Petrol.,68:151-169.

Whitford,D.A.,Nicholls,I.A. and Taylor,S.R.,1979.Spatial variations in the geochemistry of Quaternary lavas across the Sunda arc in Java and Bali.Contrib. Mineral.Petrol.,70:341-356.

Wood,D.A.,Tarney,J.,Saunders,A.D.,Bougault,H.,Joron,J.L.,Treuil,M.and Cann,J.R., 1979.Geochemistry of basalts drilled in the Northern Atlantic by IPOD Leg 49: implications for mantle heterogeneity.Earth Planet.Sci.Lett.,42:77-97.

Wood,D.A.,1979.A variably veined suboceanic upper mantle.Genetic significance of Mid-ocean Ridge Basalts from geochemical evidence.Geology,7:499-503.

Wood,D.A.,Marsh,N.G.,Tarney,H.,Joron,J.L.,Freyer,P. and Treuil,M.,1981.Geochemistry of igneous rocks recovered from a transect across the Mariana Trough,arc, forearc and trench,Sites 453-461,DSDP Leg 60.DSDP Init.Repts.,60:611-645.

Zonenshain,L.P. and Kuzmin,M.I.,1978.The Khan-Taishir ophiolitic complex of Western Mongolia and comparison with other ophiolitic complexes.Contrib.Mineral. Petrol.,67:95-109.

STRUCTURAL BEHAVIOUR OF A CONTINENTAL VOLCANIC ARC: THE MEXICAN VOLCANIC BELT

G. PASQUARE', F. FORCELLA, A. TIBALDI, L. VEZZOLI and A. ZANCHI
Dipartimento di Scienze della Terra, Università degli Studi di Milano,
20133 - Milano (Italy)

ABSTRACT

The structural evolution of South-Central Mexico during Neogene and Quaternary times clearly reveals the overlap of several tectonic phases characterized by brittle deformations. Structural lineaments linked with the southern prosecution of the Basin and Range Province were recognized, together with lineaments connected to the opening of the Gulf of California and to the convergence between the Cocos and the North-American Plates.

The Mexican Volcanic Belt consists of a volcanic arc structurally controlled by the differential development of the above mentioned lineaments, undergoing strong and repeated reactivations, contrarily to what happens in most adjacent areas. These reactivations structurally represent the unifying motive of the MVB.

The morphological and structural analyses have shown the presence of three main sectors. The western sector is characterized by a NW-SE fracture system, representing the southward prosecution of the Gulf of California structures. The central sector is defined by a composite WSW-ENE trending Graben system, probably connected with sinistral lateral shear induced by the aforementioned convergence phenomena. The Eastern sector, mainly developed within Sierra Madre Oriental, shows a reactivation of N-S Basin and Range tensional structures.

1 INTRODUCTION

Central Mexico is crossed from E to W by a Plio-Quaternary volcanic arc forming one of the outstanding geological and physiographical features of this country. Its transversal setting with respect to the cordilleran orogenic belts, to Sierra Madre Oriental belt and to the Meso-American Trench explains its name, well known in literature as Trans-Mexican Volcanic Belt. In the present paper, the name Mexican Volcanic Belt is preferred, as in Mooser (1969). Several authors tried to explain the Mexican Volcanic Belt (MVB) existence by working out geodynamical syntheses, not supported by a specific knowledge of its geological evolution (Mooser, 1969; Pal and Urrutia, 1977; Urrutia and Del Castillo, 1977;Lomnitz,1982; Anderson and Schmidt,1983; Schurbet and Cebull,1984).

Regional geological researches, directed to clarify more concretely and specifically the MVB geological history through geochemical and geochronological analysis of the volcanic products, are to be mentioned (Mooser, 1969; Demant and Robin, 1975; Thorpe, 1977; Demant, 1978; Pal et al., 1978; Robin and Tournon, 1978; Cantagrel and Robin, 1979; Demant, 1979; Gastil et al., 1979; McDowell and Clabaugh, 1979; Luhr et al., 1985; Luhr and Carmichael, 1985). But even if in some of these works (Negendank, 1972; Demant, 1985, 1978, 1979; Cantagrel and Robin, 1979; Gilbert et al., 1985; Lugo Hubp et al., 1985) outlines of regional structures of the MVB are suggested, the meaning of these remains vague.

This paper intends to propose a structural synthesis of the MVB, based on the interpretation of geological, geomorphological and structural data,collected by direct and indirect methods. The principal aim of this study is to reconstruct the Neogene and Quaternary structural evolution of the MVB, its relationships with the development of volcanic activity and with the conterminous regional structures. This work is the result of numerous geological observations carried out by the first author during many volcanological and geothermal field explorations which have occurred since 1977. Some of the other authors worked during 1984 and 1985.

2 MVB REGIONAL GEOLOGICAL FRAMEWORK

The MVB is a continental volcanic arc formed as a consequence of the subduction process of the Cocos Plate under the North-American Plate which took place in the Meso-American Trench. This arc is composed of a series of prevailingly calc-alkaline volcanic centers with subordinate alkaline products (Aguilar-Vargas and Verma, 1985; Luhr and Carmichael, 1985). The arc extends from the Atlantic to the Pacific coasts of Mexico, latitude 20°N. Most of the MVB centers are of Quaternary age; some of them are still active. Several Pliocene centers are also found, whereas the attribution to MVB of Miocene volcanic products is still doubtful.

The MVB products unconformably cover the main orogenic units of South-central Mexico. The latter are represented by a complex system of accreted terranes, forming the cordilleran orogenic belt (Campa et al., 1981; Campa and Coney, 1983) and by their foreland, known as Sierra Madre Oriental.

The cordilleran terranes accreted during several geodynamic phases and mostly during Late Triassic, Late Jurassic, Late Cretaceous and Paleocene (Laramide phase) as a consequence of the convergence between the Pacific and the North-American plates (Atwater, 1970; Tardy, 1977; Campa and Coney, 1983).

A great part of the cordilleran orogenic systems is covered by a thick volcanic sequence of Oligocene and Miocene age known as Sierra Madre Occidental. This volcanic belt largely represents the substratum of the MVB volcanic arc, with complex and still to be explained stratigraphic and structural implications between the two volcanic belts.

The cordilleran foreland is composed by a folded belt called Sierra Madre Oriental, mostly deformed during Paleocene, with eastward verging folds and overthrusts. This belt is mostly composed of Mesozoic sedimentary sequences belonging to the Mesogean paleogeographic province which extended on the whole Carribean area till Late Cretaceous (Tardy, 1977).

The structural systems deformed by the above described compressive phases are overlain by a large extensional system of faulted block which became active during Miocene and known in Western U.S.A. as Basin and Range Province. The prosecution of this system to Central-western Mexico has been suggested by Colletta and Angelier (1983), Pasquarè and Zanchi (1985), Pasquarè et al. (1985).

NW-SE trending right-lateral faults developed since Pliocene, along the Californian continental margin. They are responsible for the opening of the Gulf of California and are still active along the San Andreas fault system and the above mentioned Gulf.

3 MORPHOSTRUCTURAL ANALYSIS

In a previous paper Pasquarè et al. (1985) suggested a partition of the MVB into different morphostructural units which appear to be closely interconnected with those of the collateral geological provinces. The morphological units forming the MVB are here summarized and schematically represented in Fig. 1.

3.1 Western Sector

(1) Tepic-Guadalajara and Colima Grabens. The morphostructural development of the western sector of the MVB is essentially controlled by NW-SE faults which are locally transected by N-S and E-W faults. The NW-SE faults characterize to a great extent the morphostructural evolution of South-western Mexico, revealing an evident link with the strike-slip faults related to the opening of the Gulf

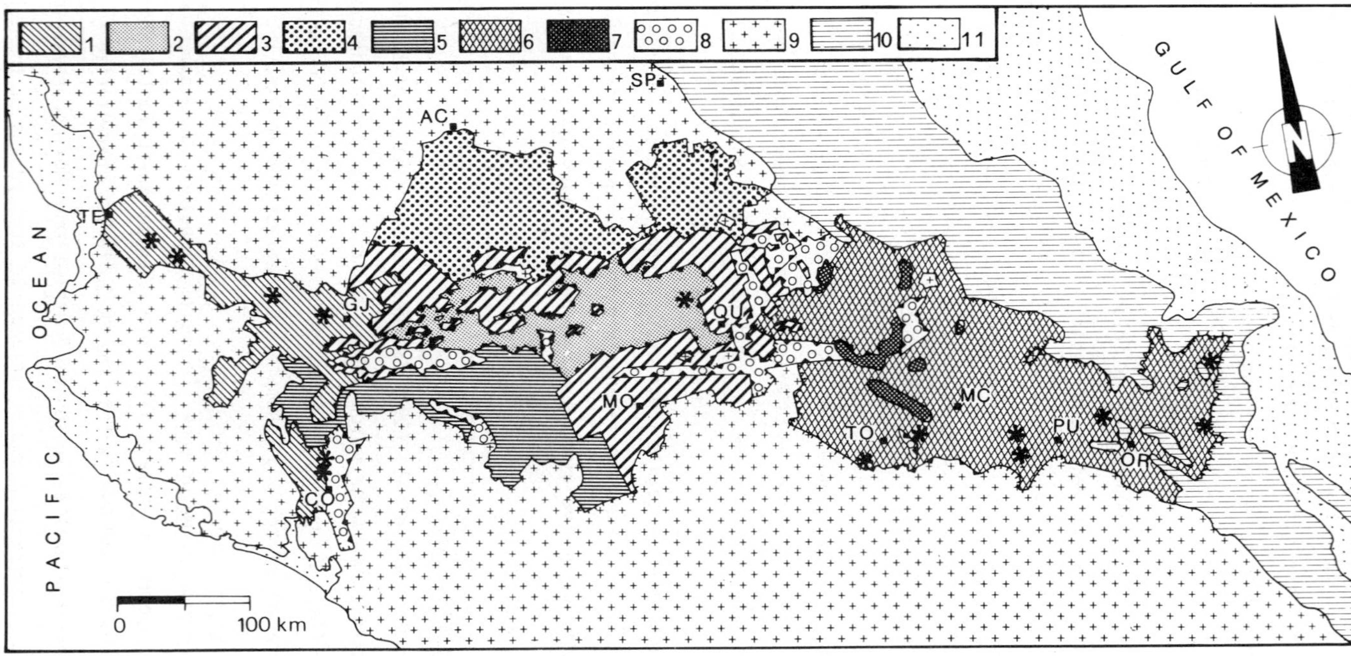

Fig. 1. Morpho-structural map of Mexican Volcanic Belt and conterminous units. Western sector: (1) Tepic-Guadalajara and Colima Grabens. Central sector: (2) Central depression; (3) ENE-WSW faulted blocks; (4) Leon-Guanajuato plateau; (5) Tarascan plateau. Eastern sector: (6) Eastern plateau; (7) Pre-Pliocene relict volcanic centers. Recent Grabens (8). Cordilleran orogenic belts (9). Sierra Madre Oriental orogenic belt (10). Coastal plains (11). Lines with barbs represent inactive faults; lines with triangles represent recent and/or active faults. Asterisks represent main volcanic centers.
AC: Aguascalientes, CO: Colima, GJ: Guadalajara, MC: Mexico City, MO: Morelia, OR: Orizaba, PU: Puebla, QU: Queretaro, SP: San Luis Potosì, Te: Tepic, TO: Toluca.

of California. These structures mostly develop along the Tepic-Guadalajara Graben system, whose marginal faults show an advanced degree of erosion. Some areas within this system were active also during Quaternary, and numerous Pleistocene volcanic centers, which are still partly active (Demant, 1979), seem to be linked with them.

In the Colima Graben, the Californian NW-SE trends appear to interfere with more recent systems of N-S faults which cause a local elongation of the Graben in that direction. The Colima Graben appears to be still active (Allan and Carmichael, 1984) and thereto are linked important Quaternary volcanic centers such as the Nevado de Colima (Luhr and Carmichael, 1985).

The E-W faults have a limited morphological importance in this sector, except for the Chapala Lake area, wherefrom a system of E-W faults deeply penetrates eastwards into the central sector of the MVB.

Both the above described Graben systems are included by Luhr et al. (1985) in a model of continental rifting presently taking place in South-Western Mexico, as a result of an eastward jump of a segment of the East Pacific Rise.

3.2 Central Sector

(2) Central Depression. This unit is represented by a tectonic depression extending from 200 km from Chapala Lake to Queretaro and mainly bordered by ENE-WSW and SE-NW fault systems. The average height of the depression is 1.800 m a.s.l. The E-W elongation of this unit, as well as the same elongation of Chapala and Cuitzeo Lakes, are related to more recent E-W faults. The eastern side of the depression is sharply interrupted by an older NNW-SSE fault system subsequently reactivated and known as Taxco-San Mighel de Allende Line (Demant, 1978). In Penjamo area, the Central Depression is crossed by a N-S Graben of very recent activation.

(3) ENE-WSW Faulted Blocks. The Central Depression is limited by ENE-WSW faulted blocks composed of Plio-Quaternary volcanic sequences connected with the MVB activity. Many volcanic centers are present along the faults which bound the blocks. Southward facing cuestas characterize the unit. The faulted blocks interact with the above-mentioned Taxco-San Miguel de Allende Line, forming a complex system of variously oriented Horsts and Grabens, herein called Queretaro Fracture Zone.

(4) Leon-Guanajuato Plateau. The Mexican Central Plateau is flanked along its southern border by a huge plateau placed at a height varying between 1.800 and

2.000 m a.s.l. The escarpment dividing the two units results on the whole extended in E-W direction, but in detail it is strongly segmented. The plateau western side shows a closely spaced dendritic pattern, hollowed in Tertiary volcano-clastic rocks. The unit is crossed by a N-S and NNE-SSW fault system, along which small recent volcanic centers grew up.

(5) Tarascan Plateau. This unit consists of a large plateau placed at an average height of 2.400 m a.s.l., located South of the Central Depression. The plateau is bounded by SE-NW and ENE-WSW fault systems and is almost entirely covered with a very closely spaced system of pyroclastic cones, lava cones and domes of Quaternary age. The volcanic activity mainly follows the ENE-WSW fractures.

3.3 Eastern Sector

(6) Eastern Plateau. The main Eastern Plateau extends for hundreds of kilometers from the Taxco-San Miguel de Allende Line to the Cofre de Perote-Pico de Orizaba N-S fracture lineament. This lineament constitutes the eastern boundary of the MVB.

The plateau is characterized by the presence of very extensive plains placed among recent volcanic ridges and ancient Sierra Madre Occidental volcanics. The depressions are flooded by recent volcano-clastic deposits coming from the largest strato-volcanoes of the MVB. The height of the plateau is generally above 2.000 m a.s.l. and reaches 2.700 m near Toluca and Puebla.

The basement of the plateau is mainly made up of Oligo-Miocene volcanics belonging to Sierra Madre Occidental volcanic province and filling up a roughly NW-SE trending tectonic depression developed during volcanic activity. A borehole near Mexico City, Pozo Texcoco, intersected the Sierra Madre Oriental sedimentary rocks after traversing 2.060 m of volcanic materials (Mooser et al., 1974).

From the structural point of view, this unit shows a lower tectonic control, compared to the recent intensive faulting affecting the western and central sectors of the MVB. Therefore, near the Queretaro Fracture Zone, the plateau is crossed by E-W faults forming recent Grabens. Also in other sectors the presence of small Grabens was recognized.

The plateau is subdivided into minor compartments by N-S and NNE-SSW Quaternary fault systems, covered by recent volcano-clastic deposits. Along the former fault set, the Popocatepetl-Ixtaccihuatl and the Pico de Orizaba-Cofre de

Perote volcanic chains have grown, while along the latter La Malinche center has developed.

In the western part of the plateau, poorly tectonized Oligo-Miocene volcanics, belonging to Sierra Madre Occidental, outcrop forming isolated ridges. The central part is characterized by the presence of Plio-Quaternary volcano-sedimentary sequences, forming the relative basement of the large strato-volcanoes. In the eastern part, extending beyond a line passing through Atlixco, La Malinche volcano and Zacatlan, the folded complexes of Sierra Madre Oriental outcrop, forming isolated ridges mainly developed in the south-eastern end of the unit.

(7) Pre-Pliocene Relict Volcanic Centers. West of Mexico City, several ridges consisting of Sierra Madre Occidental Oligo-Miocene volcanics, were recognized (Negendank, 1972; Mooser et al., 1974). They are made up of acidic lava flows and domes, poorly affected by tectonics. The absence of marginal fault scarps suggests that this unit cannot be included in Horst and Graben structures but consists of old, still preserved, volcanic centers. They are aligned along NW-SE and N-S trends.

(8) Recent Grabens. Numerous Graben structures of Late Quaternary age are evidenced within the MVB; they are characterized by probably active fault scarps, often connected with alignments of volcanic centers. The most remarkable system of Grabens extends in E-W direction along the axial part of the Central Depression. Its main elements are represented by Chapala, Cuitzeo and Acambay Grabens. The other Grabens appear to be aligned mainly in N-S and NNW-SSE directions.

4 STRUCTURAL ANALYSIS

The structural analysis is based on the interpretation of satellite images at different scales, of aerial photographs and of geological reconnaissance surveys carried out by some of the authors during numerous geological missions. The analysis of satellite images has been extended to the whole territory (Fig. 2), whereas the photogeological investigations and reconnaissance surveys have been concentrated in some test areas located within the MVB as follows:

- area between Guadalajara and Chapala Lake
- area West of Cuitzeo Lake
- area between Nevado de Toluca and Cuernavaca

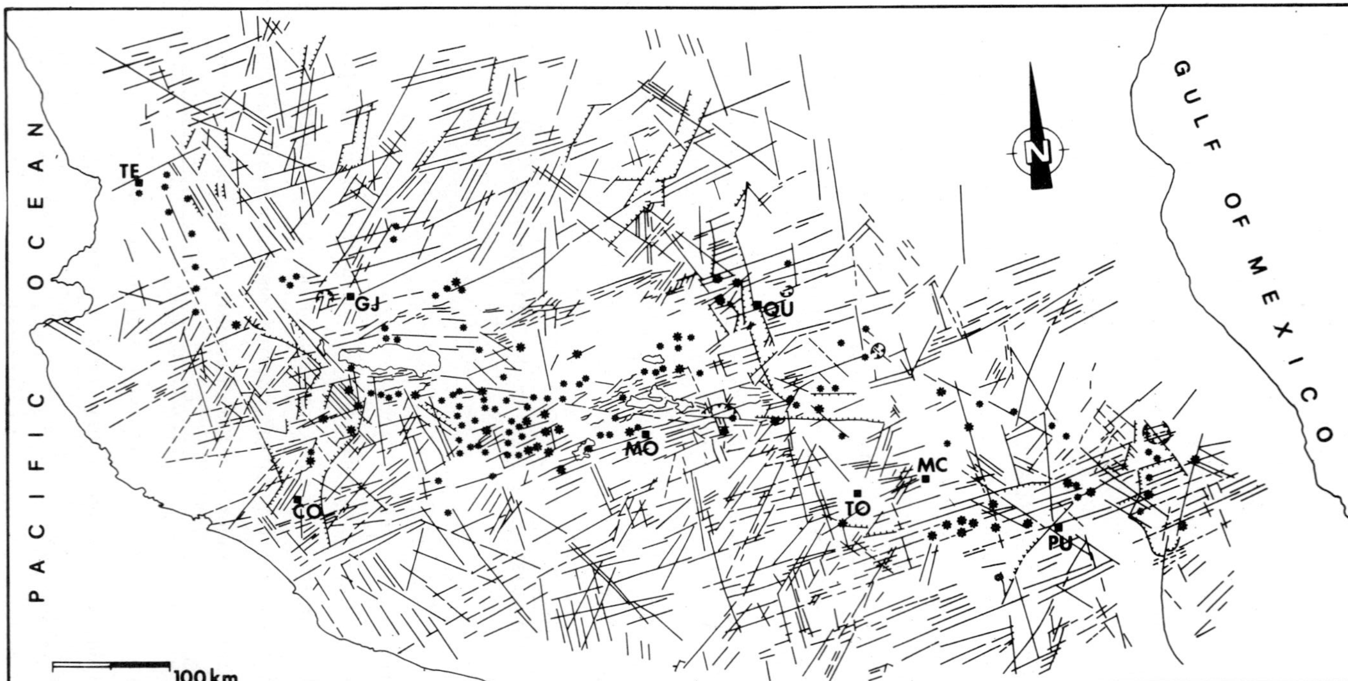

Fig. 2. Satellite fracture lineaments of south-central Mexico. Asterisks represent the main Quaternary volcanic centers, circles the main calderic structures of the Mexican Volcanic Belt.

- area between Queretaro and San Miguel de Allende
- area East of Puebla.

This analysis has been limited to the post-Laramide structures developed during Neogene and Quaternary times, affecting the growth of the MVB. The analysis of these structures has shown that they represent successive events characerized by brittle deformations greatly conditioning the present morphological features of South-central Mexico. The study of the distribution of these structures has also allowed to subdivide the area into five structural provinces, whose limits partly coincide with those of the morphostructural sectors illustrated in the previous chapter (Fig. 3). The analyzed structural elements are given by fracture lineaments showing marked systematic orientations. The results of the statistical analysis are also shown in Fig. 3.

4.1 Area 1 - Californian Province

N315°-330°. This trend has mostly characterized the geological evolution of the Californian Province during Neogene. It is represented by long and rectilinear lineaments, often accompanied by extensional fault scarps characterized by a mature morphology. In correspondence with the MVB western sector, these lineaments delimitate the Tepic-Guadalajara Graben system, within which the volcanic units of the western sector of the MVB developed since Pliocene.

N60°-75°. This set forms a system of fractures conjugated with the N315°-330° set. It is represented by frequent but generally short lineaments. This set contributes to the articulation of the Tepic-Guadalajara Graben forming depressed minor structures sub-orthogonal to the main system.

N.345°-15°. It is represented by a small number of elements, yet bearing a great importance in the formation of the Atayac Graben and particularly in the eastern sector of the Colima Graben. This fracture system seems to have been active since Quaternary time.

4.2 Area 2 - Basin and Range Province

N340°-20°. This fracture set groups the lineaments showing, despite a marked azimuthal dispersion, similar characteristics and particularly continuous variation from W to E. They mostly behave as normal faults, sometimes provoking a marked tilting of the displaced blocks. This system is densely represented and linked to a strong crustal stretching in E-W direction. The fault scarps

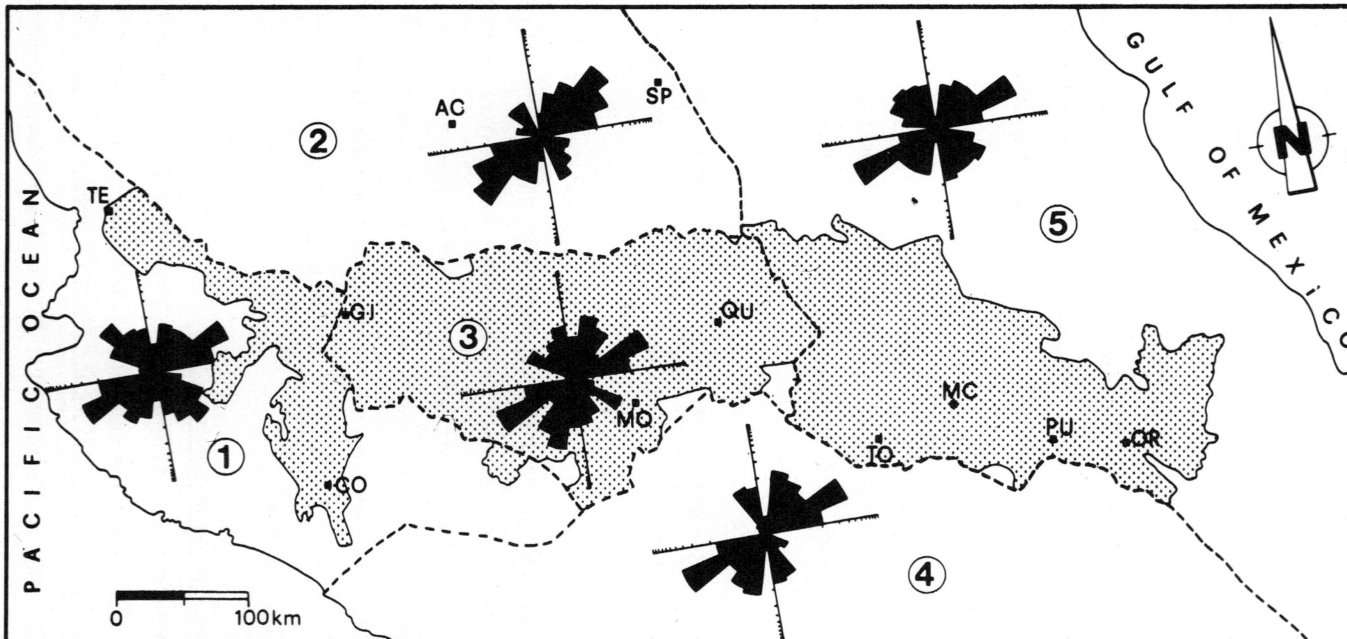

Fig. 3. Neogene and Quaternary structural map of South-central Mexico: (1) Californian Province; (2) Basin and Range Province; (3) Central Province; (4) Guerrero Province; (5) Sierra Madre Oriental Province. Dotted area: Mexican Volcanic Belt. Rose diagrams: satellite fracture lineaments, in the upper part frequence of lineaments is shown, in the lower length of lineaments is shown.

evidence an advanced degree of erosion with wide withdrawals of their rim. Some segments of these scarps suggests the presence of recent local reactivations connected with Graben formation. This is the oldest set of extensional lineaments known in the whole area.

N60°-75°. These fractures are homologous to the N60°-75° ones described in the Californian Province. They are represented by lineaments featuring a remarkable linear development, sometimes crossing the previous set of fractures. The system exerts an important control on the drainage pattern.

N300°-315°. It represents the main Californian trend, still present in the area. Its relative frequency appears to be clearly subordinate with regard to the other sets. It is represented by short and discontinuous lineaments, yet having a great morphological importance. The formation of this fracture set and of the previous one appears to be chronologically subsequent to that of the N340°-20° trending set.

4.3 Area 3 - Central Province

N65°-75°. This set is represented by swarms of faults whose distribution is not homogeneous within this area. The faults have a winding trend and a short linear development. Their distribution recalls an "en échelon" style. They give rise to a system of tilted blocks, among which numerous narrow and elongated tectonic depressions developed. The related area appears to be lowered towards the center of the depression represented by the Chapala Lake and the Celaya Graben extending between Queretaro and Salamanca. New dating makes it possible to establish that this fault system has been active from 2 to 0.3 Ma whereas the scarps do not show any sign of reactivation subsequent to the Upper Pleistocene.

N30°-50° and N300°-330°. These fractures mostly developed in the southern part of the Central Province within the Meseta Tarasca and around Morelia. The advanced morphological maturity of these lineaments and the geometrical relationships with the hereunder described sets suggest that this set is the oldest one here recognized. The intersection of these two fracture systems mark the southern limit of this MVB Province, evidenced by huge tectonic collapses. In the Meseta Tarasca, a great part of Quaternary volcanic centers appears to be linked with the reactivation of the N30°-50° set.

N80°-100°. This set consists in scarcely represented tensional faults controlling the depressions of Chapala Lake and Acambay. It is active together with the N340°-360° fractures in Late Quaternary times and results to be directly subsequent to the previous ones.

N340°-360°. It is scarcely diffused in the central area and is widely developed only West of Penjamo. A very important set of fractures is present in the eastern part of this sector, already known in literature as Taxco-San Miguel de Allende Line. It gives rise to the activation of very recent Grabens.

4.4 Area 4 - Guerrero Province

N60°-75°. This is the most frequent set in this Province and is represented by lineaments showing a remarkable continuity and morphological maturity. This set is strongly reactivated during Quaternary times in the area close to the south-eastern limit of the MVB.

N15°-40°. The lineaments belonging to this set appear to be more fragmentary and discontinuous than the previous ones. They show not univocal relationships with the previously described system, evidencing a possible contemporaneity of the two sets. This fact seems to be supported by a structural analysis carried out in the Tzitzio area (Garduño, pers. comm.). Northwards these lineaments seem to be interrupted within the MVB.

N330°-360°. This set shows an undoubtedly lower frequence than the previous one and a higher degree of fragmentation. Its geometrical relationships with the previous sets evidence an older age (Garduño, pers. comm.).

4.5 Area 5 - Sierra Madre Oriental Province

N345°-360°. These lineaments, though widely spaced, are characterized by a great linear development and are equally represented outside and within the MVB. Here they determine important alignments of Quaternary volcanic centers such as the Popocatepetl-Ixtaccihuatl range. Some important extensional faults, belonging to this system, divide the eastern sector of the MVB into altimetrically unleveled huge blocks. By comparison with the adjacent sectors we think that this fracture system is the oldest one present in Area 5, reactivated till Quaternary times.

N60°-75°. It is represented by a dense fracture system, mostly present north of the MVB. In the south-eastern sector, on the contrary, these lineaments appear to be reactivated in an extensional way and give rise to a system of blocks lowered towards SE, which separate Area 5 from Area 4.

N15°-40°. These lineaments control the distribution of some recent and big volcanic centres of MVB, such as La Malinche, the Cofre de Perote and the Pico de Orizaba Volcanoes. The fracture lineament, Cofre de Perote and Pico de Orizaba Volcanoes are aligned on, has been interpreted as the easternmost limit of the MVB (Cantagrel and Robin, 1979).

5 STRUCTURAL SYNTHESIS

The tectonic deformations accompanying the growth of the MVB during Neogene and Quaternary times are of the brittle type. These deformations are clearly subsequent to the phases of cordilleran terrane accretion and to the folding of Sierra Madre Oriental.

From the analysis carried out, it appears that these act as fracture systems affecting the whole South-central Mexico and decisively controlling the morphological features of the territory. By means of the interpretation of structural, geological and volcanological data, it has been possible to determine several deformation phases active up to Late Quaternary. The several sets of fracture lineaments that were determined, lead to the following relative sequence:

N340°-20°. This is a set of extensional probably listric faults, particularly evident in Areas 2 and 5. It is here interpreted as a southern continuation of the Basin and Range Province. It is deemed that the system was active some 10 Ma ago.

N315°-330°. This set shows an orientation and probably a kinematic meaning similar to the one of the fault system accompanying the opening of the Gulf of California since Early Pliocene. In Area 3 similarly oriented lineaments appear to be conjugated with N30°-50° fractures and are characterized by a recent strong extensional reactivation.

N60°-75°. This is the most diffused fracture set and shows different aspects and different meanings in the described structural areas. In Area 1, it can be considered as an equivalent of the previously described set. In Areas 3, 4 and 5, it should belong to the mosaic of crustal cracking caused by the reactivation of convergence processes along the Acapulco Trench from the Late Pliocene onwards. In Area 3, these fractures clearly evidence an extensional activation along an E-W trending central area bearing a transtensional character. In the other two sectors, its kinematic meaning appears doubtful.

N°15-40°. These lineaments appear to be conjugated to the previous ones and only occur in Areas 4 and 5 with a low frequency.

N80°-100°. This fracture set is scarcely represented though it clearly appears in the central area. It may locally be related with the formation of very recent Grabens, subsequent to the formation of all the above described sets.

All the aforementioned sets of fracture lineaments have been ascertained within the MVB. Moreover their areal development and kinematic meaning seem to be characerized by strongly marked variations.

In addition to this, their chronological relationships appear to be very complex and they show contrasting behaviours in adjacent areas. These observations made it possible to deduce that these lineaments have been subjected to repeated and variable reactivations within the MVB. Systems of tilted blocks and "en échelon" Grabens have been recognized in the central area of the MVB. Wide areas of downfaulting towards the depression of Balsas River were induced by an intensive reactivation of fractures along the southern edge of the MVB. Reactivations showing transcurrent components need further investigation.

Our structural observations made it possible to subdivide the MVB in 3 sectors, represented in Fig. 3, thus confirming what was observed in the morphological zoning shown in Fig. 1.

Strong and recently reactivated tensional components along Californian and N-S trends characterize the western sector. A complex intersection of fracture lineaments that repeatedly reactivated fracture lineaments with tensional and transtensional effects characterize the central sector.

This structural pattern may be attributed essentially to the lithospheric convergence phenomena acting in the Acapulco Trench. The strong and late effects of left lateral shear are probably due to oblique components in the convergence. Such structures overlap the previous block deformations of Basin and Range type, particularly evident in the Queretaro zone.

On the contrary, the eastern sector of the MVB appears to be dominated by strongly and repeatedly reactivated Basin and Range structures. They appear to be superposed by limited effects induced by the convergence. Extensional movements due to the opening of the Gulf of Mexico might have occurred in the eastern part of this sector.

6 GEODYNAMIC SPECULATIONS

The MVB represents a continental volcanic arc connected with the convergence that occurred between the Pacific Plate and the North-American Plate, within the Meso-American Trench. The arc appears as an alignment of Neogene and Quaternary volcanic centers continuous in E-W direction and oblique with regard to Meso-American Trench.

This peculiar geometry of the MVB led to the formulation of theories that differ from the plain arc-trench mechanism. For instance, the following were recalled: the continental prosecution of the Clarion Fracture Zone (Mooser and Maldonado, 1961), the south-eastern continuation of the San Andreas transform system (Gastil and Jensky, 1973) and the reactivation of an old geosuture (Mooser, 1972).

The structural analysis contained in the previous sections allow us to interpret the evolution of fracture tectonics with regard to MVB volcanism. The oldest ascertained structures belong to vertical movements of blocks directed according to N-S trends and which may be interpreted as the southern continuation of the Basin and Range Province. These structures probably reached Mexico during Late Miocene (Tardy, 1977; Mc Dowell and Clabaugh, 1979; Colletta and Angelier, 1983; Pasquarè and Zanchi, 1985); they appear to be locally reactivated until Quaternary times. The oldest volcanic activity within the MVB is probably connected with these structures. In the western part of the MVB, the Basin and Range structures were superposed by a conjugate set of NW-SE and NE-SW Californian lineaments since the Lower Pliocene. The rise of volcanic magmas in the western sector of the MVB and particularly the alignment of the volcanic centers characterizing the Tepic-Guadalajara Graben, is connected to this system.

On the contrary, a conjugate set of NNE-SSW and NE-SW fractures, probably due to the convergence of the Cocos Plate with the North-American Plate along the Acapulco Trench, developed in the central-eastern sector of the MVB. This event may support the hypothesis, formulated by Karing et al. (1978), of a change from oblique to perpendicular subduction along the Acapulco Trench during Late Miocene.

The geodynamic reorganization, begun with the opening of the Gulf of California, caused a progressive deactivation of the convergence between the Rivera Plate and the North-American Plate since the end of Pliocene (Larson,

1972). Luhr et al. (1985) correlate with this evidence an eastward jump of the East Pacific Rise segment, located between the Rivera and Tamayo fractures. The triple junction among the Tepic, Colima and Chapala Grabens may represent the new location of the spreading axis.

Therefore, in our hypothesis, continental Mexico, located NE of the Colima junction, represents a rather stable sector in the recent geodynamic evolution of the area (Fig. 4) .

Subduction processes are still active in correspondence with the Acapulco Trench where compressive stresses are transmitted to the continental lithosphere with eastward components along the central sector of the MVB. Along the contact with the stable continental Mexico, a transtensive system was activated,

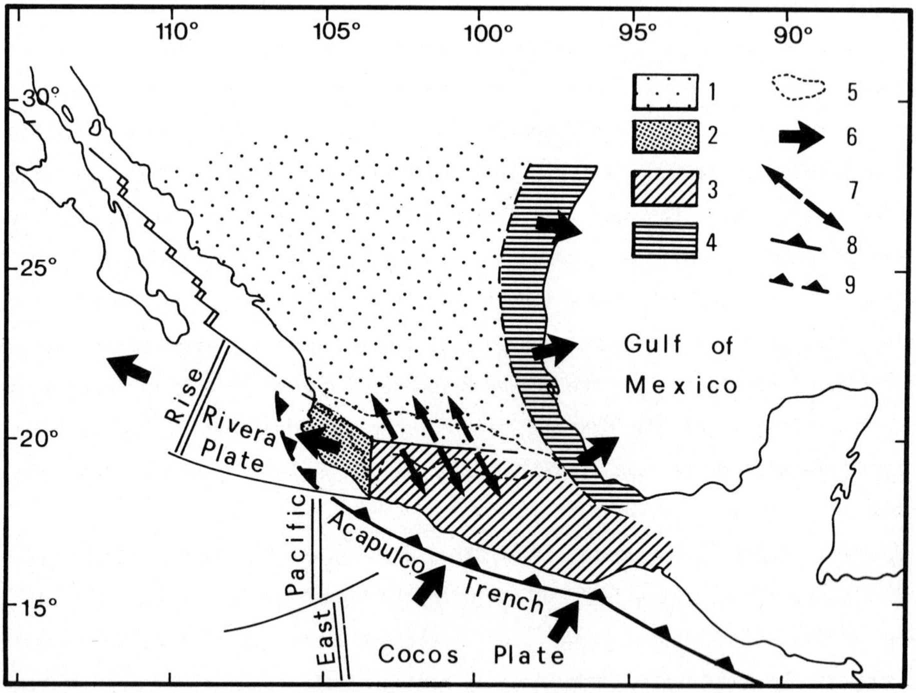

Fig. 4. Main Quaternary geodynamic features of South-central Mexico. (1) Relatively stable area; (2) Structures related to the Californian divergence; (3) Structures related to convergence along the Acapulco Trench; (4) Structures related to the Gulf of Mexico extension. (5) Boundaries of the Mexican Volcanic Belt; (6) Relative movements; (7) Main transtensional system; (8) Active trench; (9) Inactive trench.

according to a NNW-SSE trending σ_3. Such structures greatly condition the formation of volcanic centers since about 2 Ma (Pasquarè, K-Ar unpublished data). The counterclockwise rotation described by Urrutia-Fucugauchi (1981) in a sector located near the southern limit of the MVB central sector, may be considered within this hypothesis.

Moreover, the reactivation of N-S trends belonging to the Basin and Range Province contributes to the recent structural evolution of the MVB, where the main Quaternary volcanic centers are connected to this fracture system.

The problem of structural links between the MVB and the tensional system of the Gulf of Mexico still remains unsolved while the alkaline volcanic activities of the Vera Cruz area are certainly connected with the opening of the Gulf of Mexico (Cantagrel and Robin, 1979). The westward penetration of fracture lineaments connected with that system might also be present in the easternmost sector of the MVB and particularly in some anomalous lineaments such as the Pico de Orizaba-Cofre de Perote fault.

In spite of the uniformity of geochemical and volcanological characteristics found so far in the MVB, the MVB structurally appears to result from the control exerted by quite different and partly heterochronous geodynamic mechanisms.

ACKNOWLEDGEMENTS

We would like to express our appreciation for the collaboration offered by Comision Federal de Electricidad, Mexico, during field investigations. In particular, we express our special thanks to ing. Hector Alonso Espinosa, to ing. Arturo Gonzales Salazar and to ing. Antonio Razo Montiel. We gratefully acknowledge discussions with Maria Fernanda Campa Uranga, Victor Hugo Garduno Monroy and Aida Lopez Hernandez. The statistical analyses of satellite lineaments were plotted at MEAG Department, AGIP, Milan, by the good offices of dr. Emanuele Cassano.

7 REFERENCES

Allan, J.F. and Carmichael, I.S.E., 1984. Lamprophiric lavas in the Colima graben, SW Mexico: Contributions to Mineralogy and Petrology (in press).

Anderson, H.T. and Schmidt R.V., 1983. The Evolution of Middle America and the Gulf of Mexico-Caribbean Sea During Mesozoic Time. Geol. Soc. Am. Bull., 94, 941-946.

Atwater, T., 1970. Implications of plate tectonics for the Cenozoic tectonic evolution of western North America. Geol. Soc. Am. Bull., 81, 3513-3536.

Campa, M.F., Ramirez, J., Flores, R. and Coney, P.I., 1981. Terrenos tectono-estratigraphicos de la Sierra Madre del Sur, region comprendida entre los estados de Guerreros, Michoacan, Mexico y Morelos. Universidad Autonoma de Guerrero.

Campa, M.F. and Coney, P.S., 1983. Tectonostratigraphic Terranes and Mineral Resources Distribution in Mexico. Can. J. Earth Sci., 20, 1040-1051.

Cantagrel, J.M. and Robin C., 1979. K/Ar dating on eastern Mexico Volcanic rocks. Relations between andesitic and the alkaline provinces. J. Volc. Geoth. Res., 5, 99-114.

Colletta, B. and Angelier, J., 1983. Tectonique cassante du nord-ouest mexicain et ouverture du Golfe de Californie. In: M. Popoff and J.J. Tiercelin (Editors) Rifts et Fossês anciens, 433-441.

Demant, A. and Robin, C., 1975. Las fases del vulcanismo en Mexico: una sintesis en relacion con la evolucion geodinamica desde el Cretacico. Rev. Inst. Geol. UNAM, 1, 70-83.

Demant, A., 1978. Caracteristicas del Eje Neovulcanico Trans-Mexicano y sus problemas de interpretacion. Rev. Inst. Geol. UNAM, 2, 172-187.

Demant, A., 1979. Volcanologia y Petrografia del sector occidental del Eje Neovolcanico. Rev. Inst. Geol. UNAM, 3, 39-57.

Gastil, R.G. and Jensky, W., 1973. Evidence for strike-slip displacement beneath the Trans-Mexican volcanic belt. In: Proc. Conf. on Tectonic Problems of the San Andreas Fault System. Stanford Univ. Publ. Geol. Sci., 13, 171-181.

Gastil, G., Krummenacher, D. and Minch, J., 1979. The record of Cenozoic Volcanism around the Gulf of California. Geol. Soc. Am. Bull., 90, 839-857.

Gilbert, C.M., Mahood, G.A. and Carmichael, I.S.E., 1985. Volcanic stratigraphy of the Guadalajara area, Mexico. Geofisica Int., Spec. Vol. on Mexican Volc. Belt, 24, 203-216.

Karig, D.E., Cardwell, R.K., Moore, G.F. and Moore, D.F., 1978. Late Cenozoic Subduction and continental margin truncation along the Northern Middle American trench. Geol. Am. Soc. Bull., 89, 265-276.

Lomnitz, C., 1982. Direct evidence of a subducted plate under southern Mexico. Nature, 296, 235-238.

Larson, R.L., 1972. Bathymetry, magnetic anomalies and plate tectonic history of the mouth of th Gulf of California. Geol. Soc. of America Bull., 83, 3345-3360.

Lugo, H., Ortiz, P., Palacio, P. and Bocco, V., 1985. Las zonas mas activas en el Cinturon Volcanico Mexicano (entre Michoacan y Tlaxcala). Geofisica Int., Spec. issue on Mexican Volcanic Belt, 24, 83-96.

Luhr, J.F. and Carmichael, I.S.E., 1985. Contemporaneous eruptions of calc-alkaline and alkaline magmas along the volcanic front of the Mexican Volcanic Belt. Geofisica Int., Spec. issue on Mexican Volcanic Belt, 24, 203-216.

Luhr, J.F., Nelson, S.A., Allan, J.F. and Carmichael, I.S.E., 1985. Active rifting in Southwestern Mexico: Manifestations of an incipient eastward spreading-ridge jump. Geology, 13, 54-57.

McDowell, F.W. and Clabaugh, S.E., 1979. Ignimbrites of the Sierra Madre Occidental and their relation to the tectonic history of Western Mexico. Geol. Soc. Am. Bull., Sp. issue 180, 113-124.

Mooser, F. and Maldonado, K.M., 1961. Penecontemporaneus tectonics along the Mexican Pacific coast. Geofisica Int., 1, 1-20.

Mooser, F., 1969. The Mexican Volcanic Belt-structure and development. Formation of fractures by differential crustal heating. Pan-Am. Symp. on the Upper Mantle, Mexico D.F., 2, 15-22.

Mooser, F., Nairn, A.E.M. and Negendank, J.F.W., 1974. Paleomagnetic investigation of the Tertiary and Quaternary igneous rocks: VIII A paleomagnetic and petrologic study of volcanics of the Valley of Mexico. Geol. Rundschau, 63, 451-483.

Negendank, J.F.W., 1972. Volcanics of the Valley of Mexico, N. Jb. Miner. Abh., 116, 308-320.

Pal, S., Lopez, M., Perez, J. and Terrell, D.J., 1978. Magma Characterization of the Mexican Volcanic Belt (Mexico). Bull. Volc., 41, 379-389.

Pasquarè, G.and Zanchi, A., 1985. Cenzoic Volcanism and Tectonics in West-central Mexico. Rend. Acc. Naz. Lin., 95, (in press).

Pasquarè, G., Vezzoli, L. and Zanchi, A., 1986. Morphological and Structural model of Mexican Volcanic Belt. Geofisica Int., Spec. issue on Mexican Volcanic Belt, 3, (in press).

Robin, C. and Tournon, S., 1978. Spatial relations of andesitic and alkaline province in Mexico and Central America. Can. J. Earth Sc., 15, 1633-1641.

Schurbet, D.H. and Cebull, S.E., 1984. Tectonic Interpretation of the Transmexican Volcanic Belt. Tectonophysics, 101, 159-165.

Tardy, M., 1977. Essai sur la reconstitution de l'évolution paléogéographique et structurale de la partie septentrionale du Mexique au cours du Mésozoique et du Cénozoique. Bull. Soc. géol. France, (7), t. XIX, n° 6, 1297-1308.

Thorpe, R.S., 1977. Tectonic significance of alkaline volcanism in eastern Mexico. Tectonophysics, 40, 19-26.

Urrutia, J.H. and Del Castillo, L., 1977. Un Modelo del Eje Volcanico Mexicano. Bol. Soc. Geol. Mex., 38, 1, 18-28.

Urrutia-Fucugauchi, J., 1981. Paleomagnetism of the Miocene Jantetelco Granodiorites and Tepexco volcanic group and inferences for crustal block rotations in central Mexico. Tectonophysics, 76, 149-169.

THE PACIFIC ISLAND ARCS: PRODUCED BY POST-OROGENIC VERTICAL TECTONICS ?

Forese-Carlo WEZEL
Istituto di Geologia, Università di Urbino, I-61029 Urbino, Italy

Geotectonics is simpler than we thought, the gross structures stare naked from the world physical chart and things really are what they look like.

S. Warren Carey, 1958, p.179

ABSTRACT

A short review of some basic results of the IPOD drilling of the Pacific active margins is presented. From these it is deduced that the present conceptual models of tectonic processes operating at these margins do not seem to reflect the complex geological reality of these active regions.The direct data contradict the key concepts of active compression, progressive accretion and uplift of the fore-arc region. Instead, they have evidenced a generalized extensional regime since the Neogene and a rapid subsidence with submergence of Late Paleogene subaerial erosion surfaces in the fore--arc. The Neogene "molassic" slope deposits transgress regionally over block-faulted and eroded pre-Neogene orogenic belts. Particularly noteworthy are the post-orogenic collapse and extension of the back-arc area, which is generally centered over the median portions of the circum-Pacific bivergent orogens (i.e., the rigid zones of older chains, strongly folded and partly metamorphosed). These Tertiary collapse tectonics assume varying sizes and forms according to the relative evolutionary stage reached by the Pacific margin. The initial stages are represented by the central valleys of the Andes up to the great continental depressions of Basin and Range type. The successive stages lead to the marginal basins of the western Pacific, here classified by families, according to the age of the beginning of the oldest basin area opening.

The great spatial extension and depth of the transcurrent faulting subparallel to the margins of the Pacific Basin are also highlighted. It is thought possible that such a strike-slip motion may take place also in correspondence with the Benioff seismic zones.The entity of the transcurrence phenomenon has led the author to postulate that the post-orogenic collapse of the marginal basins is accompanied by tectonic rotation and great torsions which have disrupted the Pacific perimeter. The Benioff zone is considered to become steeper and deeper in relation to the increase in surface subsidence and lateral extension of the foundering reached by the marginal basins. The depth and pattern of the "oceanization" of the margin seem to indicate a broad control by events in the mantle (e.g., existence of transient "tectonospheric" waves).

INTRODUCTION

The Pacific island arcs are undoubtedly amongst the most interesting and revealing structures of the deep-seated tectonic phenomena which occur on the earth's face. These arcuate island structures have been for a long time related to the orogenic processes. Since the times of E. Haug and E. Argand the island arcs of the Sunda type have been considered a mountain chain in statu nascendi. The idea of a youthful stage of the island festoons apparently finds support in their great geological activity, manifested by the seismicity, the volcanism and the morphotectonic shape itself of the active Pacific-type margins. Such physiographic division into longitudinal swells and troughs has been thus applied by E. Argand in the reconstruction of the Alpine geosyncline in the Mesozoic, so becoming part of geologists' stock of knowledge. All these intra-geosynclinal morphologic elements are considered to be mobile and subjected to different tectonic movements (distension, subsidence and compression). Due to the successive diastrophism such paleogeographic zones become tectonic units, transforming each zone in an overthrust unit (nappe).

The plate tectonics model identifies both island arcs and mountain belts as effects of plate convergence. Most orogenic belts are considered to result from a collision event which has deformed oceanic sediments and crustal rocks. The mechanical addition of oceanic rocks, caused by a mechanism of off-scraping the upper part of the underthrusting oceanic lithosphere, constitutes the conservative process of the "subduction accretion" causing the formation of the so-called "accretionary prism" (see, Karig, 1974). Compressional stress is emphasized along active margins and the evidence cited for compression consists in the global plate motion scheme.

The formation and evolution of the modern arc systems are described as if dealing with independent phenomena not linked to the tectonic development of the surrounding coastal mountains. The geological synthesis is based almost exclusively on the analysis of a very narrow strip comprising the arc-trench region (see, e.g., Dickinson and Seely, 1979), without considering the whole wide back-arc area with its inactive "remnant" arcs and trenches. Such simplistic scientific approach originates in the conceptual belief of the adherents to plate tectonics in an inherent disorder and (temporal and spatial) irregularity of tectonic phenomena, and in an uniformitarian or actualistic view (see Sengör, 1982). Their apparent basic conviction that there was no decipherable orderliness in tectonic events led them both to the rejection of the classic concept of geotectonic cycle, considered "obsolete", and to the abandonment of the traditional terms, such as pre- syn- and post-tectonic, held to be "meaningless words" (Coney, 1970).

The aim of the present paper is that of trying to extract from the enormous literature on the Pacific a few geological facts, and as far as possible direct ones, on the circum-Pacific arc systems. The review of results from drilling of the IPOD transects across the Japan (Legs 56 and 57), Mariana (Legs 59 and 60) and middle America (Legs 66, 67 and 84) active margins, published in the relative "Initial Reports of DSDP", has been of fundamental importance. The research was carried out trying as far as possible not to be influenced by any preconceived models but to look, as "unbiasedly" as possible,exclusively at facts of direct type (drilling holes) rather than at the indirect ones of geophysics (seismic profiles). In the writer's opinion this quite preliminary panoramic look at the data, although few, is useful in so far as it permits a first valuation on the validity or not of the conceptually formulated axioms relative to the active margins. A lead concept of the writer is that both the opening of the small ocean basins and the development of the arc system are related to the previous evolutionary development of the surrounding mountain chains. This concept may be exemplified by the scheme of Fig.1, taken from the study of the Sicilian margin of the Tyrrhenian Sea.

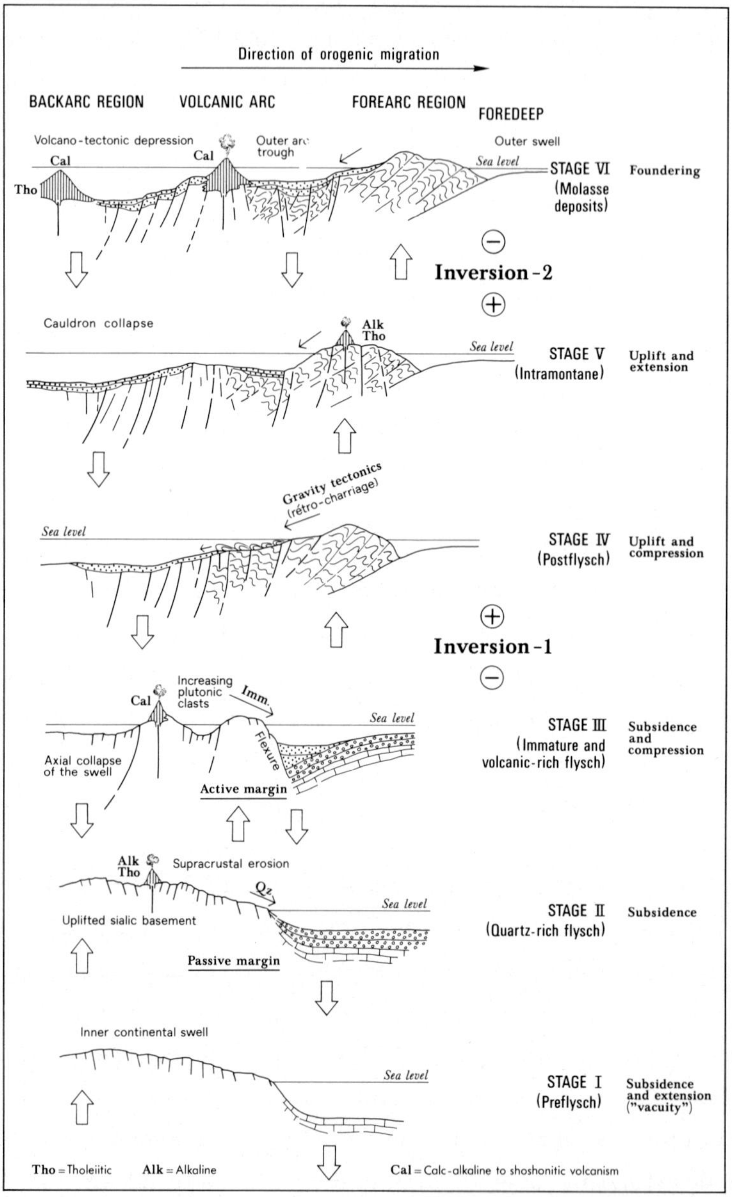

Fig. 1. Diagram illustrating the geotectonic evolution and the different tectonic regimes characteristic of each stage of a typical developmental sequence: geosyncline ⟶ mountain chain ⟶ Pacific-type active margin. The example refers to the arc system made up of Sicilian fore-arc ridge-Aeolian volcanic arc (from Wezel, 1984). For the structural framework of the Calabro-Sicilian Arc see Fig.2.

SUMMARY OF PARADIGMATIC AXIOMS OF PLATE TECTONICS

The island-arc margins should have conceptually, on the basis of plate tectonic model, the characteristic features listed below (see, e.g., Karig, 1974; Seely, 1978; Dickinson and Seely, 1979; and so forth):

1. The ocean floor seaward of the trenches is made up of deep--sea pelagic sediments deposited on the oceanic basement.

2. The accretionary prism (or subduction-accretion complex) represents a deformed belt of offscraped oceanic deposits and crustal rocks plus the turbidite trench wedges.

3. The accretionary prism is composed of imbricate tectonic slices superposed in a reverse stratigraphic order with the older at the top and the younger at the bottom of the prism.

4. The compressional accretionary prism typically grows seaward in width and vertically upward as a result of progressive tectonic thickening.

5. Volcanism in island arcs is produced by obliquely descending cold oceanic lithosphere. The slab is melted and involved in arc magmatism.

6. Inter-arc marginal basins have been formed by extensional tectonics related to plate convergence and subduction.

7. Intermediate and deep shocks along the Benioff zone indicate down-dip compression.

8. The general regional regime resulting from subduction and constant plate convergence is compressive.

THE IMPORTANCE OF VERTICAL TECTONISM OF THE LATE TO POST-OROGENIC STAGE

The kinematic analysis of the Mediterranean arc-systems, particularly the Calabro-Sicilian Arc with the Tyrrhenian inter-arc basin, has shown the large role played in the formation of the arcs by the late post-orogenic tectonics (essentially Plio-Quaternary). For this stage, which goes from the vertical uplift of the tectogenetic structures, so as to constitute the topographic high mountains,to the successive collapse and foundering of the internal zones (or Internides) of the foldbelts, the writer has coined the term krikogenesis (Wezel, 1982, 1984 and 1985; Carey, 1983 and this volume). The introduction in literature of this new term seemed to the writer a real necessity to fill both a void and cover a terminological contradiction. In fact, the elevation and the relief of mountain chains which we see today on the earth's face is not due to that process that is called orogeny (mountain-building), but to a later different type of vertical crustal movement which is in fact krikogeny.

The first krikogenic phase is represented by a crustal arching which creates the "late-tectonic basement folding" ("plis de fond" of E. Argand). As Aubouin (1981) also pointed out, the highest summits of the Alpine ranges are always located on such huge elevated swells of the continental basement. This regional undulation is therefore shown topographically in mountains and plains. To judge from the deep "litospheric roots" pointed out in the Alps and the Apennines by Mueller and Panza (see this volume) it is a question of deep-seated movements affecting the whole lithosphere. Probably the swelling is linked to the rise from the asthenosphere of "funnel-shaped diapirs" (S.W. Carey, this volume).

The last phase of krikogenic deformation is represented by the extension and collapse of the median zones of Alpide bivergent orogenic belts. The collapse zones show notable variations in size and style ranging from narrow intramontane basins to the continental depression of the Pannonian-Transylvanian Basin, and finally to the young small ocean basins such as the Tyrrhenian and the Aegean seas. The collapse basins are always superimposed on bivergent orogens which are affected by block faulting and truncation. The Plio-Quaternary collapse tectonics of the Tyrrhenian inter-arc basin are extremely fast with subsidence rates over 1 km/Ma sometimes reaching 2 km/Ma. The Tyrrhenian foundering is characterized (Wezel, 1982, 1984 and 1985) by gigantic forms of "cauldron subsidence" (e.g., the Quaternary Marsili Basin), followed by the formation of giant "ring fractures" (e.g., the Aeolian volcanic arc: see Fig.2). The whole Tyrrhenian floor is fractured like a

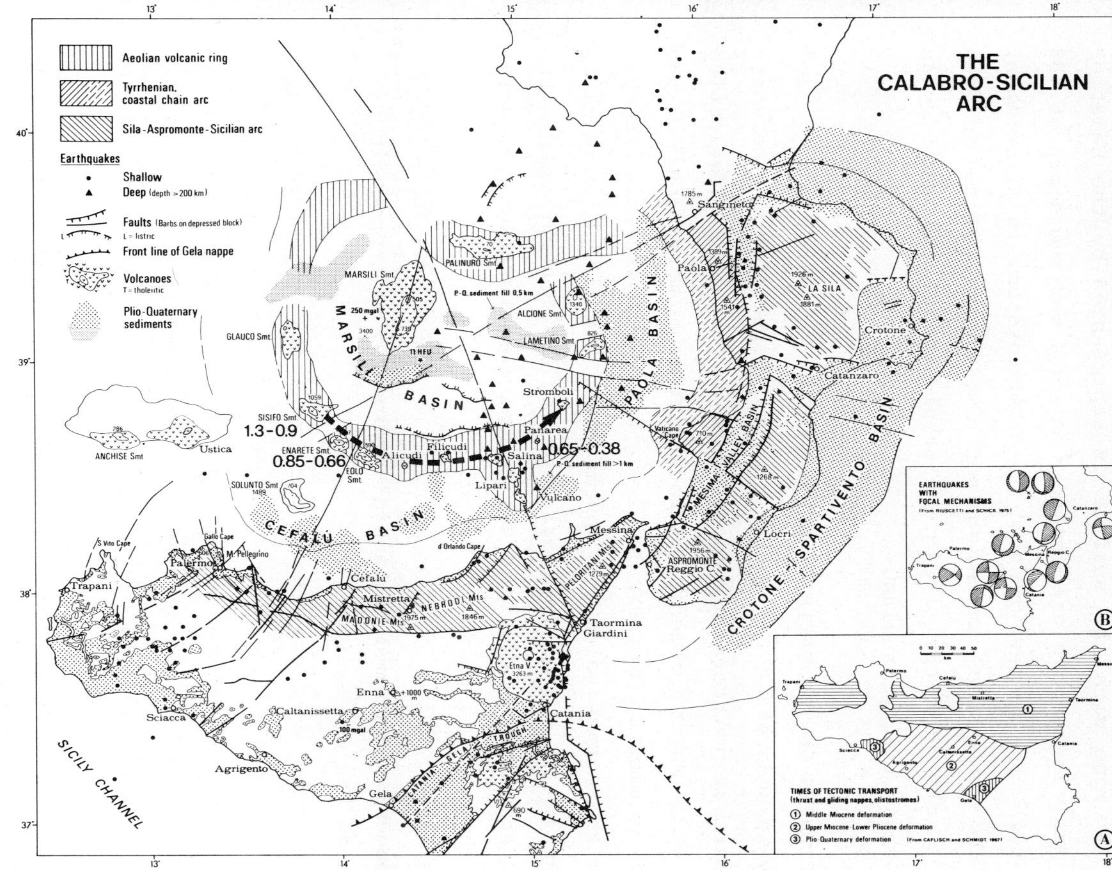

Fig. 2. Simplified geological map of the Calabro-Sicilian Arc showing the concentric regional distribution of key features. The arrow centered on the Aeolian Arc ring indicates the younging direction of the age of the calc-alkaline volcanoes (K/Ar ages based on Beccaluva et al., 1985), interpreted as manifestation of a tectonic torsion.

mosaic in different collapsed circular blocks, having a roughly constant diameter of about 100 km.This collapsing structural mosaic is found inside (concave side) a conical inclined seismic surface (Benioff zone) dipping toward the deepest central bathyal plain (Peterschmitt, 1956; Wezel, 1984 and 1985). The intersection of part of the cone with the surface is represented by the Calabro-Sicilian Arc. The shocks are recorded up to a maximum depth of 488 km, just below the deepest central basin depression. In my opinion, neither the isostatic mechanism nor the thermal anomalies can possibly account for the prodigious subsidence of the small ocean basins such as the Tyrrhenian.

MORPHOTECTONICS OF THE PACIFIC ARC SYSTEMS

To try to understand the active margins, especially in the western Pacific, it is necessary to see them as part of a wider active orogenic belt which presents characteristic geomorphological and structural elements (Fig.3). Besides it is necessary to remember that similar tectonic elements appear also, even if of quite minor dimensions, in the Alpine-Mediterranean region. Schematically, the characterizing features of the structural configuration essential to these margins are discussed below.

1) The Neogene undeformed "molassic" sediments: post-orogenic extension of the arc systems.

As was noted earlier, the arc systems are superimposed upon young mountain chains which originated during Meso-Cenozoic times. They occupy and overprint an orogenic belt of 1,000-3,000 kilometres wide. Altogether,the overprint of the orogenic system concerns both the external and internal zones plus the medial massifs of the hinterland. According to Beck and Lehner (1974) the deformation consists in phases of vertical block movement, transcurrent faulting and compressional folding, accompanied by various types of magmatic and volcanic activity (e.g., the circum-Pacific granite intrusions). The schematic regional profile by Shell (Beck and Lehner, 1974, Fig.16) from Java Trench to Borneo, 1,500 km long,

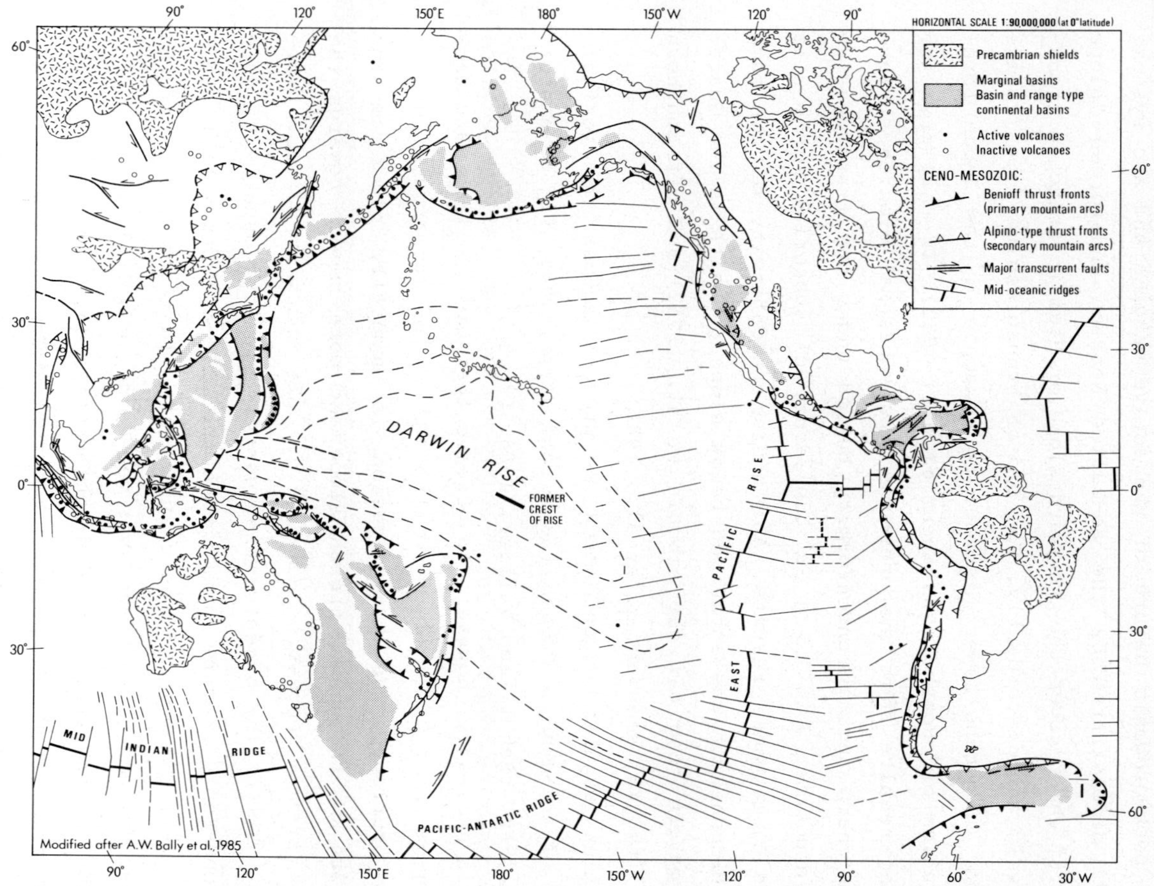

Fig. 3. Simplified tectonic map of the Pacific Ocean and adjacent continents illustrating some major features discussed in the text. The superimposition of the back-arc basins upon the Ceno-Mesozoic bivergent belts and the very long subparallel transcurrent faults are to be noted. Also highlighted is the location of the former Mesozoic "Darwin Rise".

clearly shows the Tertiary post-orogenic sediments overlying folded Mesozoic to Paleogenic sediments or basements, affected by horst and graben tectonics.

The post-orogenic nature of the active margins and arc systems is well demonstrated by the so-called Neogene undeformed "slope deposits" which cover the landward slope of Pacific trenches transgressively and discordantly. The oldest transgressive deposits are of Upper Oligocene-Lower Miocene, for example in the Japan fore--arc region off Honshu (von Huene et al., 1982). In this area the drilling results show the presence of a subaerial erosion surface with a sharp angular unconformity buried under the transgressive slope deposits. A 3 to 6 km subsidence of subaerial surface is deduced from these results. An angular unconformity of the same Oligo-Miocene age was submerged below slope deposits in the Guatemalan margin (Leg 84: see Aubouin et al., 1982) and off southern Mexico (Leg 66: see Watkins et al., 1981).

Similar angular unconformity is present across the fore-arc region of the Manila Trench (Lewis and Hayes, 1985, Fig.8) and off central Peru (Scholl and Vallier, 1983, Fig.3).

It is necessary, however, to observe that the drilling data show that there is a diachronism in the age of the submergence of the erosive surface. The transgression is generally older in the direction of the upper trench slope and more recent towards the present trench depression (see, e.g., off southern Mexico: Moore et al., 1982). The oldest transgressive deposits, usually of Late Oligocene-Early Miocene, constitute the fore-arc basin-fill sediments. Their greater thickness indicates a landward tilting of the "fore-arc ridge" (see off Honshu: von Huene et al., 1982).

Such transgressive Neogene slope deposits constitute an undeformed sediment cover which lies over the whole fore-arc region and even the trench itself. Such a situation is found in the Japan, Mariana, middle America and Peru-Chile trenches (see, e.g., Scholl and Vallier, 1983, Fig.3), while it does not seem to be found in the central Aleutian Trench, where the cover of "slope deposits" across the landward trench slope seems to be missing (McCarthy and

Scholl, 1985).

The presence of this cover of transgressive sediments even in the trench is of great importance and seems to have been until now an underestimated element: it indicates the Late Tertiary blanketing by submergence of the whole "fore-arc ridge" and, consequently, the persistence of the same tectonic regime of post-orogenic subsidence and distension across the landward trench slope and the trench floor. It shows us that the present fore-arc deformation is not of compressional type as indicated by the plate tectonic paradigm. Moreover, rather than the uplift indicated by the accretionary prism model of Seely (1978), the whole fore-arc region is foundering.

The extensional regime superimposed upon the "fore-arc ridge" is well shown by normal faulting affecting the "slope deposits" of Japan (Honza, 1981, see Fig.9), Mariana (Hussong and Uyeda, 1981, see Fig.5), Guatemala (Aubouin et al., 1982), Peru-Chile (Moberly et al., 1982). According to Aubouin et al. (1984) the entire Cordilleran margin of the Pacific shows the characteristics of an "<u>extensional</u> active margin", in the same way as the island-arc margins of Japan and Mariana. The possibility has also been put forward that even the Aleutian area off Kodiak Island belongs to this "extensional margin"-type.

The general regional transgression occurs, from Alaska to Indonesia, over folded and truncated Paleogene beds affected by a phase of block faulting. The structural picture is suggestive of a morphorogenic uplift of Mesozoic-Early Tertiary folded sediments causing their taphrogenic fragmentation and subaerial denudation, followed successively during the Neogene by subsidence phases. Evidence of Late Paleogene strong tectonic uplift of the "fore-arc ridge" with emersion, subaerial erosion, rifting and arc-related volcanism, and landward tilting comes from IPOD sites 439 and 438 of the Japan Trench,where the existence of a paleolandmass("Oyashio ancient landmass") has been hypothetized (see von Huene et al., 1982). Another paleoland has been postulated further south off the Kii Peninsula, on the basis of exotic clastics in conglomerates of

Late Cretaceous and Eocene,and of sole marks. This "Kuroshio Paleoland", with Precambrian continental crust, had emerged until the Eocene (see Suzuki and Tateishi, 1980). Also in correspondence with the Guatemalan margin the drilling data of Legs 67 and 84 show a strong Paleocene uplift with erosion of about 3,000 m of end-Cretaceous and Paleocene rocks (see Aubouin et al., 1982).

Evidence of ancient fore-arc magmatism within a few tens of kilometres of the present trench axis, not only found on the Japan but also on the Mariana transect (see von Huene, 1981) may, in my opinion, be considered indicative of proceding uplifts of the "fore-arc ridge". The same morphotectonic configuration of the island arc systems reveals the foundering of a former elevated archlike ridge, about 100 km wide. Its Neogenic-Quaternary subsidence is correlated to the recent contiguous uplift of the volcanic arc.

The circum-Pacific Neogene "slope deposits" correspond to the detrital molasse deposits of the Alpine-Himalayan chains which lie upon strongly folded successions with angular unconformity. In the Alpine system the molasse sequence originates in the marginal foredeep trough contiguous to a newly emerged and rapidly uplifted orogenic domain. Their presence is a consequence of post-paroxysmal tectonic movements of uplift and distension of the orogenic belt. It represents a late stage of geosynclinal development, successive to the phase of synparoxysmal compression, indicative of the climax of the Alpine orogeny. The molasse deposits are relatively undeformed and present a diachronism going from internal zones toward the external ones and the foredeep.

From seismic reflection profiles we deduce that the filling of the "fore-arc trough" ("interdeep") occurs due to the effect of a progressive landward tilting of the "fore-arc ridge" (see, e.g., Lewis and Hayes, 1985, Fig.7: Manila Trench). In practice it seems possible to distinguish two sequences of "slope deposits" and they are: (1) an older Oligo-Miocene sequence fed from the "fore-arc ridge" and (2) a younger essentially Plio-Quaternary sequence indicative of a clastic supply coming from the erosion of the uplifted "frontal arc". In the case of the Guatemalan margin (Leg 84: Aubouin et al., 1984, Fig.3) sequence (1) is represented by Late Oligocene to Early Miocene sandy sediments and sequence (2) by Late Miocene to Pleistocene hemipelagic muddy slope deposits. It is to be noted that Miocene sandy facies at Site 568 (midslope area) is coarser and thicker than the turbidites presently filling the trench.

2) The "accretionary prism": are we dealing with a foundered paleochain ?

On the basis of geophysical data only, the landward dipping reflections beneath the clastic wedge were interpreted as thrust surfaces inside the so-called "accretionary prism" or "subduction--accretion complex". It was considered to be made up of the mechanical addition of masses of trench turbidites, oceanic pelagic sediments and crustal rocks. The oceanic deposits were held to be tectonically accreted by the offscraping of the sediments on the underthrusting crust. The skimmed slices of rock and sediment are partially thrust beneath older underthrust slices (e.g., Seely, 1978; Dickinson and Seely, 1979). In this way, a gradual general uplift and tilting of the foldbelts of offscraped oceanic deposits results, with the formation of a tectonic pile constituted by the older skimmed slices at the top and the younger at the bottom of the "accretionary prism" (oceanward-younging slices). The sediments within each slice were to be younging landwards.

Landward-dipping seismic reflectors beneath the inner trench slope are considered to represent older uplifted trench sediments (e.g., Moore et al., 1979) and fault-bounded slices of oceanic crust (e.g., McMillen et al., 1982, Fig.3). This interpretation, however, remains quite theoretical unless the accretionary complex is directly sampled with drillings. Unfortunately, the numerous cores carried out on the inner slope of the Pacific trenches, in the DSDP/IPOD program, were not able to reach the "accretionary complex", stopping instead at the inside of the post-orogenic slope deposits. Only four drillings have been made below the slope deposits in the middle America landward trench slope off Guatemala which is so far the only trench slope drilled to basement (Sites 567, 566, 569 and 570 of Leg 84). At the penetration sites,basement was found to consist of pre-Campanian ophiolitic rocks belonging to the substratum cropping out in Central America (Aubouin et al., 1982).The Guatemalan margin is therefore underlain by pre-Campanian thrust faults bringing ophiolitic rocks above the Upper Jurassic-Upper Cretaceous Nicoya deep-sea complex. The ophiolitic basement, certainly emplaced in Cretaceous times, was recovered all along a cross section of the slope (Aubouin et al., 1982; Aubouin et al., 1984), beginning from the front of the slope. It is an ancient pre-Eocene orogenic system which has nothing to do with the present arc system, namely a submerged paleochain with a superposed new neo-tectonic regime. This contradicts in the most decisive way the concept of continuous progressive accretion and imbrication of more and more recent terrains, postulated off Guatemala by Seely and colleagues (1974). This accretionary model, widely accepted, was supported also by the multichannel seismic reflection site surveys (Ibrahim et al., 1979).

The direct data show in fact that starting from the Eocene "no net accretion has occurred in the middle America Trench off Guatemala" (Aubouin et al., 1984). The chain was uplifted at the end of Paleocene with erosion of about 3,000 m of end-Cretaceous and Paleocene rocks. A strong faulting left a sharp angular unconformity during Late Oligocene to Early Miocene. Since these times the margin has exhibited an active extensional regime.

Another indication of a subaerial emergence of a paleolandmass ("Oyashio landmass") in the Upper Oligocene comes from IPOD Sites 439 and 438, off northeastern Honshu, across the Japan Trench. Below Pleistocene to Lower Miocene mud and mudstone with ash, there are in the Lower Miocene-Upper Oligocene sand turbidites and a 48--m-thick conglomerate unit of dacitic to rhyolitic boulders (isotopically dated as 24 to 22 Ma), overlying with an angular unconformity Late Cretaceous silicified claystone (von Huene et al., 1982). In Late Cretaceous time the area was characterized by an orogenic belt, comprising a Japanese "coast range" with ophiolites and blueschists (high pressure Sambagawa metamorphic belt) and a Chinese "cordillera" intruded by granites (see Aubouin, 1984). The Cretaceous mountain chain was uplifted and deeply eroded during the Late Paleogene,when it become a significant source of sediment. Even in this case the "accretionary wedge" resulted from factual evidence to be a Mesozoic framework, not related to the present tectonic regime. This process, subduction accretion, does not seem to be documented in the circum-Pacific active margins by any DSDP drillings. Drilling off central Mexico (Leg 66: see Watkins et al., 1981), in fact has not documented the accretionary mechanism in that it did not penetrate the slope deposits in reaching the basement.

In brief, the direct data at our disposition are extremely scarce and insufficient. Drilling of the IPOD programme indicates that up to now no <u>direct</u> argument has been found in favour of the effective existence of an "accretionary prism".

3) **Plio-Quaternary vertical uplifting of island arcs.**

There seems little doubt that magmatic arcs are among the most rapidly uplifted of all tectonic features. Beck and Lehner (1974) claim that magmatic arcs show pronounced continuing vertical uplift. Evidence of such tectonic rise is provided by Quaternary raised terraces and beaches in the Aleutians and Andes (Plafker, 1972), and raised coral reefs near islands of the eastern Indonesia volcanic arc (Chappel and Veeh, 1978). According to Sakaguchi(1972) the vertical uplift of the Japan volcanic arc (Honshu), begun in the Pliocene, is not yet complete. The average uplift rate is of the order of 450 m/Ma for the Plio-Quaternary. On the same order of magnitude is the Late Quaternary uplift rate relative to the Banda volcanic arc (470 m/Ma according to Chappel and Veeh, 1978). On the other hand, the tectonic uplift and relative intense denudation since Early Holocene of the Taiwan magmatic arc (Coastal Range) has taken place at average high rates of well over ten times (5 km/Ma: see Peng et al., 1977). Such huge uplift is comparable to the present uplift rate of the Himalayan chain, about 4 to 5 mm/a (Peng et al., 1977).

Continued post-orogenic uplift of the volcanic arc may lead to erosion of the deeply buried metamorphic rocks, such as those in the Japanese islands. In this way the famous metamorphic belts of Sanbagawa (HP/LT) and Ryoke (LP/HT) of Late Jurassic-Cretaceous age crop out. The present volcanic arc of Quaternary age, results therefore located and superimposed on the internides of the Japan orogen, near the boundary of the medial massif of the hinterland region. Also the eastern Indonesia volcanic arc of Late Tertiary to Quaternary is superimposed upon the eugeosynclinal metamorphic belt of Jurassic to Permian age (Crostella, 1977).

The upheaval and block faulting of island arc mountain belts during Plio-Quaternary was accompanied by a magmatic phase. Nearly all of the presently active volcanoes in the circum-Pacific region began their explosive activity during the Quaternary, 1-2 Ma ago (see the "Cascadian episode" of Kennett et al., 1977). In most places the volcanic activity started in the Upper Miocene and was almost continuous throughout Pliocene and Quaternary times. Arc magmatism is typically centered 150-200 km landward of Pacific trenches and 100-125 km above the Benioff seismic zone.

4) Extension and collapse of the formerly elevated median area between volcanic arcs and the continent.

The small ocean basins inside the island arcs display Tertiary collapse tectonics with buried horst and graben structures. Miocene formations transgress regionally over folded, block-faulted, and truncated early Tertiary beds, a common feature of the Pacific margin from Alaska to Indonesia" (Beck and Lehner, 1974, p.386). The crust in the central area of these marginal basins is generally thin due to the lack of "layer 2". In this way undisturbed Tertiary sediments rest directly on gabbroic rocks of "layer 3". In the central portion of the Sea of Japan the Late Tertiary undeformed sediments with a thickness of 2 to 3 km overlie a basement with refraction velocities of 6 to 7 km/s, suggesting the presence of oceanic crust. This small ocean basin is located within the Hida metamorphic belt of Late Paleozoic age (medial massif).

Even a glance shows that these marginal basins (or "interarc basins" or "back-arc basins") of the Pacific are superimposed upon an ancient intermediate massif characterized by folded and metamorphosed basement, intruded by numerous granites. This means that we are dealing with areas which are relatively stable and consolidated by the previous orogeneses. There is, besides, paleogeographic evidence that such "medial massifs" have in the past abundantly provided the adjacent geosynclinal troughs with sediments. A subaerial erosion would be one of the causes of crustal thinning and progressive elimination of the "layer 2".

The presence of an ancient land area ("Asiatic continent") in the place of the present Sea of Japan Basin has been sedimentologically deduced from Late Cretaceous to Oligocene to justify the northern provenance of coarse sediments of Izumi Group (see Suzuki and Tateishi, 1980). Analogously, the presence of an "Okhotsk continental area" has been postulated as a supplier of Miocene conglomerates with gneiss and granite pebbles in the Kuril Islands (Beloussov, 1982). An ancient continental borderland, in the place of the present Banda Sea, has been shown by Silver et al. (1985). The foundering of this landmass took place in the Late Miocene and was accompanied by widespread eruption of andesites. Earlier, Umbgrove (1946, p.200 and Fig.132) had claimed, on paleogeographic bases, the existence at the end of Mesozoic and until Miocene of an extensive emergent land area which included large parts of Borneo, Malaya, the present Java Sea and South China Sea, and Sumatra. As regards the South China Sea Basin, we note the presence in this depression of a series of submarine plateaus, such as the Saigon Bench and the Reed Bank, which presumably represent subsided sectors of continental origin.

A separate chapter is represented by the zone of Melanesian and southern Pacific arcs, all around New Guinea and the Australian Cordillera (see Figs.3 and 4). This circum-Australian region is characterized by numerous and extensive submarine plateaus (e.g.,

Solomon Islands, North Fiji Plateau, Lord Howe and Chatham rises). Besides several shallow guyots are present in this area (e.g., in the Coral Basin); the tops of some are as much as 270 m below present sea level (e.g., in the Tasman Basin). This very large shoal area has, according to Menard (1964, p.146 and Fig.6.19), "the size and relief of a rise except that is chopped in pieces". The crest of this possible "Melanesian Rise" is offset right--laterally by transverse transcurrent faulting running to southeast from the Fiji Islands-New Caledonia alignment and is offset left--laterally along another fracture zone at the northern end of the Fiji Plateau. In this way the crest of the oceanic rise is subdivided in three segments: the northern of Caroline-Solomon Ridge, the middle of Fiji Plateau and the southern segment of Norfolk Island Ridge, Lord Howe Rise, New Zealand, Chatham Rise and Campbell Island Plateau. This fragmented and complicated rise is seismically active only in some places. Deep-focus earthquake belt extends over the Bismarck Basin, the Solomon Ridge, the Fiji Plateau and the South Fiji Basin.Menard noted absence of seismicity on the southern part of the crest in the area between the Tasman Sea and the South Fiji Basin. This fact suggested to him "some antiquity or decadence" (Menard, 1964, p.146), as shown also by normal heat flow. The conjectural "Melanesian Rise" has shattered the margin of Australia by longitudinal block faulting (parallel rifts) and transverse transform faulting.According to Menard (1965, p.345) the North Fiji Basin is generally shallower than 4 km and has the typical ridge and trough topography of an oceanic rise. Briefly, this marginal basin of mature type and Upper Miocene age is characterized by high mean heat flow (i.e., 2.9 HFU) and high attenuation of seismic waves (Toksöz and Bird, 1977), indicative of the upwelling of low-density mantle to shallow levels. Might we hypothetize for other marginal basins, too, the presence of ancient transient rise systems, which have successively collapsed? Can we consider the possibility that such hypothetical rises have been generated from the subsidence of the Mesozoic "Darwin Rise" of the western central Pacific Basin?

In fact, both the "Melanesian Rise" and the East Pacific Rise appear to encircle the subsided "Darwin Rise". The latter (see Fig. 3) roughly 10,000 km long, 4,000 km wide, and 2 to 3 km high, namely with the same areal dimensions as the East Pacific Rise, was hypothetized by Menard (1964 and 1965) on the basis of the actual topography of the guyots and atolls. Although "the logic behind the reconstruction of this paleogeographic and paleotectonic feature remains incontrovertible" (Schlanger et al., 1981, p.441) with the generalized acceptance of the plate-tectonic hypothesis, the concept of "Darwin Rise" "was largely forgotten or discredited". The numerous data obtained on several DSDP legs have, however, re-habilitated Menard's "Darwin Rise" (see Schlanger et al., 1981). From these data we note that the regional vertical uplift and volcanism occurred in the period between the Aptian and the mid-Maastrichtian. It interested the whole area bounded by the Nauru Basin-Marshall Islands, Mid-Pacific Mountains and Line Islands. The regional shallowing of the sea-floor reached its maximum during the Campanian-Maastrichtian times with an uplift in the range of 2.5 km in the Central Basin area where the rise crest probably passed (Schlanger et al., 1981). Since Paleocene times the "Darwin Rise" has subsided about 1.6 km in correspondence with its western limb (Nauru Basin).

The subsiding area of the fossil "Darwin Rise" is marked by a major positive anomaly in the gravity field in the surface of the geoid. A marked upheaval of the geoid is apparent with the maximum height (+81 m) centered over New Guinea, almost precisely on the equator. It represents the existence of foundering masses at depth. It is to be noted that the crest of this large mass anomaly extends from the Philippine Trench to the Fiji Island, being, that is, displaced towards SW in relation to the possible crest of the Mesozoic "Darwin Rise" and is, instead, more or less correspondent to the postulated crest of the "Melanesian Rise". Such a major positive anomaly of the Central Pacific is surrounded by a belt of great negative anomalies which lies over Ceylon (-113 m), Indian subcontinent, Central Asia and Mongolia. Such a depression of the geoid, corresponding to negative density anomalies, appears related to regions of crustal extension due to an upheaval and an updoming. The geoidal low seems to correspond to the crest of a rise possibly passing under the high plateau region of Central Asia and from the Verkoyansk Trough to the Baikal rift valley (Menard, 1964, p.151, Fig.6.1). This major continental rise, seismically traced, is superimposed on a Meso-Cenozoic fold belt having vergence towards Siberia.

The high plateaus such as Tibet and the Iranian Plateau can in fact be considered as continental equivalents of marginal basins (Toksöz and Bird, 1977). As much can be said for the whole range of medial massifs ("Zwischengebirge" or "stable median zone" or "intermediary mass") which occupy extensive areas inside the Tethys realm (e.g.,Tyrrhenian, Pannonian-Transylvanian, Rhodope-Macedonian, Anatolian, Transcaucasian intermediate massifs). Such Paleozoic massifs have constituted an important source of clasts for the adjacent geosynclinal troughs, before collapsing and being transformed into basin depressions. The belt of Tethyan median massifs corresponds roughly to the hypothetical "Lake Baikal-Spain" rise, indicated by a seismic belt from the Azores to the mouth of the Lena River (Menard, 1965, Fig.20).

The kinematic study of the Mediterranean marginal basins (e.g., Tyrrhenian Basin: see Wezel 1984 and 1985) shows that they were occupied in the Paleogene and Early Neogene by a "ridge" bounded by marginal geosynclines. Successively, the crestal area of this ridge became a zone of axial collapse generating the inter-arc basin (see Fig.1). The system shows a close analogy with the present oceanic ridges which are in fact characterized by a seismically active axial collapse zone and by a number of more or less parallel rifts that may pass through alternate periods of tectonic activity and quiescence (Beck and Lehner, 1975). The phase of updoming and outward tilting which created the geanticlinal ridge is responsible (1) for the horst and graben tectonics which have fragmented the Mesozoic-Paleogenic chain and (2) for the creation of an erosive surface and peneplainisation. It is probable also to ascribe to it the supracrustal thinning characteristic of many marginal basins. The successive phase of regional collapse of the ridge has determined (1) the general Miocene transgression over folded, block-faulted, and truncated Early Tertiary beds and (2) the formation of the circum-Pacific small ocean basins. The passage between the two phases is probably marked by voluminous outpourings of "flood basalt" lavas, fed from the parallel linear zones of fracturing.The basalt magma flooded out over the surrounding region for long distances of the order of hundreds of kilometres, with formation of a subdued new topography.

The map of the geoid indicates that the boundary between the continental belt in uplift and extension and the oceanic one in subsidence passes in Asia, just in correspondence with the western boundary of the marginal basins of Okhotsk, South China and Java seas. Therefore, all the marginal basins and island arcs of the western Pacific are included in the zone of positive anomalies of the central Pacific area, lying along the northwestern and south-western sides of the "Darwin Rise". They appear as collapse zones between the emergent continent and the subsiding ocean.

5) The Benioff zone: are we dealing with a rotating dividing boundary between the younger orogens and the ocean basin ?

One of the central points of all the geodynamics of the arc systems is the understanding of the real meaning of Benioff seismic zone. A little less than twenty years ago it was suggested that cold slabs of oceanic lithosphere descend obliquely along the Benioff zones, thereby consuming the oceanic lithosphere and producing the deep earthquakes. However, in fact, according to Freund et al.(1980) "the model of obliquely descending lithospheric slabs is not plausible". These authors propose, instead, an alter-native model in which the deep earthquakes of the Benioff zone are explained by vertical shear between separate detached blocks descending vertically. Independently, an analogous explanation has been given by the present writer for the Tyrrhenian deep seismicity, where the deep earthquakes have been related to the vertical sinking of collapsed blocks, at different stages of development, with the older more evolved blocks placed at a greater depth (Wezel, 1982, p.352). The concept of vertical descent of separate blocks is supported for the Tyrrhenian Sea also by Panza (pers. comm., 1985). Here I should like to limit myself to presenting some brief geological considerations which may perhaps help us to find a solution to this crucial problem.

The first consideration is that a close relationship exists between the belts of deep earthquakes (> 300 km deep) and the circum-Pacific marginal basins (McKerrow and Lambert, 1973). Almost all these linear belts underlie depressed areas on the surface. The majority of the deep-focus seismic belts are centered over the marginal basins of the Indonesian Archipelago and in the region of the supposed "Melanesian Rise". Only three belts are centered on the Central Pacific marginal basins (i.e.,below the seas of Okhotsk and Japan and in correspondence with the alignment Honshu-Iwo Jima Ridge-Mariana Trough). The three deep seismic zones are characterized by high seismic wave attenuation at shallow depths and high heat flow which show a warming and weakening of the lithosphere under these basins (Toksöz and Bird, 1977). They are from 300 to 1,000 km of distance from the adjacent trench.

In the Indonesian Archipelago the deep-focus earthquake belts are found below the marginal basins of Java and Banda seas and the Celebes Basin. The Sunda belt represents a very long narrow zone (about 2,800 km), located about 500-600 km north of the Java Trench. In its turn the belt underneath the Celebes marginal basin occupies the most depressed part of the basin (> 5,000 m), extending towards north in the western Mindanao until 9° N.

In the Melanesian region, the mantle earthquake belts are found below the Bismarck Basin, Solomon Islands, Fiji Plateau, Lau-Havre Trough, and the North Island of New Zealand. Even in this region the deep hypocentres lie below marginal basins, active or mature, but characterized however by high heat flow. The mature Fiji Plateau of Upper Miocene and the active Lau-Havre Trough of Pliocene show, besides, a strong attenuation of seismic waves (Toksöz and Bird,

1977). Moreover, the Fiji Plateau shows particularly high values of heat flow (1.4 to 10 HFU, average 2.9 HFU).

In short, the Pacific inclined zones of mantle seismicity ($>$ 300 km depth) seem to present the following characteristics in common:

1. they are curved narrow belts extending over long distance (300 to 3,000 km);

2. the seismic belts run subparallel to the margins of Pacific and subparallel to the general strike of young orogens;

3. they show a post-orogenic activity, superimposed on the subsiding hinterland (i.e., medial massif) of the mobile orogenic belts, behind the magmatic arc subjected to vertical uplift;

4. the deepest seismic shocks are found below the stable central part of a mobile orogenic belt characterized by a bilateral symmetry and double vergence of the thrust-fronts;

5. the inclined seismic surfaces often show a conical configuration with convexity towards the ocean basin (see, e.g., Fitch and Molnar, 1970);

6. on the internal (concave) side of the conical hypocentral surface these is high attenuation of seismic waves, indicating that the seismic surface represents a discontinuity;

7. statistically, from the literature there emerges a majority of first-arrival patterns indicative of tension (right angles to trench) and strike-slip motion (see, e.g., Tanner, 1974; Hodgson, 1957 quoted in Menard, 1964, p.109);

8. there seems to exist a relationship between the depth and steepening of Benioff zone and the thickness of the crust, in the sense that arc systems more oceanic and far from the continent are characterized by a greater depth (see, e.g., the Mariana "intra-oceanic" arc with depth of Benioff zone of 680 km);

9. an inverse correlation seems to exist between the potassium content of arc lavas and the vertical depth of the Benioff zone below. The content diminishes with the increase of the depth of the inclined seismic zone in the mantle beneath (the intra-oceanic arcs show lower potassium contents than do the continental-margin arcs).

The first three points mentioned above and also point 7 lead us to think that the active post-tectonic deformation mode is dominated by large scale trans-lithospheric transcurrent movement subparallel to the strike of the Meso-Cenozoic fold belt at the periphery of the Pacific Basin. The existence of extremely long transcurrent fault systems (megashears) around the whole perimeter of the Pacific has been put forward by various writers (e.g.,Carey, 1958, 1976 and this volume; Menard, 1964). Menard (1964, p.52) has emphasized: "According to one of the two possible interpretations of first motions of earthquakes, namely,strike-slip motion parallel to island arcs (Hodgson and Milne, 1951), more than half the perimeter is marked by wrench faults. It is difficult to escape the conclusion that all these faults have the same underlying origin".

It is worthwhile remembering that the regions of Indonesia and Melanesia are affected by major transcurrent fault zones subparallel to the general strike of the orogens. For example, the Banda volcanic arc and the Benioff zone are related by Crostella (1977) to a Recent dextral transcurrent fault. The sinistral Sorong fault zone of New Guinea, with displacement amounting to 360 km (van de Fliert et al., 1980), is related to Carey's "Melanesian Megashear", being part of the "Tethyan Sinistral Torsion" (see Carey, this volume, Figs.23 and 24). The presence along the periphery of the Pacific of major dextral transcurrent faulting with large translations gave rise, thirty years ago, to H.Benioff's and P. St. Amand's stimulating hypothesis that the inner basin of the Pacific has a counter-clockwise rotational-like movement relative to the surrounding continents, with an average slip of about 5 cm/a. Similarly, a Pacific "Dextral Counter Torsion" has been hypothetized by S.W. Carey (this volume: see Figs.15 and 16) which separates the "oceanic" and "land" hemispheres and which must thus be attributed to the interaction of rotation inertia with different moment of inertia. According to Carey (this volume) this large scale dextral shearing would explain, among other things, the dispersion of the allochthonous "suspect terranes" in the circum-Pacific fold belt systems. The fact that the Cordilleran Belt of western North America is a collage of displaced and rotated terrains, such as Wrangellia, indicates that basically it is a question of a strike-slip orogen. These facts suggest a rotational-shear relationship between the continents and the Pacific Ocean basin (i.e., the "Pacific-continent shear couple").

Points 4),5),6) and 8) are related in some way to the existence of a "deep root" below the consolidated median zone of extension and collapse of the orogenic systems, symmetrically bordered by two outer fold belts. The continent-vergent thrust front is older and presently inactive and not associated with deep-seated earthquake activity. It corresponds to Wilson's "secondary mountain arc" and Bally's "Ampferer or A-subduction zone (Alpino-type)".

DISCUSSION

The features already summarised in the previous sections seem to be plausibly interpreted in a new geotectonic picture. Given, however, the real lack of direct data, this picture can only be of a panoramic type and can pick out only a few essential aspects of the complex geological reality of the circum-Pacific. However, if my scientific analysis is sufficiently correct, it is clear that the geotectonic phenomena operating in the island arc regions are rather different from those assumed up to now.

It is evident that practically none of the basic paradigmatic axioms summarised in the second section, conceptually elaborated by plate tectonics, has resisted till now uncriticized the assault of factual data. Briefly, I recapitulate as follows (see, e.g., Uyeda, 1977 and this volume, von Huene, 1981; Aubouin et al., 1984; and others).

1. Hemipelagic rather than deep-sea pelagic sediments are widely distributed hundreds of kilometres seaward of the trenches.

2. Instead of oceanic sediments and trench-fill turbidites, rocks were recovered in the "accretionary prism" representing the offshore prolungation of tectonic complexes cropping out in the chains on land.

3. Old material (i.e., pre-Campanian ophiolites overthrust on Upper Jurassic to Upper Cretaceous units in Leg 84) instead of very

young sediments was penetrated at the trench slopes.

4. Instead of uplift during subduction, active subsidence was observed on the Japan, Mariana and middle-America trench margins.

5. to 6. According to Uyeda (1977 and this volume) the origin of volcanism and high heat flow and the origin of back-arc basins are still "basic unsolved problems" in the context of the subduction model. The frictional heat model can in fact be ruled out as a source of heat for arc volcanism. Moreover, there is still no unequivocally convincing evidence for slab melting. Thermal calculations of subduction models generally disfavor melting of slab.

7. The other of two possible interpretations of first motions of earthquakes indicates strike-slip motion on nearly vertical planes.

8. On the drilled margins the general regime resulting from subduction is tensional rather than compressional. This neotectonic regime, with no connection with prior tectonic structures, was everywhere born in Neogene.

The logical conclusion to be drawn, and disturbing as it may be, is that the tectonic processes hypothetized by the present models do not correspond to the geological reality of the arc systems. This should not cause no too much surprise in that such models "are based more on hypothetical considerations than on kinematic geologic sections established with modern geophysics and rock information" (von Huene, 1981, p.64). Without direct geological information and control the interpretation on the basis of mere indirect geophysical data (seismic reflection profiles) is very dangerous and has a good chance of being wrong. As seems in the case of the "accretionary prism" model. In fact, no accretionary process seems to work in correspondence with the Peru-Chile, middle-America, Marianas and Japan "convergent-extensional margins" (Aubouin et al., 1984).

The few data known up to now seem rather, in the writer's opinion, to belong to a quite different geotectonic interpretation of the evolution of the active margins and, thus, of the geotectonic phenomena which occur during the late-orogenic to post-orogenic

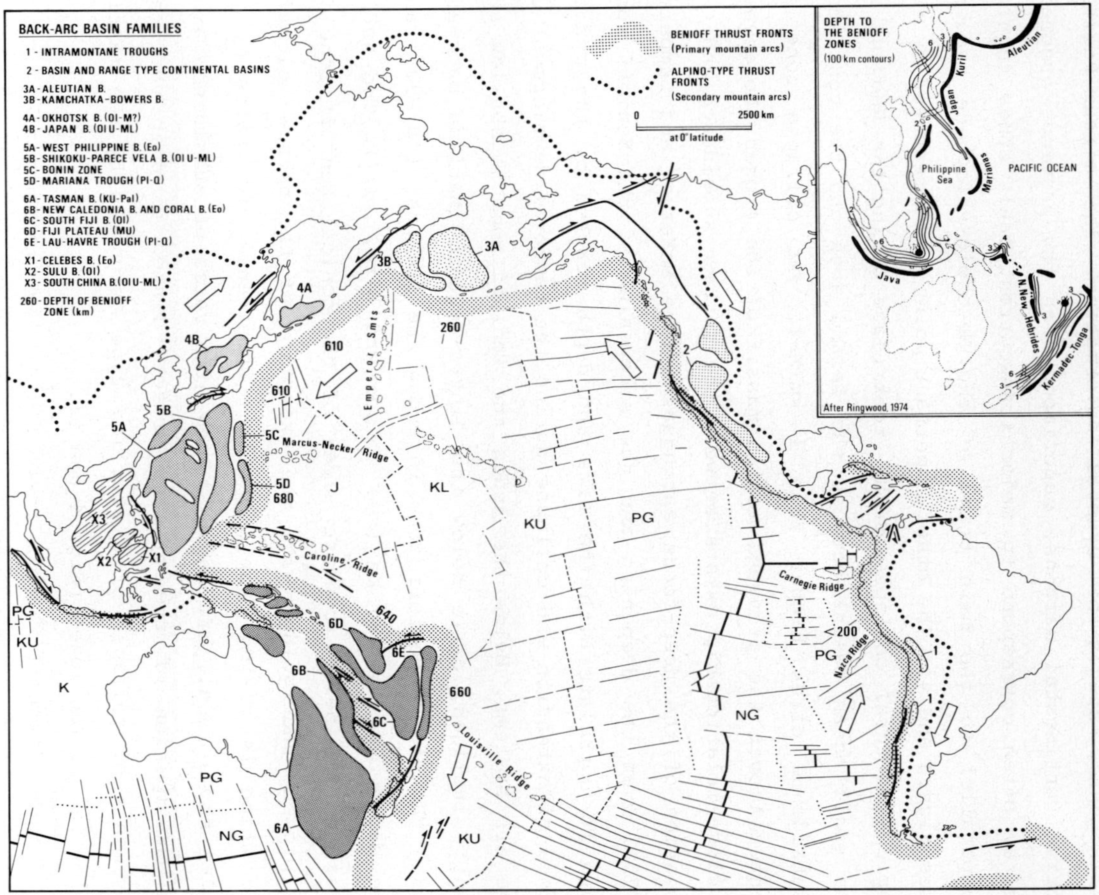

Fig. 4. Schematic map which shows that the back-arc basins are spatially distributed around the Pacific margins according to a developmental sequence, which goes from the Andes to Tasmania, from youngest to oldest. In the same counterclockwise direction can be noted a widening of the orogenic belt, an increase in the number of basins and a deepening of the Benioff zones. The inferred dextral Pacific torsion is indicated by arrows marking the "rotation-like" movements of the Pacific relative to the continents. The Indonesian Arc system is considered apart. Inset map shows the Benioff zones of the western Pacific with depth indicated by 100-km contours.

(krikogenic) stages of crustal deformation. The circum-Pacific active continental margins, superimposed on the previous orogenic belts, often correspond to major transcurrent fault zones running subparallel to the main trend of the Mesozoic-Paleogene orogens (see Figs.3 and 4). The other important point is represented by the rapid structural collapse which took place during the Neogene in the small ocean basins between the island arcs and the continent (subsidence often higher than 1 km/Ma). The collapse zone is centered over the previous median massif of the orogen. Moreover, a close relationship exists between the surface subsidence and the location of the deep-focus earthquakes (see McKerrow and Lambert, 1973). Even the deepening and the steepening of the Benioff zone seem to correlate with the size and depth reached by the median zones of extension and collapse during the temporal development of the arc systems. Below the Andes the Benioff zone is shallow and gently inclined, while below the "intra-oceanic" Mariana back-arc basin the dip of Benioff zone is among the steepest (70°-90°) and the depth reaches 680 km (see Toksöz and Bird, 1977). The whole area on the inside (concave side) of the conical Benioff zone is at present undergoing, with the exception of the narrow volcanic arc which is rising, rapid vertical subsidence with rates up to 1.7-2 km/Ma (e.g., off Japan). Such high rates of subsidence are, in the opinion of the present writer, linked to deep vertical movements of collapse of a whirling type triggered by the mantle below the median zones (Figs.4 and 5).

The development of this tectonics follows the preceding phase of krikogenic arching and vertical elevation of the orogenic systems with the formation of subaerial erosion surfaces. The up--arching creates a mountain ridge affected by extensional rifting manifested by a horst and graben morphology. The active doming is accompanied by crestal collapse of the ridge which, in the initial phases, gives rise to the formation of intramontane troughs, such as the central valleys of the Andes, or of narrow trenches like that of the Rocky Mountains. In general, the collapse zones are placed in correspondence with the medial crystalline massifs of the

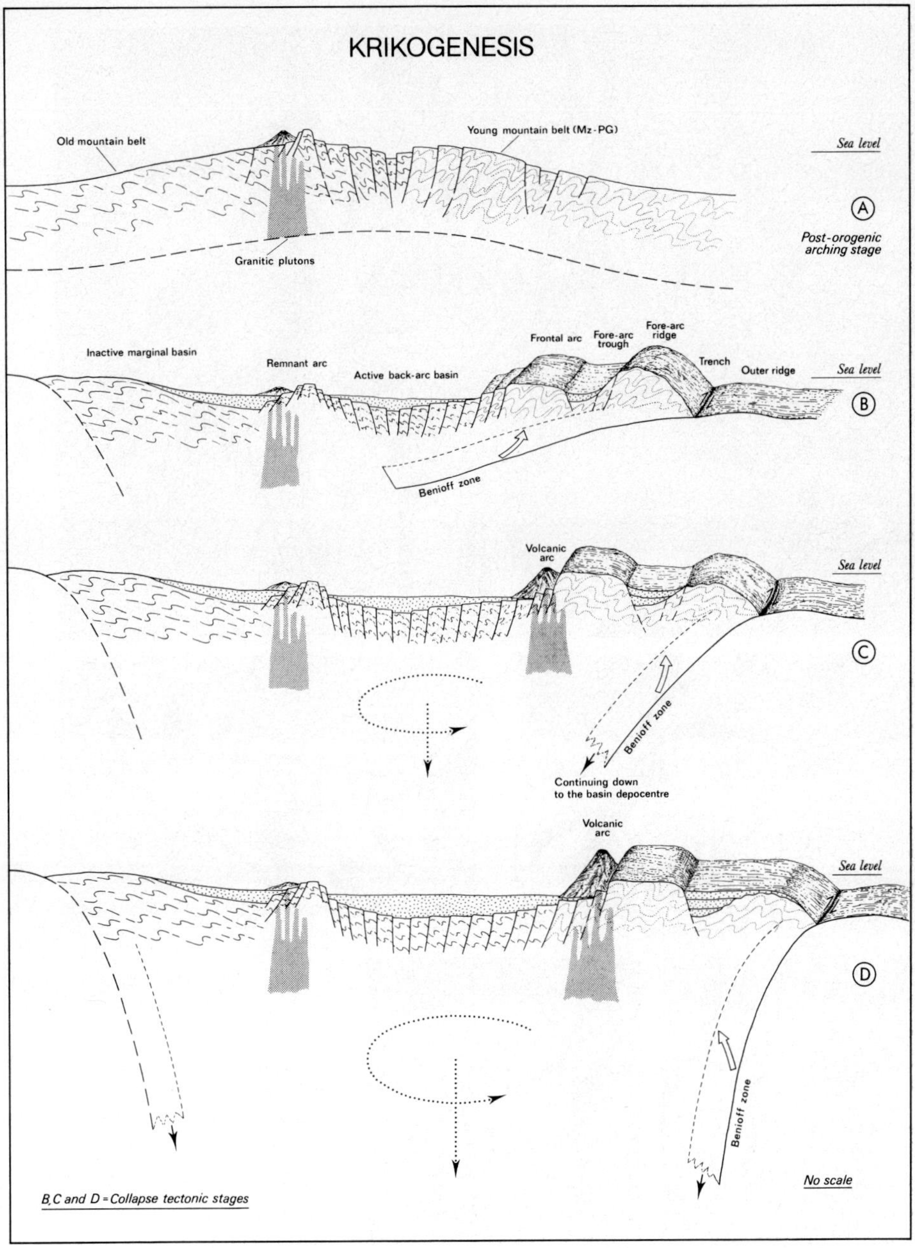

Fig. 5. Diagrammatic cross-sections through a Pacific island arc system, showing the inferred geological evolution. Note the rotational motion during the subsiding phases. For further explanation see text.

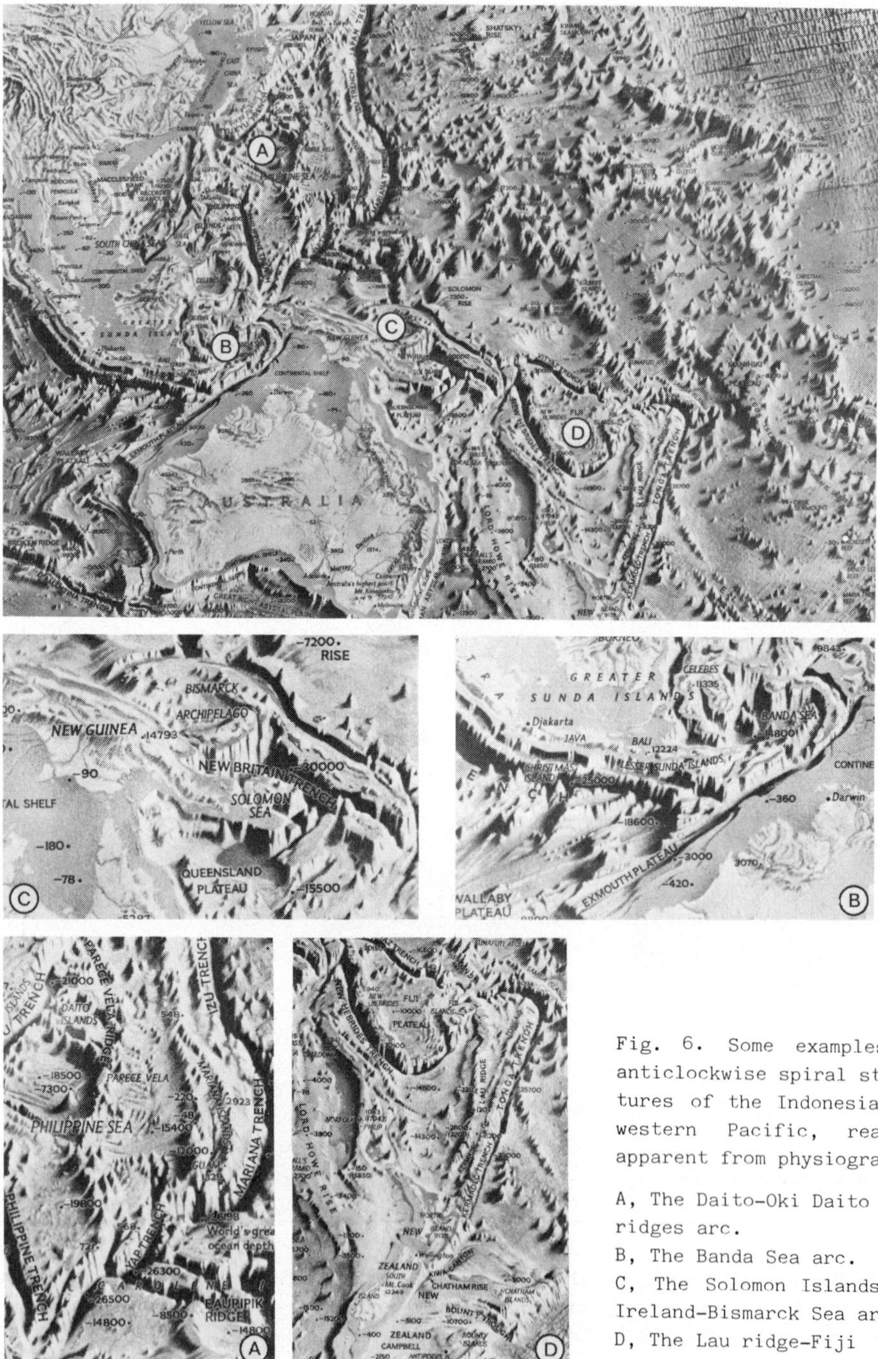

Fig. 6. Some examples of anticlockwise spiral structures of the Indonesia and western Pacific, readily apparent from physiography.

A, The Daito-Oki Daito ridges arc.
B, The Banda Sea arc.
C, The Solomon Islands-New Ireland-Bismarck Sea arc.
D, The Lau ridge-Fiji Islands arc.

mobile belt, which represent in practice the internal partly metamorphosed zones previously consolidated and stabilized during the older orogenic phases which created the opposite continent-verging folded belt (Fig.5). The post-Paleogene collapse tectonics of the median zones extend more and more laterally creating wider and wider depressions overlying block-faulted and truncated portions of the orogenic belt, mainly epi-eugeosynclinal.

The evolutionary stage reached around the Pacific by each family of back-arc basins seems to be set out by the physical map itself. Statistically, from it a younging and narrowing of the marginal basin belt in clockwise direction, from Australasia to Andes (Fig. 4), can be seen. In the same direction there is a tendency of Benioff zone to become shallower and less steep, accompanied by the tendency to shallowing of the trenches.The post-orogenic tensional tectonics tend to occupy ever narrower median sectors. The correlative thickening of the crust is accompanied by the eruption of more calc-alkaline andesitic and less tholeiitic lavas, associated with distinctly higher potassium contents. Naturally, the so-called "accretionary prisms" assume major topographic expression identifying themselves with mountain ranges (e.g., the Andes). Besides, the shallow earthquakes of great magnitude (8.0 or more) are becoming more diffuse, as for example the most destructive shocks over Alaska, Central and South America in the 19th and 20th centuries.

Moreover, the presence of some counterclockwise spiral morphostructures inside the Australasian island arcs (see, e.g., Fig.6), just in the "eye" of the Pacific system (Fig.4), is to be noticed. Incidentally, similar topographic features are present also in the Indian Ocean (see the Carlsberg Ridge), always located within the equatorial belt. Here it is enough to recall the curled shape of the Lau Colville Ridge-Fiji Islands and, more generally, of the couple Kermadec Tonga Ridge-Fiji Plateau (Fig.6D). The presence beneath the New Hebrides zone of another northeast-dipping convergent seismic surface, assures the almost complete cone closure of the area of the Fiji Plateau (inset of Fig.4). These regional bathymetric observations have been fully confirmed by paleomagnetic results indicative of a counterclockwise rotation of about 90° of the Fijian Islands during the period extending from the Late Miocene to the Present (Malahoff et al., 1982). In their turn the New Hebrides island arc has rotated 30° in a clockwise direction during the past 6 Ma.

Another well expressed spiral structure is represented by the Banda Sea arc of Eastern Indonesia (Fig.6B). The Sunda Arc in its eastern extremity, starting from the Timor and Tanimbar islands is almost folded back on itself like a hook around the Banda Sea being located more or less parallelly to the Sunda Arc at the level of the Ceram and Buru islands. The outer arc Timor-Ceram is constituted by a Middle Miocene orogenic belt, while the parallel inner arc Wetar-Damar-Banda islands-Lucipara ridge represents the Late Tertiary to Quaternary volcanic arc, superimposed over the Mesozoic Sumba metamorphic belt (Crostella, 1977; Silver et al., 1985). The volcanic arc constitutes a "curl" which almost closes in correspondence with the active volcano Gunung Api (north of Wetar). The structures of the Banda loop are suggestive of a vortex motion in counterclockwise direction (Fig.6B). In fact, large anticlockwise rotations have been confirmed paleomagnetically for Ceram and Buru (S.W. Carey, this volume). The Banda Arc curled edge corresponds at depth to the sharp curvature and sinuosity of the inclined seismic zone, which continues northward beneath Mindanao (see Fitch and Molnar, 1970). The Benioff zone forms, therefore,a deep conical surface dipping below the Banda Sea (inset of Fig.4). It is worth noting the evidence that the morphostructures, suggestive of a rotational counterclockwise movement,are situated in correspondence to the conical bends of the deep Benioff zones. The three most evident examples are represented by the Fiji, Banda (Figs.6 D,B) and Calabrian arcs (Fig.2). All three are located inside the major sinistral transcurrent fault zones, viz. in Carey's "Tethyan Sinistral Torsion" (1958, 1976, this volume).

The last considerations regard the fact that the Pacific margins derive from the destruction in post-orogenic epoch of a Meso-Paleogene chain. They certainly represent an advanced stage in the geotectonic evolution of mountain belts. If this phase is also a prelude or not to an incipient stage in a subsequent cycle of geosynclinal evolution is unknown. Rather than being mountain chains in statu nascendi the arc systems appear in their present state as regions of neotectonic destruction of young orogens. How is the position of the passive margins in this evolutionary sequence from mountain chains to island arcs? Do they represent a phase preceding or successive to that of the active margins? The direct data seem as yet insufficient for a reply.One can only note that the Atlantic passive margins are always placed and superimposed upon the medial hinterland of an old bivergent orogenic belt (Paleozoic or Precambrian). Such a structural epi-hinterland position contrasts with that of the Pacific margins which are superimposed over the frontal half of a young bivergent orogenic belt (Meso-Paleozoic), that is rather in a more external position (essentially epi-externid: see Fig.5).

A SUMMING UP

The geological kinematics lead to an interpretation of the Pacific active margins as a result of the migration towards the present continental area of a slow, gigantic transient "tectonospheric" wave. The Paleogene foundering of the huge Mesozoic "Darwin Rise" is approximately contemporary with the uplift of the rises which must have existed in correspondence with the present area of Australasian island arcs. This regional arch-shaped elevation has above all affected the Mesozoic-Paleogene orogens which have been fragmented by horst and graben tectonics and have undergone a substantial erosive and perhaps tectonic denudation.

The axial zone of the ridge, centered on the stiff, partly metamorphic, internal zones of an older opposite-vergent Alpide orogen, underwent extension and structural collapse (Fig.5). This rapid subsidence, creating the epi-hinterland marginal basins, was accompanied by intense and complex phenomena of torsion,translation and rotation of blocks of the pre-Neogene mountain chain, due to the transcurrent faulting activity. The phenomenon begins with the formation of intramontane troughs and ends with the complete disruption of the margin and the formation of a complicated series of basins and ridges of Australasian-type (Fig.4). The marginal basins may be, in short, visualized as gigantic "conical holes" which open inside the terrestrial lithosphere.

A common feature of the Pacific margin is the Neogene regional submergence of the Meso-Paleogene orogens, which begins in the Upper Oligocene-Lower Miocene with the drowning of the subaerial chain in correspondence with the "fore-arc ridge" and continues with that of the "frontal arc". Off Japan the seaward part of the margin subsided to 6 km depth with a rate of subsidence of 1.7 km/Ma. The present structural framework of the island arc system (Fig.5) reveals in itself the different stage of development reached and the beginning of the submergence of the different topographic swells, namely (1) the "outer ridge" in the Paleogene, (2) the "fore-arc ridge" in the Miocene and (3) the "frontal arc" locally subsiding in recent times. The pronounced vertical tensional uplift of the active volcanic arc is considered to be related to recent phases of collapse in the adjacent back-arc basins. Once again we note, even if on a minor scale, the shifting of a topographic wave in a landward direction, having in general (in the case of the "fore-arc ridge") an amplitude of about 6 km, half-wavelength of about 100 km and average migration rate of the order of 5 cm/a.What is most striking is the vigour of the differential vertical movements between adjacent blocks which takes place with extremely rapid rates of the order of 2 km/Ma. The importance of vertical tectonics in shaping the Pacific active margins seems to have been until now rather underestimated.

Finally, the krikogenic mechanisms of arching and rapid collapse with creation of the "conical holes" and especially the existence of a transient "tectonospheric" wave, would make one think that processes in the upper mantle and those in the fluid outer core are in some way coupled. The massive rise on the earth surface of these huge "tectonospheric" waves ("Darwin Rise" type) must have produced a global reorientation of the planet with respect to the spin axis, in such a way to place the mass excess on the equator. The reorientation probably caused the fragmentation and dispersal of continental blocks and great magmatic activity. The four main periods of intense global swelling, up to now recognized by the writer (i.e., (1) late Lower Giurassic-Middle Giurassic, (2) Aptian-Senonian, (3) Paleogene and (4) Plio-Quaternary), are manifested by a rather quiet magnetic field.

As regards the curvature of the trench, we may note that the intersection of the inclined Benioff zone with the earth's surface can produce an arc. To have, however, the curvature of the present island arcs a very shallow dip would be needed, not the steep dip that is found on some Benioff zones (e.g., in the Mariana Arc). If it dips at a steep angle from the arcuate structures, its surface must have a conical shape.

The paper is dedicated to Prof. S. Warren Carey for his 75th birthday, 30 years after the Continental Drift Symposium in Hobart and 10 years after the publication of the "Expanding Earth" volume in this same series.

ACKNOWLEDGEMENTS

The author is grateful for the English touch to Dr. David Murray and for the help given in the elaboration of the drawings to Prof. Daniele Savelli, Dr. Mario Tramontana and Mr. Franco Paolucci. Prof. Enzo Locardi read the manuscript and made helpful suggestions.

REFERENCES CITED

Aubouin, J., 1981. About mountain building. Geol.Soc.China, Memoir No.4: 33-53.

Aubouin, J., 1984. About some aspects - so called classical - of the peri-Pacific tectonics. Proc. 27th Intern. Geolog. Congr., v.7 (Tectonics), VNU Sci. Press: 93-107.

Aubouin, J., Bourgois, J. and Azema, J., 1984. A new type of active margins, as examplified by Middle America Trench off Guatemala. Proc. 27th Intern. Geolog. Congr., v.7 (Tectonics), VNU Sci. Press: 43-63.

Aubouin, J., von Huene, R. and others, 1982. Subduction sans accrétion: la marge pacifique du Guatemala: premiers résultats du Leg

84 du Deep Sea Drilling Project (janvier-février 1982). C.R. Acad. Sc. Paris, 294: 803-812.

Bally, A.W., Catalano, R. and Oldow, J.,1985. Elementi di tettonica regionale. Pitagora Editrice Bologna, 276 pp.

Beccaluva, L., Gabbianelli, G., Lucchini, F., Rossi, P.L. and Savelli, C., 1985. Petrology and K/Ar ages of volcanics dredged from the Eolian seamounts: implications for geodynamic evolution of the southern Tyrrhenian basin. Earth Planet. Sci. Lett., 74: 187-208.

Beck, R.H. and Lehner, P., 1974. Oceans, new frontier in exploration. Am. Assoc. Petrol. Geol. Bull., 58 (3): 376-395.

Beck, R.H. and Lehner, P., 1975. New geophysical data on key problems of global tectonics. Proc. 9th World Petrol. Congr.,Tokyo, v.2: 1-19.

Beloussov, V., 1982. Certain problems of structure and conditions of evolution of transition zones between continents and oceans. Geol. Rund., 71 (2): 487-511.

Carey, S.W., 1958. A tectonic approach to continental drift. In: S.W. Carey (Editor), Symp. Continental Drift, Hobart 1956: 177-355.

Carey, S.W., 1976. The Expanding Earth. Elsevier Scientific Publ. Co., Amsterdam, 488 pp.

Carey, S.W., 1983. Tethys, and her forebears. In: S.W. Carey (Editor), Expanding Earth Symposium, Sydney 1981, University of Tasmania: 169-187.

Chappel, J. and Veeh, H.H.,1978. Late Quaternary tectonic movements and sea-level changes at Timor and Atauro Island. Geol. Soc. Am. Bull., 89: 356-368.

Coney, P.J., 1970. The geotectonic cycle and the new global tectonics. Geol. Soc. Am. Bull., 81: 739-748.

Crostella, A.,1977. Geosynclines and plate tectonics in Banda arcs, Eastern Indonesia. Am. Assoc. Petrol. Geol. Bull., 61 (12): 2063-2081.

Dickinson, W.R. and Seely, D.R., 1979. Structure and stratigraphy of forearc regions. Am. Assoc. Petrol. Geol., 63 (1): 2-31.

Fitch, T.J. and Molnar, P., 1970. Focal mechanisms along inclined earthquake zones in the Indonesia-Philippine region. Jour. Geophys. Res., 75 (8): 1431-1444.

Freund, R., Kosloff, D. and Matthews, A., 1980. A dynamic model of subduction zones. In: P.A. Davies and S.K. Runcorn (Editors), Mechanisms of Continental Drift and Plate Tectonics, Academic Press: 17-39.

Honza, E., 1981. Subduction and accretion in the Japan Trench. Oceanol. Acta, Proc. 26th Intern. Geolog. Congr., Geology of continental margins symposium, Paris 1980: 251-258.

Hussong, D.M. and Uyeda, S., 1981. Tectonics in the Mariana Arc: results of recent studies, including DSDP Leg.60. Oceanol. Acta, Proc. 26th Intern. Geolog. Congr.,Geology of continental margins symposium, Paris 1980: 203-212.

Ibrahim, A.K., Latham, G.V. and Ladd, J.W.,1979. Seismic refraction and reflection measurements in the Middle America Trench offshore Guatemala. Jour. Geophys. Res., 84 (10): 5643-5649.

Karig, D.E., 1974. Evolution of arc systems in the western Pacific. Ann. Rev. Earth Planet. Sci., 2: 51-75.

Kennett, J.P., McBirney, A.R. and Thunell, R.C., 1977. Episodes of Cenozoic volcanism in the circum-Pacific region. J. Volcanol. Geotherm. Res., 2: 145-163.

Lewis, S.D. and Hayes, D.E., 1985. Forearc basin development along western Luzon, Philippines. Energy, 10 (3/4): 281-296.

Malahoff, A., Hammond, S.R., Naughton, J.J., Keeling, D.L. and Richmond, R.N., 1982. Geophysical evidence for post-Miocene rotation of the island of Viti Levu, Fiji, and its relationship to the tectonic development of the North Fiji Basin. Earth Planet. Sci. Lett., 57: 398-414.

McCarthy, J. and Scholl, D.W., 1985. Mechanisms of subduction accretion along the central Aleutian Trench. Geol. Soc. Am. Bull., 96: 691-701.

McKerrow, W.S. and Lambert, R.St.J., 1973. Deep earthquakes, surface subsidence, and mantle phase changes. Jour. Geology, 81: 157-175.

McMillen, K.J., Enkeboll, R.H., Moore, J.C., Shipley, T.H. and Ladd, J.W., 1982. Sedimentation in different tectonic environments of the Middle America Trench, southern Mexico and Guatemala. In: J.K. Leggett (Editor), Trench-Forearc Geology, Geol. Soc. London, Spec. Publ. No.10: 107-119.

Menard, H.W., 1964. Marine Geology of the Pacific. McGraw-Hill Book Co., N. York, 271 pp.

Menard, H.W., 1965. Sea floor relief and mantle convection. In: L. H. Ahrens et al. (Editors), Physics and Chemistry of the Earth, 6: 315-364.

Moberly, R., Shepherd, G.L. and Coulbourn, W.T., 1982. Forearc and other basins, continental margin of northern and southern Peru and adjacent Ecuador and Chile. In: J.K. Leggett (Editor), Trench-Forearc Geology, Geol. Soc. London, Spec. Publ. No.10: 171-189.

Moore, J.C., Watkins, J.S. and others, 1979. Progressive accretion in the Middle America Trench southern Mexico. Nature, 281: 638-642.

Moore, J.C., Watkins, J.S. and others, 1982. Facies belts of the Middle America Trench and forearc region, southern Mexico: results from Leg 66 DSDP. In: J.K. Leggett (Editor), Trench-Forearc Geology, Geol. Soc. London, Spec. Publ. No.10: 77-94.

Peng, T.-H., Li, Y.-H. and Wu, F.T., 1977. Tectonic uplift rates of the Taiwan Island since the Early Holocene. Mem. Geol. Soc. China, 2: 57-69.

Peterschmitt, E., 1956. Quelques données nouvelles sur les seismes profonds de la Mer Tyrrhénienne. Ann. Geof., 9: 305-334.

Plafker, G., 1972. Alaskan earthquake of 1964 and Chilean earthquake of 1960: implications for arc tectonics. Jour. Geophys. Res., 77: 901-925.

Ringwood, A.E., 1974. The petrological evolution of island arc systems. Jour. Geol. Soc. London, 130: 183-204.

Sakaguchi, Y., 1972. Morphogenesis of the Japanese islands and the eastern Alps. Jap. Jour. Geol. Geogr., 42 (1-4): 1-18.

Schlanger, S.O., Jenkyns, H.C. and Premoli-Silva, I., 1981. Volcanism and vertical tectonics in the Pacific Basin related to global Cretaceous transgressions. Earth Planet. Sci. Lett., 52: 435-449.

Scholl, D.W. and Vallier, T.L., 1983. Subduction and the rock record of Pacific margins. In: S.W. Carey (Editor), Expanding Earth Symposium, Sydney 1981, University of Tasmania: 235-245.

Seely, D.R., 1978. The evolution of structural highs bordering major forearc basins. In: Geological and Geophysical Investigations of Continental Margins, AAPG Mem., 29: 245-260.

Seely, D.R., Vail, P.R. and Walton, G.G., 1974. Trench slope model. In: C.A. Burk and C.L. Drake (Editors), The Geology of continental margins. Springer-Verlag, N. York: 249-260.

Sengör, A.M.C., 1982. Eduard Suess' relations to the pre-1950 schools of thought in global tectonics. Geol. Rund., 71 (2): 381-420.

Silver, E.A., Gill, J.B., Schwartz, D. and Prasetyo, H., 1985. Evidence for a submerged and displaced continental borderland, north Banda Sea, Indonesia. Geology, v.13: 687-691.

Suzuki, H. and Tateishi, M., 1980. An outline of the geology of the Shimanto terrain in the Kii Peninsula, southwest Japan. Unpubl. ms. 27 pp.

Tanner, W.F., 1974. Deep-Sea trenches and the compression assumption: reply. Am. Assoc. Petrol. Geol. Bull., 58 (12): 2525-2527.

Toksöz, M.N. and Bird, P.,1977. Formation and evolution of marginal basins and continental plateaus. In: M. Talwani and W.C. Pitman III (Editors), Island Arcs, Deep Sea Trenches and Back-Arc Basins. M. Ewing Ser.1, Am. Geophys. Un.: 379-393.

Umbgrove, J.H.F., 1946. The Pulse of the Earth. 2nd edition. M. Nijhoff, The Hague: 358 pp.

Uyeda, S., 1977. Some basic problems in the trench-arc-back arc system. In: M. Talwani and W.C. Pitman III (Editors), Island Arcs, Deep Sea Trenches and Back-Arc Basins. M. Ewing Ser.1, Am. Geophys. Un.: 1-14.

van de Fliert, J.R., Graven, H., Hermes, J.J. and De Smet, M.E.M., 1980. On stratigraphic anomalies associated with major transcurrent faulting. Ecl. geol. Helv., 73 (1): 223-237.

von Huene, R., 1981. Review of early results from drilling of the IPOD-1 active margin transects across the Japan, Mariana, and middle-America convergent margins. SEPM Spec. Publ. No.32: 57-66.

von Huene, R., Langseth, M., Nasu, N. and Okada, M., 1982. A summary of Cenozoic tectonic history along the IPOD Japan Trench transect. Geol. Soc. Am. Bull., 93: 829-846.

Watkins, J.S., Moore, J.C. and others, 1981. Accretion, underplating, subduction and tectonic evolution, Middle America Trench, Southern Mexico: results from DSDP Leg 66. Oceanol. Acta, Proc. 26th Intern. Geol. Congr., Geology of continental margins symposium, Paris 1980: 213-224.

Wezel, F.-C., 1982. The structure of the Calabro-Sicilian Arc: result of a post-orogenic intra-plate deformation. In: J.K.

Leggett (Editor), Trench-Forearc Geology, Geol. Soc. London, Spec. Publ. No.10: 345-354.
Wezel, F.-C., 1984. The Tyrrhenian Sea: a rifted krikogenic-swell basin. Mem. Soc. Geol. It., 24, pt.3 (1982): 531-568.
Wezel, F.-C., 1985. Structural features and basin tectonics of the Tyrrhenian Sea. In: D.-J. Stanley and F.-C. Wezel (Editors), Geological Evolution of the Mediterranean Basin. R. Selli Commemorative Volume, Springer-Verlag, N. York: 153-194.